POLYMER SCIENCE AND TECHNOLOGY

NEW DEVELOPMENTS IN POLYMER COMPOSITES RESEARCH

POLYMER SCIENCE AND TECHNOLOGY

Additional books in this series can be found on Nova's website
under the Series tab.

Additional e-books in this series can be found on Nova's website
under the e-book tab.

MATERIALS SCIENCE AND TECHNOLOGIES

Additional books in this series can be found on Nova's website
under the Series tab.

Additional e-books in this series can be found on Nova's website
under the e-book tab.

POLYMER SCIENCE AND TECHNOLOGY

NEW DEVELOPMENTS IN POLYMER COMPOSITES RESEARCH

STEPHAN LASKE
AND
ANDREAS WITSCHNIGG
EDITORS

nova publishers
New York

For permission to use material from this book please contact us:
Telephone 631-231-7269; Fax 631-231-8175
Web Site: http://www.novapublishers.com

LIBRARY OF CONGRESS CATALOGING-IN-PUBLICATION DATA

New developments in polymer composites research / editors, Stephan Laske and Andreas Witschnigg (Montanuniversitaet Leoben, Department Kunststofftechnik, Leoben, Austria).
pages cm
Includes bibliographical references and index.
ISBN 978-1-62948-340-5 (hardcover : alk. paper) 1. Polymeric composites. I. Laske, Stephan, editor of compilation. II. Witschnigg, Andreas, editor of compilation.
TA418.9.C6N477 2011
620.1'92--dc23

2013038302

Published by Nova Science Publishers, Inc. † New York

CONTENTS

PREFACE

The field of polymer composites is one of the most studied and published areas in polymer engineering and science. A lot of excellent research has been done and is going on, nevertheless, there are new outstanding results and findings all over the world nearly every day. The reason for this is that polymer composites offer almost endless possibilities in changing material properties. By combining at least two materials with different properties and behavior it is possible to generate materials with complete different characteristics and even to establish new material classes (e.g., wood plastic composites or nanocomposites). The ample potential and wide range of polymer composite is also observable within this book as the topics range from new processing strategies over new material combinations to matrix-filler-composite structure-property-relationships. Thereby the different chapters cover the fields of biodegradable, conducting, nano- or wood plastics composites as well as generating new (nano-) composites with sophisticated properties in terms of thermal or electrical conducting, tribology and radiation shielding.

The excellent 13 chapters within this book can be divided into material, processing and properties. Within every topic the newest developments are shown giving an impression of the comprehensive and diverse work of international scientists in the field of polymer composites. The results mark only the tip of the iceberg of polymer composites and leave a mark of what we can expect in future. After the Stone, Bronze and Iron Age we just started the age of polymer composites as material of the 21st century.

In: New Developments in Polymer Composites Research ISBN: 978-1-62948-340-5
Editors: S. Laske and A. Witschnigg

Chapter 1

EVOLUTION OF NOVEL SIZE-DEPENDENT PROPERTIES IN POLYMER-MATRIX COMPOSITES DUE TO POLYMER FILLER INTERACTIONS'

***Sheila Devasahayam*[1,*] *and Sri Bandyopadhyay*[2,†]**
[1]University of Ballarat, Australia
[2]University of New South Wales, Australia

ABSTRACT

Engineering polymers, despite experiencing rapid increases in many applications including in space, nuclear reactors and in the medical field, suffer from limitations due to lack of sufficient strength and stiffness, electrical and thermal conductivities as well as thermal stability, etc. The advances in polymer nanocomposite technology, with the filler dimensions in nanometer scales, have made inroads into engineering polymers' applications in the clean energy, semiconductor, medical and construction industries owing to their significantly larger surface to volume ratio and the highly reduced particle-particle distance compared to their micro-sized particle counterpart.

The quantum effects which begin to dominate the behaviour at the nano-scale, particularly at the lower end affect the fillers' optical, mechanical, electrical and magnetic behaviour. The scale at which discontinuity compared to bulk emerges at room temperature is different for different properties, e.g., for magnetic properties is typically ~50nm.The Bohr radius defining the optical and electronic properties is ~10nm.The critical nucleus size is 5 nm above which new crystalline phase will form, below which the clusters tend to dissolve rather than grow.

When the polymeric materials are reinforced with smaller amounts of nanofillers, due to the evolution of extraordinary range of size dependent properties on these nanofillers, the filler-polymer matrix exhibit superior barrier properties, toughness, tribological and mechanical performance, optical properties, lower melting point, increased surface area, higher specific heat, increased electrical conductivity and magnetic properties showing decreased coercive field and increased remnant magnetization, which are not possible in conventional polymer composites. For example, bulk gold appears yellow in colour while

* E-mail address: s.devasahayam@ballarat.edu.au
† E-mail address: S.Bandyopadhyay@unsw.edu.au

nanosized gold appears red in colour. Large ZnO particles scatter visible light, and appear white, while nanosized ZnO particles, due to their smaller dimensions compared to the wavelength of visible light don't scatter it and appear clear. This is attributed to the shift in optical absorption toward shorter wavelengths as the size tends to be smaller. There is also deviation in melting point from the bulk value greater than a couple of hundred degrees when the particle size goes down to below 10 nm. Fracture strength of nanostructured Cu-Fe alloys, 2.8 GPa (at ~38 nm) is about 5 times the fracture stress of iron with larger grain sizes, ranging from 50 to 150 μm. Saturation magnetization, Ms, of zinc ferrite increases significantly below 20 nm from 1 to 4.5. The coercive field of Nd2Fe14B with nanograin structure decreases significantly and the remnant magnetization increases below ~40 nm. Silica micro-particles at 16 % volume show increased energy release rate by 40 %, while silica nano particles at 5 % and 10 % volume show increased energy release rate by 60 % and 86 % respectively, i.e., from 0.162 kJ/m^2 to 0.26 and 0.3 kJ/m^2. Electrical conductivity of neat HDPE is 1×10^{-9} Sm^{-1}. However, with the addition of 10 wt % single-walled carbon nanotube (SWCNT), the electrical conductivity of HDPE increases up to 100 Sm^{-1}. Similarly thermal conductivity of neat HDPE is only 0.52 $Wm^{-1}K^{-1}$. However, it increases by 573% with an addition of 0.2 vol% SWCNT reinforcement. The elongation at break becomes 2.3 times higher when the standard composites are reinforced with nanoparticles. Impact strength increases by 1.7 times as well as the bending strength and the modulus also increase due to the incorporation of the nanofillers.

The emergence of the novel properties in polymer composites very much depend on the interfacial interaction between the polymer and these fillers. Thermodynamic and kinetic barriers inhibit the dispersal of inorganic nanoparticles with generally high surface energies within hydrophobic polymer matrices. In this chapter we discuss the effects of different types of interactions, including the intrinsic vs extrinsic properties of the nanofillers, as a result of their functionality, shape, orientation and size of the particles governing their properties.

Keywords: polymer nanocomposites, nanofillers, quantam effects, quantum confinement, size dependent-novel properties, interfacial interactions: attractive interactions, non attractive interactions, nanoconfined polymers, hierarchical morphology

INTRODUCTION

Polymer composite materials are replacing metals in various engineering applications due to their high specific strength, specific modulus, low densities [Gibson, 1994, ASM International, 1992, Zhang SW, 1998, Thomas Selvin et al. 2012, Sengupta et al. 2011], excellent friction & wear characteristics and corrosion resistance [Ning, 2011, Durand, 1985]. They have been designed and manufactured for various applications like naval ships, war planes and sea vehicles, fabrication of devices for non-linear optics, for optical data storage and for optical switching semiconductor [Reisfeld, 1996, Peyghambarian, et al., 1993].

In comparison to widely used conventional microscale particles, submicron particles have some unique features. When particle size is changed to nanoscale, properties such as melting point, fluorescence, electrical conductivity, magnetic permeability, and chemical reactivity change as a function of the size of the particle (http://www.nano.gov/nanotech-101/special). In this chapter we discuss the unique properties of the nanoscale materials, polymeric materials and the factors dominating the evolution of unique properties in polymer nanocomposites (PNCs).

POLYMER NANOCOMPOSITES

Nano composite (NC) is a conventional material reinforced by nanoscale particles or nanostructures dispersed through the bulk material. NC has at least one component in nanoscale. PNCs are nanoparticles (spheres, rods, and plates) dispersed in a polymer matrix. The principle composition of typical composite materials is illustrated in Figure 1.

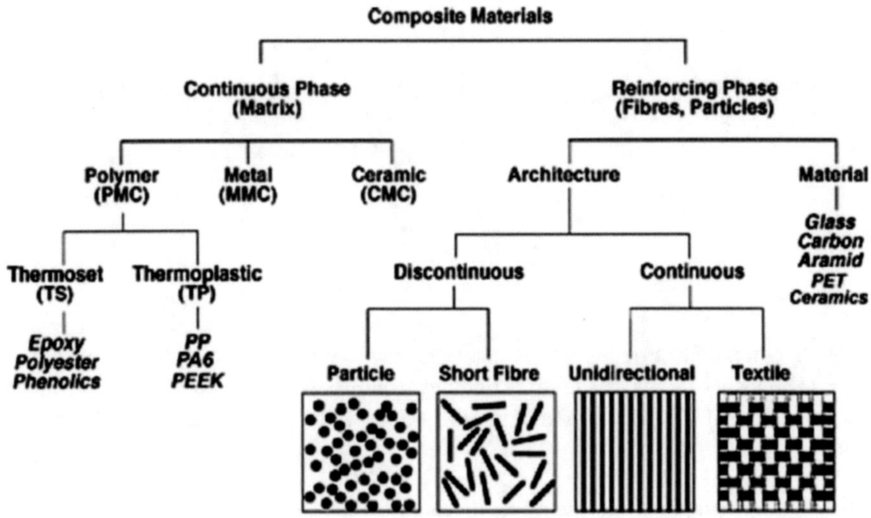

Reproduced with permission: Friedrich K , Zhang Z, Klein P : Wear of polymer composites. Stachowiak G.W (Ed) in 'Wear – Materials, Mechanisms and Practice', Wiley, Chichester, England 2005, Chapter 11, pp.269 – 290.

Figure 1. Systematic illustration of the structural components of composite materials.

In the light of global polymer production, worth 100-200 million tonnes annually, nanoparticle additions to plastics affords one of the commercially largest and diverse applications of nanotechnology. PNCs exhibit different mechanical, electrical, optical, catalytical and structural properties different from that of individual components. Nanoparticle dispersion has been shown to enhance physical (e.g., barrier, erosion resistance, and reduced flammability), thermomechanical (e.g., heat distortion temperature, thermal expansion coefficient, and stiffness) and processing (e.g., surface finish and melt strength) characteristics [Vaia and Maguire, 2007].

Performance enhancement of polymer nanocomposites capitalizes on advantages afforded by up to a 3 orders of magnitude spatial refinement of morphology relative to traditional micrometer-scale filled polymers and composites. This contrasts nanotechnology in electronics, optics, and data storage, where the nanoscale provides access to new physical processes based on quantum phenomenon. The mechanical, electrical, thermal, optical, electrochemical, catalytic properties of the nanocomposite differ markedly from that of the component materials (Figure 2). The length scale of the morphology becomes critically important as the dimensions of particle and polymer, as well as the interfacial curvature and interparticle distance, become comparable. Size limits for novel properties are typically< 5 nm for catalytic activity, < 20 nm for making a hard magnetic material soft [Banerjee et al.

2013], <50 nm for refractive index changes, and <100 nm for achieving superparamagnetism, mechanical strengthening or restricting matrix dislocation movement.

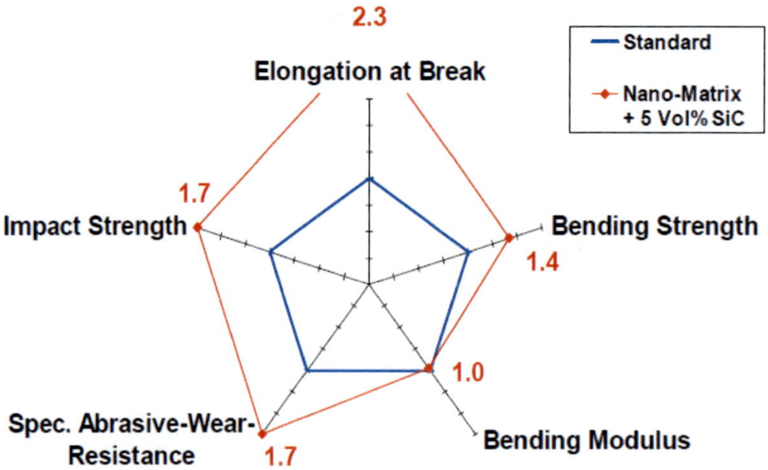

Reproduced with permission: Friedrich, K., Chang, L., Haupert, F.: 'Current and Future Applications of Polymer Composites in the Field of Tribology', in: L. Nikolais, M. Meo, E. Miletta (eds.): Composite Materials, Springer, New York, USA, 2011, http://www.springer.com/materials/special+types/book/ 978-0-85729-165-3.

Figure 2. Illustration of property improvements by nanofillers.

Factors Controlling the NC Properties

The nanoparticles lose their quantum size effect as their concentrations increase in the matrix, with the dispersion state of submicron particles determining the final properties of the PNCs. Fine tuning of size dependent properties in PNCs depends on a)dielectric function of material [Thomas et al. 2012], b) the surroundings, c) the shape and the size of the nanoparticle, d) the distance between the nanoparticles, e) the filler concentrations, and finally, f) the interface.

The NC properties are defined by three major characteristics [Anandhan and Bandyopadhyay 2011, Starr et al., 2001; Starr, et al., 2002]:

1) the nanoscale inorganic component,
2) the nanoscopically confined matrix polymer, and
3) arrangement of the inorganic component within the matrix.

1. The Nanoscale Inorganic Component

Nano-scale is defined as the size range between approximately 1 nanometer and 100 nanometer. A restriction of the definition of nanoscience and nanotechnology is that new functionalities are made available by manipulation of matter at this scale or through specificities of the nano dimensions, where the physical, chemical and biological properties of materials differ from those of the bulk matter. Nano-scale definition of 1-100 nm has incited a series of debates. It is argued that something can be described as nano-scale if it indeed shows a discontinuity, compared with the bulk, with respect to the number of phenomena emerging distinctively different from bulk behaviour [Ramsden and Freeman, 2009]. The search for discontinuities and crossovers around the size of a few nanometers revealed that the nano-scale itself is property dependent, the scale depending on external variables such as temperature and pressure. What appears indubitably as a nano-scale phenomenon at one temperature might cease to be thus distinguished when hotter or colder. Scale at which discontinuity compared to bulk emerges for different properties at room temperature is compiled in Table 1 [Ramsden and Freeman, 2009]. It is also debated that biological 'unique properties' are not constrained to sizes less than 100 nm, rather 1–1000 nm scale is recommended for the purposes of dosimetry and toxicological assessment [Teeguarden et al., 2007].

A fascinating and powerful result of the quantum effects of the nanoscale is the concept of "tunability" of properties [http://www.nano.gov/nanotech-101/special]. That is, by changing the size of the particle, a material property of interest can be literally fine-tuned. In this regard it is essential to consider the intrinsic and the extrinsic properties of the nanomaterials themselves.

Table 1. Summary of nano-scale defining lengths

Domain	formula	typical value / nm[a]
Surfaces	geometry	5
Nucleation	critical nucleus size (6)	5
Optics and electronics	Bohr radius (7)	10
Magnetism	single domain size	50
Mechanics	Griffith length (13)	50
Systems		
Electronic	electron mean free path	100
Photonic	evanescent field penetration depth	100
Fluidic	Debye length (30)	10
Mechanical	phonon mean free path	1000

[a] At room temperature.

Ramsden and Freeman, 2009.

Intrinsic versus Extrinsic Nano-Scale Properties

Intrinsic properties of manufactured nanomaterials (MNs) concern the material itself. Chemical composition leads to intrinsic properties [Thomas Selvin et al. 2012 in JAPs and JMS]. Size and shape can also underpin some intrinsic properties. The extrinsic properties are concerned with the functionality associated with the MNs. (http://2020science.org/2009/09/11/living-in-a-post-chemistry-world-the-regulatory-challenges-of-emerging-nanotechnologies/ Living in a post-chemistry world – the regulatory challenges of emerging nanotechnologies, Science, 2009, 2020). For example, the intrinsic magnetic properties of the nanomaterial do not vary considerably with size. The new magnetic behaviour different from their bulk is attributed to their extrinsic properties resulting from their interactions and is strongly dependent on their microstructure [Beke, et al., 2004; Banerjee et al. 2013].

Factors contributing to different properties of nano-scale materials compared to the macro-scale materials are discussed below [Filiponni and Sutherland, 2010]:

Gravitational Force

Gravitational force becomes negligible, as it is a function of mass and distance and is weak between (low-mass) nanosized particles, because the mass of nanoscale objects is so small.

Electromagnetic Forces

Electromagnetic forces dominate at nanoscale. Gravitational force is a function of mass and distance and is weak between (low-mass) nanosized particles. Electromagnetic force is a function of charge and distance, and is not affected by mass, so it can be very strong even when we have nanosized particles under dominance of electromagnetic forces.

Surface to Volume Ratio

Surface to volume ratio increases in nanomaterials, having a significant effect on the nanoparticles properties. Firstly, nanoparticles have a larger surface area when compared to the same volume of the material. For example, 1 kg of particles of 1 mm^3has the same surface area as 1 mg of particles of 1 nm^3. [Trantakis, 2012] As particle size decreases, a greater portion of the atoms are found at the surface compared to those inside, having profound effect on reactions that occur at the surface such as catalysis reactions, detection reactions and the reactions that require physical adsorption of certain species for the reaction to occur. For example, a particle of size 3 nm has 50 % of its atoms on its surface, at 10 nm 20 % of its atoms and at 30 nm has 5 % of its atoms on its surface, thereby 3 nm size particles leading to chemically super- highly reactive nanoparticles.

The larger fraction of the atoms on the surface influences physical properties such as melting point, which happens to be lower for the nanosized materials. This is due to the ease with which the surface atoms, which require less energy to overcome the intermolecular forces holding the atoms together, can be removed easily from bulk atoms [Burda, et al., 2005]. The atoms and molecules at the interface enhance reactivity and high surface energy, with a greater tendency to agglomerate. At nanoscale weak forces (like electrostatic and Van der Waals) result in strong interactions when the surface area is large. Materials which are inert in their bulk form, e.g., gold are reactive in their nanoscale form. Therefore a given mass of nanomaterial will be much more reactive than the same mass of material made up of large particles.

Also, surface effects become significant because of the large surface to volume ratio of very small particles [Thielsch, 1998, Brus, 1984, Steigerwald, et al., 1990, Wang, et al., 1991]. The resulting physical properties are neither those of bulk metal nor those of molecular compounds, but they strongly depend on the particle size, interparticle distance, nature of the protecting organic shell, and shape of the nanoparticles [Brust and Kiely, 2002].

Random Molecular Motion

At the macroscale level random molecular motion (molecules move due to their kinetic energy above absolute zero) is very small compared to the size of the particles. At the nanoscale these motions can be of same scale as the size of the particles thus behaving differently, a typical example being the Brownian motion.

Wave-Particle Duality

As the size of the matter decreases, electron wave-like nature dominates and exhibits tunnelling effect (tunnelling is the penetration of an electron into an energy region that is classically forbidden) where the thickness of the barrier or the energy potential is comparable to the wavelength of the particle, observed only at nanoscale.

Electrical Properties

Semiconductor nanocrystallites exhibit interesting electronic, optical and photochemical properties differing from their bulk counterpart. The size induced metal-insulator transition is observed if the metal particle is small enough (about 20 nm), when that size-dependent quantization effects can occur [Daniel and Astruc, 2004].

The nanoparticles exhibit energy separation between the valence band and the conduction band, called the band gap, unlike in bulk metals. The band gap energy increases as the size of the crystallites decreases. The increase in bandgap is caused by the quantum confinement of freely mobile electrons trapped in quantum boxes leading to discrete energy states of electrons in a nanoscale metal structure - implying more energy states are lost due to shrinking size, increasing the energy band gaps in the diameter range 1-10 nm (intermediate between the size of small molecules and that of bulk metal) [Alivisatos, 1996].

Under spatial confinement, electron energies are quantized, leading to many novel effects, collectively known as quantum size effect (QSE) [Chen, et al. 2003]. This quantum size effect arises at dimensions comparable to the Bohr-diameter of the exciton when the particles behave electronically as zero-dimensional quantum dots (or quantum boxes). The quantized energy spectrum has important consequences on electronic applications. The quantum confinement also affects the mass transport and stability of the shape of the nano-interconnect lines.

Optical Properties

The nanoscale materials show different optical properties, unlike their bulk counterpart. The increase in bandgap seen in nanoscale material due to quantum confinement leads to absorption in shorter wavelength resulting in blue shifts. When the particle radius is less than the exciton radius, photon induced transitions in exciton energy levels produce series of discrete optical absorption levels. At such crystallite dimensions the photo generated ‘electron-hole’- pairs are spatially confined if the diameter of the crystallites is smaller than the Bohr-diameter of the exciton. Freely mobile electrons trapped in such quantum boxes show a characteristic collective oscillation frequency of the plasma resonance, giving rise to the so-called plasmon resonance band (PRB) observed near 530 nm in the 5-20 nm diameter range. As a result colloids of metal nanoparticles such as gold and silver display colours different from their bulk counter parts like red, purple, or orange which depends upon the dielectric function of material and the surroundings and to the shape and the size of the nanoparticle. The PRB is also sensitive to the distance between the nanoparticles. [Pinchuk and Schatz, 2008]

Magnetic Properties

Magnetic materials such as Fe, Co, Ni, Fe_3O_4show size-dependent magnetic properties. Magnetic behaviour of a material depends on the structure of the material and on its temperature. The typical size of classical magnetic domain is ~ 1 μm. When the size of magnetic domain reaches nanoscale, these materials show new properties (Banerjee et al. 2103) due to quantum confinement like magneto resistance effect, a property of a material to change the value of its electrical resistance when an external magnetic field is applied to it. The strength of a particle's internal magnetic field and the 'coercive force' (or magnetic memory) needed to reverse an internal magnetic field within the particle is size dependent. The size of magnetic nanoparticles influences the saturation magnetization increasing significantly below a grain size of 20 nm [Chinnasamy, 2000]. For example, saturation magnetization, Ms, of zinc ferrite increases significantly below 20 nm from 1 to 4.5.

The effects of the size of the nanoparticle grain structure on $Nd_2Fe_{14}B$ indicate that in this material the coercive field decreases significantly below 40 nm and the remnant magnetization increases. This is because when the magnetic interactions extend to a distance of more than one neighbouring particle, the reversal of magnetisation becomes easier and leads to a drop in the coercivity [Manaf, et al., 1991].

Mechanical Properties

Deviations from the Hall-Petch behaviour have been observed for materials made of particles less than 20 nm in size, where the nanosize materials having smaller grain size exhibit increased yield strength. This is because the conventional dislocation-based deformation is not possible in bulk nano-structured materials with sizes < 30 nm as mobile dislocations are unlikely. The yield strength, σ_y, of a conventional grain-sized material is related to the grain size by the Hall-Petch equation:

$$\sigma_y = \sigma_0 + Kd^{(-1/2)} \quad (1)$$

where,

σ_o is the frictional stress opposing dislocation movement, or yield strength at infinite grain size,

K is a constant, and
d is the grain size in micrometers.

One method of increasing the stress at which the brittle-to-ductile transition occurs is to impede the movement of the dislocations by introducing tiny particles of another material into the lattice. This process is used to harden steel, where particles of iron carbide are precipitated into the steel. The iron carbide particles block the movement of the dislocations. Stress needed to fracture bulk glass is around 100 MPa (14500 psi), while the theoretical stress needed for breaking atomic bonds is approximately 10,000 MPa (1,450,000 psi), indicating that the fracture stress increases as the fibre diameter decreases, hence the improved uniaxial tensile strength. For nanostructured Cu-Fe alloy fracture occurs at 2.8 GPa which is about five times the fracture stress of iron having larger grain sizes, ranging from 50 to 150 μm [He and Ma, 2000].

Polymers

Polymers are very light in weight with significant degrees of strength [Anandhan & Bandyopadhyay 2011 Sengupta et al. 2011, Thomas Selvin et al. 2012]. They show low density, low coefficient of friction, good corrosion resistance and poor tensile and mechanical properties. For polymers, many bulk properties are related to the size of the polymer chain, which is characterized by the radius of gyration, R_g (~2-20 nm). The radius of gyration is the average distance from the centre of the gravity to the chain segment. Most thermoplastics are high molecular weight (M_w) polymers whose chains associate through weak Van der Waals forces (polyethylene); stronger dipole-dipole interactions and hydrogen bonding (nylon). Polymers exhibit molecular weight dependent properties such as elastic modulus and strength and melting point. The greater chain linearity and chain length tend to increase the melting point and improve the physical and mechanical properties of the polymer due to greater crystallinity. Polymers can be transparent or opaque. They are usually both thermal and electrical insulators, i.e., they do not conduct electricity and are poor conductors of heat due to low number of free electrons. Few polymers show very high electrical conductivity, about one quarter that of copper, or about twice that of copper per unit weight. When doped with electrically active impurities, they show conductivities similar to semiconductors, e.g., polyacetylene, polyparaphenylene and polypyrrole [web.utk.edu/~prack/.../Chapter%2019%20Electrical%20Properties.pdf]. Orienting the polymer chains (mechanically, or magnetically) during synthesis also results in high conductivity along oriented direction. Such polymers have applications in advanced battery electrodes, antistatic coatings, electronic devices and polymeric light emitting diodes.

2. Nanoscopically Confined Polymers

Polymers confined to nanoscopic dimensions are found in many applications including photoresists, asymmetric membranes, and nanocomposites used for automotive components. While the properties of the nanoscale fillers are extraordinary, the significant changes observed are not due to the nanofillers alone. Fig 3 is an illustration of the effect of inclusion of nanoparticles in polymer matrix, compared to micron-size particles.

Nanoconfined polymers show deviation in properties from the bulk that can result in significant enhancement in the properties, such as the creep compliance and physical ageing (structural relaxation of glassy polymers is responsible for densification and loss of mechanical properties over long time scales).

Nanoscopic confinement of a polymer matrix by nanofiller occurs when the average inter-particle distance is close or less than the dimensions of unperturbed macromolecular random coil estimated by the radius of gyration R_g. When the R_g is $<$ the interparticle distance, the nanocomposites exhibit low percolation threshold. Two effects namely: (i) size effects and (ii) interfacial effects dominate a nanoscopically confined polymer. Size effects dominate when the length scale of the confining dimension, i.e., thickness in thin films, and the length scale governing a property coincide.

The huge interfacial region resulting from the nanoscopic dimensions of the inorganic component distinguishes polymer nanocomposites from traditional composites and filled polymers, by comparison, the interfaces in conventional composite constitute a much smaller volume fraction of the bulk material.

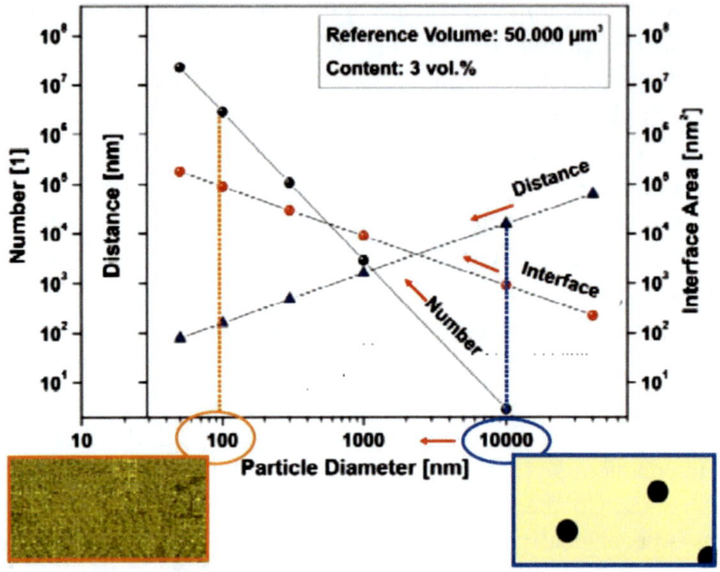

Reproduced with permission: K. Friedrich, F.Haupert, Z Zhang "Property improvements of polymer composites by spherical nanoparticles – Proceedings of ACUN-5, Developments in Composites, Advanced, Infrastructural, Natural and Nanocomposites" UNSW, Sydney, Australia, 11-14 July 2006, pages 297 – 305 Eds Sri Bandyopadhyay et al. ISBN 0 7334 2363 9.

Figure 3. Correlation among particle size, number of particles, particle surface area (interface), and surface-to-surface interparticle distance. The calculations are based on the assumption of spherical particles, cubic distribution and ideal particle dispersion within the matrix.

Given the enormous surface to volume ratio for nanoparticles, the interphase volume fraction can dwarf the properties of the nanoinclusions to percolate through the composite. Enhancing their effect is the extremely significant role that the interphase plays in these systems. The contribution of the interphase to the mechanical and functional properties of polymer nanocomposites has been extensively demonstrated by Hussain, and Hojjati, 2006. Figure 3 illustrates the interface and the number of particles increasing with decreasing particle dimensions, while the interparticle distances decrease. Interfacial effects increase significantly with an increase in the ratio of interfacial surface area to volume. It is the interplay between size and interfacial effects that results in a wealth of interesting observations including substantial increases or decreases for example, in glass transition, new non-linear viscoelastic phenomena, and their relaxation behaviour [Zhang, et al., 2011, Giannelis, et al., 1999].

Due to the high surface area of nanostructures, nanocomposites present many interfaces between the intermixed phases. The enormous amount of interface present means that most of the polymers interact directly with the inorganic component, introducing heterogeneity within the polymer matrix on the nanoscale and deforming the polymer from its bulk conformation. For example, the glass transition temperature (T_g) of 50 nm wide PMMA nanolines on silica is ~15 K lower than that of a PMMA film on silica of the same 18 nm thickness. Attractive PMMA-silica interfacial interactions increase the T_g, while free surfaces decrease the T_g of PMMA in ultrathin films relative to bulk PMMA [Mundra, et al., 2007].

Nanofiller Loadings

A homogenous dispersion of nanofillers in the polymer matrix assists in evolution of superior properties at low concentration level without sacrificing many advantages of the parent polymer, as the result of large extent of interphase region penetrating through the nanocomposite forming a percolating altered polymer network. The nanoparticles lose their QSE as their concentrations increase in the matrix. Ramanathan et al., 2007 and Reisfeld, 1996 reported that the enhancements in the T_g values of graphite/PMMA composite initially increased at low filler concentration but declined from the maximum value with increasing concentrations, achieved at higher loadings. Cho and Daniel, 2008, demonstrated that the glass transition temperature of a composite increased only modestly with nanotubes which were somewhat agglomerated, while the T_g was increased by 39 °C in a composite where better nanofiller dispersion was achieved. Attractive interfacial interaction leading to T_g enhancement, is attributed to high crosslink densities leading to higher M_w, reduced chain mobility, and crystalline molecular assemblies [Devasahayam, 2006, Tsui, et al., 2001, Wang and Zhou, 2002]. These phenomena are hypothesized to result from decreases in nanoparticle dispersion achieved at higher loadings.

Filler addition causes structural variations in the polymeric network, significantly influencing the degree of crystallinity and producing defects in the polymeric matrix as seen in polyethyl methacrylate (PEMA) films doped with up to 15.0 wt % of $CoCl_2$ and LiBr, having potential applications in photonics, bio-sensing, microlithography, microelectronics, packaging [Abdelrazek, et al.,2012, Zidan, 2003]. Consequently there is a decrease in optical absorption edge and optical band gaps (both direct and indirect) as the dopant concentration increases. These defects produce localized states in the optical band gap, decreasing the optical gap resulting in an increase in the degree of disorder in the films [Abdelghany, et al., 2012]. In other words, the agglomeration of nanoparticles remarkably reduces the volume fraction of interphase and impedes the formation of a percolating interphase network inside the polymer matrix. Therefore, the ex-relaxation dynamics of nanocomposites with clustered nanoparticles show less change from the pure polymer relative to nanocomposites with well-dispersed nanoparticles. Enhancement in T_g in nanocomposites varies with varying degrees of dispersion, and decay above a threshold volume fraction of nanofillers [Qiao, et al., 2011].

Polymers in general are non-conductive. Electric conductivity of metal–polymer composite predominantly depends on the electric conductivity, the shape and size, and the volume fraction and the spatial arrangement of the metal particles. The percolation threshold decreases rapidly with increasing metal particles axial ratio and decreasing the metal particle size. While HDPE is hardly electrical conductive, the dielectric strength and electrical conductivity being respectively 31.2 KV/mm and 1×10^{-9} Sm^{-1}, the HDPE nanocomposites exhibit improved electrical properties [Shah et al., 2009]. The dielectric strength of HDPE was reportedly doubled with an addition of 5wt % nanoclay improving the volume resistivity and surface resistivity as 1000 times as neat HDPE. With an addition of 10 wt % SWCNT, the electrical conductivity increased to 100 Sm^{-1}.

The conductivity of mechanical processed Graphene/Cobalt-containing PEO and PMMA polymer nano composites shows a typical percolation-controlled conductivity profile with a resulting steep drop at low metal loadings, Figure 4, [Luechinger et al., 2008], the composites displaying conductivities over 50 S cm^{-1} at metal loadings of 40 volume % and higher.

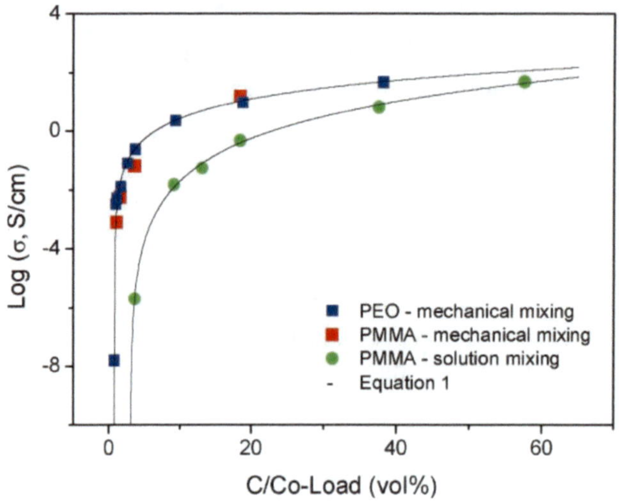

Luechinger et al., 2008.

Figure 4. The conductivity of C/Co-containing composites.

The electrical conductivities σ, over the C/Co volume content, ϕ of the composites follow the classical expression derived from percolation theory (eq 2) for C/Co contents above the percolation threshold ϕ_c (t is a critical constant) [Stauffer, 1985., Zallen, 1983].

$$\sigma \alpha \, (\phi - \phi_c)^{t}: \quad (2)$$

The percolation limits of mechanically prepared composites demonstrate a material independent behaviour both in PEO and PMMA, the percolation behaviour indicating that the C/Co 3D network connectivity controls the conductivity at room temperature [Benoit et al., 2002]. The mechanical blending initially generated a very phase segregated heterogeneous material with metal preferably found on the surface of the initially used polymer micro particles (Figure 5a). The electrical conductivity of this material was high (Figure. 5b; zero deformation cycles) due to the poorly mixed metal-rich domains proceeded throughout the whole sample, offering percolating conductance pathways. After four subsequent deformation cycles, the C/Co nanoparticles are well-dispersed in the polymer matrix (5d), with corresponding sharp decrease in electrical conductivity (5b). Even a single deformation cycle reduced the conductivity by two orders of magnitude confirming the successful integration of the metal particles into the polymer. Subsequent deformation cycles yielded a composite of much lower conductivity but a high degree of particle dispersion.

Polymers exhibit diamagnetic susceptibility and the magnetic resonance. The magnetic susceptibility is negative and small and temperature independent for diamagnetic materials. Paramagnetism can occur in a polymer only when it contains transition group elements or unpaired electrons in free radical or triplet state. Super paramagnetic particles are known to display strong interactions with external magnetic fields leading to large saturation magnetizations, optoelectronics and optical computing for neural networks applications.

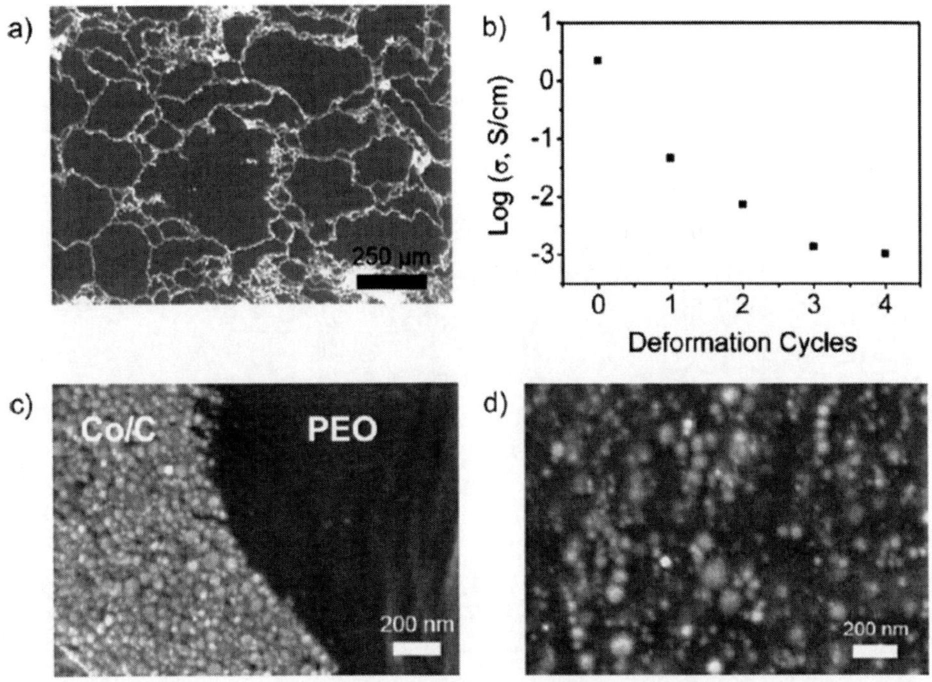

Reproduced with permission: Norman A. Luechinger, Norman Booth, Greg Heness, Sri Bandyopadhyay, Robert N. Grass, and Wendelin J. Stark , Surfactant-Free, Melt-Processable Metal–Polymer Hybrid Materials: Use of Graphene as a Dispersing Agent, Adv. Mater. 2008, 20, 3044–3049.

Figure 5. a, b, c, d) Scanning electron microscopy images and b) volume conductivity of a 40 wt % C/Co containing PEO nanocomposite. The heterogeneous microstructure of a hot compacted pill (a) shows distinct regions of pure C/Co nanoparticles and polymer (c). After 4 deformation cycles, the C/Co nanoparticles were well-dispersed in the polymer matrix (d), which stays in line with a sharp decrease in electrical conductivity (b).The heterogeneous microstructure of a hot compacted pill (a) shows distinct regions of pure C/Co nanoparticles and polymer (c). b) volume conductivity of a 40 wt % C/Co containing PEO nanocomposite.

Pronounced ferromagnetism is reported for a nanocomposite containing 90 wt % C/Co with potential application in organic spintronic devices, with logic, communications and data storage. A very high saturation magnetization of 158 emu g^{-1} is measured for the nanocomposite containing 90 wt % C/Co, which compares favourably with existing melt-processable magnetic polymers, as seen in Figure 6. Low coercivity of both materials is a result of the small particle size of C/Co which is below the magnetic domain size of cobalt. Co shows a saturation magnetization in the range of 162 emu g^{-1}. The saturation magnetization values (105 and 158 emu g^{-1}, respectively) allow a saturation magnetization of 175 emu g^{-1} for pure cobalt, about 7 % higher than the theoretical value for bulk cobalt. Respaud, et al., 1998, reported increased saturation magnetization for ultrafine cobalt particles. The very low coercivity of both materials is a result of the small particle size of C/Co which is below the magnetic domain size of cobalt [Ning et al., 2011].

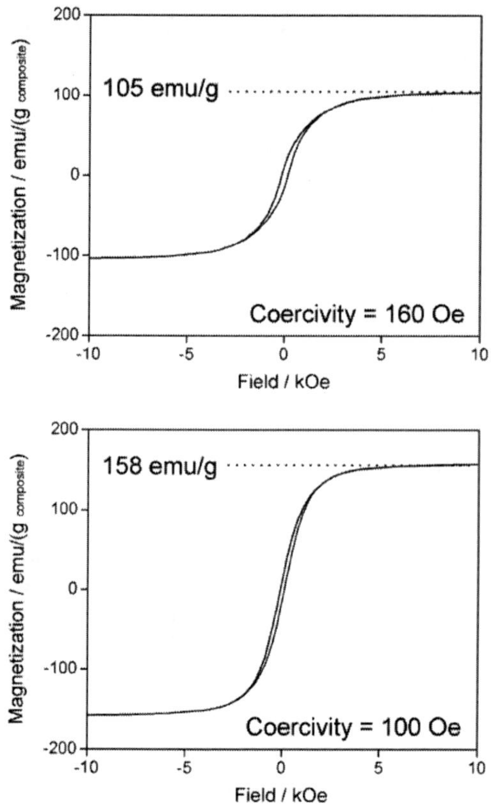

Reproduced with permission: Norman A. Luechinger, Norman Booth, Greg Heness, Sri Bandyopadhyay, Robert N. Grass, and Wendelin J. Stark , Surfactant-Free, Melt-Processable Metal–Polymer Hybrid Materials: Use of Graphene as a Dispersing Agent, Adv. Mater. 2008, 20, 3044–3049.

Figure 6. The magnetization of PMMA-based C/Co composites with 60 wt % (top) and 90 wt % (bottom) cobalt gives access to bulk magnetizations of up to 158 emu g-1, suggesting melt-processable composites as an alternative to currently used magnetic materials. The coercivity of both composites stayed low (<160 Oersted).

A substantial room-temperature saturation magnetization of about 0.65 emu g^{-1} was reported in a crystalline blend film of conjugated polymer poly(3-hexylthiophene) (P3HT) and phenyl-C61-butyric acid methyl ester (PCBM), requiring no transition metals, due to nanoscale phase separation. Here, the origin of organic ferromagnetism was associated with the crystallization of P3HT and charge transfer between the P3HT and the PCBM. The majority of magnetization is attributed to the intrinsic properties of the P3HT: PCBM blend film. The nanoscale magnetic domains are correlated with the crystallization of the P3HT blending of the P3HT and PCBM [Yang et al., 2012].

Voitenko et al., 2007, reported that concentrations of quantum dots have significant effects on the wavelength of the photoluminescence in polymer blends doped with PbSe quantum dots. At low concentrations of quantum dots, the energy transfer appears to be from excitation of the polymer which is transferred to the quantum dot. At higher concentrations, it appears that quantum dot to quantum dot interactions are important and are mediated by the polymer matrix [Voitenko, et al., 2007].

Mechanical

As illustrated in Figure 2, incorporating nanoparticles significantly improves the mechanical properties of the polymer matrix. Making materials with nano-sized grains has the potential to provide significant increases in yield stress, and has many useful applications such as stronger materials for automobile bodies [Poole and Owens, 2003]. This is attributed to much higher specific surface area of submicron size particle promoting stress transfer from the matrix to submicron particles. The required amounts of submicron particles in polymer matrices are usually much lower than those of the corresponding microparticles [Zhang and Friedrich 2005., Schwartz and Bahadur S, 2000., Jawahar and Balasubramanian, 2006]. A crack provides a mechanism whereby a weak external force can break stronger bonds one by one. In general, the crack initiation is followed by stable crack propagation as shown in Figure 7. When the crack size of specimen exceeds a critical value, a_c, an unstable fracture occurs. According to the Griffith theory (Krausz and Krausz, 1988), the total energy for crack extension is derived from the surface energy, U_s and the stored elastic energy, U_e. They obey an energy equilibrium until fracture, that is, $\Delta U_{total} = \Delta U_s + \Delta U_e$. As displayed in Figure 7, at critical crack size the total energy reaches a peak and the change in total energy is zero.

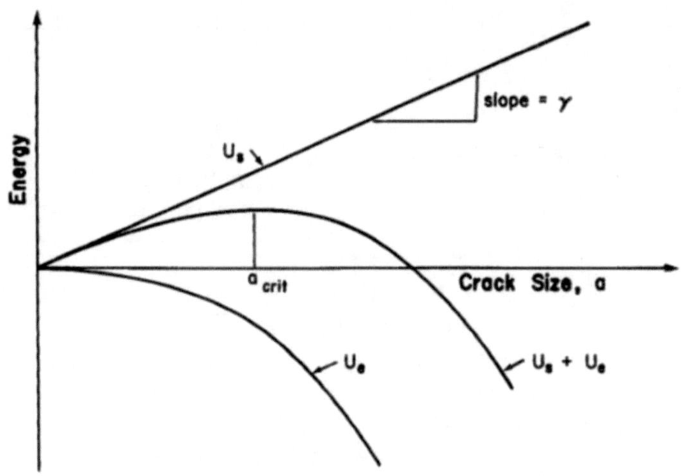

Krausz and K. Krausz, Fracture Kinectics of Crack Growth. Kluwer Academic Publishers, Dordrecht, 1988, pp. 2-6.

Figure 7. Variation of the surface energy Us, stored elastic energy Ue, and total energy Us + Ue A. S.

The fracture stress increases as the fibre diameter decreases, hence the higher uniaxial tensile strength. This is because the high specific surface area of submicron size particle promotes stress transfer from the matrix to submicron particles. The required amounts of submicron particles in polymer matrices are usually much lower than those of the corresponding microparticles (Zhang and Friedrich, 2005, Schwartz and Bahadur, 2000, Jawahar and Balasubramanian, 2006).

Composite stiffness depends significantly on particle loading, not necessarily on particle/matrix adhesion since the fillers have much larger modulus than the matrix. There is also a critical particle size, usually in the nanoscale, below which the composite stiffness is greatly enhanced due to the significant effect of the particle size, probably caused by the much larger surface areas imparting a ‘‘nano”-effect [Fu, et al., 2008].

Improved barrier properties and wet adhesion in epoxy-steel system have been reported due to precipitation of nano organometallic crystals contributing to trilayer interface in laminar composites when concentration exceeds solubility product [Devasahayam, 2006[a]]. A trilayer interface due to formation of fibre-like nano-crystallites within the organic layer during slow curing process shows a gradient and a higher value of Young's modulus than a bilayer interface system exhibiting higher fracture energy and barrier properties [Devasahayam, 2006[a], Devasahayam, 2006[b], Devasahayam, 2004]. Figure 8 illustrates the effect of volume fraction on the impact energy of the nanoparticles. It is obvious that the smaller the particles are, the more pronounced is the improvement of the impact energy at already lower filler volume fractions [Wetzel, 2006; Wetzel, et al., 2002].

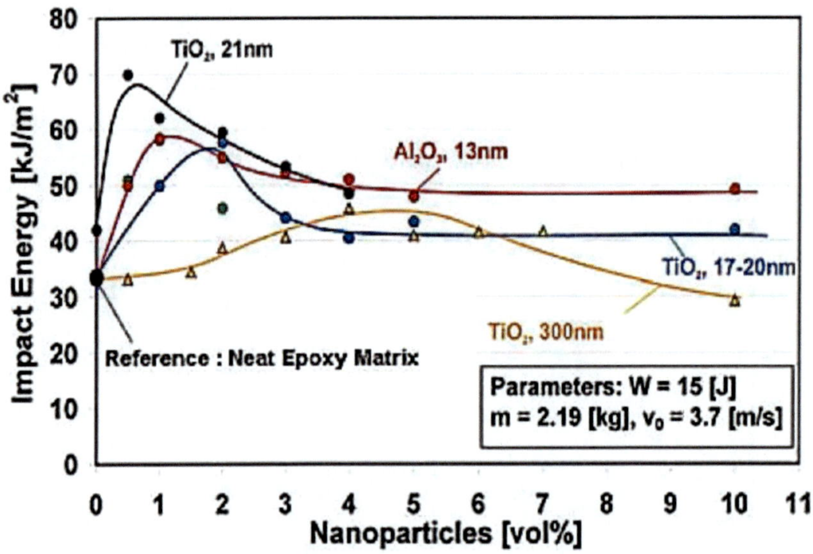

Reproduced with permission: Wetzel, B.: Mechanische Eigenschaften von Nanokompositen aus Epoxydharz und keramischen Nanopartikeln, Dissertation Technische Universität Kaiserslautern, Germany, February 2006; Wetzel, B., Haupert, F., Friedrich, K., Zhang, M. Q., Rong, M. Z.: Impact and wear resistance of polymer nanocomposites at low filler content, Polymer Engineering and Science 42, (2002) 1919-1927.

Figure 8. Impact energy of EP-resin composites, filled with different amounts of various types of Nanoparticles.

Great improvement in thermal conductivity of polymers after nanoparticle inclusions has also been reported. Thermal conductivity of neat HDPE is only 0.52 $Wm^{-1}K^{-1}$. However, it increased by a factor of about 6 with an addition of 0.2 vol % SWCNT) reinforcement [Mark, 2009].

Wavelength tunability of the response by varying the size of the NCs is reported for the conjugated polymer poly[2-methoxy-5-(2'-ethylhexyloxy-*p*-phenylenevinylene)] (MEH-PPV) sensitized with PbS nanocrystals (~5 nm diameter). The PNC shows photoconductivity at wavelengths where only the nanocrystal component is absorbing. This photo response corresponds to infrared sensitization of the polymer by the NCs. The photocurrent is attributed to the absorption in the nanocrystals with subsequent hole transfer to the polymer [McDonald, et al., 2004]. The increased absorption of the blend in the 300 – 400 nm region, compared to the pure polymer film, is also due to the NCs. significant photocurrent, indicating infrared sensitization of the polymer by the NCs. Excitonic structures have been observed only in large gap semiconductor particles such as ZnS and CdS where exciton binding energy is strong and the band gap changes with size. In narrow gap semi-conductors, such as PbS, the large exciton radius and the strong screening can give small exciton binding energy and the absence of excitonic structures in the absorption edges. Bulk PbS has a direct gap at 0.41 eV at 300 K. As the cluster size decreases, the absorption edge shifts to higher energy, from 3200 nm for the bulk to 530 nm for very small clusters, with the band gap increase covering a range of 2 eV. Wang and Herron, 1992 reported the first demonstration of photoconductivity in a polymer doped with semiconductor nanoclusters (CdS). Wang et al.,1987 explained the transition of PbS from molecular to bulk form in polymer films based on the effect of band nonparabolicity, where, as the particle size decreases, the band gap shifts to the blue and eventually approaches the transition energy of the first allowed excited state, $X\text{-}\rightarrow A$, of a PbS molecule, The band gap approaches the value of the first allowed excited state at around 20 A. implying that the clusters become "molecular-like" in this size regime, as far as their optical properties are concerned.

The Role of Nanoparticle Shapes in Dispersion

Given the same volume, the extent of surface area depends on the shape of a nanomaterial. For example, for a sphere and a cube having the same volume, the sphere will have a larger surface area than the cube. Orientation of nanofillers in a matrix plays crucial role in enhancing the mechanical properties due to the alignment of for example, the montmorillonite along the carbon-fibre surface or montmorillonite and polymer crystallites in a Nylon –montmorillonit enanocomoposites, where improvements in transverse and axial strength in the final fiber-reinforced structure due to the alignment are reported [Gusev and Rozman, 1999].

Carbon nanotube bundles are difficult to disintegrate into single tubes due to strong Van der Waals interactions inhibiting the incorporation of carbon nanotubes into a polymer matrix. In contrast, the spherical geometry of the materials facilitates dispersion, as the contact area between two spheres is much lower than between two tubes. It is reported incorporating spherical nanoparticles in an epoxy matrix results in a simultaneous improvement in the impact energy and the elastic modulus. The effective electric conductivity increases rapidly with the increase of the metal particle axial ratio before it reaches the axial ratio threshold [Xue, 2004]. The percolation threshold of metal–polymer composites, e.g., tin–polypropylene composites decreases rapidly with increasing the metal particle axial ratio and decreasing the metal particle size. Using metal particles with larger axial ratio and smaller size helps to reduce the minimum metal content required to reach a certain electrical conductivity value [Xie et al., 1996].

Anisotropic nanoparticles influence the permeability of gases through a polymer. The increased aspect ratios of the nanoparticle exhibiting a better barrier properties at lowest volume fractions of particles are illustrated in Figure 9, with potential application in Cryogenic tanks, durability to diffusion species. Incorporation of high aspect ratio conductive particles such as CNT, graphite flake, metals, etc. as percolated network in resin between conductive fibres find application in adhesives, coatings, gap fillers and electronic enclosures [Krishnamoorti, 2010].

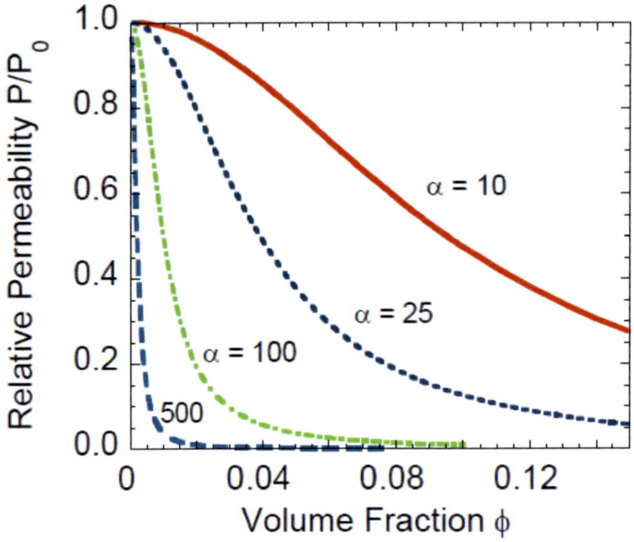

Figure 9. E.L. Cussler, Stephanie E. Hughes, William J. Ward III, Rutherford Aris J. Memb Sci., 1998, Volume 38, Issue 2, August 1988, Pages 161–174. α = aspect ratio.

Interaction between the Nano Fillers and Polymers

Properties in NC are affected by different interactions. They include a) local packing of the polymer chains, b) monomer density profile, c) relaxation behaviour, d) fracture energy and toughness, e) viscosity, f) aggregation and dispersion of nanoparticles, g) glass transition temperatures, h) Young's modulus, i) electrical conductivity and j) magnetic properties.

Types of Interaction

Two types of interactions namely, a) attractive interactions due to chemisorption /grafting (anchoring by chemical bonds) and b) non-attractive interactions due to physisorption (chain attachment by Van der Waals interactions) dominate in PNCs.

Dispersion

The dispersion state of nanoparticles in a polymer matrix affects the properties of the resulting nanocomposite. For example, the CNTs need to be dispersed on nanometer scales to impart outstanding mechanical and electrical properties onto polymer matrices. A well dispersed nanofiller-polymer systems indicates superior interaction between the particle-matrix, resulting in thermodynamically stable phase. Major challenge is to achieve increased performance (e.g., thermomechanical) through dispersion at the single-particle level determined by the thermodynamic miscibility of the nanoparticles with the polymers. According to Flory theory, this miscibility is comparable in difficulty with mixing two large polymers, where the enthalpic effects dominate this situation with entropy playing only a weak role [Kumar and Krishnamoorty, 2010].

Interactions between the nanoparticles and/or interactions between nanoparticles and the surrounding polymer matrix determine the aggregation/dispersion in a nanocomposite. The strong adhesion forces between metal particles due to their high surface energy, within the polymer matrix having a lower surface energy is the primary cause for dispersion problems in NCs (see Table 2 for a representative overview). Strong Van der Waals interactions inhibit the incorporation of carbon nanotubes into a polymer matrix and reflect the large contact area of two adjacent carbon nanotubes in an aggregate (Luechinger et al., 2008). CNT-PNC comprising an immobilized polymer phase or interphase is composed of polymer chains surrounding the CNTs or other nanoparticles. The polymer chains are strongly attracted by the CNT surfaces because of Van der Waals and/or π-π interatomic stacking forces. As a result, they behave quite differently from bulk polymers. In addition, they feature good interfacial adhesion, filler/filler entanglement, and interfacial strength (Prado, 2011). For the particles to disperse, the monomer-particle interaction strength needs to be only slightly stronger than the monomer-monomer interactions (Figure 10) [Starret al., 2004].

Table 2. Surface energy of materials

Solid material [a]	γ_s [mJ m^{-2}]
carbon fibers	40–46
carbon nanotubes	28–45
PTFE	20–21
PE	30–33
PMMA	41–43
copper	1800
iron	2400
cobalt	2500

[a]PTFE: Poly(tetrafluoroethylene); PE polyethylene; PMMA: poly(methylmethacrylate).
Luechinger et al., 2008.

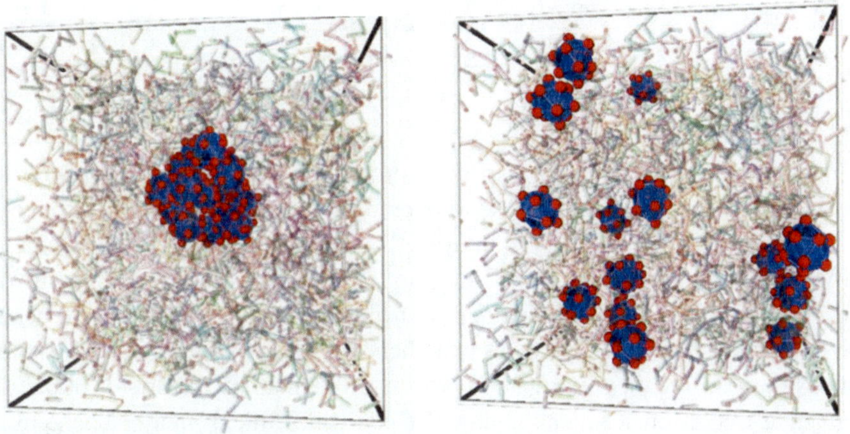

Reproduced with permission: Starr, Francis W. and Glotzer, Sharon C., "Science and Engineering of Nanoparticle-Polymer Composites: Insights from Computer Simulation" (2004). Division III Faculty Publications. Paper 249. http://wesscholar.wesleyan.edu/div3facpubs/249.

Figure 10. Clustered nanoparticles for a relatively weak Monomer-particle interaction strength = 1.0 dispersed nanoparticles for Monomer-particle interaction strength = 1.5.

The free energy of adhesion, W_a per area between two surfaces of the same material acting in a liquid medium is given by eq 3:

$$W_a = -2\gamma_s \tag{3}$$

where,

W_a = free energy of adhesion, per area between two surfaces of the same material in a liquid medium, and γ_s= interfacial tension between the solid and liquid phase.

A low interfacial tension γ_{sl} (equation 3) leads to low adhesion forces between particles, breaking apart nanoparticle agglomerates and causing dispersion of primary particles facilitated (Luechinger et al., 2008,). Contact angle measurements and the surface energies of the solid and liquid phase (γ_s and γ_l,) as well as the interfacial energy γ_{sl} (eq 4) are good indicators of the dispersability of nanoparticles in polymers.

$$\gamma_s = \gamma_s - \gamma_l cos\theta \tag{4}$$

where θ is the wetting contact angle.

Low-surface-tension liquids such as hydrophobic polymer melts (with low-surface-tension) are able to wet most solid materials that is $\theta < 90$ (Table 3).

Table 3. Wetting contact angles of polymers on solid materials

Liquid polymer [a]	Solid material	Contact angle
PE	aluminum	30°–55°
PP	aluminum	17°–28°
PP	carbon fiber	17°
PP	carbon nanofiber	18°
PEO	carbon fiber	26°
PEO	carbon nanofiber	36°

Luechinger et al., 2008.

Different approaches to solve the dispersion problem have been reviewed by Grossiord et al., 2006. At present the use of surfactants to overcome the gap between the different surface energies of filler and matrix is widely employed. The major processing routes which are used for incorporating nano and micro-particles into different types of polymer matrices are illustrated in Figure 11. High shear energies are necessary to break up the agglomerates with the aim to separate the individual particles, wet them with the polymer fluid and stabilize their separated positions during the solidification process of the matrix [Friedrich, 2006].

A surfactant-free mechanical entrapping of metal nanoparticles in a graphene like carbon shell to facilitate the contact between the polymer and the particles to overcome dispersion has been reported by Luechinger et al., 2008. This involved mechanically incorporating graphene coating of 1-2 nm of the metal particle like Cobalt nanoparticles of 10-50 nm size into a polymer matrix. This allowed filler loading up to 90%.

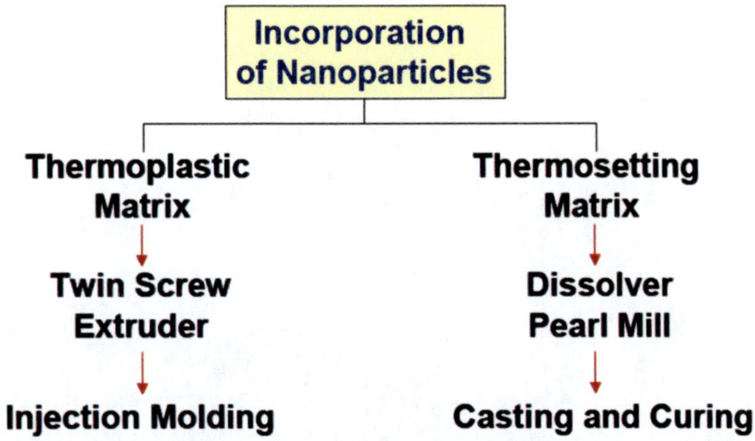

Reproduced with permission: K. Friedrich, F.Haupert, Z Zhang "Property improvements of polymer composites by spherical nanoparticles – Proceedings of ACUN-5, Developments in Composites, Advanced, Infrastructural, Natural and Nanocomposites" UNSW, Sydney, Australia, 11-14 July 2006, pages 297 – 305 Eds Sri Bandyopadhyay et al. ISBN 0 7334 2363 9.

Figure 11. Processing routes of polymer nanocomposites applied in this study.

3. Arrangement of the Inorganic Component within the Matrix

Arrangement of nanoparticles is an important aspect to fine tune the properties of the nanocomposites. Filler distribution and the interphase determine the morphology of polymer nanocomposites. Polymer chains pack differently in the vicinity of a surface or interface of particle, with the Interface acting as a reflecting plane, and the chain trajectory reflecting back into the polymer melt. The reflection does not perturb projection of radius of gyration parallel to the plane, R_g, and is comparable to bulk R_g (Brown and Russell, Macromolecules 1996, 29, 798-800). However, it is reduced in perpendicular direction, $R_g(\perp)$, consequently, the coil-coil interpenetration decreases resulting in a loss of entanglement at interface (Figure 12) [Baumchen et al., 2012]. The R_g ($\perp$) normal to the interface is reduced by a factor of $(1 - 2/\pi)^{0.5} = 0.603$ after reflection. Consequently, the volume pervaded by a given chain length is smaller close to the interface. The pervaded volume is reduced by 0.603 after reflection. The reduced interpenetration of neighbouring chains in favour of self-penetration leads to, a) weakening of the polymer network, b) decreased resistance to melt flow and c) reduced entanglement density near the interface leading to decreased entanglement molecular weight, M_e (Molecular weight of the polymer between cross-link points, at the interface). Larger is the M_e, longer is the time needed to relax. Failure properties, such as toughness, depend critically upon the ratio of the M_w of the polymer to its M_e. Consequently, the coil-coil interpenetration decreases resulting in a loss of entanglement at the substrate/melt interface.

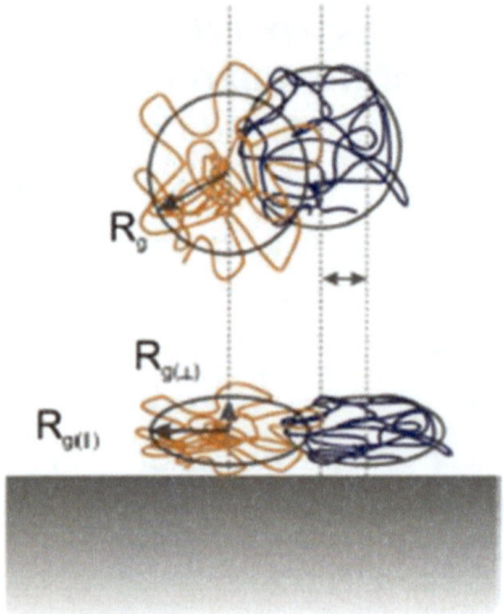

Reproduced with permission: Baumchen et al., 2012, Inspired by Soles C L and Ding Y 2008, Nanoscale polymer processing Science 322, 689.

Figure 12. Polymer conformation in a bulk melt (upper Figure) and at the substrate/melt interface (lower Figure): Two neighbouring coils interpenetrate each other and may form entanglements. At the interface, the radius of gyration parallel to the substrate Rg(‖) is comparable to bulk Rg, while it is reduced in perpendicular direction (Rg(⊥)).

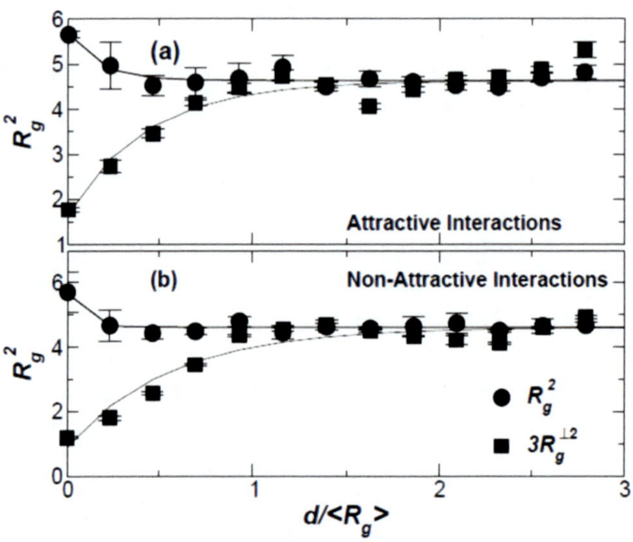

Reproduced with permission: Starr, Francis W. and Glotzer, Sharon C., "Science and Engineering of Nanoparticle-Polymer Composites: Insights from Computer Simulation" (2004). Division III Faculty Publications. Paper 249. http://wesscholar.wesleyan.edu/div3facpubs/249.

Figure 13. R_g as a function of distance $d/<R_g>$,of center of mass of a chain from nanoparticle surface- Polymers become slightly elongated near the surface, and flatten significantly, orienting with the particle surface.

The behaviour of R_g^2 and $(R_\perp)^2$ is nearly identical for attractive, and non-attractive interactions, the chains becoming increasingly elongated and "flattened" as the surface of the nanoparticle is approached (Figure 13).

Monomer Density Profile ρ(d)

Monomers form well-defined layers surrounding the nanoparticle. In the attractive case, the monomer density in the first layer is enhanced due to the relatively strong monomer-particle attraction (Figure 14). The peak location changes weakly with temperature since there is a preferred distance due to the attractions between the nanoparticle and the monomers [Starr, et al., 2001; Starr, et al., 2002]. The splitting in the first peak at lowest T occurs when the nanoparticle is not spherical. In the case of non-attractive interactions, there is also an enhancement in the density in the first layer, attributable to monomers far from the nanoparticle "pressing" the inner monomers toward the interface with the nanoparticle. Since there is no preferred distance for the inner-most monomers, the location of the first peak is more strongly T dependent.

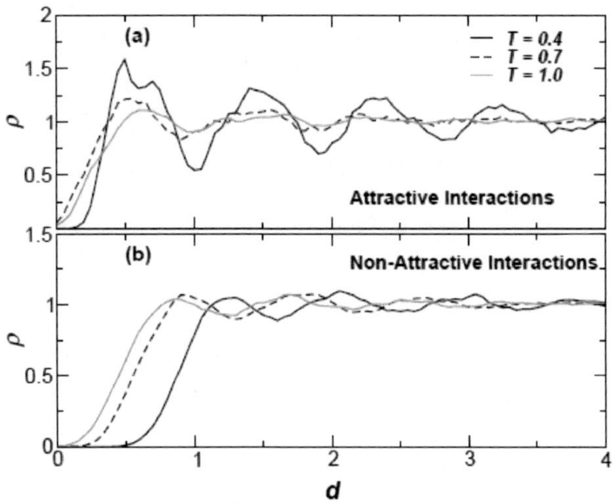

Reproduced with permission: Starr, Francis W. and Glotzer, Sharon C., "Science and Engineering of Nanoparticle-Polymer Composites: Insights from Computer Simulation" (2004). Division III Faculty Publications. Paper 249. http://wesscholar.wesleyan.edu/div3facpubs/249.

Figure 14. Monomer density profile ρ(d) as a function of the distance from the nano-particle surface at several temperatures for both (a) attractive and (b) non-attractive monomer-particle interactions.

Relaxation Behaviour in Nanocomposites

Fracture properties of Polymer and NCs depend on Interfacial chemistry via the relaxation time. The relaxation of the monomers closest to the nanoparticle surface is slowed by roughly 2 orders of magnitude when the monomers and nanoparticle have an attractive interaction (Figure 15). Conversely, when there is no attraction between monomers and the nanoparticle, the relaxation of surface layer monomers is significantly enhanced by roughly one order of magnitude near the surface compared to the bulk. The relaxation time of the outer-most layer in both cases nearly coincides with the relaxation time of the pure system [Starr, et al., 2004].

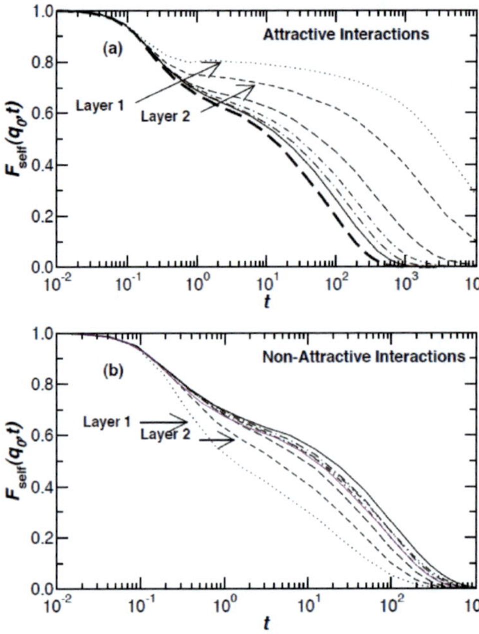

Reproduced with permission: Starr, Francis W. and Glotzer, Sharon C., "Science and Engineering of Nanoparticle-Polymer Composites: Insights from Computer Simulation" (2004). Division III Faculty Publications. Paper 249. http://wesscholar.wesleyan.edu/div3facpubs/249

Figure 15. Relaxation in attractive and non-attractive interactions ((Fself (q0, t), real-space time-dependent spatial-correlation function).

The concomitant energy dissipation arising from the extension and the relaxation of the interfacial bonds contributes a significant resistance to fracture. When the energy barriers separating the metastable states are significantly larger than the characteristic vibrational and thermal energies of the system, the interface does not relax within any experimentally observable time frame. On the other hand, when the energy barriers are comparable to thermal and vibrational energies, the contact area exhibits a time dependent relaxation with a concomitant dissipation of strain energy. Significant drag to crack propagation arises from rate-dependent processes occurring right at the interface. Fracture energy can be virtually independent of the molecular weight if the relaxation time is very small (Figure 16). When interfacial interaction is primarily due to dispersive forces the relaxation time is on the order of a microsecond or less. The relaxation time of bond dissociation for the breaking of polymer chains (case I) is about 10^{13} s. When relaxation time < 1 s chain pull out dominates while for relaxation time ~ 10^{13} s [Ghatak, et al., 2000], chain scission /bond scission dominates. Figure 17, [Ghatak, et al., 2000; Chaudhury, 1999] shows fracture energy, G increasing logarithmically with rate of crack propagation for attractive interactions except when the dispersion forces dominate the interfacial interaction. Three types of adhesive interfaces are studied here. Case I corresponds to a covalent interaction between glass and a silicone elastomer. In this case, significant chain scission occurs at the glass-polymer interface. Case II corresponds to H-bonding interaction between silica and end-tethered silicone polymers of various molecular weights Case III corresponds to the dispersion interaction between a silicone elastomer and ahexadecylsiloxane monolayer adsorbed on silica.

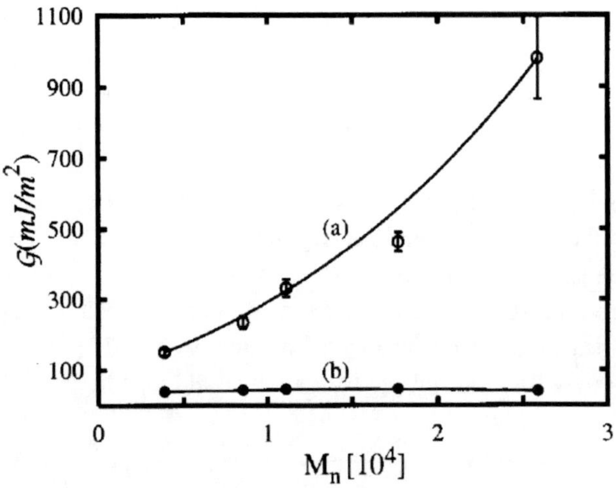

Reproduced with permission: AnimangsuGhatak, Katherine Vorvolakos, Hongquan She, David L. Malotky, & Manoj K. Chaudhury, J. Phys. Chem. B, Vol. 104, No. 17, 2000, 4018-4030; Chaudhury, M. K. J. Phys. Chem. B 1999, 103, 31, 6562.

Figure 16. Dependence of the fracture energy on the molecular weight of polymers. Dependence of the fracture energy on the molecular weight of polymers. (a) Represents the H-bonding interaction of PDMS and silica. (b) represents the dispersion interaction of PDMS with hexadecylsiloxane-coated silica.

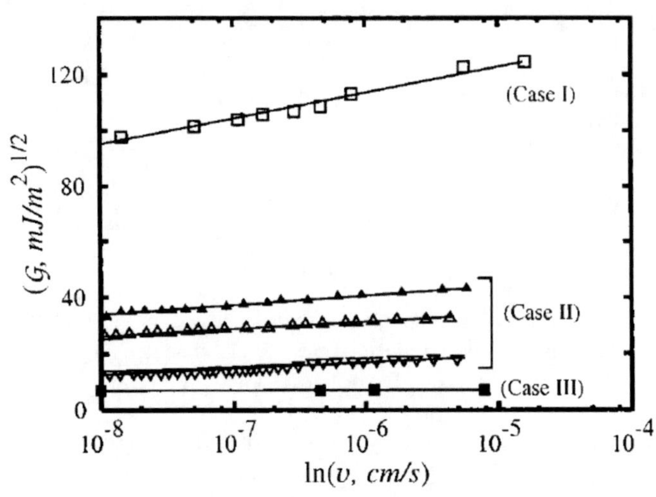

Reproduced with permission: Animangsu Ghatak, Katherine Vorvolakos, Hongquan She, David L. Malotky, & Manoj K. Chaudhury, J. Phys. Chem. B, Vol. 104, No. 17, 2000, 4018-4030; Chaudhury, M. K. J. Phys. Chem. B 1999, 103 (31), 6562

Figure 17. Dependence of fracture energy, G on the rate of crack propagation, Case I: covalent interaction between glass and a silicone elastomer. Case II: H-bonding interaction between silica and end-tethered silicone polymers of various molecular weights. ▲,Δ, ∇: molecular weights of 25.8 kD, 17.7 kD, and 3.8 kD, respectively. Case III: dispersion interaction between a silicone elastomer and ahexadecylsiloxane monolayer adsorbed on silica.

PNCs exhibit higher viscosities in presence of attractive interactions due to increased dispersion. In a well dispersed configuration, surface area increases linearly with the number of particles. For a clustered configuration, the surface area increases sub-linearly with the number of particles. Since change in viscosity is proportional to the amount of polymers in contact with the nanoparticles, the viscosity is expected to be larger for well dispersed configurations. It follows that for *attractive* interactions, the relaxation times of surface monomers increase as viscosity increases (Starr, et al., 2004). Similar property improvements of polymer composites by spherical nanoparticles are illustrated in Figure 18 where improvement in both the fracture toughness K_{IC} as well as the fracture energy G_{IC} is reported (Friedrich, 2006). Note the difference in G value between Figures 16 and 18. This is due to the fact that Figure 16 represents typical bond (covalent, etc) fracture energy; whereas Figure 18 represents the measured crack propagation energy based on linear elastic fracture mechanics, reflecting new fracture surface formation (Bandyopadhyay, 1990; 1993).

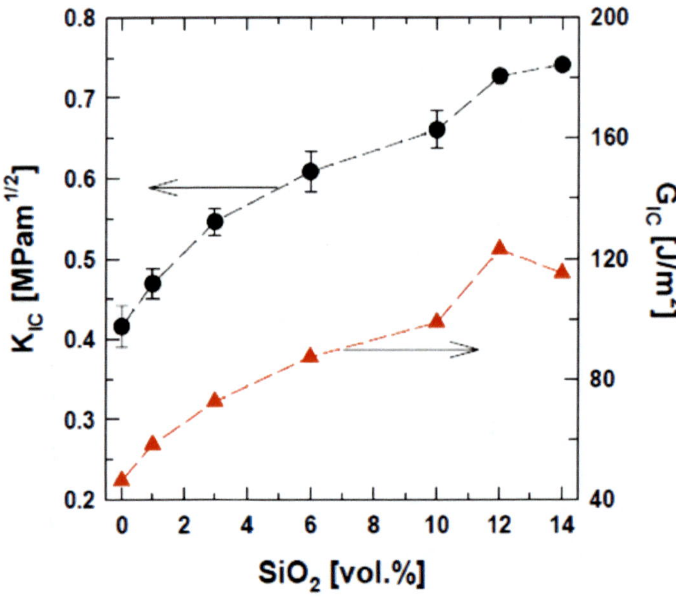

Reproduced with permission: Zhang, H., Zhang, Z., Friedrich, K., Eger, C: Property improvements of in situ epoxy nanocomposites with reduced interparticle distance at high nanosilica content, ActaMaterialia (2006) Volume 54, Issue 7, April 2006, Pages 1833-1842.

Figure 18. Fracture toughness and fracture energy in epoxy-SiO2 (nano particles) thermoset system.

Hierarchical Morphology

While arrangement of nanoparticles is an important aspect to fine tune the properties of the nanocomposites, random arrangements of nanoparticles will not provide optimized electrical, thermal, or optical performance for many potential high- technology applications, such as dielectric underfills for electronic packaging, printed flexible electronics, engineered aerospace structural components, reconfigurable conductive adhesives and optical gratings [Vaia and Maguire, 2007] The hierarchy of morphology in a nanocomposite is more important than in micrometer-scale filled polymers due to the extreme aspect ratio of many nanoparticles, e.g., exfoliated clays and carbon nanotubes. Gusev and Rozman, 1999, showed that comparable shear moduli could be obtained at only half the volume fraction of particles if a weblike morphology could be generated rather than random or hexagonal arrangement.

Two general approaches to invoke hierarchical morphology control are *external-in* (directed patterning of nanoparticle dispersions) and *internal-out* (mesophase assembly of nanoparticles MANP). Relying on the ability to tailor interparticle interactions, both particle-particle and particle-matrix result in thermodynamically stable (and defined) mesophases. Materials like CNTs often exhibit hierarchical structures that are composed of several levels. Although CNT diameters are in the nanometer range, larger structures are formed because of re-agglomeration of the nanoparticles. The nanospheres can self-assemble in different ways, including as densely packed structures and linearly packed structures (necklace -like), which can further build up different configurations, such as spider-web-like networks or particulated structures. The resulting configuration tends to be far more complex for CNTs than for polymer chains because of the tubular structures of the former. This enables a larger number of possible morphologies [Prado, 2011].

Directed patterning of nanoparticle dispersions using external electric fields relies on the creation of a multidimensional morphology by an external means to transform a random distribution to a prescribed ordered construction. For example, particulate additives, such as oxides or metals, under an electric field, can reversibly form fibrous structures within a non-conducting fluid. These controlled aggregates parallel to the applied field and can increase the viscosity of an electro-rheological fluid by up to 10^5. Similar morphologies are found for magneto-rheological fluids where particle-particle interactions, which arise from induced magnetic dipoles, lead to chaining and subsequently long range, periodic ordering of the particle chains parallel to the lines of magnetic flux [Vaia, et al., 2007]).

Conclusion

This chapter has reviewed emergence of the novel properties in polymer matrix nanocomposites arising from nano inclusions and the factors dominating the evolution of these unique properties.

Intrinsic and extrinsic nano-scale properties of the nanomaterials predominated by the interfacial interactions and the quantum effects associated with nanodimensional structures are analysed.

The nanoinclusions offer improvement in barrier properties, mechanical aspects, thermal nature, optical features and lower melting point in PNCs. The role of hierarchical morphology and patterned morphology that offers optimized electrical, thermal, or optical performance to PNCs is assessed.

Acknowledgment

The authors are grateful to Nova Science Publishers for giving us the opportunity to write on the very important topic of nanocomposites – their fundamental and application orientated aspects. We would also like to thank the authors who provided permissions allowing us to present their figures/diagrams and tables in the chapter to make the correlation clearly understandable. Also the authors thank their respective universities (U Ballarat, and UNSW) for the encouragement provided by the management in compiling this chapter.

REFERENCES

Abdelghany, A. M., El Batal, H.A., Mari, L., Optical and shielding behaviour studies of Vanadium Doped Lead Borate Glasses, *Radiation Effects and Defects in Solids*, 2012, 167, (1) 49-58.

Abdelrazek, E. M.,Abdelghany, A. M., Orabyand, A.H., Asnag, G.M., Investigation of Mixed Filler Effect on Optical and Structural Properties of PEMA Films, *International Journal of Engineering & Technology IJET-IJENS* 2012, 12 No:04, 98-102.

Alivisatos, A. P., Semiconductor Clusters, Nanocrystals, and Quantum Dots, *Science* 1996, 271, 933-937.

Anandhan, S., and Bandyopadhyay, S., "Chapter 1: Polymer Nanocomposites: from Synthesis to Applications", pages 3 – 28, *Nano composites and Polymers with Analytical Methods*, Editor: John Cuppoletti; In Tech Publishers, Croatia, ISBN 978-953-307-352-1, 2011.

Animangsu Ghatak, Katherine Vorvolakos, Hongquan She, David L. Malotky, & Manoj K. Chaudhury, *J. Phys. Chem. B*, 2000, 104, 17.

ASM International. ASM handbook, *Material park*, USA; 1992.

Bandyopadhyay S., "Correlation of Macroscopic Fracture Behaviour with Microscopic Aspects of Deformation in Toughened Epoxies", Chapter 9, in "*Toughened Plastics I: Science and Engineering*", Eds. K.C. Riew and A.J. Kinloch, American Chemical Society, 1993, Chapter 9, pp. 211-58.

Bandyopadhyay S., "*Review of the Microscopic and Macroscopic Aspects of Fracture Behaviour of Unmodified and Modified Epoxy Resins*", Materials Science and Engineering, 1990, A125, pp. 157-184.

Banerjee, Shilpi., Hajra, Partha., Mada, Mykanth Reddy., Bhaumik, Asim., Bandyopadhyay, Sri., Chakravorty Dipankar., "Exchange Bias effect in Nickel Zinc Ferrite-Mesoporous Silica Nanocomposites" *J. of Magn. andMagn. Mater.*, 2013, 332, 98–102, http://dx.doi.org/10.1016/j.jmmm.2012.12.023.

Baumchen, O., Fetzer, R., Klos, M., Lessel, M., Marquant, L., Hahl, H., and Jacobs, K., arXiv:1202.1002v1 [cond-mat.soft] 5 Feb 2012, http://arxiv.org/pdf/1202.1002.pdf.

Beke, D.L, Szabo, S and Kis-Varga, M., Intrinsic and domain magnetism in nanomaterials, *Advances in condensed matter and materials research*, 2004, 5, 77-112.

Benoit, J. M., Corraze, B., Chauvet, O., Localization, Coulomb inter-actions, and electrical heating in single-wall carbon nanotubes/polymer composites, *Phys Rev* 2002, B65:241405 (R).

Brus, L.E., *Journal of Luminescence*, 1984, 31/32, 381.

Brust, M.; Kiely, C. J. Some Recent Advances in Nanostructure Preparation from Gold and Silver: A Short Topical Review, *Colloids Surf. A:Physicochem. Eng. Asp*. 2002, 202, 175-186.

Burda, C., Chen, X., Narayanan, R., El-Sayed, M. A., Chemistry and properties of nanocrystals of different shapes, *Chem. Rev*. 2005, 105, 1025-1102.

Chaudhury, M. K., *J. Phys. Chem*. B. 1999, 103 (31), 6562.

Chen, D., Yamada, T., Quantum Size Effects in Nanoscale Metallic Structures, *IEEE*, 0-7803-7976-4/ 2003.

Chinnasamy, C. N., *J. Phys. Condens. Matter*, 2000, 12, 7795.

Cho, J., Daniel, I. M.,*Scripta Mater.*, 2008, 58, 533–536.

Daniel, M. C., and Astruc, D., Gold Nanoparticles: Assembly, Supramolecular Chemistry, *Chem. Rev*. 2004, 104, 293−346.

Das, M., Roy, D., Bandyopadhyay, N.R., Banerjee, A., and Bandyopadhyay, S., Thermogravimetric and Resistivity Study of Ex-situ and In-situ PMMA/-COOH Functionalized MWCNT Composites", *Journal of Applied Polymer Science*, 2011, Vol.120, Issue 5, pp.2954–2961, DOI 10.1002/app.33320.

Devasahayam S., *J Polym Sci, Part B: Polym Phys.* 2004, 42, 3822.

Devasahayam S., *Journal of Applied Polymer Science*, 2006, 99, 2006, 3318.

Devasahayam, S., *Journal of Applied Polymer Science,* 2006, 99, 3318.

Devasahayam, S., *Journal of Applied Polymer Science*, 2006, 99, 5, 2052-2061. doi:10.1002/app.22588.

Durand, J. M., Vardavoulias, M., Role of reinforcing ceramic particles in the wear behaviour of polymer-based model composites. *Wear* 1995, 181–183, 833–9.

Friedrich, K., Chang, L., Haupert, F., Current and Future Applications of Polymer Composites in the Field of Tribology, in: L. Nikolais, M. Meo, E. Miletta (eds.): Composite Materials, Springer, New York, USA, 2011, http://www.springer.com /materials/special+types/book/978-0-85729-165-3.

Filiponni, L and Sutherland, D., *Unique properties at the nanoscale: Nanoscience: the basic sense behind nanoscience* Sources 2010: http://www.bc.pitt.edu/prism-logo.gif; http://www.physics.umd.edu/lecdem/outreach/QOTW/picsk3-06.gif.

Friedrich, K., Haupert, F., Zhang, Z., "Property improvements of polymer composites by spherical nanoparticles" – *Proceedings of ACUN-5, Developments in Composites, Advanced, Infrastructural, Natural and Nanocomposites UNSW,* Sydney, Australia, 11-14 July 2006, pages 297 – 305 Eds Sri Bandyopadhyay et al. ISBN 0 7334 2363 9.

Friedrich, K., Property improvements of polymer composites by spherical nanoparticles, Lecture for the MATS3574 Polymer Engineering Course at the UNSW, Sydney, Australia, 7. September 2006.

Friedrich, K, Zhang Z, Klein P: Wear of polymer composites. Stachowiak G.W (Ed) in 'Wear – Materials, Mechanisms and Practice, Wiley, Chichester, England 2005, Chapter 11, pp.269 – 290.

Gibson, R.F., Principles of composite material mechanics. Philadelphia, USA, 1994, 190–67.

Fu, S.Y., Feng, X. Q., Lauke, B., Mai, Y. W., Effects of particle size, particle/matrix interface adhesion and particle loading on mechanical properties of particulate–polymer composites, *Composites*: Part B: 2008, 39, 933–961.

Giannelis, E. P., Krishnamoorti, R., Manias, E., Polymer-Silicate Nanocomposites: Model Systems for Confined Polymers and Polymer Brushes, *Advances in Polymer Science*, 1999, 138, 107-147.

Grossiord, N., Loos, J., Regev, O.,Koning, C. E., *Chem. Mater*. 2006, 18, 1089.

Gusev, A. A., Rozman, M. G., *Comput. Theor. Polym. Sci.* 1999, 9, 335.

He, L., Ma, E., *J. Mater. Res*. 2000, 15, 904.

Hussain, F., Hojjati, M., Review Article: Polymer-Matrix Nanocomposites, Processing, Manufacturing, and Application: An Overview, *Journal of Composite Materials*, 2006, 40, No. 17, 1511-1575.

Jawahar, P. R., Balasubramanian, M., Tribological behaviour of clay–thermoset polyester nanocomposites, *Wear* 2006, 261, 835–40.

Krausz, A. S., and Krausz, K., *Fracture Kinectics of Crack Growth*. Kluwer Academic Publishers, Dordrecht, 1988, pp. 2-6.

Krishnamoorti, R., *Advanced Polymer Nanocomposites: Novel Properties and Applications*, University of Houston, 2010, http://cnst.rice.edu/uploadedFiles/Resources/Nano%20103_RK(1).pdf

Kumar, S. K., and Krishnamoorti, R., Nanocomposites: Structure, Phase Behavior, and Properties., *Annu. Rev. Chem. Biomol. Eng*. 2010, 1, 37–58.

Luechinger, N. A., Booth, N., Heness, G., Bandyopadhyay, S., Grass, R. N., and Stark, W. J., Surfactant-Free, Melt-Processable Metal–Polymer Hybrid Materials: Use of Graphene as a Dispersing Agent, *Adv. Mater*. 2008, 20, 3044–3049.

Luis Antonio Sanchez de Almeida Prado, *Morphological and structural aspects of polymer nanocomposites,* 011 Society of Plastics Engineers (SPE), http://www.4spepro.org/view.php?source=003504-2011-01-29.

Manaf, A., Buckley,R.A., Davies, H.A., Leonowicz, M., Enhanced magnetic properties in rapidly solidified Nd-Fe-B based alloys, *Journal of magnetism and magnetic materials,* 1991, 101, 1-3, 360-362.

Mark, J. E., *Polymer Data Handbook*, Second Edition. Oxford University Press, New York, 2009, pp. 646.

McDonald, S. A., CyrP. W.,Levina, L., and Sargent, E. H., Photoconductivity from PbS-nanocrystal/semiconducting polymer composites for solution-processible, quantum-size tunable infrared photodetectors, *Appl. Phys. Lett*., 2004, 85, No. 11, 13.

Mundra, M. K., Donthu, S.K., Dravid, V.P, and Torkelson, J.M., "Effect of Spatial Confinement on the Glass Transition Temperature of Patterned Polymer Nanostructures, "*Nano Letters* 2007, 7, 713-718.[Link].

Ning, Z., Li, X.C., Yan, S.Y., Ye, H., Zong, F.W., Effects of load and sliding distance on the wear properties of hybrid friction material. *Mater Sci Forum* 2011, 686, 401–5.

Peyghambarian, N., Hanamura, E., Koch, S.W., Masumoto, Y., Wright, E.M., in A.S. Edelstein, H. Hahan (Eds.), *Nanostructured Materials: Synthesis, Characterization and Uses*, Adam Higler, Bristol, 1993, 173.

Pinchuk, A. O., Schatz, G. C., Nanoparticle optical properties: Far- and near-field electrodynamic coupling in a chain of silver spherical nanoparticles, *Materials Science and Engineering*, 2008, B 149 251–258.

Poole, Jr, C P., Owens, F. J., *Introduction to Nanotechnology*, Wiley-Interscience, New Jersey, 2003.

Qiao, R., Deng, H., Putz, K. W., Brinson, L.C., Effect of Particle Agglomeration and Interphase on the Glass Transition Temperature of Polymer Nanocomposites, *Journal of Polymer Science Part B: Polymer Physics* 2011, 49, 740–748.

Ramanathan, T., Stankovich, S., Dikin, D. A., Liu, H., Shen, H., Nguyen, S. T., Brinson, L. C., *J. Polym. Sci. Part B: Polym. Phys*., 2007, 45, 2097–2112.

Ramsden, J.J., Freeman, J., Thenano-scale, *Nanotechnology perceptions*, 2009, 5, 3-25.

Reisfeld, R., *Sol–Gel Glasses and Modern Application*, Vol. 85, Springer-Verlag, Berlin/New York, 1996, 99.

Respaud, M., Broto, J.M., Rakoto, H., Fert, A. R., Thomas, L., Barbara, B., Verelst, M., Snoeck, E., Lecante, P., Mosset, A., Osuna, J., Ely, T. O., Amiens, C., Chaudret, B., Surface effects on the magnetic properties of ultrafine cobalt particles, *Phys. Rev. B* 1998, 57, 2925.

Schwartz, C. J., Bahadur, S., Studies on the tribological behavior and transfer film counterface bond strength for polyphenylene sulfide filled with nanoscale alumina particles. *Wear* 2000, 237:261–73.

Sengupta, R., Bhattacharya, M., Bandyopadhyay, S., and Bhowmick, A.K., "A review on the mechanical and electrical properties of graphite and modified graphite reinforced polymer composites", *Progress in polymer science*, Volume 36, Issue 5, pp.638-670 (2011).

Shah, K. S., Jain, R.C., Shrinet, V., A. K. Singh, A. K., High Density Polyethylene (HDPE) Clay Nanocomposite for Dielectric Applications, *IEEE Transactions on Dielectrics and Electrical Insulation*, 2009. 16 (3), 853-861.

Starr, F. W., and Glotzer, S. C., "*Science and Engineering of Nanoparticle-Polymer Composites: Insights from Computer Simulation*" Division III Faculty Publications. 2004, Paper 249. http://wesscholar.wesleyan.edu/div3facpubs/249.

Starr, F.W., Schrøder, T.B. and Glotzer, S.C., *Macromolecules*, 2002, 35, 4481.

Starr, F.W., Schrøder, T.B. and Glotzer, S.C., *Phys. Rev.* 2001, E 64, 021802.

Stauffer, D., *Introduction to Percolation Theory*, Taylor and Francis, London 1985.

Steigerwald, M.L., and Brus, L.E., *Accounts Chemical Research*, 1990, 23, 183.

Teeguarden, J, G., Hinderliter, P. M., Galya Orr., Brian D., Thrall, and Joel G. Pounds, Review: Particokinetics In Vitro: Dosimetry considerations for in vitro nanoparticle toxicity assessments, *Toxicological sciences*, 2007, 95, 300–312.

Thielsch, R., Bohme, T, Reiche, R., Schlafer, D., Bauer, H. D., and Bottcher, H., Quantum-Size Effects Of PbS Nanocrystallites In Evaporated Composite Films, *Nano Structured Materials,* 1998, 10. No.2.131-149.

Thomas Selvin.P., Abdullateef, A A., De S K., Bandyopadhyay, S., Basfar, A A., Atieh, M A., Effect of phenol functionalization of carbon nanotubes (CNT) on properties of natural rubber nanocomposites, *Journal of Applied Polymer Science,*2012, 124, 2370–2376.

Thomas Selvin P., Abdullateef Adedigba A., Atieh, Muataz A., De S K., Mostafizur Rahaman, Chaki, T. K., Khastgir D., Bandyopadhyay, S**.,** Electrical properties of natural rubber nanocomposites: effect of 1-octadecanol functionalization of carbon nanotubes, *Journal of Materials Science,* 2012, 47, 3344-3349.

Thomas Selvin. P**.**., Thomas, Sabu., Zafeiropoulos, N. E., Bandyopadhyay, Sri., Wurm, Andreas., Schick, Christoph., Polystyrene/Calcium Phosphate Nanocomposites: Morphology, Mechanical and Dielectric properties, S, *Polymer Engineering & Science*,2012, 52, 689–699.

Trantakis, I., *Toxicity of Food Related Nanomaterials, Current Toxicology* – FS2012, p 1-27, http://www.toxicology.ethz.ch/education/Toxicity_of_Food_Related_Nanomaterials_Curr_Tox.pdf.

Tsui, O. K. C., Russell, T. P., Hawker, C. J., *Macromolecules,* 2001, 34, 5535.

Vaia, R. A., Maguire, J. F., Polymer Nanocomposites with Prescribed Morphology: Going beyond Nanoparticle-Filled Polymers, *Chem. Mater*., 2007, 19(11), 2736–2751.

Vaia, R., Koerner, H., Lu, W., Manias, E., Polymer Nanocomposites with Prescribed Morphology: Going Beyond Nanoparticle-Filled Polymers, *Proceedings Published 2007 by the American Chemical Society*, http://www.plmsc.psu.edu/~manias/PDFs/POLY_inv2_mar07.pdf.

Voitenko, I., Muth, J. F., Gerhold, M., Cui, D., Xu, J., Tunable photoluminescence of polymer doped with PbSe quantum dots, *Materials Science and Engineering*, 2007, C 27, 1078–1081.

Wang, X., Zhou, W., *Macromolecules,* 2002, 35, 6747.

Wang, Y., and Herron, N., *Chem. Phys. Lett.* 200, 71, 1992.

Wang, Y., and Herron, N., *Journal of Physical Chemistry,* 1991, 95, 525.

Wang, Y., Suna, A., Mahler, W. and Kasowski, R., *Journal of Chemical Physics*, 1987, 87, 7315.

Wetzel, B., Haupert, F., Friedrich, K., Zhang, M. Q., Rong, M, Z., Impact and wear resistance of polymer nanocomposites at low filler content, *Polymer Engineering and Science*, 2002, 42, 1919-1927.

Wetzel, B., Mechanische Eigenschaftenvon Nanokompositenaus Epoxydharz und keramischen Nanopartikeln, *Dissertation* Technische Universität Kaiserslautern, Germany, February, 2006.

Xie, P., Gu, P., Beaudoin, J.J., *J Mater. Sci.* 1996, 31, 4093.

Xue, Q., The influence of particle shape and size on electric conductivity of metal–polymer composites, *European Polymer Journal* 2004, 40, 2004, 323–327.

Yang B, et al., Room-temperature organic ferromagnetism in the crystalline poly (3-hexylthiophene): PhenylC61-butyric acid methyl ester blend film, *Polymer*, 2013, 54 490-494. http://dx.doi.org/10.1016/j.polymer.2012.12.024.

Zallen, R., *The Physics of Amorphous Solids*, Wiley, New York 1983.

Zhang, C., Guo, Y., and Priestley, R. D., Glass Transition Temperature of Polymer Nanoparticles under Soft and Hard Confinement, *Macromolecules* 2011, 44, 4001–4006;

Zhang, S.W., State-of-the-art of polymer tribology. *Tribology Int.*, 1998, 31, 49–60.

Zhang, Z., Friedrich, K., Tribological characteristics of micro and nanoparticle filled polymer composites. In: *Polymer composite from nano to macroscale*. Berlin: Springer: 2005, 169–85.

Zidan, H. M., Filling level effect on the physical properties of $MgBr_2$ and $MgCl_2$ filled poly (vinyl acetate) films, *J. Polym. Sci.: Part B: Polym. Phys.* 2003, 41, 112-119.

In: New Developments in Polymer Composites Research ISBN: 978-1-62948-340-5
Editors: S. Laske and A. Witschnigg

Chapter 2

NEW DEVELOPMENT IN POLYMER COMPOSITES RESEARCH AND APPLICATIONS OF POLYMER-CEMENT MATRICES FOR SOLIDIFICATION/STABILIZATION OF RADIOACTIVE WASTES

S. B. Eskander[1,*] and M. E. Tawfik[2]
[1]Radioisotopes Department, Atomic Energy Authority, Egypt
[2]Department of Polymers and Pigments,
National Research Center, Dokki, Egypt

ABSTRACT

Composite materials combine and maintain two or more distinct phases to produce materials that are far superior to either of the base components. Polymer composite, is multi-phase material produced by combining polymer resins, e.g., polyester, vinyl ester, and epoxy with filler and/or reinforcing additives to produce a bulk product with properties better than those of individuals. Fillers are often used to provide bulk to the polymer, reduce cost, low bulk density or aesthetic features.

Polymer-Cementitious composite, as an example of that group, had enjoyed wide spread uses in various fields of our life. In construction industry they were applied for many years in non-critical applications. Rapid expansion of their uses in road paving and dental repair were also reported.

One of the most recent applications of the polymer-cement composite is in the field of immobilization of radioactive wastes. Various workers have been reported for solidification/stabilization of some low and intermediate radwastes based on polymer-cement composite. The main advantages of these matrices are high strength and stiffness, low weight, high resistance for the hazardous radionucleides leachability, acceptable deformation in addition to their favorable physical and thermal stabilities.

* E-mail address: samireskander11@yahoo.com (Corresponding author)

Keywords: PET waste, chemical recycling, radioactive borate waste, cement composite, solidification/stabilization, Cesium-137, Cobalt-60, leaching

INTRODUCTION

Composite materials, also called *composition materials* or shortened to *composites*, are materials made from two or more constituent materials with significantly different physical or chemical properties, that when combined, produce a material with characteristics different from the individual components. The individual components remain separate and distinct within the finished structure. In other words, in its most basic form a composite material is one which is composed of at least two elements working together to produce material properties that are different to the properties of those elements on their own. In practice, most composites consist of a bulk material, the "matrix", and a reinforcement of some kind, added primarily to increase the strength, environmental sustainability and stiffness of the matrix. Today, the most common man-made composites are the polymer –cement composites. Cement-polymer composites are being developed and tested as a replacement for traditional cement. Polymer-cement composites have been reported as popular construction materials because of their excellent properties in comparison with ordinary cement. Polymers have been used for improving mechanical properties, adhesion with substrates, or waterproofing properties of mortars and concretes. The literatures agree that the properties of polymer modified mortar and concrete depend significantly on the polymer content or polymer cement ratio, that is, the mass ratio of the amount of polymer solids in a polymer-based admixture to the amount of cement in a polymer-modified mortar or concrete [1].

The combination of advanced polymer chemical modification and advanced formulations of cement/polymers could lead to improved thermally insulated, self-extinguishing, and moisture semi-permeable materials. These novel materials will result in significant energy and cost savings as well as affordable and robust housing. Anticipated benefits and potential commercial applications of the research and development include low-cost, ultra-lightweight cement-polymer composites with improved strength, structural stability, and reduced thermal conductivity for zero-energy buildings.

Polymer-modified cement is prepared by mixing either a polymer or monomer in a dispersed, powdery, or liquid form with fresh cement mortar mixtures, and subsequently curing, and if necessary, the monomer contained in the mortar or concrete is polymerized *in situ*. Several types of polymer-modified mortars and concretes that is, latex-redispersible polymer powder, water-soluble polymer, liquid resin, and monomer modified mortars and concretes, are produced by using the polymers and monomers and applied widely in the construction fields [2-6].

Polymer and Polymer Wastes

Around 4 per cent of annual petroleum production is converted directly into plastics from petrochemical feedstock (British Plastics Federation 2008) [7]. As the manufacture of plastics also requires energy, its production is responsible for the consumption of a similar additional quantity of fossil fuels. Therefore, the objective of designing a composite structure, with improved or optimized characterizations and based on the recycled plastics wastes, has been involved extensively.

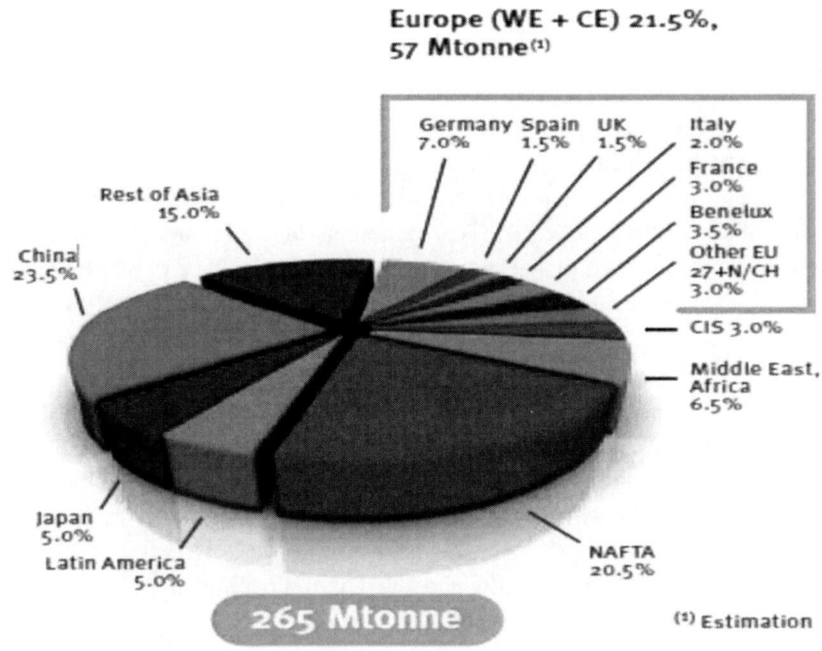

Image source EUPC, 2011.EUPC, (2011) Plastics – the Facts: An analysis of European plastic production, demand and recovery for 2010, 1-32, version 3.0, September 2011.

Figure 1. Overview of the world plastic production in 2010.

Photo: Md. Ibrahim: one of the winning entries from the 2007 WHO ICF.

Figure 2. 'Waste management' in developing countries.

During the past 30 years, the global production of plastics has increased by over 500 percent. Currently, at a global level, the annual plastic production is estimated over 230 million ton, (figure 1), and is expected to reach 400 million ton in 2020 with a further increase 5 % annual growth rate [8-10].

Polymers cover very large, broad classes of compounds, both natural and synthetic, with a wide variety of properties. They play essential and ubiquitous roles in everyday life, from those of familiar synthetic plastics and others materials of day to day work and home life, to natural biopolymers that are fundamental to biological structure and function.

Plastics are composed of different potentially toxic chemicals that are introduced during the production phase of their life cycle (i.e., monomers and oligomers, bisphenol-A (BPA), phthalate plasticizers, flame retardants and antimicrobials) [11].Once plastic loses its utility it is normally discarded and sent to either a recycling plant (in the best case scenario) or to a dump sit.

The waste-oriented society is standing at the crossroads between deciding to ignore serious environmental damage caused by waste and taking radical measures. Waste generation continues to increase worldwide in tandem with growth in consumption. Economic globalization facilitates the global transfer of wastes. Waste composition changes rapidly with new production technologies where waste is generated as a by-product. In this way, human beings are developing transnational waste streams and new waste types that may have serious unknown impacts on both human health and the environment. Illegal waste practices and transboundary transport have become very problematic because of many accidents. Wastes are often disposed of in the 'cheapest' countries without environmental liability (Figure2) [12].

Owing to these reasons, international waste management conventions striving for regulation of the international waste streams have been established. The life cycle of a product in connection with the solid waste stream from generation to disposal, including recycling, several waste recovery and treatment methods, and landfilling, stand in the forefront of this discussion. (Figure 3).

Source: http://www.sustainabilityed.org/pages/example2-2.htm.

Figure 3. Solid waste management hierarchy.

Landfill is the conventional approach to waste management, but space for landfills is becoming scarce in some countries. All disposal techniques are more or less energy-intensive and are connected with further distinct environmental damage [13]. Managing waste in an efficient way while minimizing environmental threats in the form of released pollutants requires international control of waste management and has attracted the attention of researchers to find new ideas to solve the problem.

Every year, large quantities of waste polymers are produced from industrial, agricultural and household activities. It has been reported that plastics make up more than 12% of municipal solid waste stream, a dramatic growth from 1960, when plastics were only 1% of the waste stream [14]. The widespread applications and non- biodegradability of these polymers create huge amounts of waste and disposal; what results in a serious environmental problem [13]. At present, most waste plastics are disposed by reclamation and incineration. Such a disposal of waste plastics has become a major environmental problem all over the world. Traditional waste plastic disposal, in fact, could lead to the release of hazardous substances either by weathering and natural drying or by incineration processes. Reclamation could cause the release of plastic additives such asphthalates and various dyes polluting ground water. Incineration is an alternative to reclamation disposal of waste plastics, but this practice could result in the formation of unacceptable emissions of gases such as nitrous oxide, sulfur oxides, dusts, dioxins and other toxins. Especially, waste plastics containing PVC pollute the environment and shorten the life of incinerator as generating hazardous HCl gas, dioxins containing chlorides, etc. [15, 16].

The new environmental, economic, and petroleum considerations have induced the scientific communities to increasingly deal with polymer recycling. The efficient treatment of waste polymers is still difficult challenge. The recycling rate for different types of plastic significantly varies, leading to an overall recycling rate of only 8% or 2,4 million tons in 2010 [17].

There are various techniques for recycling of waste polymers including energy recovery, mechanical and chemical recycling. The recycling process is the best and commercially important way to manage the waste of polymers because it converts the waste materials to useful added-value products [18-23]. It saves money, energy, raw materials, and land space, in addition to reducing pollution. Some recycling processes reuse materials for the same application purposes. Other turned old materials into entirely new products. Old tires, for instances, are shredded and turned into rubberized road surfacing. Newspapers become cellulose insulation.

However, there are some main problems for polymer recycling such as separation before handling [24].

The increases in the amount of plastics wastes have contributed to various environmental problems. Now research for new and innovative uses of waste materials are continuously advancing. These research efforts try to match society's need for safe and economic disposal of waste materials by preparing composite materials from different wastes polymer including plastic, rubber, agro-wastes....[25].

Polyethylene Terephthalate Waste (PET)

Polyethylene terephthalate (PET) is a thermoplastic polymer resin of polyester family. PET consists of polymerized units of the monomer ethylene terephthalate, with repeating $C_{10}H_8O_4$ units. Because of its structural regularity, PET crystallizes readily above its transition temperature. The presence of a benzene ring in the main chain leads to a stiffer chain with the result that the polyester has both a higher glass transition temperature (80° C) and crystalline melting point (254°C). Polyethylene terephthalate has inherently poor chemical resistance to alkalis but good resistance to ultraviolet light. PET is one of the versatile engineering plastics that are used to manufacture films and bottle for packaging. Examples include soft drink bottles, detergents, products of edible oils and others.PET does not create any direct hazard to the environment, but due to the increasing in the plastics waste and its high resistance to the atmosphere and biological agents, it could be considered as pollutant materials. The major problems are in terms of resource utilization and disposal places [26]. Therefore, recycling of PET will not only serve as a partial solution to the solid waste problem but also contribute to the conservation of raw petrochemical products and energy. In other words, product made from recycled plastics can results in, at least, half energy saving [27].

Post-consumer PET is often an attractive material for recycling. Unlike other polymers, recycled PET can be produced that is suitable for contact with food. PET can also be used in applications such as carpet fibers, geo-textiles, packaging and fiber fill. PET can be converted into polybutylene terephthalate (PBT) resin, which can be a valuable material for injection and blow-moulding applications [28].

Poly-ethylene terephthalate (PET) waste is collected in scrap textiles and beverage bottles after use in many locations, (figure 4), and recycled by different techniques to be used in various applications. PET waste is a potential source for the production of new chemical reagents such as unsaturated polyester resins, polyurethane foams……etc [29].

Figure 4. Bales of crushed PET bottles.

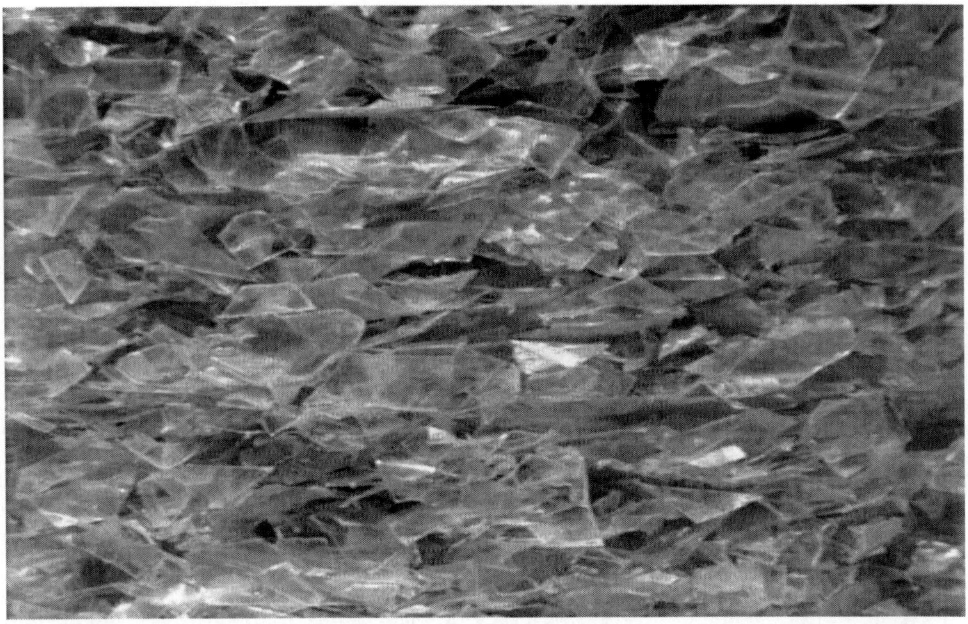

Figure 5. Flakes of PET waste.

Recycling of PET

Plastic recycling is very important for, at least, two main reasons: *firstly*, to reduce the increasing volumes of plastic waste and *secondly*, to generate value-added materials from low cost sources by converting them into valuable materials.

In addition to the landfilling and thermal processes, there are two global methods for recycling of PET waste, namely, mechanical and chemical methods [30].

Mechanical Recycling

In mechanical method, generally, the beverage bottle can be recycled to the usage cycle by chipping them into very fine scraps, fig 5, heating, melting and remolding to new products. Frequently in the mechanical methods the reground scrap PET is blended with virgin PET, the other polymer and also mineral to achieve useful materials for different purposes [31-33].

Chemical Recycling of PET Wastes [34]

PET is chemically processed by total depolymerization into monomers or partial depolymerization into oligomers and other products. The monomers are then re-polymerized and the regenerated or the reconstituted polymer is formed into new article.

Different recycling processes for chemical reprocessing or chemolysis of PET waste are available (35-55) and most of them consist of esterifying the polyester with an excess of reactant such as diols, diamines, alcohols, or water, Scheme 1.

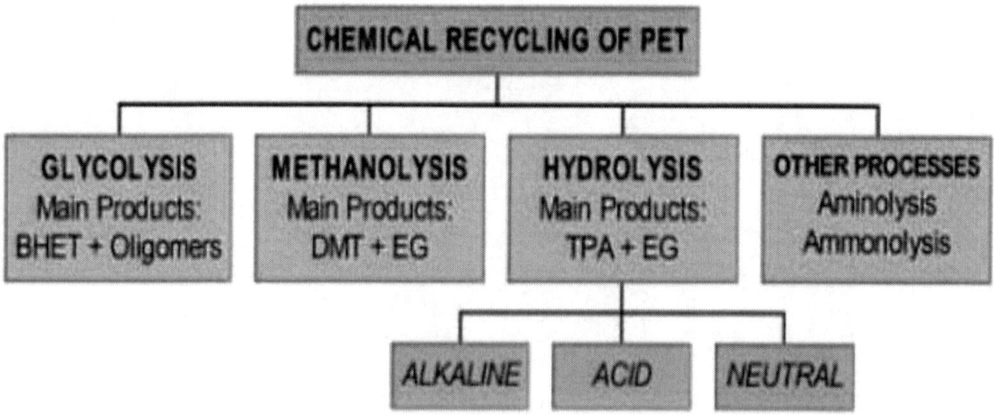

Source: Chong, M. and Kim, G., Cluster 8, COSMOS 2007.

Scheme 1.Some chemical recycling processes.

Methanolyticde polymerization of polyethylene terephthalate (PET) was carried out in a stainless stirred autoclave at temperatures of 523–543 °K, pressures of 8.5–14.0 MPa, and with a weight ratio of methanol to PET from 3 to 8. The solid products mainly composed of dimethyl terephthalate and small amounts of methyl-(2-hydroxyethyl) terephthalate, bis (hydroxyethyl) terephthalate, dimers and oligomers were also carried out [56].

Chemical recycling of Polyethylene terephthalate (PET) was undertaken through catalytic glycolysis method and the obtained oligomers were used to prepare several unsaturated polyester resins. PET in the presence of diethylene glycol (DEG), triethylene glycol (TEG) and zinc acetate as a catalyst was depolymerized, followed by purification of the products [57].

This chapter is based on a large body of research carried out at Inorganic and Applied Chemistry Unit, Radioisotope Department, Egyptian Atomic Energy Authority for nearly 5 decades. The Egyptian National Research Center plays an important complementary role in these studies. The published work, in addition, to that contained in literatures reviewed the development and applications of polyester-cement composite (PCC), based on the chemically recycled PET,in the field of solidification/stabilization of radioactive wastes.

PET Waste Aminolysis

Tawfik and Eskander [21] studied the aminolysis of post consumer PET bottle where a predetermined weight of the PET flakes was depolymerised at 190°C under atmospheric pressure using ethanolamine (EA) in presence of different concentration of dibutyl tin oxide (DBTO) as a catalyst. At the end of a chosen definite reaction time (i.e., 1, 2, 4 h), distilled water was added in excess to the reaction mixture with continuous vigorous agitation, followed by filtration to separate the white precipitate from the reaction medium. The filtrate mainly contained unreacted ethanolamine and some water-soluble PET degradation products. For purification, the precipitate obtained was dissolved in boiled distilled water, and then small fibres were produced through evaporation of excess water and chilling. The yield was dried, and weighed before subjecting to characterization.

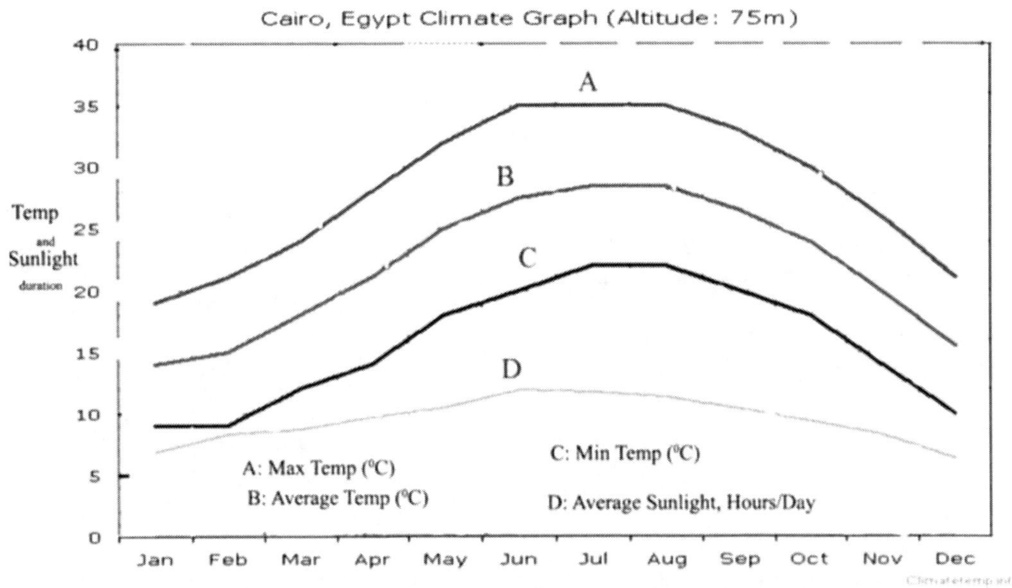

Figure 6. Temperature and average sunlight duration in Cairo, Egypt, in June, July, and August.

A series of experiments covering various reaction conditions such as PET: EA ratios, catalyst concentrations and reaction time were studied systematically to optimize the aminolysis process.

Tawfik et al. in 2010 [13] had introduced an alternative green procedure for the aminolysis of PET postconsumer wastes based on the use of sunlight as a renewable clean source of energy for the reaction instead of a conventional heating system and catalytic reagents that were carefully selected to be superior in their role. The chemicals used were commonly available and low in cost e.g., sodium acetate (SA), cetyltrimethyl ammonium bromide (CAB), and dibutyl tin oxide (DBTO). EA was used for the aminolysis of the small pieces of PET waste (1: 4 PET: EA w/w) in a properly sealed, round-bottom flask. The reaction was carried out in the presence of one of the three catalysts (i.e., DBTO, CAB, or SA) for 21 days of degradation time. The catalyst was added at a ratio relative to the weight of the PET.Glass flasks filled with the polymer waste, EA, and the catalyst were dipped in a sand bath and exposed to sunlight in an open-air area. The aminolytic process was undertaken during June, July, and August. These months are the summer season in Egypt and record the hottest temperatures all over the year. The recorded maximum and minimum temperatures and sun duration through these months are represented in Figure 6. At the end of the predetermined period, each flask was withdrawn and distilled water was added in excess to the reaction mixture with vigorous agitation to precipitate out the white product. The filtrate mainly contained the white precipitate in addition to the unreacted EA and small quantities of water-soluble PET degradation products. The precipitate obtained was filtered and redissolved in distilled water by boiling for about 30 min. The degradation white crystalline powder of BHETA was obtained through concentration of the filtrate by boiling and chilling. The end product was further purified by recrystallization in water, dried in an oven at 80oC, and then weighed to estimate the yield percentage relative to the unreacted part of the PET waste. This unreacted part of the PET waste was thoroughly washed with water and then dried in a vacuum oven for 2 h at 60oC, and its weight was followed until a constant value was reached. Similar degradation experiments were performed in the presence of increasing concentrations of DBTO as a catalyst (0.5–2 wt % of the PET waste) for 21 days under the same conditions. On the basis of the percentage yield obtained, a DBTO concentration of 1.5 wt % of PET was chosen for the processing of the plastic waste under the previously described conditions (i.e., dipping in a sand bath and exposure to sunlight), where the aminolysis reaction was studied for various time periods (7, 14, 21, 30, 45, and 60 days). At the end of each time interval, the white precipitates were separated from the unreacted PET waste, and the weight of the white crystalline BHETA was determined.

PET Waste Glycolysis

Polyethylene terephthalate bottle wastes were shredded and glycolyzed in propylene glycol [PG] (50 wt%) under reflux at 473°K for 8 h. Zinc acetate (0.5% by weight of the PET) was added as a catalyst to obtain oligomers with a hydroxyl value of 727mgKOH/g following the method described in previously work[58].

From the glycolysis process, unsaturated polyester (PC) was obtained with hydroxyl to carboxyl ratio 1.1: 1 mol/mol. The PET oligomers produced were charged in a reaction vessel with both maleic anhydride (MA) and adipic acid (AA) in a molar ratio of 1.1: 0.5: 0.5 and heated gradually at 353–413°K for 1 h. Then the temperature of the mixture was increased at a rate of 10 °K/h until it reached 483 °K. The total reaction time persisted for about 8 h and the reaction was allowed to continue until a determined acid number was reached.

To evaluate the properties of PC obtained from recycled PET, a control one based on virgin materials was also prepared. Bis-(2- hydroxy propyl terephthalate) (BHPT) was prepared from DMT and PG in the molar ratio (1: 2.3) according to the method described previously in literature [59].

Unsaturated polyester was obtained by the polyesterification of the BHPT with MA and AA in a molar ratio of 1.1: 0.5: 0.5 following the same procedure used for preparing that based on virigin polymer.From spectroscopic and physicochemical characterizations, it was found that, the prepared UP based on the recycled PET bottles and that based on virgin materials are quite similar [20].

Also, PET waste was depolymerized by diethelylene glycol (DEG)/propylene glycol (PG) mixture at a ratio of 1:0.5:0.5 by weight, in the presence of lead acetate as a catalyst (0.5% by weight of PET) at the reflux temperature of the glycols. The mixture was charged into a four-necked round bottom flask, which was fitted with a stirrer, reflux condenser, nitrogen inlet and thermometer. The flask was immersed in oil bath, and the temperature was brought up to 200°C for 5 hours. In order to check the extent of depolymerization, the glycolyzed product was analyzed for hydroxyl number value, and it was found to be 617 mg KOH/g [60].

Polyester is a category of polymer which contains ester functional group in their main chain (at least 85% by the weight). Although there are many types of polyesters, the term "polyester" as a specific material most commonly refers to polyethylene terephthalate (PET). Polyesters include naturally occurring chemicals, such as in the cutin of plant cuticles, as well as synthetics through step-growth polymerization such as polycarbonate and polybutyrate. Natural polyesters and a few synthetic ones are biodegradable, but most synthetic polyesters are not. Depending on the chemical structure, polyester can be a thermoplastic or thermoset;

however, the most common polyesters are thermoplastics (61). Polyesters as thermoplastics may change shape after exposing to heat. While combustible at high temperatures, polyesters tend to shrink away from flames and self-extinguish upon ignition. Saturated polyesters refer to that family of polyesters in which the polyester backbones are saturated. Unsaturated polyesters refer to that family of polyesters in which the backbone consists of alkyl thermosetting resins characterized by vinyl unsaturation. However, unsaturated polyester resin (UPR) is thermosetting material. There are several reasons for the importance of polyester such as the relatively easy accessible raw materials, the very well understood and described simple chemical process of polyester synthesis, the low toxicity level of all raw materials and side products during polyester production and processing, the outstanding mechanical and chemical properties of polyester and the recyclability. Polyester is very durable: resistant to most chemicals, stretching and shrinking, wrinkle resistant, mildew and abrasion resistant. Polyester is hydrophobic in nature and quick drying.

The production of unsaturated polyester resins based on recycled poly (ethylene terephthalate) (PET) is very interesting and possible. The resins made with the recycled PET can be used to produce materials composites of good quality. These polyesters prepared from PET wastes are a potential source of valuable materials with fairly good mechanical integrity and could be used for many practical applications among them the solidification and stabilization of radioactive wastes.Water extended polyester (WEP) based on the recycled poly (ethylene terephthalate) prepared from consumed soft drink bottles was mixed with Portland Cement (PC) to incorporate borate waste solution simulates [60, 62-65].

Cement

Cement has been described as adhesive substance capable of uniting fragments of solid matter to compact solid blocks (66). There is a wide range in composition of cement materials used for the solidification of radioactive wastes and in the construction of various barriers in a repository. Their common feature is that the main constituents are lime (CaO), silica (SiO_2) and alumina (Al_2O_3). Portland cement (OPC)is the most commonly used type in radioactive waste management applications (67). It has lime as principal constituent which is obtained by mixing calcareous and argillaceous or other silica-alumina and iron oxide-bearing materials. The principal compositions and oxides percentages of Portland cement used in our studies are represented in Table 1. Portland cement is a complex mixture of several phases that can vary appreciably in composition and hence reactivity. The composition of OPC had been changed largely over the last two centuries. The actual compositions vary according to the composition of the raw materials and the fuel used, but typically the following substitutions should occur:

Alite:	3-4% of other oxides, including Mg, Al, Fe
Belite:	4-6% of other oxides, notably Al, Fe, K, Mg, SO_3
Aluminate:	can be up to 20%, notably Fe, Si, Mg, Na, K, Ti
Ferrite:	part of a series varying between $Ca_2Fe_2O_5$ and $Ca_2A_{11}.4Fe_2O_5$

The typical composition in Portland cement has lower iron content than that of C_4AF [68].

Table 1. Chemical and mineral composition (%w/w) of Portland cement

Chemical compositions, %		Compounds compositions,%	
SiO_2	19.84	C_3S	53.11
Al_2O_3	4.74		
Fe_2O_3	4.0	C_2S	16.89
CaO	61.01		
MgO	2.5	C_3A	5.81
K_2O	0.6		
SO_3	2.4	C_4AF	12.16
Insoluble residue	0.95		

-Loss on ignition = 3.96%.
-Lime saturation factor = 96% by wt.

Polymer–Cement Composite

Cement hydration in polymer modified material is influenced by the presence of the polymer particles and polymer film in the fresh state, during hydration as well as in the hardened state. The properties of the fresh mixture are influenced to a large extent by the surfactants, present at the surface of the polymer particles. The cement particles are better dispersed in the mixture and amore homogeneous material is formed. The hydration of the cement is reflected in the strength evolution of the material [69].

Immediately after mixing, the cement particles and polymer particles are dispersed in the water. The first hydration of the cement takes place, which results in an alkaline pore solution. This is indicated as step 1, Figure 7. The second step is presented in Figure 8. A portion of the polymer particles is deposed on the surface of the cement grain and the aggregate. The polymer-cement ratio determines the amount of polymers present in the pore solution and present at the aggregate surface. Part of the polymer particles may coalesce into a continuous film. This preferably takes place at the surface of the cement hydrates where extra forces are exerted on the polymer particles due to the extraction of water for cement hydration. The polymer film can partly or completely envelop a cement grain, which results in a retardation or even a complete stop of the hydration of the cement grain. The following step, Figure 9, consists of cement hydration, polymer flocculation and possibly polymer coalescence into a film. The processes that take place depend on the curing conditions. If no dry curing period is included, the overall film formation is retarded and the influence on the properties of the fresh mixture is limited at this stage. If a dry curing period is included, polymer film formation takes place during this step, which influences the cement hydration process as well as the strength development at early ages. In the bulk liquid phase, hydrate precipitations are present, which form a combined inorganic and organic product. The fractions of the different types of product formed depend on the polymer/cement ratio used.

The final step, Figure 10 includes further hydration and final film formation. Through the cement hydrates, a continuous polymer film forms as water is further removed from the pore solution. The part of the polymer particles, that is still present in the dispersion, is restricted to the capillary pores and at the interface of the aggregates and the bulk polymer-cement phase. It is this part which contributes the most to the elastic and final strength properties. The continuity of the polymer phase through the binder matrix is more pronounced in the case of a higher polymer\cement ratio.

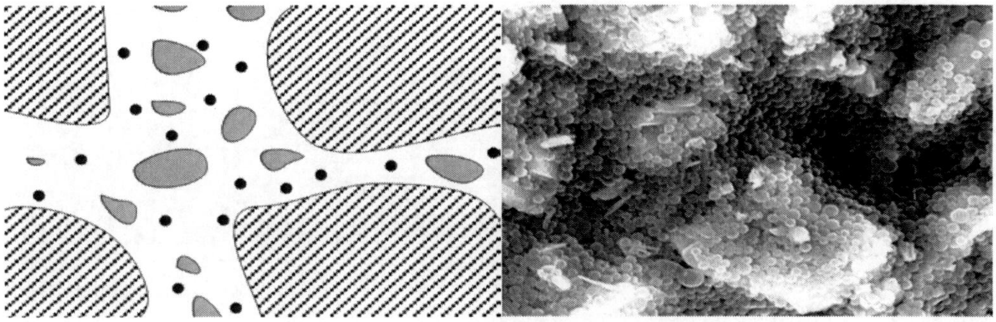

Figure 7. Step 1, immediately after mixing, aggregates, cement particles, polymer particles and mixing water – small ettringitte needles are formed.

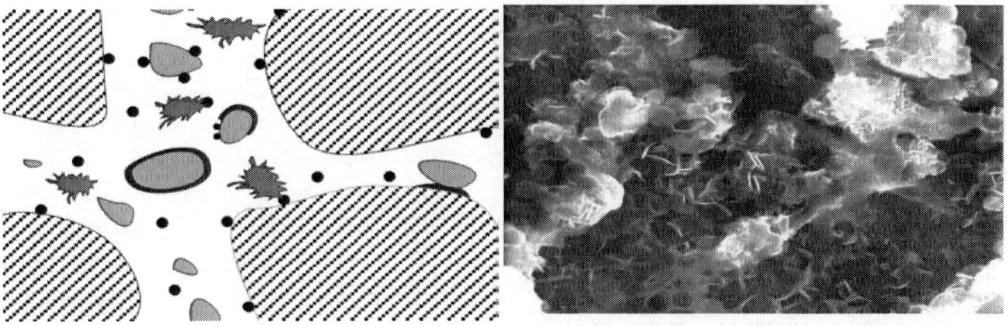

Figure 8. Step 2, after mixing, the polymer particles interact with the cement particles and the aggregates. In the case a dry curing period is introduced, a continuous film may be formed polymer particles flocculate together, on restricted places; no coalescence has taken place at this stage.

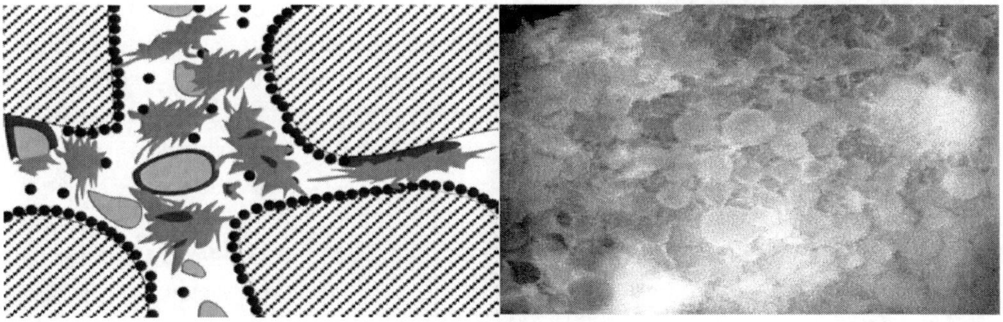

Figure 9. Step 3, cement hydration proceeds, polymer film formation starts on specific spots –polymer particles coalesce together into a continuous film.

Also, Machovič et al. found that during the alkaline hydrolysis of PET fibres with sodium hydroxide at an elevated temperature, morphological changes occur on the fiber surface. Their surface increases as well as the relative abundance of hydroxyl groups on the surface [72]. These changes lead to increasing fiber adhesion to the cementitious matrix during the hydration of the cement composite [73].

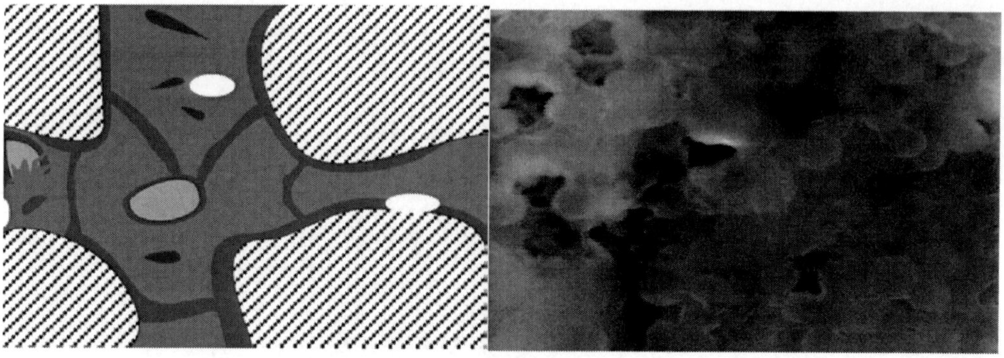

Figure 10. Final step, cement hydration continues, the polymer particles coalesce into a continuous film – cement particles are hydrated [70-71].

Polymer-Cement Matrices for Solidification/Stabilization Radioactive Wastes

Radioactive waste is defined as material that contains, or is contaminated with radionuclides at concentrations or activities greater than clearance levels as established by individual countries' regulatory authorities, and for which no use is foreseen.

Radioactive wastes that generated from a wide range of sources, including the power industry besides the medical and scientific research institutions, presenting a range of challenges in dealing with a diverse set of radionuclides of varying concentrations. There are three categories describing these wastes; Low Level Waste (LLW), Intermediate Level Waste (ILW) and High Level Waste (HLW). The main distinctions between these are the nature and level of the radiation. HLW produces significant quantities of heat during storage whereas ILW and more-so, LLW produce virtually no heat or only relatively low levels of heat evolution.It should be noted, however, that in some establishments, e.g., in the USA, only two categories, HLW and LLW, are adopted and the ILW being essentially incorporated into the latter.

Conditioning technologies are essential for the encapsulation and immobilization of these radioactive wastes, forming the initial engineered barrier required for their transportation, storage and disposal. The need to ensure the long term performance of radioactive waste forms is a key driver of the development of advanced conditioning technologies.

The conversion of liquid and wet solid wastes into a stable monolithic form by solidification minimizes the potential for radionuclide release to the environment during interim storage, transportation and disposal. The solidified waste form significance, especially during transportation, in order to reduce the probability of waste form fracture and breakage which may increase the rate of radionuclides release due to the increased surface area exposed to the environment. High thermal and radiolytic stability ensure that the waste form will retain its original properties over time and not pressurize the waste container.

The immobilization of radioactive waste is an integral part of the cycle for efficient use of nuclear energy and during the past twenty five years, a great deal of research into suitable matrix and methods has resulted in the deployment of a variety of successful processes on an operational scale. These are dependent on the nature of the waste itself and in particular, it's associated radioactive properties [74].

Different matrices, ranging from cement, bitumen, polymers and composites, are being used for immobilization of low level liquid radioactive wastes concentrates.

Cements have several attributes that make them suitable for consideration as an encapsulation medium for radioactive waste. Thus, they

- are readily available and relatively inexpensive
- can be formed into a fluid grout
- are durable
- can incorporate many ions into solid solution
- maintain a high pH, which decreases radionuclide solubility
- provide radiation shielding that is not degraded by the radiation
- can be modified to suit a particular purpose through selection of appropriate constituents
- are tolerant to a wide variety of waste forms
- It can be relatively easily processed in a remotely operated plant

It has, however, major disadvantages for the encapsulation of radioactive waste in cement material which is the extent and rate of heat evolution, besides the relative porosity of the cemented waste forms.

Polymers are generally inert, permit high waste loadings compared with conventional cementitious encapsulation, and durable providing excellent waste form characteristics such as low leachability and good compressive and tensile strength, an important consideration for hydrophilic wastes [75-77]. Both unsaturated styrenated polyester (SPE) and polymethyl methacrylate (PMMA) were used to fill the pores in cement containers that used for disposal of radioactive wastes. Two different techniques were adopted for the addition of organic polymers based on their viscosity. The low density PMMA was added using impregnation technique. On the other hand high density SPE was mixed with cement paste as a premix process [78].

Recycled polymers were studied for use as encapsulating agents in order to potentially improve process. Economics, provide a valuable new market for non-biodegradable plastics, reduce the usage of valuable petroleum resources and lessen the municipal waste burden. The recycled plastics tested, were all post-consumer or post-industrial and included low-density polyethylene (LDPE), linear low-density polyethylene (LLDPE), high-density polyethylene (HDPE) and polypropylene (PP). Recycled plastics were used as a direct substitute for virgin resin or in a selected proportion with virgin resin based on expected processing characteristics. Impacts on processing parameters and final waste form performance were examined [79].

Polymer impregnated cement (PIC) is a precast cement- composite, containing liquid or solid radioactive waste, which has been impregnated with a monomer that is subsequently polymerized [80]. In this way void spaces within the cement are filled, reducing leachability and improving mechanical properties.

Polymer modified cement materials offer considerable improvement in properties and performance compared with traditional mortars [81-82].

Polymer cement composites using unsaturated polyester resins based on recycled poly-(ethylene terephthalate) were investigated and are considered of importance for their potential applications in the nuclear field [83]. The mechanism of interaction of polymers with the hydration products of Ordinary Portland Cement (OPC) has been also studied [84-86].

Improvement in the properties of cement waste forms-styrenated polyester based on poly (oxy-diethylene maleate) as a polymer additive was, also, investigated by Salehetal at 2012. The composite material of cement/polymer was used as a matrix for the immobilization of borate waste simulates [87].

Cement-Water Extended Polyester (WEP) Composite Based on Recycled PET

Preparation of Unsaturated Polyester (UP)

Unsaturated polyester (UP) was prepared by the esterification of the glycolized PET oligomers with maleic anhydride (MA) and adipic acid (AA) at a molar ratio of 1.1:0.7:0.3 for hydroxyl to carbonyl groups, respectively. The polyesterification reaction was carried out in a four-necked round bottom flask, having a stirrer, reflux condenser, an inert gas inlet and thermometer. A Dean–Stark trap was used for measuring the water amount, which was removed throughout the reaction. The reactants were heated in oil bath, and its temperature rose to 150°C. The temperature was then increased to 210°C at a rate of 10°C/h until the acid number value reached nearly 45 ± 2 mg KOH/g. Table 2 described some physical and mechanical properties of the prepared UP resins.

Table 2. Some physicomechanical properties of the prepared UP resins

Physicomechanical properties	
Specific gravity at 298 °K	1.29
Acid number (mg KOH/g)	42
Hydroxyl number (mg KOH/g)	60
Number average molecular weight (Mn)	1100
Weight % of styrene monomer in resin	40
% of cross linking	92
Elongation at break of the neat cast (%)	30
Tensile strength of the neat cast (MPa)	18
Appearance	yellow

Preparation of Water Extended Polyester (WEP)

The obtained UP was dissolved in a styrene monomer at a ratio of 60:40% by weight at room temperature (25 ± 5°C). WEP emulsion was prepared by dispersing with continuous stirring of 20% water by weight, containing 0.2 mmole DEA/g prepolymer as an emulsifier, into 10 g of the obtained styrenated polyester. The stable emulsion was cured by addition of 2% from the prepared WEP by weight benzoyl peroxide (Bz_2O_2) as a catalyst. The obtained water extended polyester mixture was mixed with cement paste at various water/cement ratios to prepare either the final waste form by incorporating a predetermining weight of borate waste simulate or polyester-cement composite container.

Characterization of Unsaturated Water Extended Polyester (WEP) Based on Virgin Raw Ingredient

Both IR and NMR analyses of the prepared UP from its virgin raw materials illustrate the expected formula of the polyester given in the literature and determine the main characteristic groups of the prepared polyester used as an additive to the cement waste forms (Figs. 11, 12) [88].WEP is used as an organic additive to improve the properties of cement waste forms. Nuclear magnetic resonance (NMR, ppm) analysis for the UP based on PET was, also, carried out using a Bruker AC-400 spectrometer and deuterated chloroform as a solvent. The spectrum represented in Figure 13 shows the obtained signals for the UP analysis and their corresponding assignments represented the following groups; at 8.07–8.08 (4H, C_6H_4), at 6.85–6.89 (olefinic protons in the CO-CH=CH=CO chromophore of maleic and fumaric acid residues), at 5.2–5.6 (methine proton of PG residue), at 3.58–4.68 (-O-CH_2-CH_2-O and –O-C-CH_2-O), at 1.63–2.14 (adipic acid residue), and at 1.27–1.48 (methyl group in PG residue) [60].

The IR spectrum of unsaturated polyester (UP) obtained from recycled soft drink bottles of PET shows completely similar to that based on the virgin materials. The broad band that appeared around 3500 cm^{-1} is due to the stretching vibrations of the OH terminal in the hydroxyl group in the hard polymer. The band at 3020 cm^{-1} may be assigned to the aromatic CH while bands at 2950 cm^{-1} and 2880 cm^{-1} may refer to the stretching of CH in methylene and methyl olefinic groups, the stretching frequencies of the ester carbonyl group (C=O) in the polymer appeared near 1720 cm^{-1} and 1640 cm^{-1} while that for olefinic -C=C- appeared at 1580 cm^{-1}. The band in the region 1450–1380 cm^{-1} may be attributed to the bending vibration of CH_2 and CH_3 in propylene glycol. The curve shows also the bending in trans -CH=CH- groups, which appears at 980 cm^{-1}.

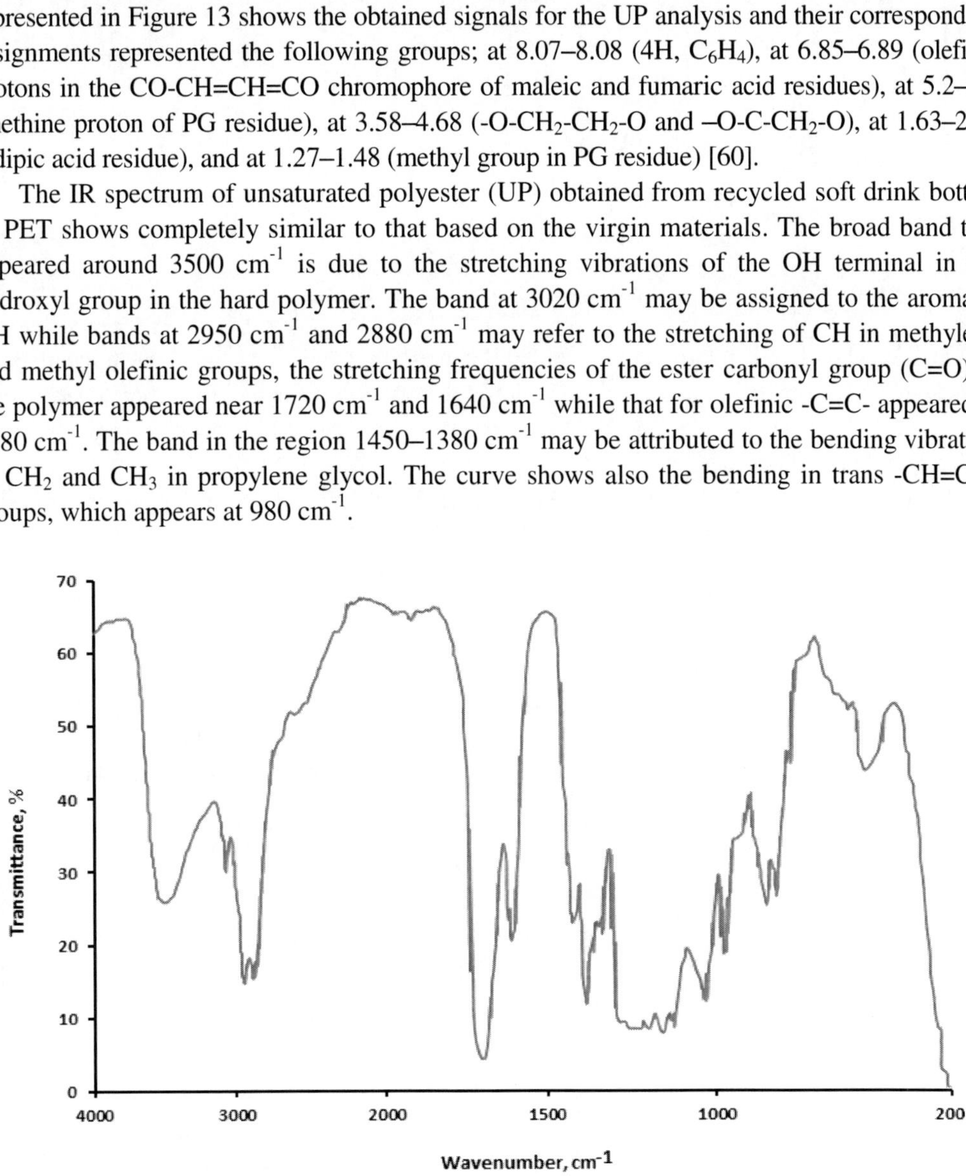

Figure 11. FT-IR spectrum of the unsaturated polyester prepared from virgin raw ingredients.

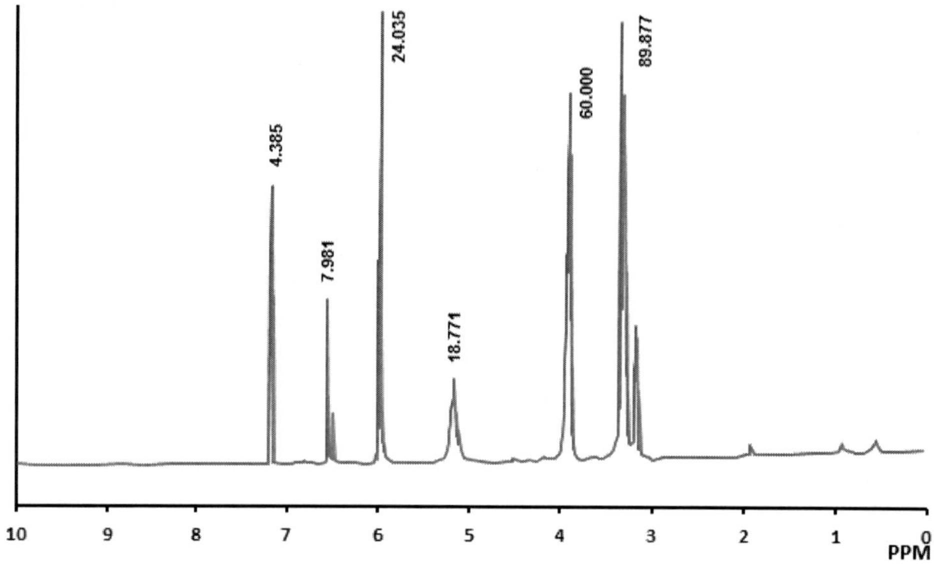

Figure 12. NMR spectrum of the unsaturated polyester prepared from virgin raw ingredients.

On the other hand, bands at 880 cm^{-1}, 750 cm-^{1}, and 650 cm^{-1} may be attributed to cisolefinic and aromatic residues [62]. It could be definitely stated that there were non-significant differences between the WEP based on the recycled PET and that prepared from its virgin raw ingredients.

However, it is worth mention that the economical and environmental aspects favored the preparation of WEP based on the recycled PET.

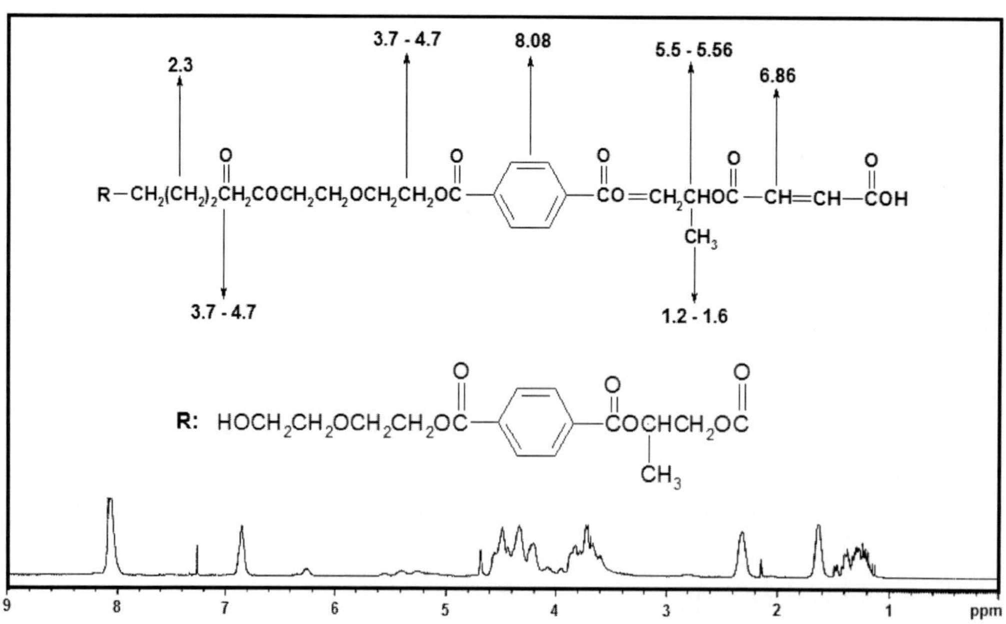

Figure 13. NMR spectrum of polyester based on glycolized PET waste.

On the other hand, styrenated unsaturated polyester products based on both the recycled PET and poly (oxy-diethylene maleate) mixed with styrene monomer showed mechanical and physical characterizations closely similar to each other as reported in Table 3.

Table 3. Chemical, physical and mechanical characterizations of WEP

Composition	WEP based on recycled PET	WEP based on virgin raw ingredients
Styrene monomer	30 %	30 %
Specific gravity at 25°C	1.45	1.45
Water of emulsion	50 %	50 %
Initiator (Benzoyl peroxide)	2 %	2 %
Compressive strength, MPa	9.57	9.95

Evaluation of the Final Waste Form

Pressurized water reactors (PWRs) are controlled, in part, by the manipulation of the concentration of boric acid in the primary coolant. Boron has very high neutron captures cross section, making it a convenient method for adjusting the reactivity, which in turn controls a reactor's power output. Thus; PWRs produce radioactive liquid waste that usually contains relatively high concentrations of boric acid.

Transformation of that borate wastes from pressurized water reactor into homogeneous monolithic solid waste forms was achieved using cement and polymer-modified cement as an inert matrices. The aim was to reduce the potential migration and dispersion of radionuclides from waste forms to the environment, the primary objectives were to provide possible options helping in choosing the proper immobilization media, to improve the quality of the final waste forms and process technology.

Polyester-Cement Waste Form (PCC)

The final waste forms were prepared by mixing predetermined weight of powdered radioactive borate waste simulates with polymer-cement composite paste. This mixture was poured in a polyethylene moulds and allowed to set and harden at room temperature in its humid atmosphere for 28 days. At the end of curing time, the obtained blocks were demoulded giving solid cylindrical blocks of 60 ± 2 mm height and 31 ± 0.5 mm diameter. Various factors that may affect the physicochemical stability of the formed composite specimens were studied. The resulting PCC products assumed to show a good mechanical integrity, enhanced physical properties, acceptable radiation stability, low leachability of radionuclide, and withstand relatively high temperatures. However, it could be stated that the characters of the prepared PCC depend mainly on the properties of UP used.

Compressive strength is generally given for the description of structural stability of the final waste forms. To evaluate the mechanical integrity of the solid PCC containing the borate waste simulate, compressive strength measurements were carried out for PCC solid blocks under variable factors that may affect e.g., polymer/cement ratio, water content, or borate waste simulate concentration. Remarkable changes in compressive strength values for PCC samples containing an increasing ratio of WEP premixed with cement paste having different water/cement ratios were recorded, as shown in Figure 14. Premixing cement with WEP affected the final properties of the final waste forms in two contradicting ways depending on the WEP/PC ratio used. In the case of low polymer contents (up to 3% of hydrated cement), the compressive strength values were increased by increasing the WEP content. On the other hand, increasing the polymer content, more than 3 weight percent, the mechanical integrity of the final waste forms diminished clearly. The improvement in mechanical integrity of the blocks containing low WEP concentration may be attributed to the positive effect of the organic WEP on the final waste forms. When premixing the polymer with the cement and water, it becomes uniformly dispersed within the fresh paste. On hardening, the polymer particles coalesce and gradually build up together to form a three-dimensional cross-linked network that contributes to the toughness and durability of the PCC-system [60]. In other words, the WEP additives act as binding materials improving adhesion of the different heterogeneous cement components. Moreover, the WEP dispersion during hardening seals the pores of the cement matrix and, hence, diminishes the porosity of the products reflecting improvement of the mechanical integrity of the final waste forms. Also, the WEP may act as a filler material that has a plasticizing effect and results in a detectable increase in the durability of the final waste forms. On the other hand, the high WEP contents negatively affected on the mechanical properties of the final solid blocks, as shown in Figure 14. This may be due to the fact that the dispersion of the excess organic polymer materials in the paste restricts the access of water in cement grains by forming a thin coating skin over them and also enhanced the ettringitte formation. In addition, the hydration process of cement could also be negatively affected by the interaction of excess polymer with Ca^{++} in the cement materials. The carboxylate anions (COO^-) produced by the hydrolytic reaction of esters combine chemically with Ca^{++}cations released from cement grains during the hydration of cement material to form a molecular structure that prevents the scission of the polymer chain.

Basically, in the cementation process, the anhydrous tricalcium silicate (C_3S) and dicalcium silicate (C_2S), which contribute with the most abundant constituents of PC, commences the hydration reaction very rapidly. Water reacts chemically with the cement constituents to form hydrate silicate and aluminates compounds, which normally contribute to setting and hardening of cement [60]. Figs. 14 and 15 describe the effect of water content on the mechanical properties of PCC. Two sets of PCC solid blocks were prepared having different water contents. In the first set, the water of emulsion was included in the water of hydration, while in the second set the water of emulsion was not included in the water of hydration. It is clear from Figure 14 that when the water of emulsion was not included in cement hydration water, solid blocks with higher compressive strength values were obtained, compared to that of the solid blocks in which the water of emulsion was included in the cement hydration water. This may be attributed to the fact that in the second set, part of the water was used for emulsion formation and at this low water content; the hydration reaction of cement may be stopped in the presence of significant amounts of non-reactive cement materials due to the lack of sufficient amounts of water needed for the hydration process.

Therefore, it is recommended to calculate the cement hydration water separately from the polymer emulsion water. It is worth mentioning that the same trend was exhibited for two different w/c ratios, i.e., 35 wt% or 45 wt% which was not including the water of emulsion. The ratio of 35 wt% of higher mechanical integrity was obtained compared to that at the ratio 45 wt%, but the workability at the w/c ratio 45 wt% was better than that at 35 wt% (Figure 15) [60].

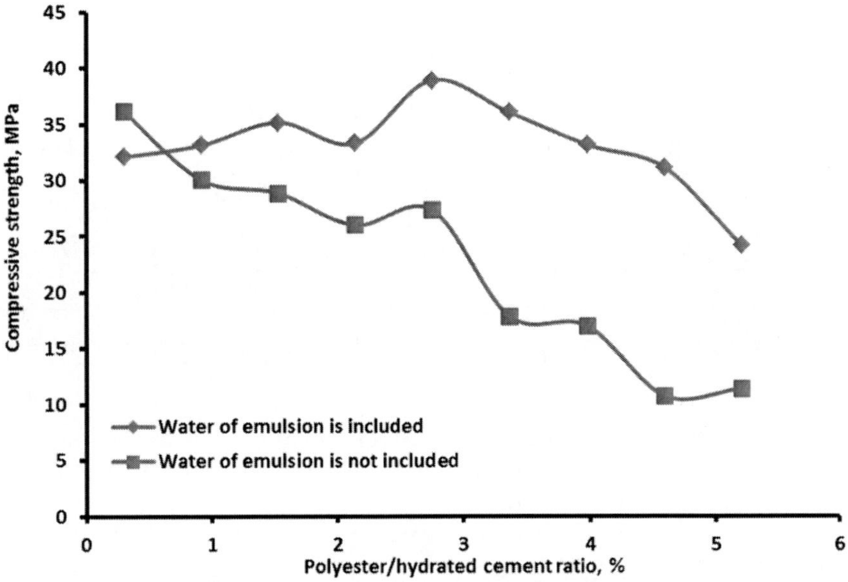

Figure 14. Mechanical integrity of the final polymer-cement composite waste form; Effect of the water content on the compressive strength of the final waste form.

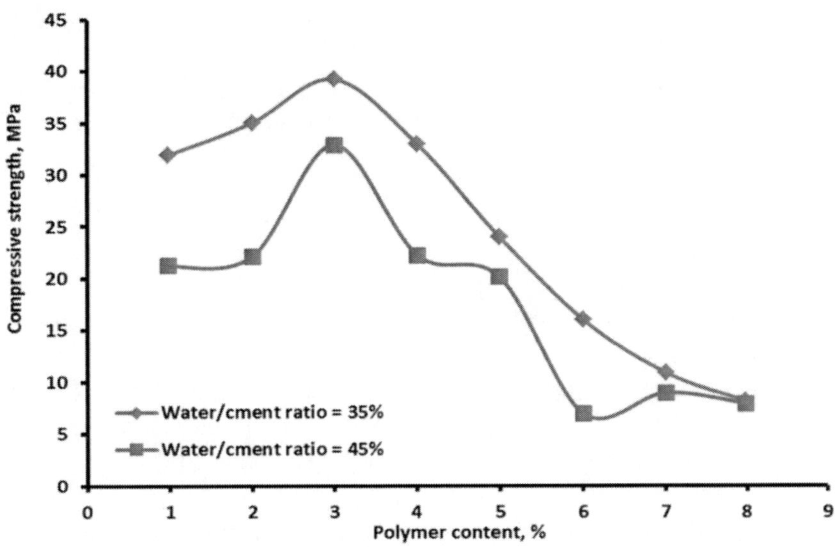

Figure 15. Mechanical integrity of the final polymer-cement composite. Effect of polymer content on the compressive strength of the final waste form.

The effect of borate content on the mechanical integrity of PCC final waste forms were followed by increasing the waste simulate percentages that premixed with WEP/PC mixture containing 3% polymer (WEP) and at water/cement ratio 40%. After the curing period, the compressive strength measurements were performed for the solid blocks and the data obtained were represented in Table 4. Increasing the borate content premixed with the PCC caused a remarkable decrease in the mechanical integrity of the final waste forms.

Table 4. Compressive strength values of the final PCC waste forms containing different percentages of borate waste simulates

Borate, wt.%	Compressive strength of blocks, MPa	
	Containing 3%, by wt WEP	Containing 6%, by wt WEP
0	33.6	32.74
1	30.06	28.74
2	19.0	0.38
3	20	brittle
4	2.5	brittle
5	brittle	—

Borate waste simulate mainly includes a large percentage of boric acid H_3BO_3 (i.e., 180 g/1) and some borates, e.g., BO_3^{-3} and $B_4O_7^{-2}$ are known as cementation inhibitors. Boric acid reacts with calcium hydroxide during cementation as follows:

$$2H_3BO_3 + Ca\,(OH)_2 \rightarrow Ca\,(BO_2)_2.2H_2O + 2H_2O$$

Calcium borate is sparingly soluble and is assumed to create a barrier at the cement particles and the solution interface. This barrier retards the diffusion of water and colloidal particles.

Acceptable compressive strength values for solid blocks containing up to 3% borate waste simulate were achieved for PCC waste forms containing 3% WEP and w/c ratio 40%. At higher borate levels, a sharp deterioration in the mechanical integrity was detected [60]

Chemical Characterization

Development of leach resistant waste forms is a major area of research in the field of radioactive waste management in order to minimize the environmental impact through the back release of the hazardous materials.

The long-term chemical stability of the polymer-cement composite waste (PCC) aged for 7 years under normal weather conditions and then exposed to different leachants was studied. Cumulative fraction leached of radiocontaminants and mass changes of the waste form during leaching periods were addressed [64].

The leaching mechanisms of simulated low and intermediate level radioactive waste forms immobilized in polymer-cement composite are being determined as a support for the development of their chemical stability under different leaching parameters such as static and dynamic conditions, various dipping media, environment temperatures, etc. However, it should be noticed, and to the best of our knowledge, that no data have been published which provides a basis and direct comparison of the leaching behavior of the candidate final waste form aged for 7 years before subjected to leachants flooding under previously stated conditions, which is supposed to play a role in the long-term durability of the prepared composite.

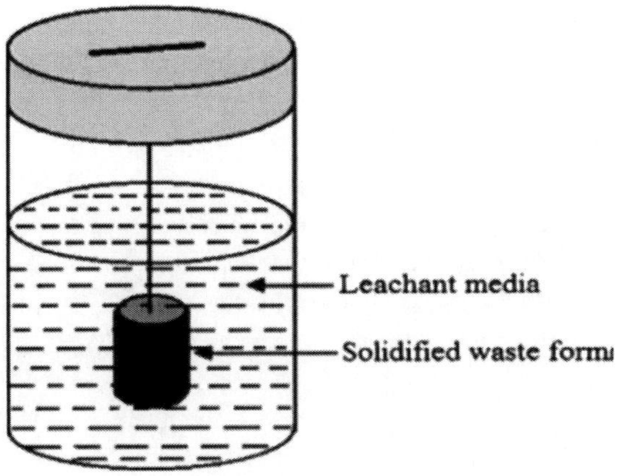

Figure 16. Schematic diagram of suspended solidified waste form during leaching test.

Static and Dynamic Leaching Conditions

PCC samples spiked with ^{137}Cs and ^{60}Co were suspended in the center of a closed leaching jar containing groundwater (leachant volume of 150 ml) under static conditions (i.e., no agitation, no movement and the leachant was not replaced by fresh one during the whole test period). The whole volume of the leach solution was periodically withdrawn out of the leaching jar, (Figure 16), subjected to counting and then returned back again. While in dynamic experiments, similar PCC blocks were treated in the same manner, except, the leaching solution was replaced periodically by fresh one at time intervals parallel to the sampling in a static test.

The leachability, of ^{137}Cs and ^{60}Co radionuclides from the PCC monolith waste form, was sum of all fractions leached during the time intervals ΣAn/Ao. S/V, was plotted versus the square root of time in days ($t_{1/2}$) where; An is the activity of radionuclides released during the particular leaching period with index n, Ao is the initial activity of radionuclides present in the borate waste immobilized in PCC and tn is the elapsed time of the leaching period n in days. S and V are the surface area and volume of the tested block, respectively [64]. The counts of radionuclides were determined using multichannel analyzer, PCAP, USA. From the plot of the cumulative fraction leached as a function of the square root of time in the diffusion region of leaching, slopes were calculated and applied to determine the diffusion coefficient using the following formula:

$$De = \pi (mV/2S)^2$$

where; De is the apparent diffusion coefficient (m^2/s), m is the slope of the curve demonstrating the ΣAn/Ao versus the square root of leaching time in seconds, V is the sample volume (cm^3) and S is the sample surface area (cm^2). On the other hand, leach index (Lx) which denotes a direct measure of the extent of leaching of radionuclides from the candidate polymer-cement composite, was computed based on the relation,

$$Lx = \log (1/De).$$

For all leaching curves, it should be notified that the leaching behavior of radionuclides from the monolith waste forms revealed two distinguished performances. The first initial period characterized by a high leachability of radionuclides relative to the second final one. This may be explained by the fact that up to nearly 25 days from the starting time, the high release may be attributed to the washout of loose radioactivity from the surface of PCC waste form. This can be due to the relatively rapid washing process occurred when the blocks were firstly exposed to the leaching solution. This was followed by nearly a steady state diffusion controlling periods that persist. Leaching solution velocity (v) is defined by the volume (v) of the leachant contacted with solid waste per unit of surface area (S) and unit of time (t) given as:

$$v=v/ (S. t).$$

At low leaching solution velocities, i.e., static condition, where the leachant was not replaced by fresh one, the amount leached approaches the saturation limit and leaching solution allowed to equilibrate with waste forms. On the other hand, under dynamic conditions where the leachant was periodically replaced maximum saturation limits were not obtained [64] and consequently, higher leach rate was reached as presented in Figure 17a, b. In the forthcoming leaching experiments, the dynamic condition will be performed.

Chemical Behavior of PCC in Various Leaching Media

The long-term durability of the polymer-cement composite under the imposed disposal conditions is a key factor controlling whether the proposed PCC can be accepted as a solidification matrix for borate waste or not. One of the major concerns that meet the cementitious waste form products in the disposal site can be the aggressive flooding environment. The leaching test was used to predict the long-term durability of the waste forms when exposed to aqueous media. Polymer-cement waste forms are alkali in nature, and they are vulnerable to acid attack either from acid rains, seawater or groundwater that may be reached the disposal environment.

Seven years aged PCC blocks, spiked with ^{137}Cs and ^{60}Co, were suspended in separate jars each of which contains one of the following leachants, 0.1 M H_2SO_4, groundwater, seawater or tap water. Cumulative fractions leached of the two radionuclides in the different leaching solutions were plotted as a function of leaching time and the data obtained are shown in Figure 18a, b. Both ^{137}Cs and ^{60}Co recorded the highest leachability in 0.1 M H_2SO_4 followed by seawater and then groundwater while the leachability in tap water was the lowest. In sulfuric acid medium, the dissolution of calcium hydroxide crystals and the highly decalcified calcium silicate hydrate (CSH) gels due to the acid attack result in increasing porosity and enlarged capillary pores in the leached specimens. This causes self-accelerating leaching [64]. Due to exposure to sulfuric acid for 260 days, the 7 years aged PCC solidified waste forms showed some swelling and corrosion due to the formation of a gypsum layer on the surface. It is worth mentioning that the visual appearance of the blocks, shown in Figure 19 is in consistence with the leaching rate results and confirms the data obtained for the high leachability of both radionuclides in sulfuric acid (Figure 18a, b).

The attack caused by solutions containing sulfate, e.g., seawater or groundwater can have the same diverse effects on the cementitious-base waste form. The effect of sulfate in both media may be associated mainly with the expansion of the gypsum and ettringitte that can be formed as a result of contact with such leachants, as shown in the following formula:

$$\underset{\text{Portlandite}}{Ca(OH)_2(s)} + H_2SO_4\,(a) \rightarrow \underset{\text{gypsum}}{CaSO_4.2H_2O}$$

$$\underset{\text{Gypsum}}{3(CaSO_4 \cdot 2H_2O)\,(s)} + \underset{\text{hydrogarnet}}{3CaO \cdot Al_2O_3 \cdot 12H_2O(s)} + 14H_2O(a) \rightarrow \underset{\text{ettringitte}}{3CaO \cdot Al_2O_3 \cdot 3CaSO_4 \cdot 32H_2O\,(s)}$$

The expansion of these two abrasive salts can cause cracking and consequently, increase in the exposed surface area of the PCC blocks to the leaching solution and accordingly results in higher leach rates [64]. Due to the absence of the most aggressive components (e.g., sulfate or chloride) in tap water, the ^{137}Cs and ^{60}Co leachability represented the lowest rate even after more than 260 days (Figure 18a, b). Hence, when PCC present in places flooded by tap water or non-acidic rain water, reasonable leaching performance could be expected.

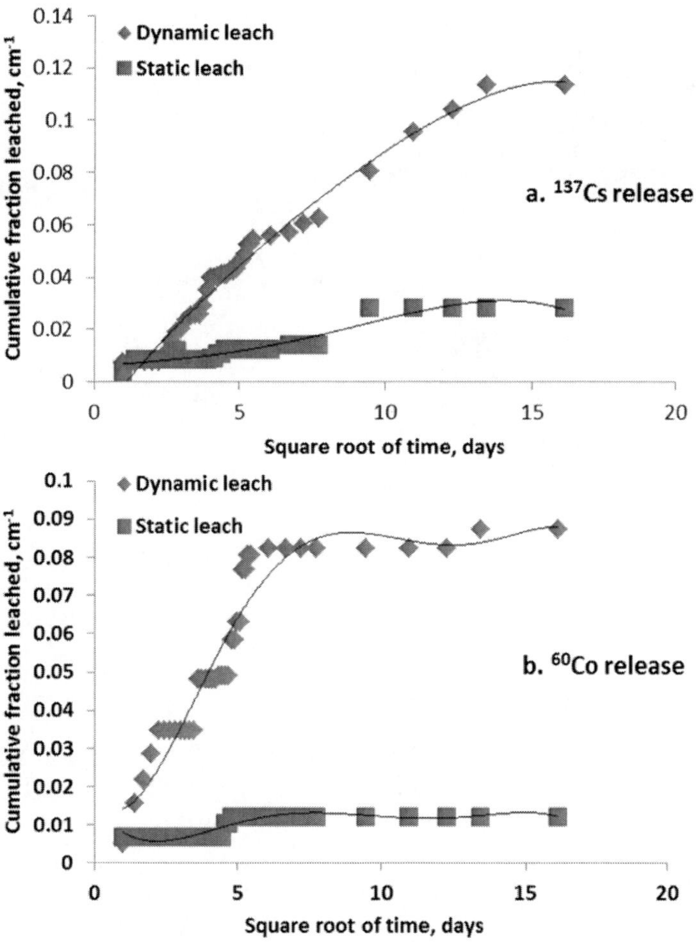

Figure 17. Cumulative fraction leached of ^{137}Cs and ^{60}Co released in groundwater understatic and dynamic leach conditions at room temperature (25 ± 5°C).

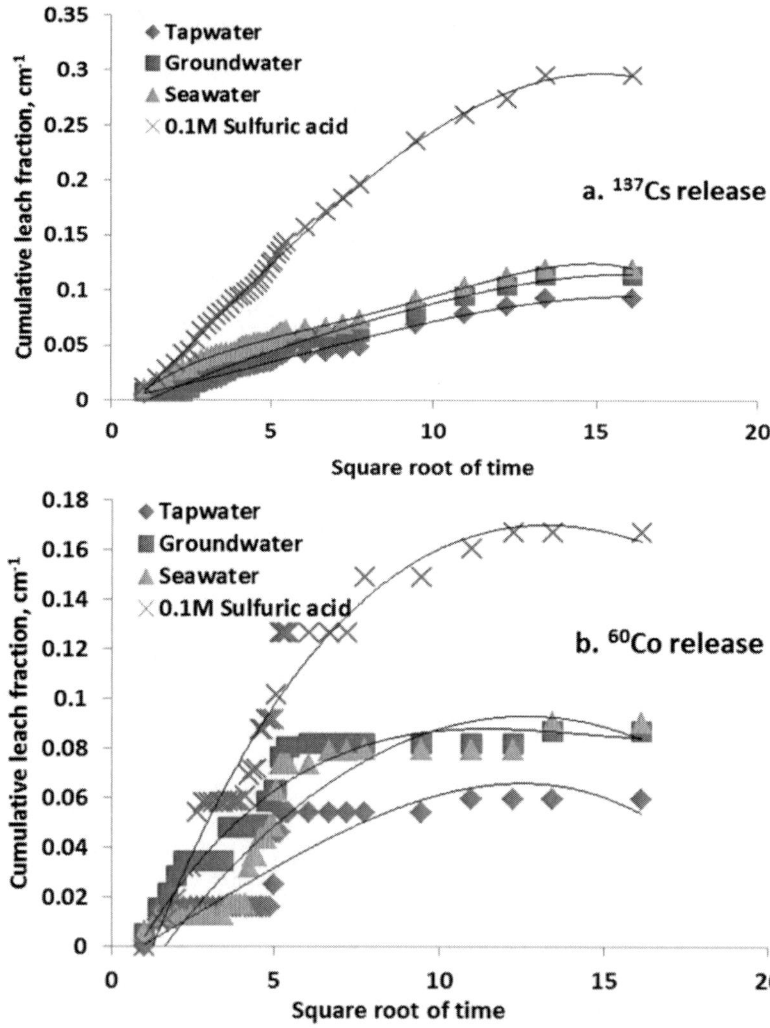

Figure 18. Cumulative fraction leached of ^{137}Cs and ^{60}Co released in different leachants at room temperature (25 ± 5°C).

Diffusion Coefficient and Leach Index

It is clear from all leaching curves that, after nearly 30 days a plateau of stabilization of the cumulative fraction leached is reached for the PCC blocks under all various studied conditions, which shows that the only leaching mechanism of the two radionuclides is controlled by diffusion mechanism.

Based on the diffusion coefficient (De), the leachability index (Lx) is calculated to characterize the efficiency of the candidate matrix to solidify the problematic borate waste concentrate. The value 6 of leach index is the threshold value to accept a given formulation as adequate for certain solidification/stabilization of nuclear wastes [64]. The higher values of the leachability index represent the lower release of the radionuclides from the monolith waste forms. The diffusion coefficients (De) in (m^2/s) together with leach indices data for the various leaching conditions are represented in Table 5. Cesium, being the most mobile waste soluble species in the nuclear waste stream, is poorly adsorbed in the basic environment of cement [64]. On the other hand, cobalt may form insoluble hydroxide at this higher pH condition. Therefore, the leachability of ^{137}Cs was greater than that of ^{60}Co from the PCC-waste form immersed in the same leachants. The data in Table 5, indicated that all leach indices of either ^{137}Cs or ^{60}Co in various leachants and under different conditions were usually greater than the value of 6 [64]. The relative high Lx values may be attributed to the role played by polymers within the PCC in reducing the porosity of the final waste forms and hence retarding the release of radiocontaminants. Recently, it was reported that more than 60% of cesium activity was released from immersed PCC-waste form cured for 28 days only [64]. This value can be easily compared to 11.3% of radiocesium leached after 260 days from the aged PCC under investigation. This assures the task achieved by the polyester based on the recycled PET waste in increasing the chemical stability of the proposed PCC even after 7 years of aging followed by immersion for 260 days.

Before leaching **After leaching**

Figure 19. Polymer-cement composite waste form (PCC) before and after immersion in 0.1 M H_2SO_4 for 260 days.

Table 5. Diffusion coefficients and leach indices at different leaching conditions

Leaching Conditions	^{137}Cs		^{60}Co	
	De (m^2/s)	Lx	De(m^2/s)	Lx
0.1M H_2SO_4acid*	1.02E-13	8.99	2.105E-14	9.67
Sea water*	1.29E-14	9.89	7.255E-15	10.14
Groundwater*	1.24E-14	9.91	2.786E-16	11.56
Tap water*	9.49E-15	10	2. 7.32E-15	11.65
Static condition*	1.03E-15	11	232E-16	10.14

* Leachant volume/surface area of waste form (V/S) =3.

Based on all leach tests data obtained, it could be stated that the candidate polymer-cement composite is an acceptable solution for immobilizing the radioactive borate waste even after 7 years of aging and under a breakthrough of water into the disposal site.

CONCLUSION

The recycling process is the most recommended and commercially sustainable way to manage the polymeric wastes, because it converts those materials to useful added-value products. It saves money, energy, raw materials, and land space needed for disposal and mainly reducing the environment pollution. The production of unsaturated polyester resins based on recycled poly (ethylene terephthalate) (PET) is very interesting and practice widely. These polyesters prepared from PET wastes are a potential source of valuable materials with fairly good mechanical integrity and could be used for many practical applications among them is solidification and stabilization of radioactive waste. Water extended polyester (WEP) based on the recycled poly (ethylene terephthalate), (PET), prepared from consumed soft drink bottles, was mixed with Portland Cement (PC) in a composite to incorporate borate waste solution simulate. Based on the all experimental data obtained, it can be stated that, the reached cement–polymer composite is an additional acceptable recommended solution for immobilizing the radioactive borate waste.

REFERENCES

[1] Soufiane Benosma, A., Mouli, M., Taibi, H., Belbachir, M., Senhadji, Y., Behlouli, I., Houivet, D., (2012), "Mineralogical Study of Polymer-Mortar Composites with PET Polymer by Means of Spectroscopic Analyses". *Materials Sciences and Applications*, 3, 139-150.

[2] Sakai, E.,Sugita, J., (1995), "Composite mechanism of polymer modified cement" *Cement and Concrete Research*, 25, 1, 127–135.

[3] Jansen, D., Goetz-Neunhoeffer, F., Neubauer, J., Haerzschel, R., Hergeth W.D.(2013), "Effect of polymers on cement hydration: A case study using substituted PDADMA". Cem*ent and Concrete Composites*, 35, 1, 71–77.

[4] Silva, R.V., De Brito, J., Saikia, N., (2013), "Influence of curing conditions on the durability-related performance of concrete made with selected plastic waste aggregates", *Cement and Concrete Composites*, 35, 1, 23–31.

[5] Ferreira, L., De Brito, J., Saikia, N., (2012), "Influence of curing conditions on the mechanical performance of concrete containing recycled plastic aggregate". *Construction and Building Materials*, 36, 196–204.

[6] Kim, S.B., Yi, N.H., Kim, H.Y., Kim, J-H.J., Song, Y-C., 2010, "Material and structural performance evaluation of recycled PET fiber reinforced concrete".*Cement and Concrete Composites*, 32, 3, 232–240.

[7] British Plastics Federation Oil consumption (20 October 2008) http://www.bpf.co.uk/ Oil_Consumption.aspx.

[8] PLASTEX, (2010)http://www.plastexonline.com/Press/PressDetailes.aspx?AID=19.

[9] Braun, D., (2004), "Poly (vinyl chloride) on the way from the 19th century to the 21st century" *Journal of Polymer Science Part A: Polymer Chemistry*, 104, 1839-1844.

[10] Simoneit, B. R. T., Medeiros, P. M., Didyk, B. M., (2005), "Combustion products of plastics as indicators for refuse burning in the atmosphere" *Environmental Science and Technology*, 39, 6961-6970.

[11] Lithner, D., Larsson, A., Dave, G., (2011), "Environmental and health hazard raking and assessment of plastic polymers based on chemical composition" *Science of the Total Environment*, 409, 3309-3324.
[12] Ustohalova, V., (2011)," Management and Export of Wastes: Human Health Implications" *Encyclopedia of Environmental Health*, 603–611.
[13] Tawfik, M. E., Mohamed, N. A., Eskander, S. B., (2011), "Aminolysis of poly(ethylene terephthalate) wastes based on sunlight and utilization of the end Product [Bis(2-hydroxyethylene) Terephthalamide] as an ingredient in the anticorrosive paints for the protection of steel structures". *Journal of Applied Polymer Science*, 120, 2842–2855.
[14] http:/www.epa.gov/osw/conserve/materias/plastcs.htm.
[15] Chul-Hyun Park, Ho-Seok Jeon, Jai-Koo Park,(2007)," PVC removal from mixed plastics by triboelectrostatic separation". *Journal of Hazardous Materials*, 144, 470–476.
[16] Rebeiz,K.S., Craft,A.P. (1995),"Plasticwaste management in construction: technological and institutional issues". *Resources Conservation and Recycling*, 15, 245–257.
[17] Winter, W., (1992), De Die Makromol. *Chem., Macromolecular Symposia*, No. 57, 253
[18] Zhang, Y.A.,Broekhuis, A.Picchioni, F.,(2009), "Thermally self-healing polymeric materials: the next step to recycling thermoset polymers." *Macromolecules* , 42, 1906-1912.
[19] Tawfik, M. E.,(2003), Preparation and characterization of water extended polyester based on recycled polyethylene terephthalate. *Journal of Applied Polymer Science*, 89, 13, 3693- 3699.
[20] Tawfik, M. E., Eskander, S. B.,(2006), Polymer composite from marble wastes and recycled polyethylene terephthalate. *Journal of Elastomers and Plastics*, 38, 1, 65-79.
[21] Tawfik, M. E., Eskander, S. B., (2010), Chemical recycling of polyethylene terephthalate waste using ethanolamine. *Sorting of the end products. Polymer Degradation and Stability*, 95, 187-194.
[22] Eskander, S. B., Tawfik, M. E.,(2011), Polymer-Cement composite based on recycled expanded polystyrene foam waste. *Journal of Polymer Composites* 32:1430–1438.
[23] Nawar, G. A. M., El Sabbagh, S. H., Tawfik, M. E.,(2012), "Preparation of plasticizer material for rubber (dibutyl phthalate) from polyethylene terephthalate waste and its effect on the physicomechanical and rheological properties of rubber loaded with lignin as a filler" *Eg. Patent* 258/2009, (12/2012 vol 736).
[24] Burat, F, I., Güney, A., Olgae Kangal, M., (2009), Selective separation of virgin and post cosumer polymer (PET and PVC) by flotation methods. *Waste Management* 29, 1807- 1813.
[25] Siddique, R., Khatib, J., Kaur, I., (2008), Use of recycled plastic in concrete: A review. *Waste Management* 28, 10, 1697-2030.
[26] Won, Jong-Pil; Jang, Chang-il; Lee, Sang-Woo; Lee, Su-Jin; Kim, Heung-You,(2010), *Construction and Building Materials*, May 1.
[27] Suh, D. J., Park, O. O., Yoon, K. H., (2000). "The properties of unsaturated polyester based on the glycolyzed poly (ethylene terephthalate) with various glycol compositions," *Polymer*, vol. 41, no. 2, pp. 461–466.
[28] European Commission (DG Environment) "Plastic waste in the environment ", Final Report April 2011.

[29] Rahimi, A., Shokrolahi, P., (2001)," Application of inorganic polymer materials:1. Polysiloxanes". *International Journal of Inorganic Materials* 3, 843-847.

[30] Vakili, M. H., Haghshenas Fard M. (2010),"Chemical Recycling of Poly Ethylene Terephthalate Wastes". *World Applied Sciences Journal* 8 (7): 839-846.

[31] Frounchi, M., Mehrabzadeh, M., Ghiaee, R. (1997), "Studies on Recycling of Poly (ethylene terephthalate) Beverage Bottles" *Iranian Polymer J.,* 6: 269-272.

[32] Fraisse F., Verney, V., Commereuc, S., Obadal, M., (2005), "Recycling of polyethylene terephthalate)/polycarbonate blends", *Polymer Degradation and Stability*, 90: 250-255.

[33] Navarro, R., Ferrandiz, S., Lopez, J., Segui, V.J. (2007), "The influence of polyethylene in the mechanical recycling of polyethylene terephthalate", *J. Materials Processing Technol.*, p 1-7.

[34] Jankauskaite, V., Macijauskas, G., Lygaitis, R.,(2008),"Polyethylene Terephthalate Waste Recycling and Application Possibilities, a Review" *Materials Science* (MEDŽIAGOTYRA) 14, 2.

[35] Karayannidis, G. P., Achilias, D. S.,(2007),"Chemical Recycling of Poly (ethylene Terephthalate)" *Macromolecular Materials and Engineering* 292 (2), 128 – 146.

[36] Subramian, P. M., (2000), "Plastics Recycling and Waste Management in the US Resources", *Conservation and Recycling*, 28, 253 – 263.

[37] Guclu, G., Yalcinyuva, T., Ozgumus, S., Orbay, M., (2003), "Hydrolysis of Waste Polyethylene Terephthalate and Characterization of Products by Differential Scanning Calorimetry" *Thermochimica Acta*, 404, (1), 193 – 205.

[38] Grause, G., Kaminsky, W., Fahrbach, G.,(2004), "Hydrolysis of Poly (ethylene Terephthalate) in Fluidised Bed Reactor" *Polymer Degradation and Stability*, 85, 1, 571-575.

[39] Vakili, M.H. and Haghshenas Fard, M., (2010), "Chemical Recycling of Poly Ethylene Terephthalate Wastes", *World Applied Sciences Journal*, 8, (7). 839-846.

[40] Rosmaninho, M., Jardim, E., Flavia, M. C. C., Ferreira, G., Thom, V., Yoshida, M., Araujo, M. H., Lago, R. M.,(2006), "Surface Hydrolysis of Postconsumer Polyethylene Terephthalate to Produce Adsorbents for Cationic Contaminants" *Journal of Applied Polymer Science*, 102, (6), 5284 – 5291.

[41] US Patent 6649792: "*Method of Chemical Recycling of Polyethylene Terephthalate Waste*", 2003.

[42] Patent WO/2001/068581: "*The Method of Chemical Recycling of Polyethylene Terephthalate Waste*", 2001.

[43] Carta, D., Cao, G., D´Angeli, C., (2003), "Chemical Recycling of Poly (ethylene Terephthalate) (PET) by Hydrolysis and Glycolysis" *Environmental Science and Pollution Research,* 10 ,6, 390 – 394.

[44] Goto, M., Koyamoto, H., Kodama, A., Hirose, T., Nagaoka, S., (2002), "Depolymerization of Polyethylene Terephtha-late in Supercritical Methanol" *Journal of Physics: Condensed Materials*, 14, 44, 11427 – 11430.

[45] Kurokawa, H., Ohshima, M.-A., Sugiyama, K., Miura, H.,(2003). "Methanolysis of Polyethylene Terephthalate (PET) in the Presence of Aluminum Tiisopropoxide Catalyst to Form Dimethyl Terephthalate and Ethylene Glycol" *Polymer Degradation and Stability*, 79, 3, 529 – 533.

[46] Genta, M., Iwaya, T., Sasaki, M., Goto, M., (2007), "Supercritical Methanol for Polyethylene Terephthalate Depolymerization: Observation Using Simulator" *Waste Management*, 27, 9, 1167 – 1177.

[47] Spychaj, T., Fabrycy, E., Spychaj, S., Kacperski, M., (2001), "Aminolysis and Aminoglycolysis of Waste Poly (ethylene Terephthalate)" *Journal of Material Cycles and Waste Management*, 3, 1, 24 – 31.

[48] Shukla, S. R., Harad, A. M., (2006), "Aminolysis of Polyethylene Terephthalate Waste" *Polymer Degradation and Stability*, 91, 8, 1850 – 1854.

[49] Japan Patent 03A0649522: "*Chemical Recycling of Poly (ethylene terephthalate). 3. Preparation of N, N-dibutyl terephthalamide and N, N'-diallyl terephthalamide via One-pot-two-step degradation process*", 2001.

[50] Awaja F, Pavel D., (2005), "Recycling of PET". *Eur. Polym. J.*, 41,1453-77.

[51] Chen Ch.-H., Chen Ch.-Y., Lo Yu-W., Mao Ch.-F., Liao W.-T. ,(2001), "Studies of Glycolysis of Poly (ethylene Terephthalate) Recycled from Postconsumer Soft-drink Bottles. I. Influences of Glycolysis Conditions" *Journal of Applied Polymer Science*, 80, 7, 943 – 948.

[52] Pardal, F., Tersac, G. ,(2007), "Kinetics of Poly (ethylene Terephthalate) Glycolysis by Diethylene Glycol. Part II: Effect of Temperature, Catalyst and Polymer Morphology" *Polymer Degradation and Stability*, 92, 4, 611 – 616.

[53] US Patent 6410607: "Glycolysis Process for Recycling of Post-consumer PET", 2002.

[54] Kosmidis, V. A., Achilias, D. A., Karayannidis, G. P., (2001), "Poly (ethylene Terephthalate) Recycling and Recovery of Pure Terephthalic Acid. Kinetics of a Phase Transfer Catalyzed Alkaline Hydrolysis" *Macromolecular Materials and Engineering*, 286, 10, 640 – 647.

[55] Chen Ch.-H., Chen Ch.-Y., Lo Yu-W., Mao Ch. F., Liao W. T. "Studies of Glycolysis of Poly (ethylene Terephthalate) Recycled from Postconsumer Soft-drink Bottles. II. Factorial Experimental Design" *Journal of Applied Polymer Science* 80 (7) 2001, 956 – 962.

[56] Yong Yang, Yijun Lu, Hongwei Xiang, Yuanyuan Xu, Yongwang Li, (2002), "Study on methanolytic depolymerization of PET with supercritical methanol for chemical recycling". *Polymer Degradation and Stability*, 75, 185–191.

[57] Tahvildari, K., Mozafari, Sh., Tarinsun, N., (2010),"Chemical Recycling of Poly Ethylene Terephthalate to Obtain Unsaturated Polyester Resins", *J. Appl. Chem. Res.*, 12, 59-68.

[58] Tawfik, M.E., (1997), *PhD Thesis*, Cairo University, Egypt.

[59] Tawfik, M.E., Elsabee, M.Z., (2002), "Preparation and Characterization of Polyester Based on bis-(2-hydroxypropyl terephthalate)", *Journal of Polymer Research*, 9, 129.

[60] Tawfik, M. E. , Eskander, S. B., Bayoumi, T. A.,(2005), "Immobilization of Borate Waste Simulate in Cement-Water Terephthalate Waste 1. Mechanical Properties of the Final Waste Forms Extended Polyester Composite Based on Polyethylene". *Polymer-Plastics Technology and Engineering*, 44, 7, 1355-1368.

[61] Rosato, D. V., Rosato, D.V., Roa\sato, M. V.,(2004), "*Plastic Product Material and Process Selection Handbook*", Elsevier,Oxoford, UK.

[62] Eskander, S. B., Tawfik, M. E., Bayoumi, T. A., (2006), "Immobilization of Borate Waste Simulate in Cement-Water Extended Polyester Composite Based on Poly

(ethylene Terephthalate) Waste. 2-Frost resistance of the polymer modified cement composite"*Polymer-Plastics Technology and Engineering*, 45, 939- 949.

[63] Tawfik, M. E., Eskander, S. B., (2007), "Immobilization of Borate Waste Simulate in Cement-Water Extended Polyester Composite Based on Poly (ethylene Terephthalate) Waste. 3- Behavior of solid waste form during a scenario of flooding accident in the disposal site."*Journal of Applied Polymer Science*, 104, 6, 4113-4120.

[64] Saleh, H.M., Tawfik, M. E., Bayoumi,T.A., (2011), "Chemical stability of seven years aged cement-water extended polyester based on recycled PET composite containing radioactive borate waste simulates", *Journal of Nuclear Materials*, 411, 185–192.

[65] Eskander, S.B., Bayoumi, T.A., Saleh, H.M., (2012), "Performance of aged cement–polymer composite immobilizing borate waste simulates during flooding scenarios". *Journal of Nuclear Materials*, 420, 175-181.

[66] Holcomb, W.F., (1978)," A survey of the available methods of solidification for radioactive wastes ", University of Michigan Library, Washington.

[67] Peter C, H., (2003)," Lea's Chemistry of Cement and Concrete", Fourth Edition, Butterworth-Heinemann, Oxford, UK.

[68] Taylor, H.F.W, (1997),"*Cement Chemistry*", 2nd. Ed. Thomas Telford, London.

[69] Van Gemert, D., Czarnecki, L., Łukowski, P., Knapen, E., 2004, *A report from 11th ICPIC Congress*, Berlin.

[70] Beeldens, A., Van Gemert, D., Schorn, H., Ohama, Y., (2004), "Integrated model for microstructure building in polymer cement concrete", *Proceedings of 11th Congress on Polymers in Concrete*, Berlin, 1-10.

[71] Tabor, L. J., (1987), "Dispersed polymers", revision of concrete society technical reportno.9 - polymer concrete, chapter 7, *Contribution to the Fifth International Congress on Polymers in Concrete at Brighton*, UK, 83-94.

[72] Machovic, V., Lapcak, L., Borecka, L., Lhotka, M., Andertova, J., Kopecky, L., Miskova, L., (2013),"Microstructure of interfacial transition zone between PET fibres and cement paste". *Acta Geodyn. Geomater.*, 10, 1 (169), 121–127.

[73] V., Andertová, J., Kopecký, L., Černý, M., Borecká, L., Přibyl, O., Kolář, F., Svítilová, J., (2008), "Effect of aging of PET fiber on the mechanical properties of PET fiber reinforced cement composite", *Ceramics-Silikáty*, 52, 3, 172–182.

[74] Sharp, J. H., Hill, J., Milestone, N. B., Miller, E. W., (2003), " Cementitious systems for encapsulation of intermediatelevelwaste" *Proceedings of ICEM 03: The 9th International Conference on Radioactive Waste Management and Environmental Remediation September 21 – 25, 2003.* Examination School, Oxford, England, ICEM 03-4554.

[75] Ghattas, N. K, Ikladious, N. E., Eskander, S. B., (1983), "Poly (methyl methacrylate) as incorporation medium for spent ion exchange resins II- simulated resins", *Journal of Applied Polymer Science*, 28, 1779-1786.

[76] Ghattas, N. K,Ikladious, N.E. Eskander, S. B., (1986),"Studies on incorporation of simulated spent ion exchange resins into phenol formaldehyde polymer", *Radioactive Waste Management and Nuclear Fuel Cycle*, 7, 1, 81-96.

[77] Ikladious, N. E., Ghattas, N. K. Eskander, S. B., (1986), "Incorporation of radioactive waste into styrenated polyester", *Nuclear and Chemical Waste Management*, 6, 101-105.

[78] Ghattas, N. K., Eskander, S. B., Bayoumi T. A., Saleh, H. M., (2009), "Cement-Polymer Composite Containers for Radioactive Wastes Disposal", *J. Rad. Res. Appl. Sci.*, 2, 1, 103-118.

[79] Lageraaen, P. R., Kalb P. D.,(1997),"Use of recycled polymers for encapsulation of radioactive hazardous and mixed wastes final" Report Brookhaven National Laboratory Environmental & Waste Technology Center .USA- BNL-66575 Informal Report.November 1997.

[80] Donato, A., (1976),"Incorporation of radioactive wastes in polymer impregnated cement," INS Management of Radioactive; Wastes From the Nuclear Fuel Cycle. Volume II, International Atomic Energy Agency, Vienna, Austria.

[81] Ohama, Y., (1997),"Recent Progress in Concrete –Polymer Composites) *Advanced Cement Based Materials*, 5,2, 31-40.

[82] Kardon, J.B., (1997)," Polyester –Modified Concrete: Review", *Journal of Materials in Civil Engineering*, 9(2), 85-92.

[83] Sumathy, C.T., Dharakumar, M., Devi, M.S., Saccubai, S., (1997), "Modification of Cement Mortar by Polymer Latex", *Journal of Applied Polymer Science*, 63(10), 1251-1257.

[84] Rebeiz, K. S., Fowler, D. W., (1996) "Flexural Strength of Reinforced Polymer Concrete Made with Recycled Plastic Wastes", *ACI structural Journal* 93, 5, 524-530.

[85] Larbi, J.A. and Bijen, J.M., (1990), "Interaction of Polymers with Portland Cement during Hydration: A Study of the Chemistry of the Pore Solution of Polymer Modified Cement Systems "*Cement and Concrete Research*, 20(1), 139-147.

[86] Sugama, T. and Kukacka, L. E., (1982), "Products formed in aged Ca-polyester complexed polymer containing cement" *Cement and Concrete Research*, 12(6), 789-796.

[87] Saleh, H.M., Bayoumi, T.A., Shatta, H.A., (2012), "Mechanical and chemical characterizations of polyester modified cement immobilizing nuclear wastes", *Advances in Chemical Science* (ACS), 1, 1, December2012.

[88] Shatta, H.A., (1996), "Applications of polymer-impregnated cement in the field of radioactive wastes", *Ph.D. Thesis*, Science and Education, Chemistry Department. University of College of Women for Arts, Cairo, Egypt.

In: New Developments in Polymer Composites Research ISBN: 978-1-62948-340-5
Editors: S. Laske and A. Witschnigg

Chapter 3

NEW FABRICATION TECHNIQUES OF CONDUCTIVE POLYMER/INSULATIVE POLYMER COMPOSITE FILMS FOR PRINTED ELECTRONICS

Mitsuyoshi Onoda
University of Hyogo, Kobe, Japan

ABSTRACT

Studying the fabrication techniques of composite films has led to new scientific understandings and exciting application devices. A way to overcome the lack of processibility of conductive polymers and to improve their electrical stability in air is to prepare composite materials based on conventional thermoplastic materials. That is, by some method, for example, by mixing and combining or connecting more than two kinds of materials with different characteristics, the fault of a single material can be compensated, and such a combined material that has characteristics not obtainable with a single material is called "a composite", *In situ* polymerization, involving the impregnation of a suitable insulating polymer with a catalyst, gives homogeneous conductive polymer/insulative polymer composites.

The chapter is organized as follows. The background of study is explained as Introduction in section 1. The introduction in section 2 gives an outline of organic electronics based on conductive polymers and their application, then what the composite films are, etc. Section 3 includes the electrolysis polymerization technique. Principles of molecular self-assembly process are described in section 4. Section 5addresses the new fabrication techniques of conductive polymer/insulative polymer composite films based on a sol-gel method, electrolysis polymerization method, molecularly self-assembly method, electrophoretic deposition technique, etc., including characteristics and device application of their obtained composite films. By the last sentence the paradigm shift to printed electronics is summarized.

Keywords: Conductive polymer, electrolysis polymerization, molecular self-assembly process, sol-gel process, *in situ* polymerization, electrophoretic deposition

1. INTRODUCTION

Conductive polymers with extensive π electron conjugated systems in their main chain are considered to be organic semiconductors with relatively small bandgaps, and are being intensively studied as new materials in the field of physical chemistry. To date, various applications of conductive polymers, such as in semiconductive devices and optoelectronic devices, have also been proposed. Unfortunately, except for rechargeable batteries and capacitors, these applications have not yet been developed to the stage of practical use [1, 2]

Major advantages of conductive polymers as semiconductors against conventional inorganic semiconductors may be in their unique properties such as mechanical flexibility and solubility. Especially, the latter feature enables the preparation of semiconductor films under the atmospheric pressure by wet-processes, motivating a number of researches on the application of conductive polymers for light-emitting devices [3-5], photocells [6, 7], field-effect transistors [8] and other electronic devices of "printed electronics".

In order to fully exploit the potential offered by polymers, the device material must have good environmental stability and be processable in its polymeric form, while also having excellent electrical, optical, mechanical and other properties. However, many of these conductive polymers, which consist of rigid conjugated double, show strong interchain interaction and generally are difficult to process due to their nonfusibility and insolubility. This is a serious problem in their practical utilization. However, by appropriate modification of the molecular structure of these polymers, such as the introduction of long alkyl side chains, the investigation of the polymerization process, and other developments, they have been made soluble in any solvent, and even fusible at relatively low temperatures, and reveal good environmental stability. Accordingly, the study of conductive polymers have now reached a new stage.

Since the spin-coating technique only requires simple and cheap apparatus, this technique widely used for the laboratory-scale preparation of the polymer electronic devices. However, other than the incompatibleness with the patterning, this technique seems to possess some disadvantages for the commercial production. For example, the material efficiency is poor because most of materials place on the substrate is blown out during the spinning. Moreover, the polymer solution for this technique must be relatively thick if one targets this type of polymer films suitable for sandwich-type electronic devices. To our typical experience, a chloroform solution containing several g/l of a conductive polymer is appropriate to obtain a 100 nm-thick polymer films. Although the concentration required depends on the polymer and the solvent, a polymer solution containing less than 1 g/l may be generally useless for the spin-coating in the case of typical conductive polymers.

A way to overcome the lack of processibility of conductive polymers and to improve their electrical stability in air is to prepare composite materials based on conventional thermoplastic materials. That is, by some method, for example, by mixing and combining or connecting more than two kinds of materials with different characteristics, the fault of a single material can be compensated, and such a combined material that has characteristics not obtainable with a single material is called "a composite". In situ polymerization, involving the impregnation of a suitable insulative polymer with a catalyst, has been reported to give homogeneous conductive polymer/insulative polymer composites [9, 10].

Since spin-coating technique only requires simple and cheap apparatus, this technique widely used for the laboratory-scale preparation of the polymer electronic devices. However, other than the incompatibleness with the patterning, this technique seems to possess some disadvantages for the commercial production. For example, the material efficiency is poor because most of materials placed on the substrate is blown out during the spinning. Moreover, the polymer solution for this technique must be relatively thick if one targets this type of polymer films suitable for sandwich-type electronic devices. Recent progress in the field of ink-jet printing technology allowing precise positioning of very small amount of ink is going to overcome these disadvantages [11, 12].

However, for the production of flat and uniform films without patterning, which are suitable for lighting applications for example, this method seems to be too sophisticated. The electrodeposition techniques have a long history and are practically used for painting and coating of industrial products such as automobiles. The electrophoretic deposition technique [13], which deposites the material over the entire electrode surface at once by using the electric field, may be one of the most important candidates for this type of deposition [14-16]. We have shown that the electrophoretic deposition of a soluble conductive polymer with polymer suspensions containing relatively large amount of good solvent(solvents having excellent solubility)for the polymer can yield flat and uniform films applicable to the electronic devices such as light-emitting devices [17]. From the viewpoint of the required concentration of solution to make suspensions, this result is important. For example, while the suspension consisting of 90 % of poor solvent (solvents having inferior solubility), 10 % of good solvent and 0.1 g/l of the polymer is made by mixing a unit of the polymer solution containing 1.0 g/l of the polymer with 9 units of the poor solvent, the suspension consisting of equivalent volumes of the poor and good solvents with the same polymer content requires a solution containing just 0.2 g/l of the polymer.

On the other hand, in the coating technology based on the colloidal suspension, a liquid which is poor solvent of polymer itself and containing polymer as solid nanoparticles, the stage of film formation is separated from the solidification. The drying of the film is conducted in the nonsolvent vapor. Therefore, this class of processes is expected to yield polymer films with novel microscopic structures [14, 18].

And then we report the electrophoretic deposition of some conductive polymers and their mixtures from colloidal suspensions. The atomic force microscope (AFM) observation clearly revealed that this classic, but new for conductive polymer deposition, technology yields nanostructured films. Additionally, this unique deposition process is found to be quite useful to make mixtures. In other word, the nanostructured composites films can be fabricated by using the electrophoretic deposition technique.

This chapter is organized as follows. The introduction in section 2 gives an outline of organic electronics based on conductive polymers and their application, then what the composite films are, etc. That is, a review is given on recent progress of the conductive polymer and its application to optoelectronic devices. New types of functional applications have also been proposed Conductive polymers with highly extended conjugated electron systems in their main chains have bandgaps ranging mostly between 1.0 eV and 3.5 eV, resulting in the characteristic color depending on molecular structure. The optical property of a conductive polymer changes drastically in association with an insulator-metal transition upon doping and also with conformation change. Interaction of light with conductive polymers is also unique.

The applications of these novel characteristics in optoelectronics, such as color switching (electrochromism), optical recording and memory (photochromism), thermochromism, solvatochromism, and gelchromism, etc., are introduced and discussed. Section 3 includes the electrolysis polymerization technique. In section 4, principles of molecular self-assembly process are described. Section 5 addresses the new fabrication techniques of conductive polymer/insulative polymer composite films based on a sol-gel method, electrolysis polymerization method, molecularly self-assembly method, electrophoretic deposition technique, etc., including characteristics and device application of their obtained composite films. By the last sentence, section 6, the paradigm shift to printed electronics is summarized.

2. Novel Optical Property of Conductive Polymer and Its Application [1, 2]

2.1. Introduction

Polymers have been considered to be insulative and also colorless. However, in such electrically and optically non-active polymers, polymer main chains are composed of saturated σ-bonds. On the other hand, conductive polymers whose main chains are composed of unsaturated π-bonds have been synthesized and found to exhibit high conductivity. This type of conjugated polymer is named a conductive polymer and has attracted much attention from both the fundamental and practical viewpoints. That is, a reversible insulator-metal transition was found in this new category of polymer upon doping and various new concepts such as the soliton, polaron and bipolaron have been introduced to explain these unique characteristics.

Various applications of conductive polymers such as electronic elements, optical switching elements, sensors, etc., have been proposed, which also stimulated syntheses of various new conductive polymers. The conductive polymer has also been found to demonstrate novel optical properties.

In this chapter, we report the structure and properties of conductive polymers and discuss various possibilities for optical applications of the novel optical properties of conductive polymers.

2.2. Structure and Property of Conductive Polymers

Figure 1 indicates the molecular structure of typical conductive polymers. As is evident in this figure, conductive polymers are divided into two groups; (a) linear one-dimensional conductive polymers and (b) rwo- or three-dimensional polymers. The first group of polymers are mostly insulative or semiconductive and the latter metallic. This chapter mainly discusses polyacetylene, polythiophene and poly(p-phenylene vinylene) and their derivatives, whose molecular structures are shown in Figure 2. Properties of conductive polymers are modified by the introduction of side chains and also of hetero atoms because they have influence on the planarity of the conjugated system and therefore electronic band scheme. Properties, especially electrical properties, are also strongly influenced by the morphology of polymers.

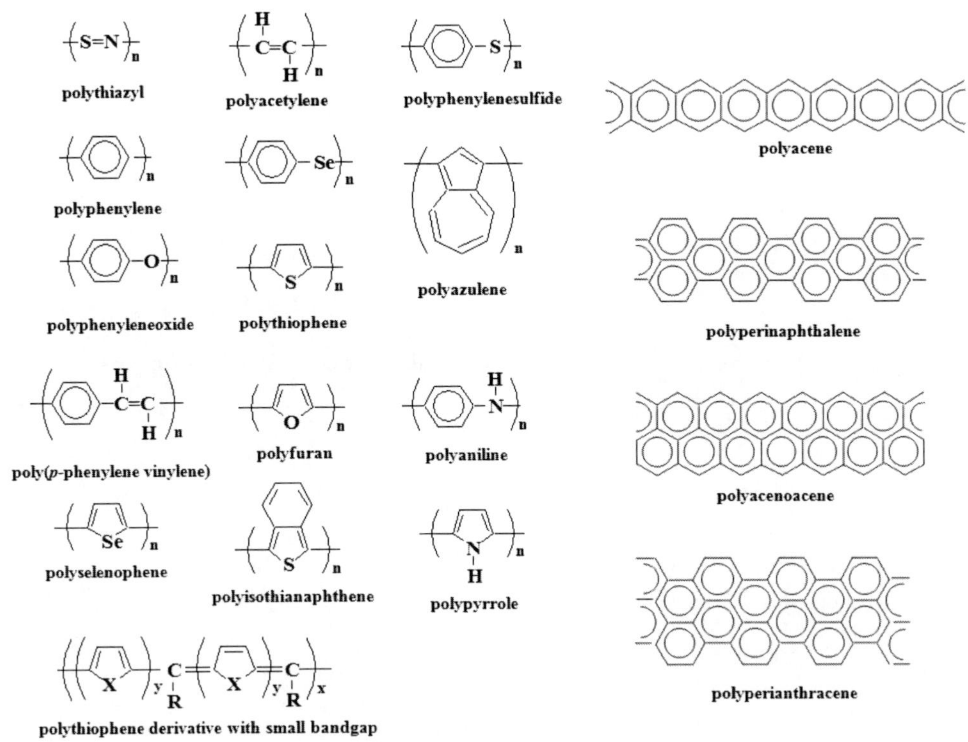

Figure 1. Structures of various conductive polymers.

The bandgap energy and therefore color of conductive polymers are influenced strongly by the molecular structure. In most conductive polymers, bandgaps range between 1.0 and 3.5 eV, but appropriately designed conductive polymers exhibit bandgaps of less than 0.5 eV. Even in polyacetylene with the simplest molecular structure, *trans* and *cis* isomers exhibit different bandgaps, of about 1.5 eV and 1.9 eV, respectively. As is also evident from the absorption spectra in Figure 3, substituted polyacetylene such as polyphenylacetylene (PPA) and polychlorophenylacetylene (PCPA) indicate the much larger bandgaps of 2.2 eV and 3.1 eV, respectively. Therefore, the former is red in color but the latter yellow. This difference should originate in the effective conjugation length.

The inter-chain distance also influences the property of conductive polymers. For example, the lattice constant (a) of poly(3-alkylthiophene) is found to increase remarkably with increasing alkyl chain length (16.70 Å and 25.89 Å in poly(3-hexylthiophene) and poly(3-dodecylthiophene), respectively), which also influence the optical characteristics such as luminescence. Energies of the valence band and the conduction band also depend on the molecular structure. For example, introduction of a substituent with donating characteristics elevates the top of the valence band. In poly(3-methylthiophene) with a donating nature of CH3 substituent, the energy of the top of the valence band is higher than that of non-substituted polythiophene, as is evident from the photoemission spectra shown in Figure 4 [19].Property is also strongly dependent on the dimensionality of the molecular structure of polymers. In the liniear conjugated system so far discussed, a bandgap exists, and the polymers are semiconductive or insulative.

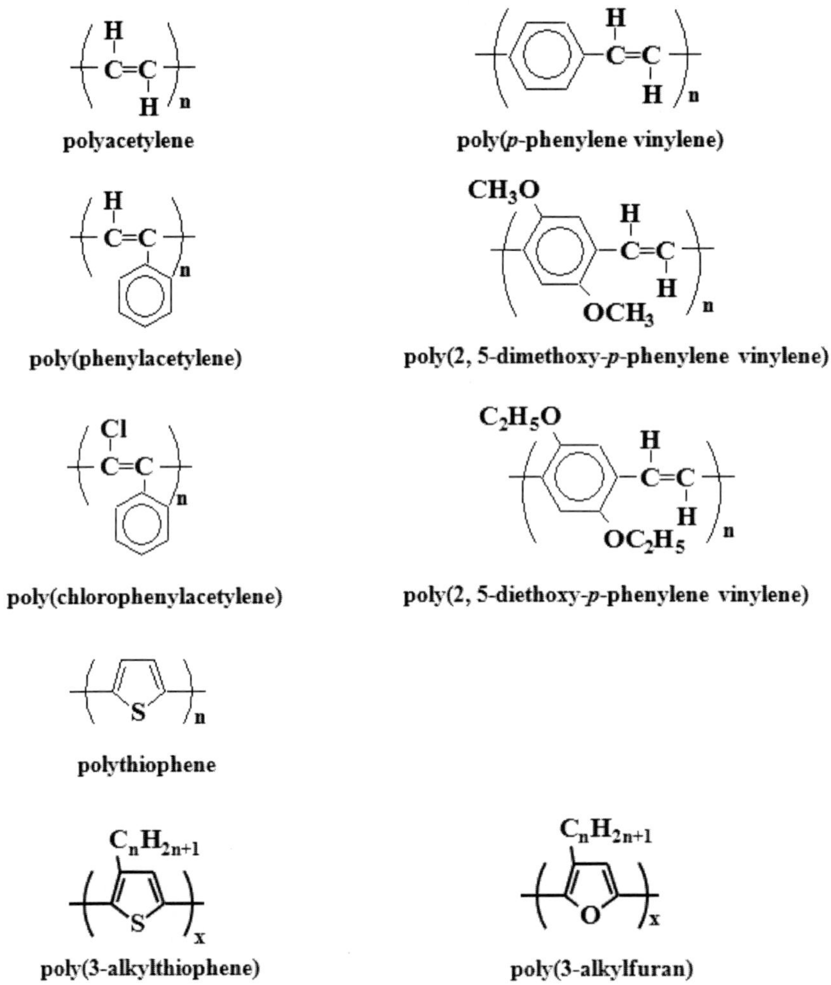

Figure 2. Structures of conductive polymers.

However, materials with strong interchain interactions such as $(SN)_x$and a 2-dimensional conjugated system such as polyacene demonstrate metallic characteristics. Graphite is the extreme case of the latter group. It should also be noted that $(SN)_x$ even exhibits superconductivity [20]. In metallic conductive polymers too, plasma reflection can clearly be observed [20].

It should be stressed that even linear conductive polymers demonstrate insulator-metal transition upon doping. As shown in Figure 5, for example, electrical conductivity of polythiophene increases by more than 10 orders of magnitude upon doping of a small amount of dopants. Similar enhancement of conductivity upon doping was found in most conductive polymers. The highest conductivity attained in the doped state is of the order of 5×10^5 S/cm in polyacetylene at this stage, which is comparable to the value of copper metal. In association with this insulator-metal transition, optical and magnetic properties also change drastically. As is evident in Figure 6, the absorption spectrum of polythiophene changes drastically with doping.

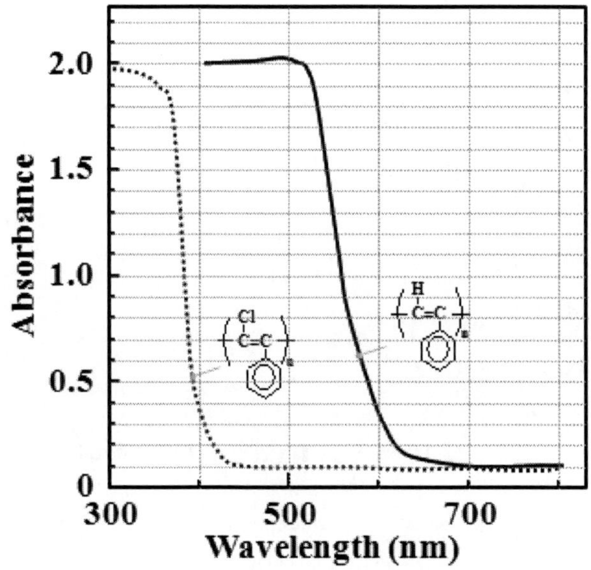

Figure 3. Optical absorption spectra of poly(phenylacetylene) and poly(chlorophenylacetylene).

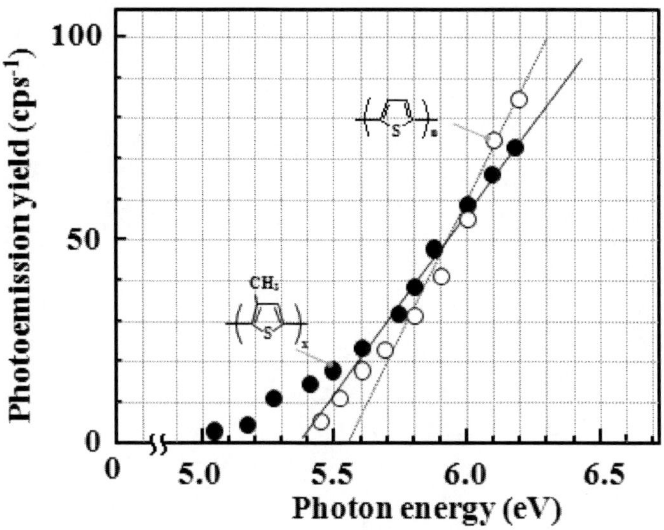

Figure 4. Photoelectron emission spectra of undoped polythiophene and poly(3-methylthiophene) in air atmosphere at room temperature.

ESR linewidth and magnetic susceptibility also demonstrase drastic change upon doping, as shown in Figure 7 for poly(2,5-dimethoxy-*p*-phenylene vinulene) [21].

In the heavily doped state, truly metallic behavior was confirmed, such as high conductivity, plasma reflection (Figure 8 for polypyrrole), Pauli paramagnetism and small linearly temperature dependent thermoelectric power.These novel characteristics of conductive polymers have been explained in terms of sliton, polaron, and bipolaron models.

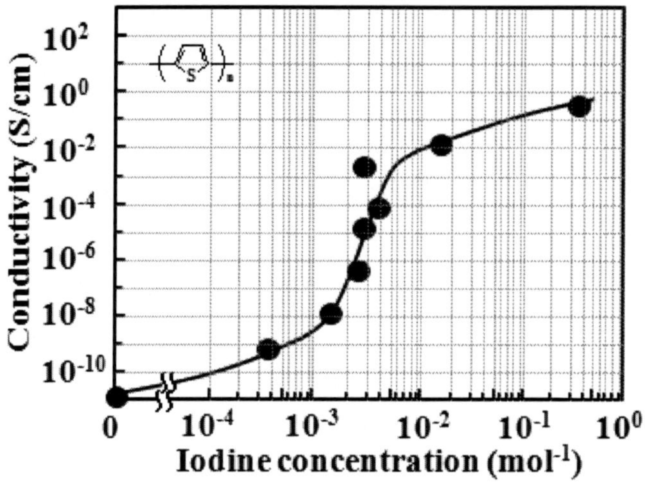

Figure 5. Iodine-concentration dependence of the electrical conductivity for polythiophene.

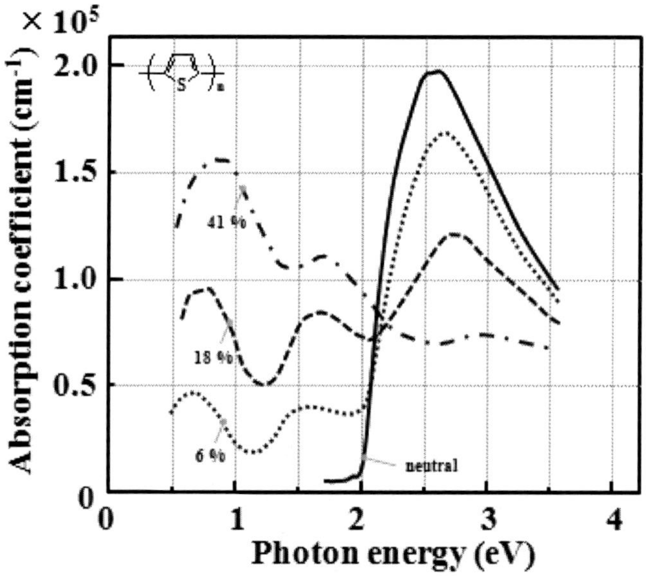

Figure 6. Optical-absorption-spectral change of polythiophene as a function of the dopant concentration.

2.3. Applications of Optical Properties of Conductive Polymer

2.3.1. Color Switching (Electrochromism)

We expected drastic color change of conductive polymer upon doping associating with the insulator-metal transition. That is, the transition between interband absorption in the undoped state and metallic reflection in the doped state is expected to induce reversible color switching of a conducting polymer upon doping and undoping.

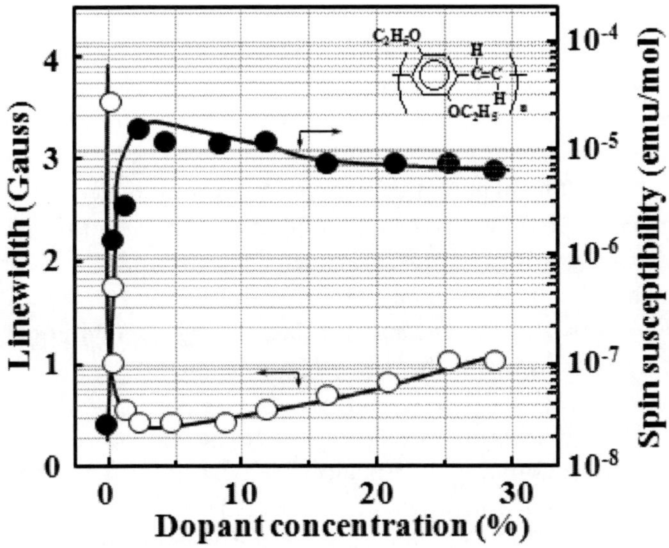

Figure 7. Dependence of the spin susceptibility and linewidth estimated from *in situ* ESR measurements of poly(2,5-diethoxy-*p*-phenylene vinylene) on dopant concentration.

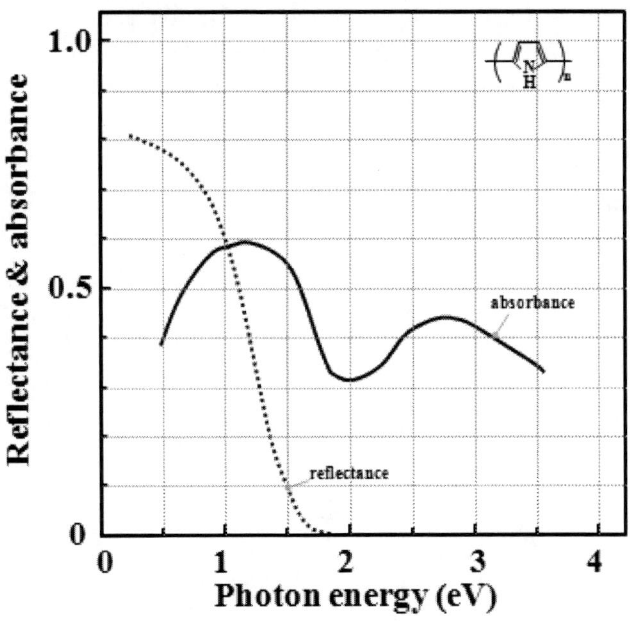

Figure 8. Reflectance and absorption spectra of doped polypyrrole.

However, it was confirmed later that the evolution of polaronic states also plays an important role in the color change. For example, color changes between red and blue in polythiophene, yellow and blue in polypyrrole, and yellow and purple in polyalkylfluorene were confirmed [22]. In some samples such as polyaniline the entire visible spectral range was covered by the transition, as shown in Figure 9 [23].

The switching speed was comparable to that of nematic liquid crystal elements. That is, a switching time of several msec was realized. However, speed depends on the preparation method and condition of the conductive polymer film, the thickness, and the electrolyte and solvent. A cycle life of the order of 10^6 cycles can be relatively easily attained. However, for practical applications life cycles of more than 10^8 cycles are desired.

At least in this polymer color switching element, there are several advantages compared with liquid crystal elements. (1) Large size of the element can be easily obtained. (2) Optical polarizer is not used, resulting in brightness and a wide viewing angle.

This color switching element also can be used as a voltage-controllable color filter and can be applied to various types of sensors, monitors, and cameras.It should also be mentioned that colorless conductive polymers can be established upon doping in the conductive polymer of a bandgap less than ca. 1.5 eV, because the interband absorption in the visible range is suppressed and absorption evolves in the infrared upon doping.

2.3.2. Optical Recording and Memory Element (Photochromism)

When an insulator-metal transition is induced by light irradiation, optical recording and memory elements can be established. Therefore several possible optically induced insulator-metal transitions exist: (1) change of inter-carbon distance of the main chain to suppress the bond alternation, (2) change of dimensionality from one dimension to two or three dimensions by enhanching interchain interaction, which also suppresses Peierls transition, and (3) optical induced doping.As an example of application of optical induced doping, we have prepared conductive polymer films such as polythiophene, containing appropriate photosensitive dopants such as triphenyliodoniumtetrafluoroborate. In this case, upon irradiation of light BF_4^- is released and becomes an effective dopant for conductive polymer.

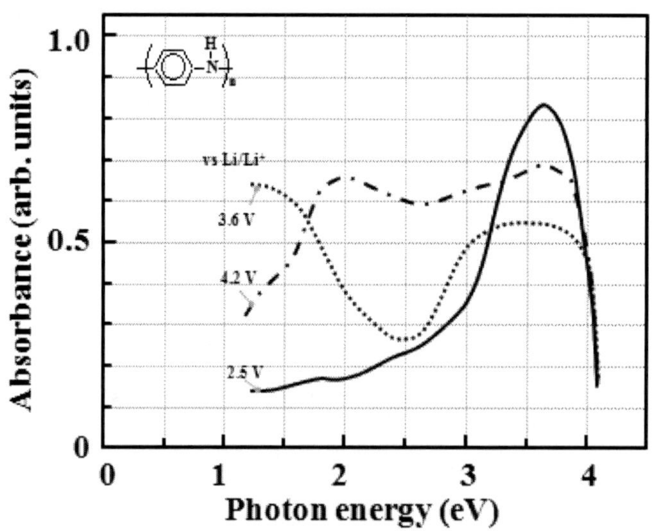

Figure 9. Optical-absorption-spectral change of polyaniline depending on applied potential (vs. Li/Li^+).

As shown in Figure 10, the color changes from red to blue by light irradiation of the composite film. Optimum design of the dopant species and its concentration is expected to produce a much faster response. The changed color was stable for a long time.

By sandwiching the conductive polymer-optically sensitive layer (polymer sheet containing photo-sensitive material) composite between two ITO (In-Sn oxide) glass electrodes, the erasable optical recording element is established as shown in Figure 11. In this case doping was optically induced and undoping was electrochemical.

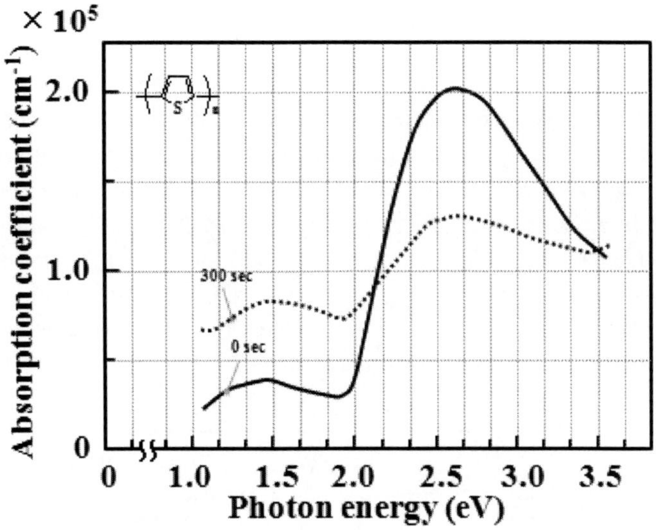

Figure 10. Optical-absorption-spectral change of polythiophene containing photosensitive dopant depending on light irradiation time.

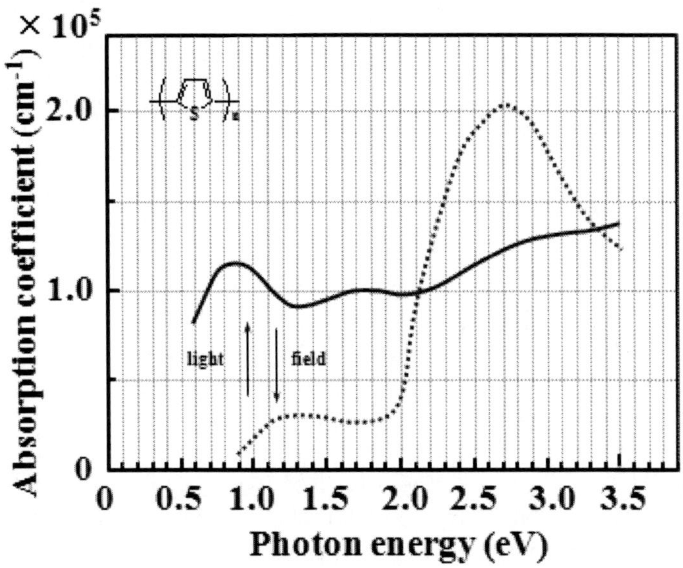

Figure 11. Optical-absorption-spectral change of a photochromic device constructed with photosensitizing polymer and polythiophene.

Optical induced isomerization can also be used for the optical memory element. We have demonstrated that the absorption spectrum of *cis*-PPA changed to that of *trans*-PPA upon irradiation of light.

In the case of conductive polymers with substituted side chains which can be isomerized by photons such as the azoxy group, the polymer conformation and also the interchain distance can be changed upon light irradiation. These phenomena are also expected to be used for optical recording and memory.

2.3.3. Thermochromism

Conductive polymer has been considered to be nonsoluble and also nonfusible. However, poly(3-alkylthiophene)s of long alkyl side chains were found to be soluble in conventional solvents and even fusible at relatively low temperatures.

After this finding, several conductive polymers such as poly(3-alkylfuran), poly(3-alkylselenophene) [24] and polyalkylfluorene were also confirmed to be soluble and fusible.

Poly(3-alkylthiophene)s were also confirmed to demonstrate novel optical properties upon a change in temperature [25]. That is, reversible color change with temperature, thermochromism, was observed in these compounds.

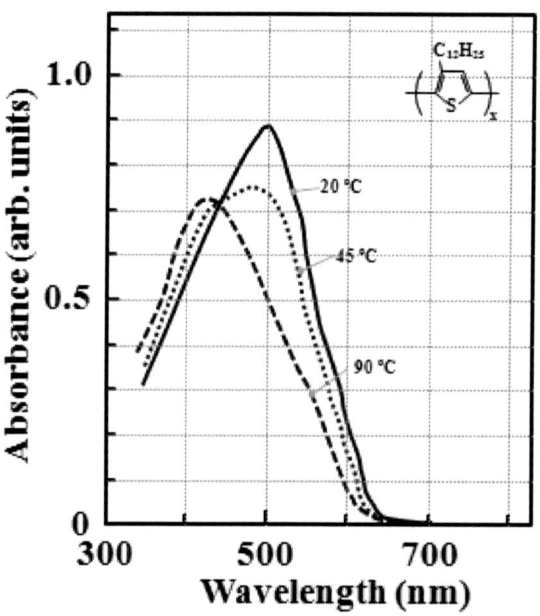

Figure 12. Optical-absorption-spectral change of poly(3-dodecylthiophene) at various temperatures.

As shown in Figure 12, the absorotion spectrum of poly(3-alktlthiophene) changes drastically with increasing temperature and returns to the original color with decreasing temperature. The temperature at which the spectrum changes drastically decreases with increasing alkyl chain length. This thermochromism in poly(3-alkylthiophene) is interpreted in terms of change of effective conjugation length due to the introduction of torsion of bonds between thiophene ringson account of the steric hindrance between S atom and alkyl chain.

That is, with increasing temperature the alkyl side chain undergoes Trans-Gausch conformation change. The Gausch conformation is bulky and induces a steric hindrance effect between S atoms. Therefore, at high temperature the absorption spectrum demonstrates blue shift due to the shortening of conjugation length and increased bandgap.

In the case of poly(3-alkylthiophene), color changes from red to yellow. The appropriate design of the molecular structure facilitated a much wider spectral change.

The detailed spectral change of this thermochromism is a little complicated. In the pre-melting region only the absorption peak shifts to higher energy; the absorption edge remains constant. However, at the melting point, the absorption edge also demonstrates step wise blue shift. Figure 13 indicates alkyl chain length dependence of the melting point and glass transition point. The spectral change may be initiated at around the glass transition point.

All the substituted conductive polymer does not necessarily demonstrate thermo-chromism. For example, even in the same oxygen group of heterocyclic conductive polymers, thermochromism was not observed in poly(3-alkylfuran) and poly(3-alkylseleno-phene). Differences in size, polarizability or electronegatively between S, O, and Se atoms may explain these differences in the temperature dependence of optical properties.

It should be noted that not only the film of poly(3-alkylthiophene) but also poly(3-alkyl-thiophene) contained in matrix polymers such as polystyrene and polymethylmethacrylate also demonstrated thermochromism.

It should also be mentioned that this film and composite matrix film were used as optical recording media. That is, the spectrum changed upon irradiation of Ar ion laser light. In these cases, it is desirable to irradiate the sample in an oxygen-free state.

The introduction of appropriate molecular groups containing electrical or optical sensitive units in the long side chains makes possible the control of thermochromism by voltage application or light irradiation.

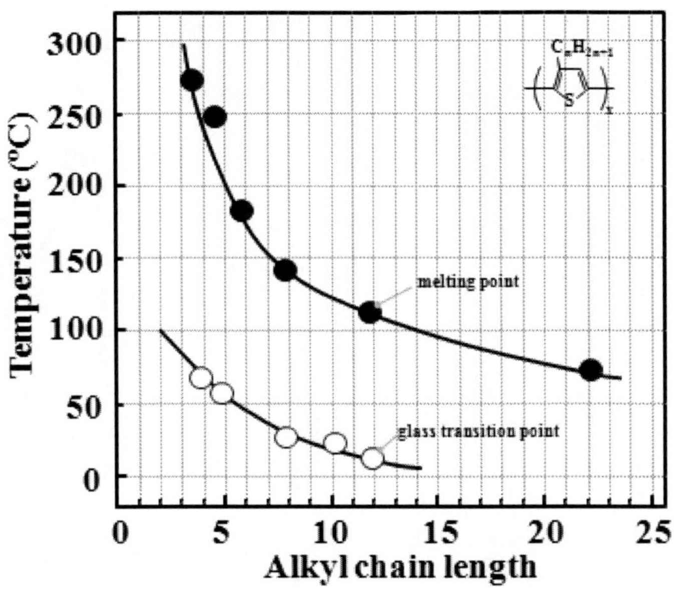

Figure 13. Plots of melting points and glass transition temperature of various poly(3-alkylthiophene)s as a function of alkyl chain length.

2.3.4. *Solvatochromism*

The color of these conductive polymers also changes drastically with a change of solvent. As is evident in Figure 14, the absorption spectrum of poly(3-alkythiophene) in dichloromethane shifts to a longer wave length than that of the sample in chloroform. It should also be mentioned that in the former case, the spectrum also showed blue shift with increasing temperature. This spectral shift upon changing solvent is named solvatochromism and can be explained as follows. Conformation of the alkyl side chain depends on solvent. For example, in dichloromethane at 20 ºC, docosyl side chain of poly(3-docosylthiophene) is in *Trans* conformation.

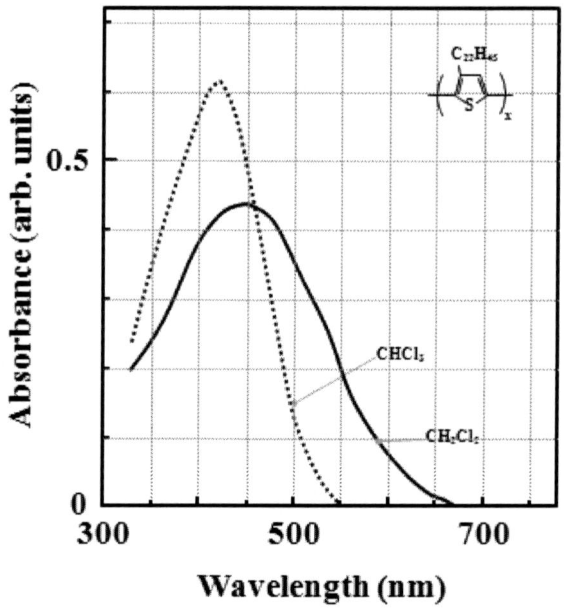

Figure 14. Optical absorption spectra of poly(3-docosylthiophene) in various solvents.

However, in chloroform at room temperature it is bulky *Gaush* conformation. Therefore, in chloroform, the steric hindrance between *Gaush* from of the alkyl chain and S atom induces torsion on bonds between thiophene rings, resulting in an effectively short conjugation length and large bandgap compared with the case of dichloromethane, in which steric hindrance is not remarkable due to *Trans* conformation of docosyl chain. However, even in dichloromethane at higher temperature the conformation of docosyl chains transformed into *Gaush* form, resulting in the blue shift of absorption.

The absorption spectrum is influenced not only by this steric effect but also by the interaction between solvent molecule and conjugated electron system in the main chain [26]. Such solvatochromism was also observed in poly(3-alkylfuran) and poly(alkylfluorene), as shown in Figure 15.

If the conformation of the side chain is controlled by either application of electric field or light irradiation, the solvatochromism can be controlled and used for various applications. For such a purpose the introduction of dipoles and also molecular units into side chains which show photoisomerization, such as the azoxy group, may be effective.

It should also be mentioned that in good solvents in which planarity of conjugated main chains is preserved, the interchain interaction should be strong and should tend to aggregate with lowering of temperature.

Therefore in such a case light scattering will also be influenced. That is, the control of aggregation at the µm and sab-µm level could open the possibility of optical applications utilizing non-scattering – scattering transition.

Ultrasonic irradiation may also have interesting effects.

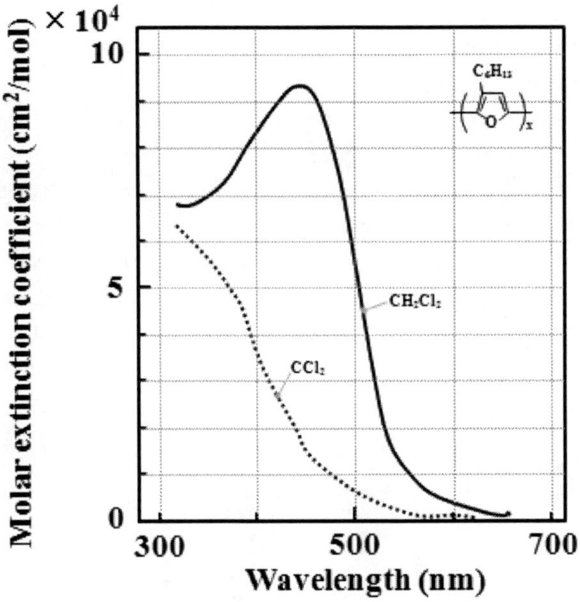

Figure 15. Optical absorption spectra of poly(3-hexylfuran) in various solvents.

2.3.5. Optical Recording by Optically Induced Stability

Conductive polymers such as polythiophene and its derivatives are relatively stable even under light irradiation.

However, we found that nonsoluble poly(3-alkylthiophene) films immersed in some solvent were even dissolved upon irradiation of light with photon energy exceeding the bandgap, if the solvent was not free of oxygen gas [27, 28]. When the oxygen was removed from the solvent, the film was not influenced by light.

This phenomenon can also be used for etching, cutting, and hole burning in the film. Figure 16 is the photograph of a hole made in nonsoluble poly(3-hexylthiophene) in air-saturated chloroform, upon irradiation of Ar ion laser for 1 minute. By using a focused laser beam much smaller holes, of the order of µm in diameter, can be formed in a very short time. Dye laser light of 50 nsec pulse width excited by Nd-YAG laser can also be used for hole formation.

Figure 16. Photograph of light-etched pattern on poly(3-hexylthiophene) film irradiated with Ar ion laser (488 nm).

In other words, this phenomenon can be used for optical recording. It is also possible to use it not only in the wet system but also in the dry system. In such a case, evaporation is also effective for removal of the sample in the irradiated area.

2.3.6. Optical Recording by Light Irradiation on Precursor Polymer

Some conductive polymers such as polydiacetylene, poly(*p*-phenylene vinylene), and their derivatives have been obtained from the precursor insulating films by heat treatment, light irradiation, or γ-ray irradiation.

In such materials, the information can be recorded easily in the nonsoluble polymer area by intense light irradiation upon the precursor film and subsequent washing of the non-irradiated area with solvent. The precursor of poly(*p*-phenylene vinylene) can be selectively polymerized only in the light irradiated area.

On the other hand, as shown in Figure 17, after irradiation on the precursor film in oxygen atmosphere, only the non-irradiated area can be polymerized . Perhaps the pre-irradiated area is deteriorated so that polymerization reaction does not take place. Therefore, by washing with alkaline-solution, a hole can be made at irradiated spots [28, 29].

Figure 17. Optically recorded pattern on poly(2,5-dimethoxy-p-phenylene vinylene) by light irradiation upon a precursor film and subsequent heat treatment and washing.

Similarly, polythiazyl, $(SN)_x$ film which is prepared by the evaporation of $(SN)_x$ crystal and exhibits gold metallic reflection can be selectively removed by the irradiation of intense light or electron beam, which can also be used for recording media.

It should be mentioned that such materials and processes as mentioned above can also be applied to the photolithography technique.

2.3.7. Gelchlomism

We have found that lightly cross-linked poly(3-alkylthiophene) exhibits gel characteristics [30]. That is, the volume changes by more than 10-100 times with a change of solvent and also in temperature, as shown in Figure 18 for example. Along with this volume change, the color of the polymer gel was found to change drastically, as shown in Figure 19 for example, in poly(3-hexylthiophene). These gel characteristics are interpreted in terms of reversible cross-linking influenced by a change in interchain interaction due to the change of torsion of polymer main chains triggered by the conformation change of alkyl chain depending on the conditions. Change of size, conductivity, and color of the conductive polymer gel was also successfully controlled by doping, as shown in Figure 20 for example. Electrochemical doping makes electrical control of the gel possible [31].

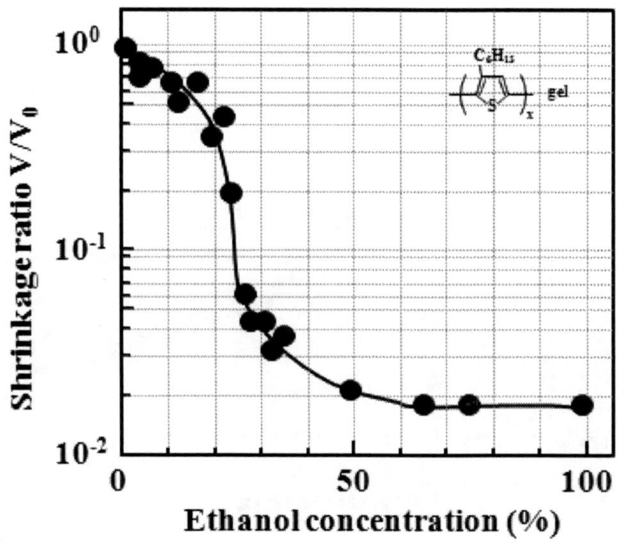

Figure 18. Plots of shrinkage ratio of poly(3-hexylthiophene) gel as a function of concentration of ethanol in immersed ethanol/chloroform solution.

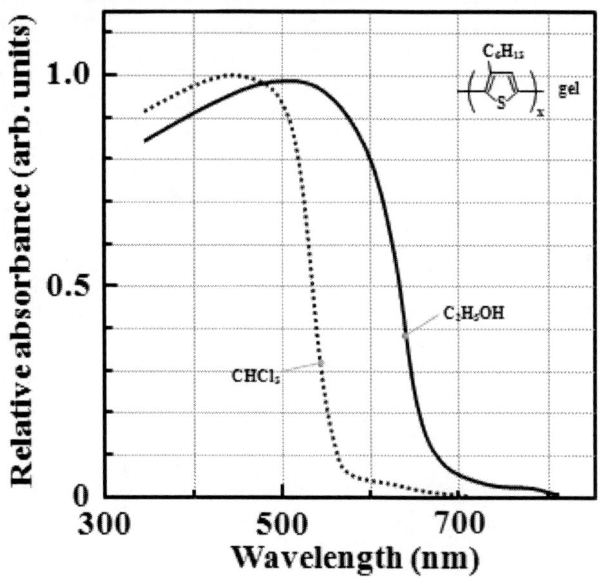

Figure 19. Optical absorption spectra of poly(3-hexylthiophene) gel in chloroform, $CHCl_3$ (expanded state) and ethanol, C_2H_5OH (shrunken state).

The optical control of gel characteristics should be possible when the conformation of a side chain was influenced by light irradiation. A change in gel characteristics such as size and color can also be controlled by irradiation of light in the conductive polymer gel-optically sensitive dopant composite.

These phenomena can also be used for optoelectronic elements and devices.

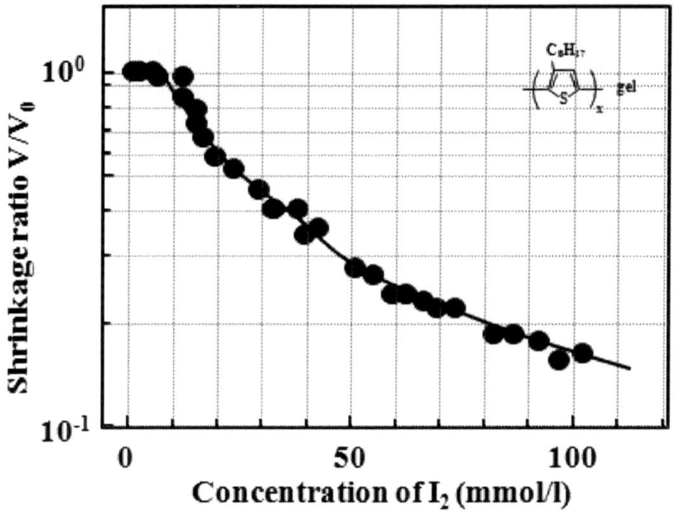

Figure 20. Plots of shrinkage ratio of poly(3-octylthiophene) gel against concentration of iodine in the immersed solution.

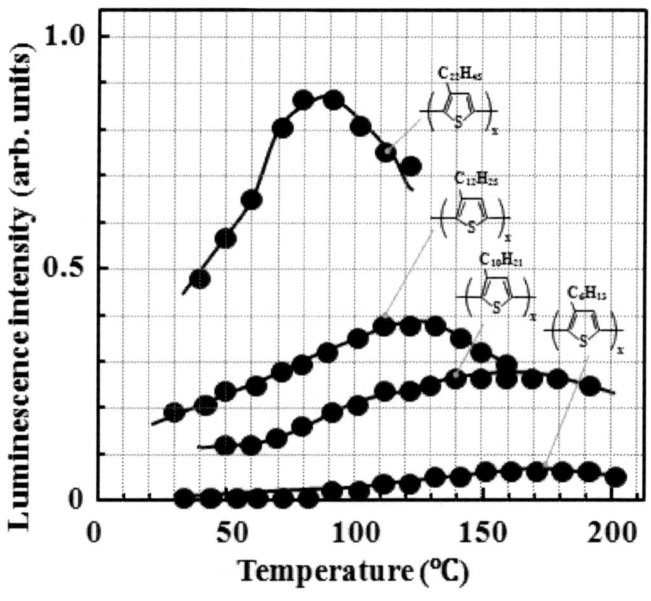

Figure 21. Temperature dependence of luminescence intensity of poly(3-alkylthiophene) with various alkyl side chain length.

2.3.8. Anomalous Luminescence of Conductive Polymer

Figure 21 shows the temperature dependence of luminescence in conductive poly(3-alkylthiophene)s [32].

As is evident in this figure, luminescence intensity was enhanced with increasing temperature. Then, after indicating peak intensity at the melting point, the luminescence intensity again decreased at higher temperatures. It should also be noted in this figure that poly(3-alkylthiophene) with a longer alkyl chain exhibits higher intensity, and the peak of luminescence was observed at lower temperatures. These results can be interpreted in terms of the confinement of excited species in the photoexcited area and the higher probability of radiative recombination due to the introduction of bond torsion between thiophene rings and the decreased conjugation length.

One-dimensionally is more remarkable in the sample of longer alkyl chains because of longer interchain distance and the smaller possibility of transfer of photoexcited species from one chain to the neighboring chains.

It was also confirmed that a variety of junctions can be made utilizing conductive polymers, such as diodes, transistors, and field-effect transistors (FET). Utilizing these poly(3-alkylthiophene)s, remarkably temperature-dependent diode characteristics were realized [33].

It should also be noted that some conductive polymers exhibit intense luminescence. Therefore, in future, some light-emitting diode or laser may be achievable utilizing conductive polymers, whose intensity is higher at higher temperatures.

2.3.9. Nonlinear Optical Effect in Conductive Polymers

The existence of high-density π-electrons suggests that a conductive polymer is highly electronically polarizable. Therefore nonlinear optical responses were also expected in these conductive polymers. Indeed, a large third-order nonlinear effect was confirmed in various conductive polymers such as polydiacetylene, polyacetylene, polythiophene, and poly(3-alkylthiophene) [34, 35].

As a second-order nonlinear optical effect, optical second harmonic generation (SHG) in conductive polydiacetylene was studied in detail. For SHG, the center of inversion must be lost in the structure, and appropriately designed conductive polymer exhibits large second-order nonlinear susceptibility.

Third-order optical nonlinearity, such as third harmonic generation (THG) and nonlinear refractive index, is expected in most conductive polymers, because in this case the lack of the center of inversion symmetry is not a necessary condition. Ultrafast nonlinear response was indeed confirmed in the observation of phase conjugation in conductive polymers [36].

2.3.10. Neural Functional Component

We have found that a learning effect analogous to neural function is manifested in conductive polymers [37]. For example, as shown in Figure 22, the diffusion coefficient of dopants in chemically prepared poly(3-alkylthiophene) became larger with repeating doping-undoping cycles. This reaction is dependent on the solvent and electrolyte used.

This result means that doping speed becomes faster with repeating cycles. That is, when a constant width voltage pulse is applied, a conductive polymer which has experienced pre-doping can be switched to the metallic state but a fresh one cannot be switched. Therefore, not only electrical conductivity but also color change due to the insulator-metal transition should depend on cycles, and a learning effect should exist. This is considered analogous to learning in the neural network and also can be interpreted as one type of memory.

In conclusion with spectral change due to the insulator-metal transition, optical sensitive neural functional component should also be possible.

In other conductive polymers such as poly(*p*-phenylene vinylene) and poly(2,5-diethoxy-*p*-phenylene vinylene), similar learning effects were confirmed.

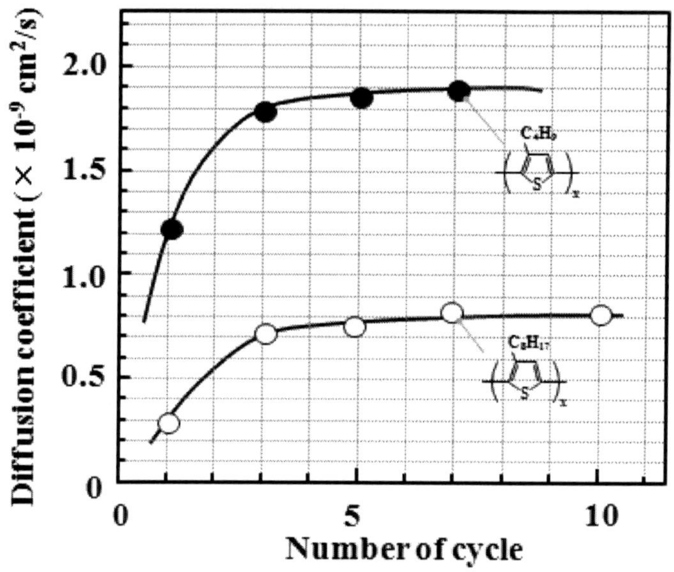

Figure 22. Learning effect on diffusion coefficient of dopant within chemically prepared poly(3-butylthiophene) and poly(3-octylthiophene) films.

The learning effect was explained in terms of the change of morphology by doping-undoping cycles at this stage.

It should also be mentioned that the accumulation of dopants by repeating cycles should be used for realization of the learning effect.

2.4. Summary

We have explained that conductive polymers have novel optical characteristics which are quite different from the conventional insulative polymers. Though the mechanism of the unique characteristics has not been fully understood, there are a variety of possible applications to optoelectronic elements and devices.

In this chapter, the possibilities for using conductive polymers as optoelectronic materials, elements, and devices were discussed. It is clear that new types of conductive polymers will be developed and that completely new properties will be discovered, which can also be used for new types of optoelectronic elements and devices.

The study of conductive polymers can also be interpreted as the first step in realizing molecular electronics and molecular optoelectronics devices.

3. Electrolysis Polymerization of Conductive Polymers

3.1. Electrolysis Polymerization Technique

In general, electrolysis polymerization technique is performed with the electrolysis polymerization equipment shown in Figure 23. If the aromatization compound monomer that is to be polymerized is dissolved into a solvent containing a suitable supporting electrolyte and an appropriate voltage is applied to the electrode pair immersed into this solution, the monomer is oxidized or reduced on the surface of the anode or cathode, respectively, and it polymerizes in the form of powder or a film and at times may show arborization [38].The case in which the monomer is oxidized and polymerized on the anode surface is called an electrochemical oxidation polymerization.

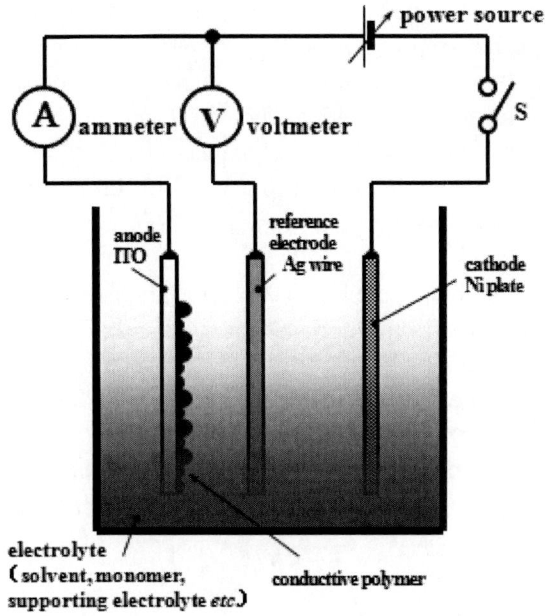

Figure 23. Equipment of electrolysis polymerization.

On the other hand, the case in which the monomer is reduced and polymerized on the surface of the cathode is called electrochemical reduction polymerization. The reference electrode might be immersed in the solution if necessary.

The most important point in this polymerization method is that the composition of the electrolyte, that is, the kind of solvent, the kind of supporting electrolyte, and the monomer, and the difference in the concentrations excert a strong effect on the polymerization reaction. Even when the same monomer is used, if the composition of the electrolyte is different, the form of the product might also differ greatly, and at times, the product might mot be obtained at all. Besides this as a predominant factor of such an electrolysis polymerization reaction, it is likely that the applied voltage, the current density and the polymerization temperature, and depending on the circumstances, the material of the electrode and the distance between electrodes, might also be periodic strong effects.

In order to obtain good-quality conductive polymer films by the electrolysis polymerization technique, it is necessary to examine the above predominant factors in detail and to understand the optimal electrolysis polymerization condition.

Moreover, the conductive polymer can be reversibly electrochemically doped and dedoped and the dopant concentration can be widely adjusted by limiting the voltage, enabling a film has an arbitrary electric conductivity to be obtained.

3.2. Mechanism of Electrolysis Polymerization Reaction

The reaction mechanism of conductive polymer synthesis by electrolysis polymerization has not been clarified, because various factors, such as the composition of the electrolyte and the electrolytic conditions, interact very complexly with the electrode reaction. Therefore, the polymerization reaction condition differs for each individual conductive polymer and the optimum conditions are determined through trial and error. However, qualitatively, the following reaction mechanisms are accepted in general. The dynamic movement of the molecule, which takes part in the monomer, and the electrolytic ion formed by the transfer of an electron from the solvent are caused at the interface between the electrode and the electrolyte in the electrolysis polymerization technique.

Usually, 2-2.5 electrons are consumed in the electrolysis polymerization reaction, and ultimately, two pieces of electrons are used in the polymerization reaction and the remainder is used in the doping.

As a result, the same number of protons (H+) that as the number of electrons used for the polymerization will accumulate in the polymerization liquid. Therefore, the polymerization reaction occurs as a result of the removal of the electron from the monomer, and the coupling reaction, in which the generated radical cation (positive ion) is assumed to be the activated species, and the deprotonation reaction are thought to progress repeatedly. The polymerization reaction is thought to be either reaction 1, the electrophilic substitution coupling reaction, or reaction 2, the radical coupling reaction (Here, M indicates a monomer.), as shown in Figure 24.

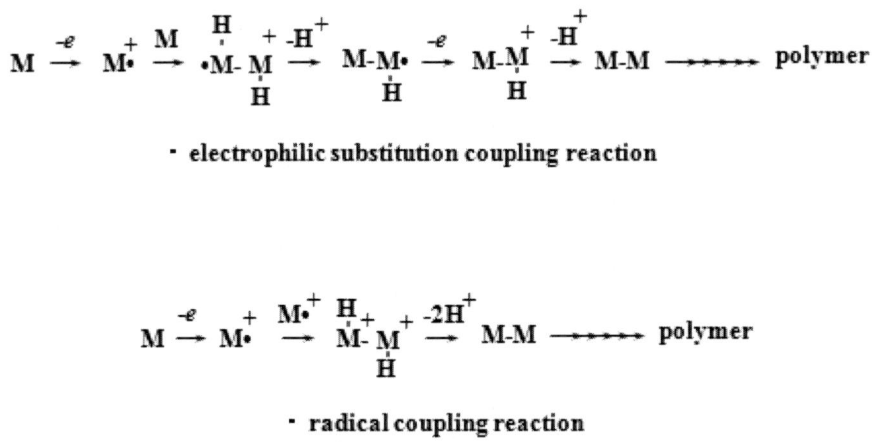

Figure 24. Electrolysis polymerization reaction.

However, the structure analysis of the obtained polymer is difficult because the polymer is nonfusible and nonsoluble. Then, the polymerization reaction mechanism itself becomes very complex and a united opinion is not obtained, because electrolysis polymerization is a reaction of a nonuniform system the progresses in a limited area in the neighborhood of the electrode, and it is necessary to consider the supply of monomers and electrolytic ions to that area. That is, the electrolysis polymerization is thought to begin with the oxidizing or reducing reaction of the monomer in the electrolyte upon applying potential, followed by the generation of the radical cation or the radical anion in the aromatic compounds, and subsequently, the polymerization progresses repeated the coupling and the deprotonation reaction.

It is necessary to prevent the solvent from undergoing the electrochemical reaction with the monomer at a low voltage, because the effects of the various factors such as the type of solvent, the supporting electrolyte, the polymerization voltage, and the polymerization temperature, on the electrolysis polymerization reaction are not yet fully clarified.

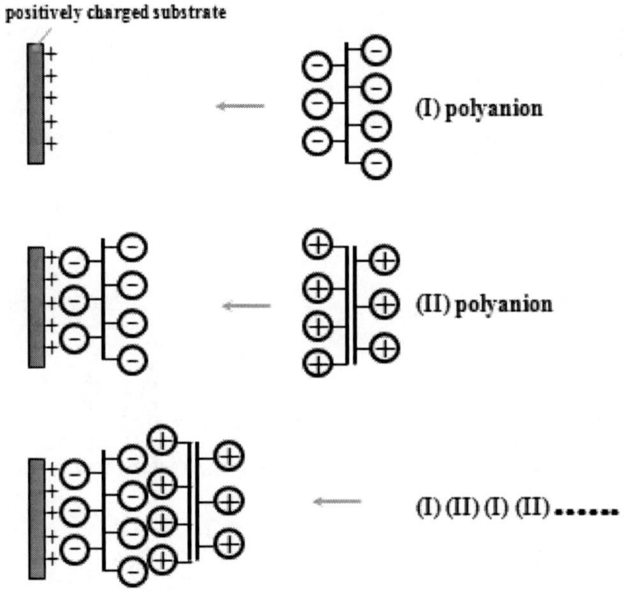

Figure 25. Molecular self-assembly process involving the alternate deposition of polyanions and polycations. In this case, the substrate surface is positively charged.

4. Principles of Molecular Self-Assembly Process

The basic process used to create alternating layer thin films involves dipping a substrate into a dilute aqueous solution of a polyanion, rinsing the substrate with water and then dipping it into a dilute solution containing a polycation.

Thus, in the first dip, a monolayer thick film of a negatively charged conjugated polyion is deposited onto the surface of the substrate whereas in the second dip, a monolayer thick film of a positively charged conjugated polyion is deposited.

This process can be repeated as many times as desired to build multilayer thin films in which each layer deposited is only about 0.5-2 nm in thickness (depend on such parameters as solution concentration, pH, and ionic strength). Each complete cycle of deposition can take as little as 30 s to complete, thereby making it possible to easily and quickly form electrically conducting thin film coatings with complex multilayer organizations and controllable functionality.

For example, the self-assembly process as described by Decher et al., is illustrated in Figure 25 [39]. In this case, a positively charged glass substrate, created by suitable silane chemistry, is first immersed into a dilute solution of a polyanion followed by immersion in a dilute solution of a polycation. The repetition of this cycle produces a multilayer thin film comprised of alternating layers of polyanions and polycations. The thickness and conformation of each polymer layer deposited are determined by the chemistry of the deposition solution.

Figure 26 shows molecular structures of typical conjugated polyions that we can successfully utilize to fabricate multilayer thin films via this process.

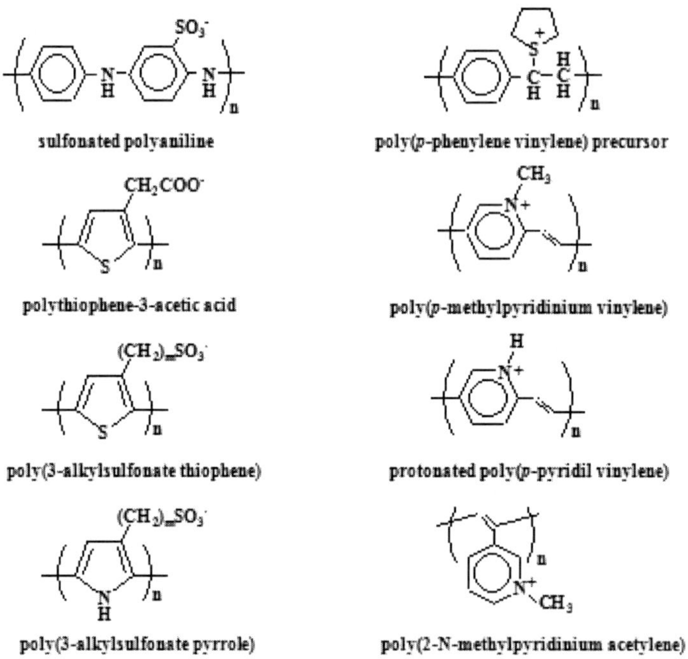

Figure 26. A list of the conjugated polyions used to fabricate multilayer thin films via self-assembly process.

5. New Fabrication Techniques of Polymer-Composite Films

5.1. Self-Assembled Multilayer Heterostructures of Electroactive Polymers

5.1.1. Introduction

In the 20th century, professors Alan J. Heeger, Alan G. MacDiarmid, and Hideki Shirakawa have been among the Nobel prize awardees in chemistry for their pioneering works on the discovery and development of conductive polymers [40-42]. Conductive polymers are generally becoming more important as semiconducting materials in new applications. Recent progress has been made in the fabrication of electronic devices based on metal-semiconductor Schottky junctions, where conductive polymers are used as active semiconducting materials [43].

Particularly exciting are the reports of light-emitting diodes (LEDs), based on the poly(p-phenylene vinylene) (PPV), and its derivatives [4, 44-48]. Since then, a number of polymer

LEDs have been reported, using different conductive polymers emitting in various portions of the visible spectrum [49-51].

The emission of light in the range from green-blue to orange-red can be produced with the proper chemical modifications of PPV [46]. Furthermore, fascinating properties due to the polymeric nature of the semiconducting material have been demonstrated: light sources processing mechanical flexibility and the capability of emitting polarized light and large LEDs are already being developed [46, 52]. Although inorganic blue electroluminescent (EL) diodes consisting of III-V compound semiconductors have substantially developed, a polymer emitting blue light is particularly desirable[50, 51].

On the other hand, ultrathin organic films are currently gaining interest in many areas such as integrated optics, sensors, electronics, and so on [53].

It is most interesting and challenging to contract films with a supermolecular architecture in which the individual organic molecules are macroscopically oriented and in which molecules with different functionally can be incorporated into individual layers.

For example, the Langmür-Blodgett (LB) technique can be utilized fabricate multilayer heterostructures from a wide variety of relatively diverse conductive polymers. Although the LB technique provides a powerful processing vehicle for the fabrication of multilayer thin films, it is limited in terms of the types of materials that can be used, the shape, and geometry of the substrates coated and the time it takes to fabricate complicated multilayer thin films [54].

About 20 years before, a new technique of constructing multilayer assemblies by consecutively alternating adsorption polyelectrolyte was developed. Rubner et al., have found that it is possible to extend this exciting new technique to the manipulation of conductive polymers [55]. That is, recent advances in the molecular level processing of conductive polymers have made it possible to fabricate thin film multilayer heterostructures with an unprecedented level of control over the type and thickness of the deposited layer. Since this new layer-by-layer deposition process is carried out with dilute aqueous solutions of the charged polymers, it is an extremely simple and effective way to produce thin film structures that are controllable at the molecular level. Then, as the dimensions of the individual layers approach molecular scales, it may also be possible to realize quantum effects in these multilayer contracts.

The layer-by-layer molecular self-assembly technique of conductive polymers is based on the spontaneous adsorption of a charged polymer onto a surface from a dilute polymer solution. Accordingly, the delocalized positive charges (polarons and/or bipolarons) of in situ polymerized conductive polymers such as polypyrrole (PPy), polyaniline (PAn) can be used to self-assemble multilayer thin films when alternated with a suitable polyanion such as poly(styrene sulfonic acid).

That is, the basis of this new process is the electrostatic attraction developed between the ionized functional groups of a polyanion and the delocalized positive charges along the backbone of *p*-type doped conductive polymers.

We have already reported that the nature of the substrate surface plays an important role in the early stages of the PPy growth process[56]. That is, there is a linear correspondence between the dipping time and the amount of PPy deposited onto the negatively charged substrate, however for surfaces that are positively charged, there is a rest period of 20-30 min during which no PPy is deposited. After this rest period, the growth of the PPy on the surface is significantly less than that observed on a negatively charged surface. In particular, it is very interesting from fundamental and practical viewpoints to investigate what happened during the rest period.

In this section, we describe the fabrication of self-assembled multilayer heterostructures of conductive polymers and report the basic properties of electronic and optical applications such as LEDs, photovoltaic cells, actuators, and so on, consisting of a multilayer thin film. Furthermore, we discuss that the delocalized positive charges such as polarons and/or bipolarons of *in situ* polarized conductive polymers can be used to self-assemble thin films. We discuss the initial deposition process of PPy onto the positively charged substrate.

5.1.2. Nature of the Substrate Surface

It should be noted that the fabrication of high optical quality and multilayer thin films with useful electrical properties can be accomplished with a variety of different substrates, including substrates that are either hydrophobic or hydrophilic. As long as the first layer deposited is a polycation, it is not necessary to use substrates that have been chemically treated to render them charged. However, the nature of the substrate surface influences the deposition process and molecular organization of the deposited monolayer.

In this study, indium-tin-oxide (ITO)-coated conductive glass with positively charged surfaces were used as substrates for the absorption process. That is, ITO glass substrates were first conditioned successively with solutions of methanol, methanol/toluene (1/1), and toluene each for 10 min.

These substrates were then treated with 5 vol-% TMS (N-(2-aminoethyl-3-aminopropyl)trimethoxysilane) in toluene for 15 h. The TMS-coated substrates were then heated in a warm toluene both for 1 h to form a physical bond between the substrates and TMS. After the heat treatment the substrates were cleaned in toluene, methanol/toluene (1/1), and methanol, each for 10 min. Finally, the substrates were rinsed extensively with deionized water to remove residual TMS and dried in air [56].

5.1.3. Experimental Procedure

5.1.3.1. Preparation of Polyions

In this section, we discuss the use of PPV polycation precursors to fabricate multilayers of PPV with sulfonated polyaniline (SPAn) as a polyanion. PPV-precursor/SPAn multilayer films can be successfully built with a remarkable reproducibility of the amount of material delivered in each deposited layer. The precursor present in this film can also be easily thermally converted to its fully conjugated form, thereby creating films with electroactive PPV layers as shown in Figure 27.

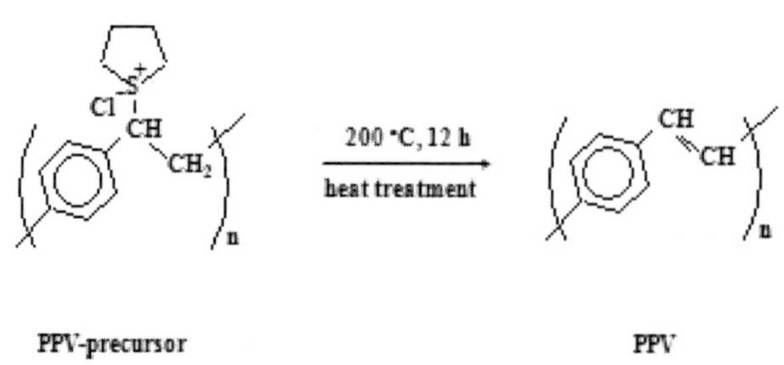

Figure 27. Thermal conversion of the PPV-precursor to PPV.

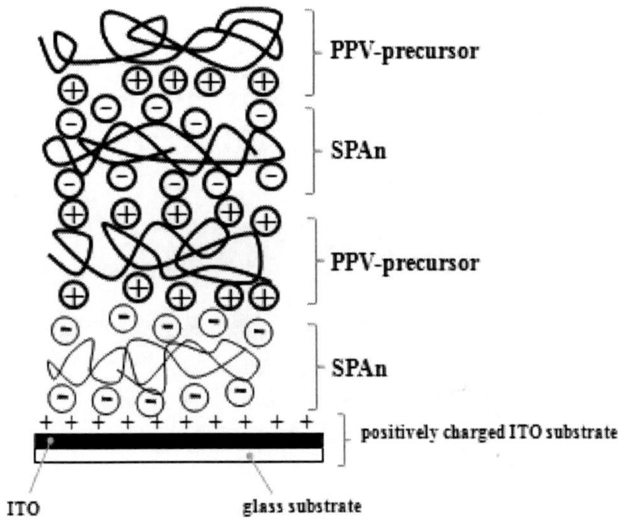

Figure 28. Schematic illustration of a multilayer heterostructure fabricated via the self-assembly process.

PPV-precursor was synthesized by tetrahydrothiophene (THT) route [57]. SPAn was synthesized using the procedure described by Yue et al. [58]

The SPAn solution was prepared by dissolving the polymer on a 0.1 mol/l NaOH. The acidity of the solution was then adjusted with HCl to give a measured pH of 4.5. All the solutions were prepared in the concentration range of 0.001-0.1 mol/l and filtrated through a 1.0 μm filter prior to use.

5.1.3.2. Fabrication of SPAn/PPV Multilayer

The multilayer films were built up by first immersing the ITO glass substrate in the SPAn solution for 15 min, rinsing with deionized water and air drying. The subsequent dipping in the PPV-precursor solution for equal time followed by the rinsing and the drying procedures puts down another layer. This cycle can be repeated as many times as desired.

Figure 28 shows a schematic illustration of a self-assembled film fabricated with alternating layers of two different polymers: SPAn (polyanion) and PPV-precursor (polycation). The thermal conversion of the SPAn/PPV-precursor to SPAn/PPV multilayer was performed in an oven at 200 °C in vacuo for 12 h.

5.1.3.3. Fabrication of SPAn/PPV Multilayer EL Device

The construction of typical SPAn/PPzV multilayer EL device is illustrated in Figure 29. A high work function metal such as ITO (about 4.8 eV) serves as the anode and is semitransparent. The self-assembled multilayer heterostructures of SPAn/PPV were deposited as a thin film on the ITO electrode and a low work function such as Al (about 4.2 eV) is evaporated on to the polymer surface by vacuum metal vapor deposition. Active areas of SPAn/PPV multilayer LED were 2 mm^2.

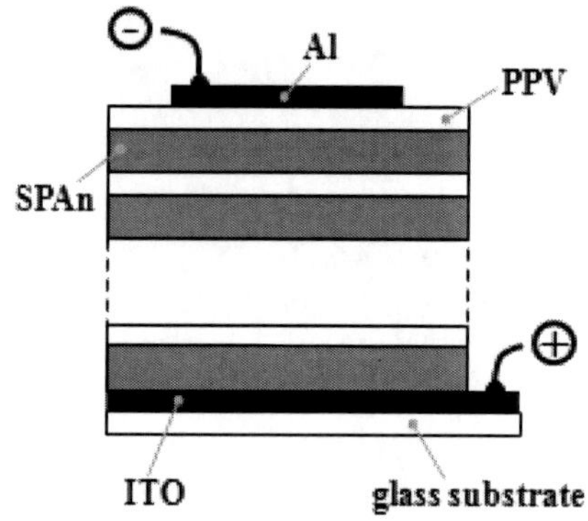

Figure 29. Structure of the SPAn/PPV multilayer light-emitting diode.

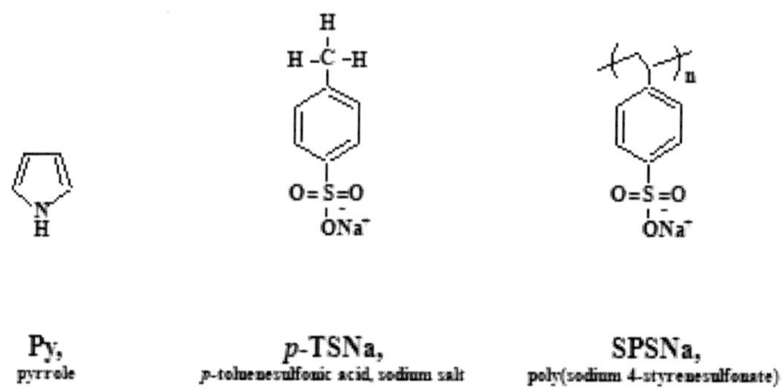

Figure 30. Molecular structures of the main chemicals used in this study: pyrrole (Py), *p*-toluenesulfonic acid, sodium salt (*p*-TSNa), and poly(sodium 4-styrenesulfonate) (SPSNa).

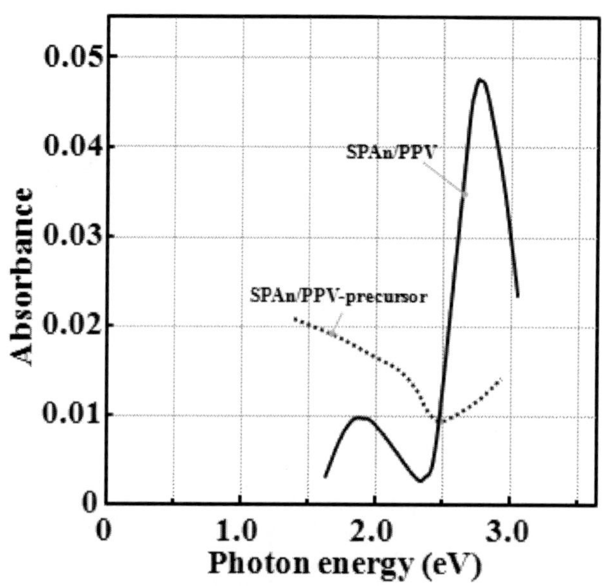

Figure 31. Optical absorption spectrum of a SPAn/PPV multilayer thin film. Optical absorption spectrum of a SPAn/PPV-precursor multilayer thin film is also shown for reference. The multilayer structure consists of 10 periods of 20 Å thick SPAn/PPV.

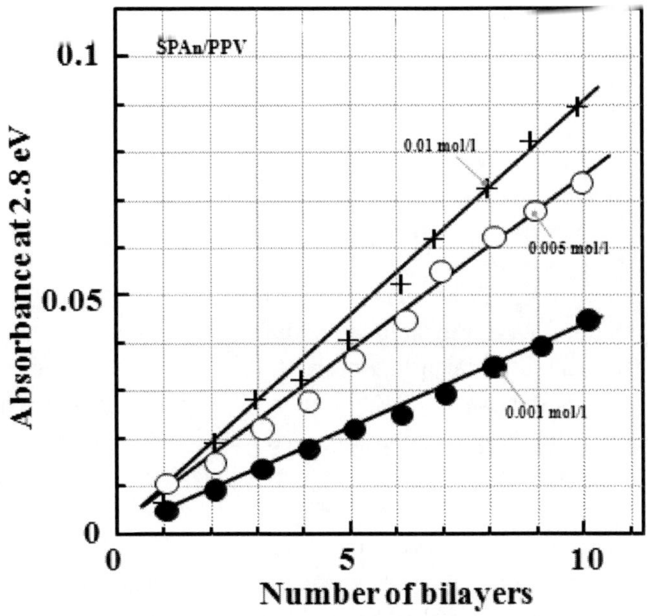

Figure 32. Dependence absorbance at 2.8 eV of a 10 bilayer SPAn/PPV multilayer thin film on the number of bilayers deposited from solutions with different polyion concentrations.

The assembled multilayer films were monitored by UV-vis adsorption spectroscopy using a Hitachi 330. The thickness of the layer was determined by optical absorption measurements (i.e., the optical absorbance of the film showed a linear relation with the film thickness asshown in Figure 32.

The detailed explanation is discussed later in section 5.1.4.1.). The EL emission spectra and light intensity properties were measured by using Nikon P-250 spectrometer with a photomultiplier and a Si photodiode. For photoluminescence (PL) spectra, the polymer was excited with light at 450 nm. Current-voltage curves were obtained with Keithley 990 digital multimeter and Metronics 5244 DC voltage source. All the measurements were performed in air at room temperature.

5.1.3.4. Preparation of the Substrate Surface

Glass slides with positively and negatively charged surfaces were used as substrates for the adsorption process. The substrates were all cleaned by first placing them in a hot H_2SO_4/H_2O (7/3) bath for 1 h and then in a $H_2O/H_2O_2/NH_3$ (5/1/1) bath for 30 min. The substrates were extensively rinsed with deionized water after each cleaning step. This procedure produces hydrophilic glass slides.

Positively charged surfaces were made using a solution treatment of precleaned hydrophiric slides with TMS. The precleaned hydrophilic substrates were first conditioned successively with a solution of methanol, methanol/toluene (1/1), and toluene each for 10 min. The preconditioned slides were then immersed in a TMS/toluene (5 vol.-% of TMS) solution for 15 h. The TMS gel coated substrates were then heated in a warm toluene bath for 1 h to form a physical bond between the substrates and TMS. After the heat treatment, the substrates were each cleaned in toluene, toluene/methanol (1:1), and methanol for 10 min.

Finally, the substrates were rinsed extensively with deionized water to remove residual TMS and dried with compressed air.

Negatively charged surfaces were made by a solution treatment of precleaned positively charged slides. That is, the positively charged substrates were immersed in a poly(sodium 4-styrenesulfonate) (SPSNa) solution (pH 2.5, 0.01 mol/l) for 20 min to deposit a single layer of negatively charged SPS^- anions onto the surface.

Figure 30 shows molecular structures of the main chemicals used in this study.

5.1.3.5. Preparation of *In Situ* Polymerization Electrolyte

PPy dipping solutions were prepared by adding pyrrole (Py) monomer to an aqueous ferric chloride ($FeCl_3$) solution which had been pH=1 using concentrated hydrochloric acid (HCl). To enhance conductivity, *p*-toluenesulfonic acid, sodium salt (*p*-TSNa) was also added to the $FeCl_3$ solution. The dipping solution was aged for 30 min and filtrated using a 0.45 μm filter prior to use. PPy dipping solutions were comprised of 0.04 mol/l Py monomer, 0.02 mol/l *p*-TSNa, and 0.005 mol/l $FeCl_3$. To insure controlled deposition, dipping solutions were typically only used for 3-5 h. To prevent the oxidation of Py, PPy dipping solutions were sufficiently shielded from light. The deposition of PPy was monitored by UV-visible absorption spectroscopy using a spectrophotometer (Hitachi, Model U-3410). The thickness of the sample film was determined using the spectrophotometer and the absorbance-thickness curves, which were prepared using a multiple interference microscope. The surface states of PPy were evaluated using an atmospheric low-energy photoelectron spectroscopy instrument (Rikenkeiki Co. Ltd., Model AC-1).

5.1.4. Results and Discussion

5.1.4.1. Self-Assembled Multilayer Heterostructures of Electroactive SPAn/PPV

Although the PPV-precursor does not absorb light in the visible region, it is possible to readily monitor the growth process as the SPAn absorbs in the visible region. Figure 31 shows the optical absorption spectra of a 10 bilayer film fabricated from the SPAn/PPV-precursor using 0.001 mol/l polyion solutions before and after thermal conversion. As the SPAn thin films are doped by the HCl of the polyanion dipping solution (pH=4.5), the spectrum of the SPAn/PPV-precursor film shows a characteristic expected for a doped SPAn film. It is suggested that the charges on the SPAn polymer are actually in the form of delocalized polarons and/or bipolarons. After the thermal conversion, the spectrum of the SPAn/PPV shows a small broad peak at around 2.0 eV due to the thermal dedoping of the SPAn and a strong peak which corresponds to the interband transition of the newly formed PPV film appears at around 2.8 eV. The absorption peak at 2.0 eV has been assigned to the optical absorption of polarons in partially doped SPAn. It has been reported that the interband transition peak of the SPAn appears at around 3.8 eV [59]. Accordingly, the multilayer film containing alternating layers of the SPAn and the PPV is essentially seen to be a superposition of these two spectra, indicating that both materials are successfully transferred into multilayers and suggesting that no ground state interaction is taking place between these two materials. The bandgap of the SPAn/PPV multilayer film was evaluated to be about 2.6 eV from the $(\alpha h\nu)^2$ versus the $h\nu$ plot of the absorption edge of the interband absorption peak of PPV, where α, h, and ν are the absorption coefficient, Plank's constant and the frequency of light, respectively. This bandgap is compatible with that of the PPV film formed by spin coating [29].

The heterostructure thin films formed with these different polymers were of the same high quality as the films formed from the individual polyions and also exhibited a perfectly linear relationship between the absorbance of the conductive polymer and the number of layers deposited. Figure 32 shows the absorbance of the SPAn/PPV multilayer film measured at 2.8 eV as a function of the number bilayers deposited from solutions with different concentrations. The linearity of these plots clearly demonstrates that each deposited monolayer contributes an equal amount of material to the thin film on average. Thus, alternating polyanion (SPAn)-polycation (PPV-precursor) bilayers can be successfully assembled from conjugated-conjugated polyions and it is possible to fabricate novel SPAn/PPV multilayer heterostructures by thermal conversion. In the case of our experimental condition, preliminary results showed that the thickness contributed by each bilayer of SPAn/PPV varies from about 20 Ä (0.001 mol/l solutions) to 42 Å (0.1 mol/l solutions) when dipping times of 15 min are used for each layer. We determined that if the dipping time is more or less than 15 min, the films could be made to grow thicker or thinner than the above thickness, respectively. Thus, by simply controlling the dipping sequence and the types of polyions deposited, it is possible to fabricate diverse collection of composition modulated heterostructures that are controllable at the molecular level.

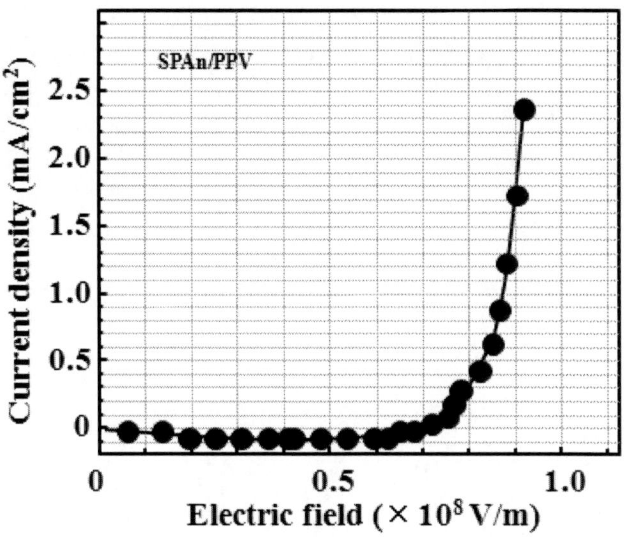

Figure 33. Current density-electric field haracteristic for an electroluminescent (EL) device having a SPAn/PPV multilayer thin film as an emitting layer. The multilayer structure device consists of 10 periods of 20 Å thick SPAn/PPV. Only the forward bias is shown.

5.1.4.2. LEDs Using SPAn/PPV Multilayer Thin Film

Figure 33 indicates the typical dark current density-electric field characteristic of the EL device with 10 bilayers of the SPAn/PPV thin film as an emitter material. A pronounced rectification effect was observed in this device, where the forward bias corresponds to positive on the ITO electrode. That is, the forward bias current of this smalldevice increases nonlinearly at electric fields higher than about 0.7×10^8 V/m, which is nearly identical to the threshold of light emission. This threshold electric field means that substantial charge injection into an emitting layer occurs. It should be noted that the EL emission was detected only in the forward-bias and the reverse-bias current of this device remained small; the rectification ratio was approximately 10^2-10^3. Furthermore, the EL intensity increased drastically with the increasing electric field and was approximately linear with injected current. High forward-bias voltage is required in order to achieve charge injection.

However, reducing the thickness lowers the device operating voltage. It was clearly demonstrated that the current-voltage curves depend on the electric field. This clearly points to a tunneling model for carrier injection in which one or both of the carriers (electrons and holes) are electric field emitted through a barrier at the electrode interface.

A schematic energy diagram of the SPAn/PPV EL device is illustrated in Figure 34. The ionization potential of the PPV and the SPAn films are about 5.7 and 5.6 eV, respectively. In having the bandgap of 2.6 eV, the bottom of the conduction band is estimated to be located at an energy higher than a Fermi level of Al by about 1.1 eV. Furthermore the top of the valence band of the SPAn is located at an energy lower than a Fermi level of ITO by about 0.8 eV. In the SPAn/PPV EL diode, the energy of about 0.8 eV is required for holes to be injected from the ITO electrode into the SPAn film. Since the conduction band of PPV is located at about 3.1 eV below the vacuum level, the energy barrier of about 1.1 eV exists between the Al electrode and the conduction band. This means that the electron injection into PPV from the Al electrode is somewhat difficult compared to the hole injection into SPAn from the ITO electrode. It was assumed that the top of the valence band of SPAn is located at a higher energy state than that of the PPV by about 0.1 eV. The larger bandgap of the SPAn by about 0.9 eV compared with that of the PPV indicates that the bottom of the conduction band of the SPAn is located at a higher energy than that of the PPV by about 1.0 eV. Accordingly, it follows that the accumulation of holes and electrons at the SPAn-PPV heterojunction interface is expected.

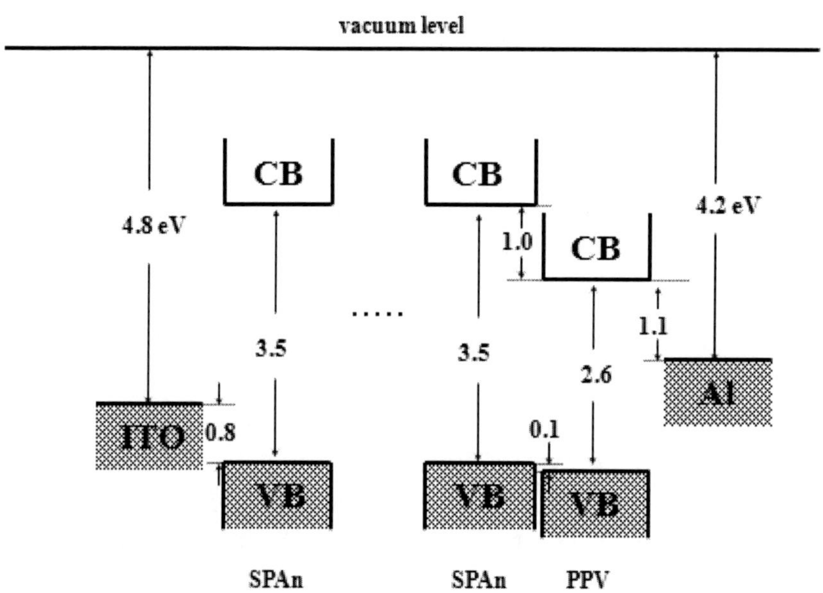

Figure 34. Schematic electron level structure for SPAn/PPV multilayer heterostructure EL device.

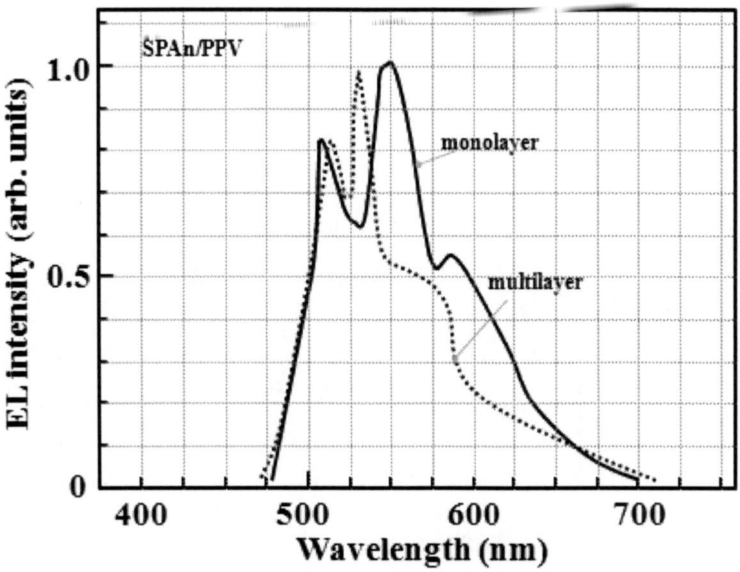

Figure 35. EL spectrum for light-emitting diode (LED) with the SPAn/PPV multilayer and monolayer structure. The multilayer structure LED consists of 10 periods of 20 Å thick SPAn/PPV. The monolayer LED consists of 1 period of 120 Å SPAn/PPV.

The emission spectrum of a 10-bilayer (10 periods of 20 Å thickness SPAn/PPV LED device at room temperature is shown in Figure 35. The EL spectrum of the SPAn/PPV monolayer film (1 period of about 120 Å thickness) formed by spin coating is also shown for comparison. The driving condition (injection current 0.6 mA/cm^2) is the same. Although the emission intensity of the SPAn/PPV multilayer LED is weak compared with that of the SPAn/PPV monolayer LED, the emission intensities of the LEDs are normalized. The emission properties are hardly affected by the absorption of the SPAn having very low optical densities at about 2.0 eV.

However, the presence of adjacent monolayers of the SPAn may slightly weaken the EL emission of SPAn/PPV LED due to a weak reabsorption of the SPAn at 2.0 eV. The EL spectrum of the LED with monolayer structure is just similar to that reported Burroughes et al., with a peak near 555 nm (2.23 eV) [4].Therefore, this device emits light in the greenish-yellow range of the spectrum and can be easily seen under normal lighting even at low driving voltage. On the other hand, the peak of the EL emission of the LED with mukltilayer structure is at about 536 nm corresponding to the photon energy of 2.31 eV; the bluish-green emission can be seen with the naked eye. The light emitted from this device is also visible under ordinary room lighting. These SPAn/PPV heterostructure LEDs show an emission spectrum characteristic of a single polymer layer, in this case PPV. There is no contribution to the emission spectrum from excitons decaying in the SPAn. The EL spectrum separates into three peaks. This separation of the EL spectrum is considered to be due to the Stokes shift based on the absorption and emission of phonon. Similar vibronic structure is also observed in the EL spectrum of PPV LEDs [4, 48].

There is a small shift of about 20 nm (0.08 eV) between the two spectra as can be seen in Figure 35. This can also be observed by the eye; the EL of the device with monolayer structure appears greenish-yellow while that with multilayer appears bluish-green. Figure 36 shows the shift of peak emission energy of the EL emission specrrum of the SPAn/PPV multilayer LED from that of the SPAn/PPV monolayer LED as a function of bilayer thickness.

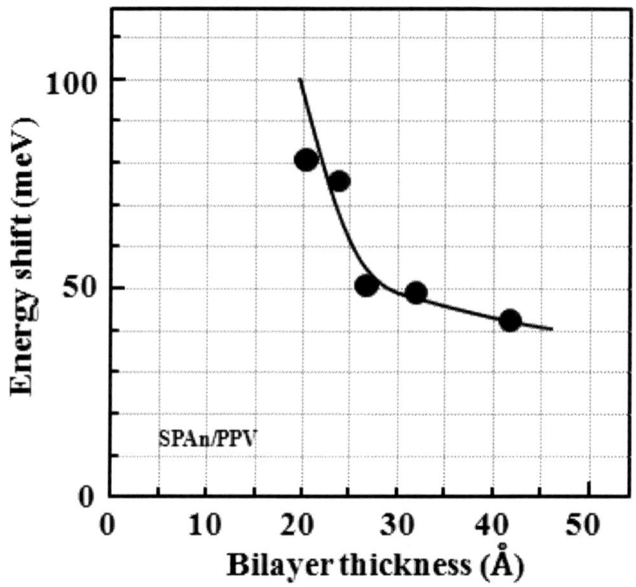

Figure 36. Relation between peak energy shift of EL emission spectrum of the LED with multilayer structure and bilayer thickness. Solid line was estimated from Krönig-Penny model.

The origin of this shift is not clear at present. Usually, the EL emission is generated by the recombination of the electrons and the holes injected from the opposite electrodes of the device. The EL emission of the SPAn/PPV heterostructure LEDs can be tentatively interpreted by taking the dynamics of oppositely charged polarons generated by the injection of electrons and holes into consideration.

In the energy band scheme shown in Figure 34 schematically, the energy barrier for the electrons in the PPV was estimated to be 1.0 eV, whereas that for the holes in SPAn is 0.1 eV. Therefore, it is considered that the electrons in a PPV layer are sandwiched by the SPAn energy barrier, and holes in the SPAn layer by the PPV, separately. The SPAn/PPV multilayer LEDs seem to be of Type I' superlattice structure similar to an inorganic compound semiconductor. Accordingly, the confined electrons in the PPV layer lovalized at a higher energy state and since the state densities at higher energy are larger than those at lower energy, the recombination of carriers from the higher energy state is expected and should be large. That is, a small emission energy shift can be interpreted in terms of the confined carriers in the multilayer structure. Then the slight narrow emission spectrum can also be explained since the quantized energy level has a steplike state density. The energy shift was evaluated using the Krönig-Penny model for the localized electrons. As can be seen in Figure 36, experimental results exhibit a good fit to the solid line by the Krönig-Penny model which devices an energy-wave number vector relationship in a rectangular type potential profile by assuming $m^*=0.8m_e$, where m^* and m_e are effective electron mass and electron mass, respectively. Quantized energy levels are indicated schematically with a dashed line in Figure 37. However, at this stage, we cannot conclude that the phenomena that we observed in the EL diodes consisting of the SPAn/PPV multilayer structure arises from the same quantum size effect as PPV discussed in inorganic compound semiconductors. Because the nature of SPAn-PPV interface is not yet sufficiently clear and the quantized energy states may not exist due to the layer fluctuations and the imperfect crystallinity, and so on.

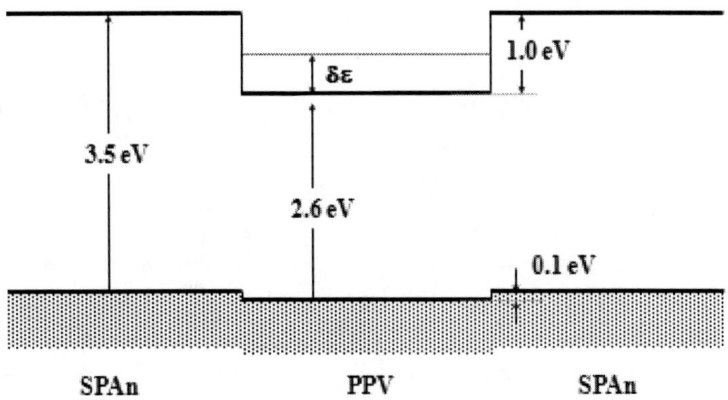

Figure 37. Simple quantized energy level diagram of SPAn/PPV multilayer structure.

Although it cannot be explained satisfactory at this stage that the EL emission of SPAn/PPV multilayer LED comes only from the PPV layer, we hypothesize as follows. The generation of polaron excitons is expected to take place very close to the SPAn-PPV heterojunction.

Accordingly, it is considered that the EL emission occurs when the holes tunnel through the heterojunction barrier and polaron excitons were formed into the PPV layer of SPAn/PPV hetrostructures. That is, SPAn carriers the roles of assisting the injection of holes from the ITO and the transport of the injected holes to the boundary of heterojunctions.

We observed novel optical effects in the LED device with the periodic multilayer structure of SPAn/PPV thin film.

Of particular interest are multilayer thin films in which layers of functionally different materials are used to control the injection, transport, and the recombination of electrons and holes in Schottky barrier based LEDs.

5.1.5. Summary

We have investigated the fundamental physical properties of SPAn/PPV multilayer thin film LED fabricated by self-assembly technique. The results of the experimental study can be summarized as follows.

1. The total thickness of the multilayer film assembly increases linearly with the number of deposited SPAn-PPV bilayers.
2. The injection properties of carriers in this device can be explained in terms of the Fower-Nordheim tunneling theory and the EL emission is due to the radiative recombination of polaron excitations formed by injection of electrons and holes.
3. The EL emission of device with SPAn/PPV monolayer structure appears greenish-yellow white that with multilayer appears bluish-green and there is a slight shift between the two spectra of these devices. It was assumed that the EL emission peak

shift of SPAn/PPV multilayer LED is due to the confinement effect of carriers in the superlattice structure consisting of SPAn/PPV multilayer system.

4. The simple electronic band structure of SPAn/PPV multilayer LED was deduced, with which the mechanism of the shift of EL emission spectra can be explained, was deduced.

5.2. Conductive Polypyrrole-Silica Glass Composite Films Prepared by Electrolysis Polymerization

5.2.1. Introduction

Conductive polymers with extensive π-electron delocalization are intensively studied as materials for use in electronics and optics [1-3]. Polypyrrole (PPy), whose molecular structure is shown in Figure1, is readily produced by electrochemical procedures and is attractive because of its good environmental stability and relatively high electrical conductivity. To date, most syntheses have focused on the use of an organic solvent system, with relatively little attention being paid to aqueous systems. However, it was found recently that PPy prepared electrochemically in aqueous solutions also possesses high electrical conductivity. Furthermore, improvements in processibility, mechanical properties, and atmospheric stability of conducting polymers have also been made by forming polymer-polymer composites [60, 61]. The electrochemically polymerized composites of PPy have attracted much attention due to their excellent stability and flexibility.

The sol-gel process offers a new approach to the synthesis of oxide materials. Starting from molecular precursors such as metal alkoxides, an oxide network is obtained via inorganic polymerization reaction [62]. Low-temperatureprocessing allows the synthesis of metastable oxide phases or even mixed organic-inorganic materials [63].Our laboratory has found that conductive PPy-silica glass composite films can be simply prepared by electrolysis polymerization of PPy simultaneously with the synthesis of silica glass by sol-gel processing.

In the work reported here, the electrochemical behavior of the exchange process for dopant anions in PPy-silica glass composite film was investigated by means of cyclic voltammetry. That is, the electrochemical doping-undoping process in PPy-silica glass composite films has been activated by repeated doping-undoping processes, similar to a learning process in neural networks. These learning phenomena are attributable to the increase of the diffusion coefficients of dopant anions caused by structural changes of the polymer layers during doping-undoping cycles. Application of this new type of PPy-silica glass composite film in an ionic sensor which responds to specific ions has also been proposed.

5.2.2. Experimental Procedure

Pyrrole was distilled and stored under a nitrogen atmosphere in the dark. Various electrolytes – tetra-*n*-butylammonium chloride (Bu_4NCl), tera-n-butylammonium tetrafluoroborate (Bu_4NBF_4), tetra-*n*-butylammonium hexafluorophosphate (Bu_4NPF_6), tetra-*n*-butylammonium perchlorate (Bu_4NClO_4), and tetra-*n*-butylammonium *p*-toluenesulfonate (Bu_4N*p*-TS) – were purified and dried before use. The conductive PPy-silica glass composite films were prepared by carrying out the electrolysis polymerization of PPy simultaneously with the synthesis of silica glass by sol-gel processing.

Cyclic voltammetric studies were performed using a Hokuto Denko Ltd. HA-303 potentiostat/galvanostat generator. Acetonitrile solutions containing various electrolytes were used under a nitrogen atmosphere in these studies. A concentration of 0.25 mol/l was used for the electrolyte solutions. An indium tion oxide (ITO)-coated conductive glass substrate working electrode with PPy-silica glass composite deposited on its surface, a platinum plate counter electrode, and an Ag reference electrode were used in a three-electrode cell system. All electrode potential values reported in this paper are referenced to Ag/Ag+. A potential sweep rate of 15 mV/s was used. The resulting films were then washed with water and acetonitrile to ensure removal of any residual monomer and electrolyte.

Elemental analyses were carried out using a Nihon Denshi Ltd. TN-422J X-ray microanalyzer.

5.2.3. Results and Discussion

5.2.3.1. PPy-Silica Glass Composite Film Preparation

The electrolysis polymerization was carried out in a one-component cell using an HA-303 potentiostat/galvanostat equipped with a Hokuto Denko Ltd. HF-201 coulombmeter. That is, the composite film was deposited onto an ITO-coated conductive glass substrate at a constant potential of 2.3 V versus Ag/Ag^+ in aqueous solution containing 1.0 mol/l pyrrole, 1.o mol/l sodium-*p*-toluenesulfonate (*p*-TSNa), 8 g tetraethyl orthosilicate [$Si(OC_2H_5)_4$], 10 ml ethanol, 0.2 g HCl (concentration 35 %), and 2 ml *N*, *N*-dimethylformamide (a crack-preventing agent). Deposition was carried out until a fixed charge had passed through the electrode. This film was heated at 150 °C in a vacuum of about 10^{-3} Torr for 10 h to obtain the PPy-silica glass composite film. This film contained some dopant (*p*-TS^-) and was blue. To compare the doping effect of PPy-silica glass composite film with that of PPy film, the latter was also prepared by electrochemical oxidation of the pyrrole monomer.

Figure 38 illustrates the morphology of PPy film and PPy-silica glass composite film. For PPy film the surface is covered with many wrinkles [64].

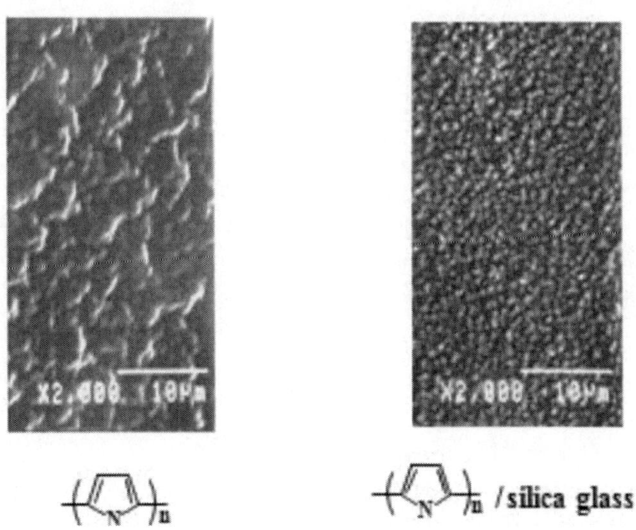

Figure 38. Scanning Electron Microscope (SEM) photos (left) PPy film and (right) PPy-silica glass composite film.

On the other hand, for PPy-silica glass composite film the surface is covered with many smaller protuberances and is relatively smooth.

It is well known that by the the silanol groups [$Si(OH)_4$] formed by the hydrolysis of $Si(OC_2H_5)_4$ adsorb various molecules and bind with them because of their high polarity, weak acidity, and high reactivity [65].

Accordingly, it is considered that pyrrole adsorbed on the silanol groups polymerizes electrochemically and the sol-gel process of silanol groups advances slowly.

Figure 39 shows the X-ray microanalysis spectra of PPy film (a) and PPy-silica glass composite film (b). Figure 39 (a) shows a single peak S with an energy of 2.31 keV, which indicates the existence of some dopant (p-TS^{-}).

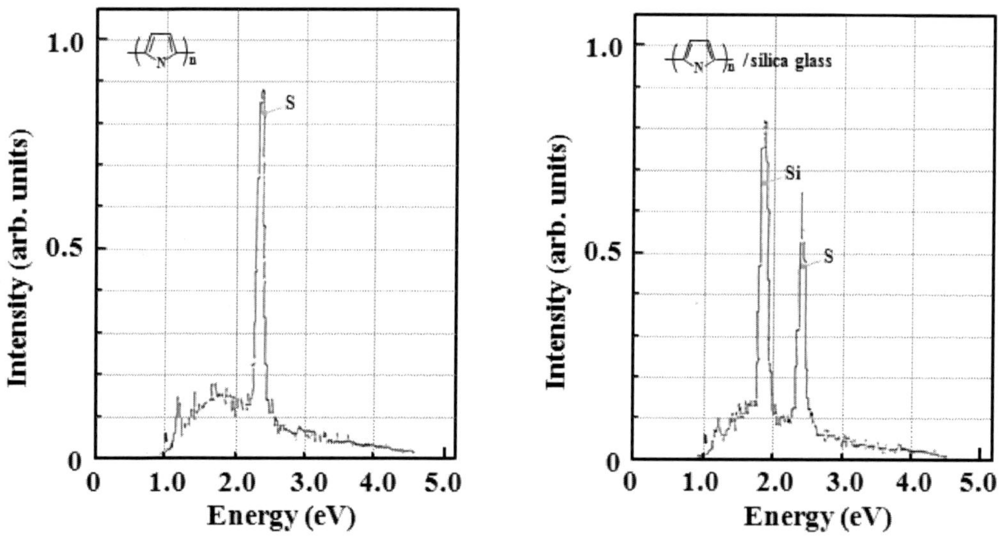

(a) (b)

Figure 39. X-ray microanalyzer spectra of (a) PPy film and (b) PPy-silica glass composite film.

Figure 39 (b) shows another peak Si appearing near 1.74 keV. The reaction process involves the following two steps. Reaction (1) and (2) occur simultaneously in the PPy-silica sol-gel densification process, resulting in the PPy-silica glass composite film.

5.2.3.2. Doping Properties of PPy-Silica Glass Composite Film

A special feature of conductive polymers is doping effects. Figure 40 shows typical cyclic voltammograms of PPy film (a) and PPy-silica glass composite film (b) doped with p-TS^{-}, which were taken in acetonitrile solutions containing 0.25 mol/l NBF_4 electrolyte anions.

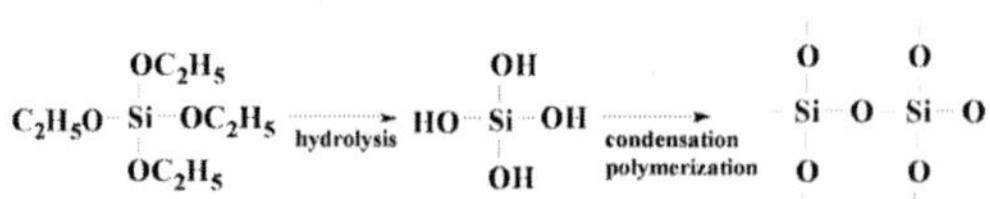

(1)

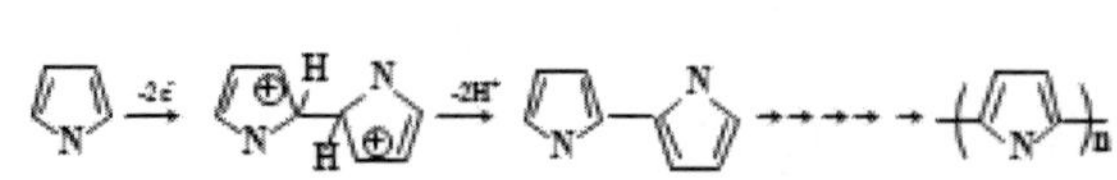

(2)

Both films could be cycled repeatedly between oxidized and neutral states with no appreciable decomposition of materials.

The redox reaction was accompanied by reversible color change. Cyclic voltammograms of PPy-silica glass composite films are quite different from the typical voltammograms of PPy films. That is, the cyclic voltammograms of PPy films became saturated after repeating redox procedures, as shown in Figure 40 (a).

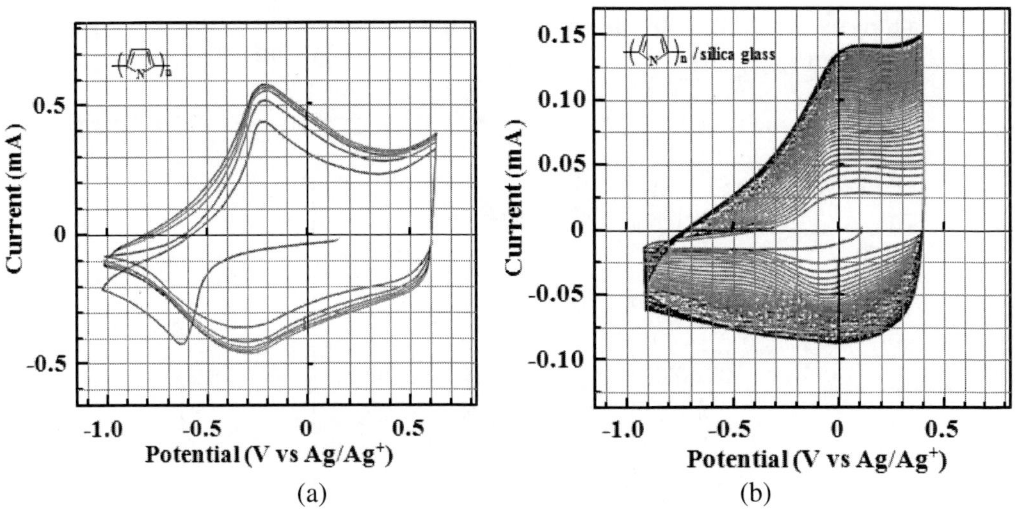

Figure 40. Cyclic voltammograms of (a) PPy film and (b) PPy-silica glass composite film.

However, the cyclic voltammograms of PPy-silica glass composite films gradually became larger as the number of measurements increased as shown in Figure 40 (b). The redox process was activated by repeated electrochemical doping-undoping cycles and saturated with currents after multiple repeated redox procedures.

This activation behavior was previously unknown. Plotting of cyclic voltammograms of PPy-silica glass composite films between -0.9 and 0.4 V against Ag/Ag^+ revealed a broad oxidation wave with a current maximum at about 0.1 V, and a reversible reduction wave reaching a maximum at about 0 V. It should be noted that the doping threshold of BF_4^- dopant is located at about -0.7 V against the Ag reference electrode. Although the current levels increased with the number of cycles, presumably owing to the continuous redox reactions of the PPy-silica glass composite film, the coulombic recovery was greater than 90 %, indicative of a high degree of electrochemical reversibility.

Figure 41 shows temperature dependences of the electrical conductivities in PPy film and PPy-silica glass composite film. For both films the electrical conductivities show the thermal activation behavior and the activation energy was evaluated to be about 0.01 eV.

Accordingly, it is considered that the observed electrical conductivity should be restricted by defects or the interchain carrier transport of PPy films and PPy-silica glass composite films in the conduction mechanism.

Figure 42 plots the evolution of the optical absorption spectra of a PPy-silica glass composite film taken *in situ* during electrochemical BF_4^- doping as a function of the doping potential in Bu_4NBF_4/acetonitrile ranging from the nearly undoped state to the oxidized state. The absorption spectrum of PPy-silica glass composite film showed only a single peak in the neutral state due to interband excitation at about 3.2 eV. The bandgap of the PPy-silica glass composite film evaluated from a plot of $(\alpha h\nu)^2$ vs $h\nu$ for the absorption edge of the interband absorption peak was about 2.7 eV, in which α and $h\nu$ are the absorption coefficient and photon energy, respectively. As the doping proceeded, the interband absorption peak was suppressed, but two new absorption peaks appeared simultaneously, at around 1.0 and 2.3 eV, respectively. As mentioned above, the electrical and optical properties of PPy-silica glass composite films prepared by this method were nearly similar to those of PPy films prepared by the electrolysis polymerization method. However, novel electrochemical properties in PPy-silica glass composite films were observed, as shown in Figure 40.

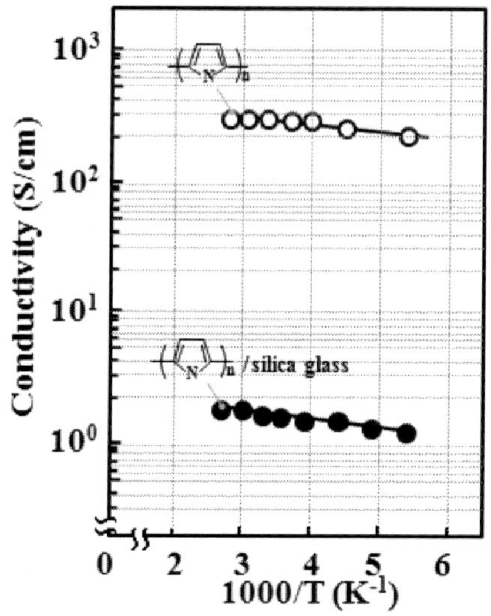

Figure 41. Temperature dependence of electrical conductivity in PPy film andPPy-silica glass composite film.

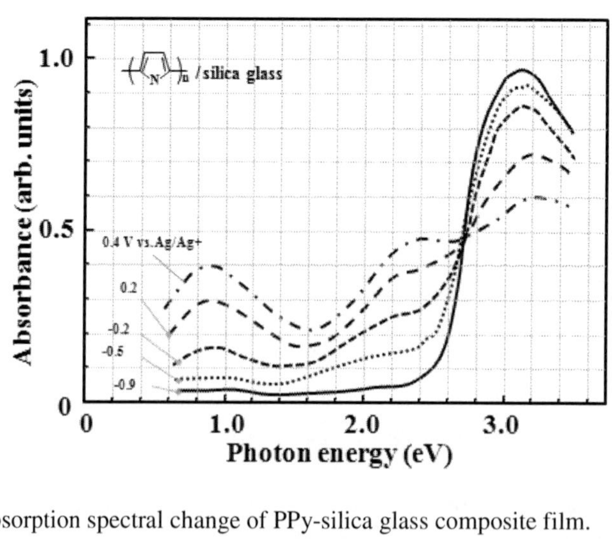

Figure 42. Optical absorption spectral change of PPy-silica glass composite film.

It has been reported that the electrochemical properties of PPy films depend on the kind of dopant and electrolyte, but not on the size of the electrolyte anion [66]. In this experiment, it can be speculated that p-TS^- in PPy-silica glass composite films was gradually exchanged with the BF_4^- anions during the electrochemical operation. To verify this speculation, we examined the electrochemical behavior of PPy-silica glass composite films in various electrolyte solutions.

5.2.3.3. Electrochemical Behavior of PPy-Silica Glass Composite Films in Various Electrolyte Solutions

Figure 43 shows several cyclic voltammograms of PPy-silica glass composite films doped with BF_4^- in acetonitrile solution containing various electrolytes. In all cases, the oxidation-reduction current values gradually increased with repeated potential sweeping. Figure 44 shows the relation between the number of cycles up to saturation of the oxidation-reduction current and δ_r [ionic radius of dopant (BF_4^-) – ionic radius of electrolyte anion].

For PPy films the number of cycles up to saturation of the oxidation-reduction current does not depend on δ_r and the redox reaction saturated several times.

However, for PPy-silica glass composite films it should be noted that when the size of electrolyte anions becomes smaller than that of the dopant (BF_4^-), the number of cycles up to saturation of the oxidation-reduction current increases. The ionic radii of p-TS^-, ClO_4^-, PF_6^-, BF_4^-, and Cl^- are about 9.8, 6.2, 5.9, 5.3, and 3.3 Å, respectively.

Therefore, it is considered that the size of electrolyte anions plays an important role in causing this activation behavior. This electrochemical behavior is characteristic of PPy-silica glasscomposite films and cannot be observed in PPy films. Electrolyte anions transfer in the redox process of PPy-silica glass composite films based on cyclic voltammetry measurements, and the oxidation-reduction current values are in proportion to the diffusion rate of the electrolyte anions. This diffusion rate depends on the radius of the electrolyte anion and is in proportion to the osmotic swelling pressure of the PPy layers.

From these viewpoints, this activation phenomenon is thought to be induced by the electrochemical doping-undoping reaction. The origin of the increasing tendency of the doping-undoping current in the voltammetric response is investigated by potential step anions. chronoamperometry, a well-known method of determining diffusion coefficients of the dopant ion within the PPy layer of PPy-silica glass composite films during electrochemical processes. The observed undoping current is nearly inversely proportional to the square root of the time t, suggesting that the Cottrell equation is applicable to this case. The Cottrell equation, which is known to be usable for control of kinetics with linear semi-infinite diffusion phenomena [67], is

$$i=\pi^{-1/2}SCFD^{1/2}t^{-1/2}$$

where S, C, F, and D are the apparent electrode surface area (cm^2), the concentration of electrolyte anion (mol/liter), the Faraday constant (9.6485×10^4 C/mol), and the apparent diffusion coefficient of electrolyte anion (cm^2/s). D was evaluated from the gradient of the Cottrell plot, that is, i versus $t^{-1/2}$ plot, using the C value which was determined from the undoping electricity measured in each constant-potential undoping process. As shown in Figure45, the D value increases with the number of cycles, tending to saturation for small electrolyte The saturated D of PPy-silica glass composite film increased with decreasing size of electrolyte anion. Furthermore, the D of PPy film was evaluated to be about 2 to 4×10^{-9} cm^2/s, regardless of the size of the electrolyte anions and the number of cycles.

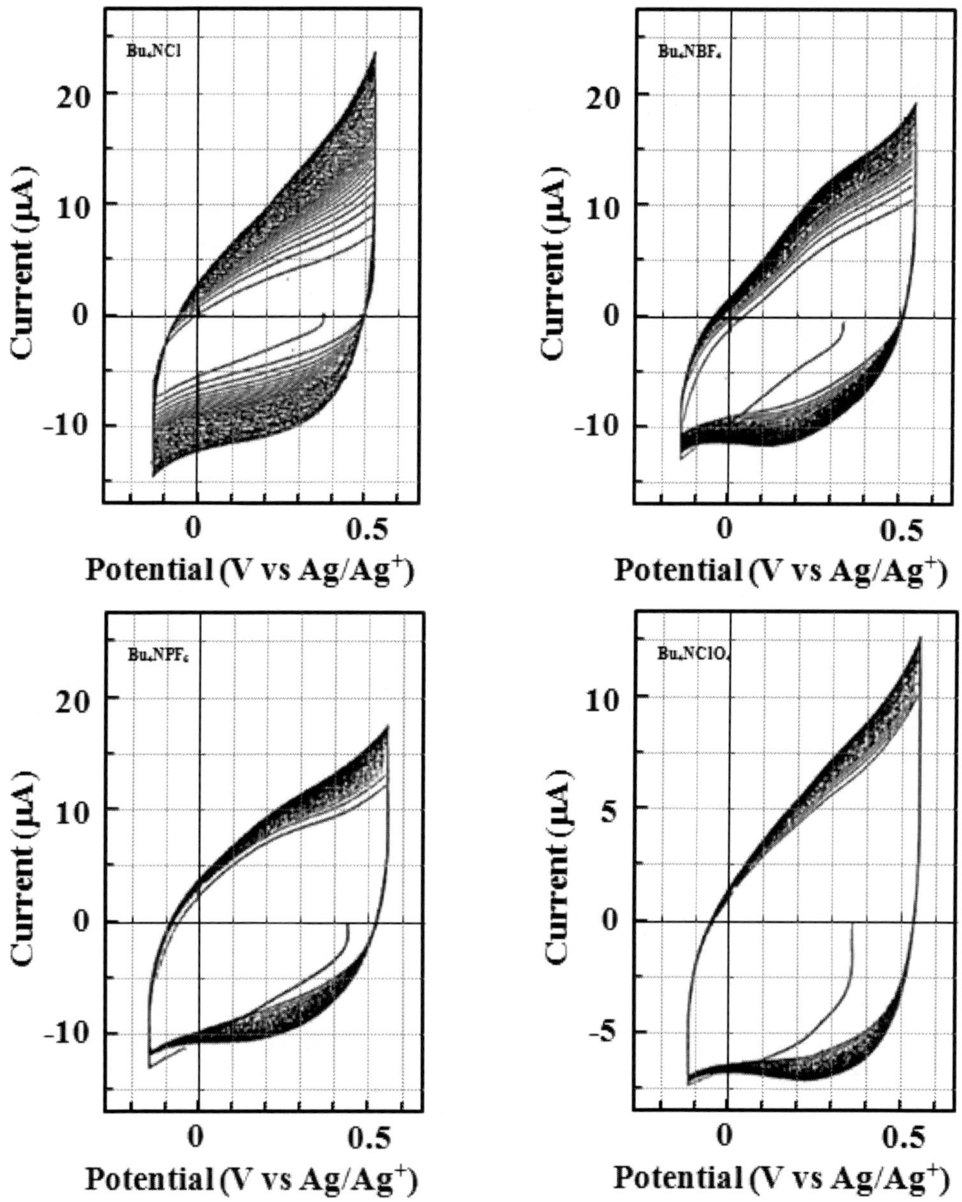

Figure 43. Cyclic voltammograms of PPy-silica glass composite film in acetonitrile solution containing various electrolytes.

Although the origin of this anomalous electrochemical behavior of PPy-silica glass composite film has not been conclusively determined, the following speculation has been made. In the redox process at the beginning of the potential sweep, the dpant anion (BF_4^-) contained in the PPy-silica glass composite film play an important role.

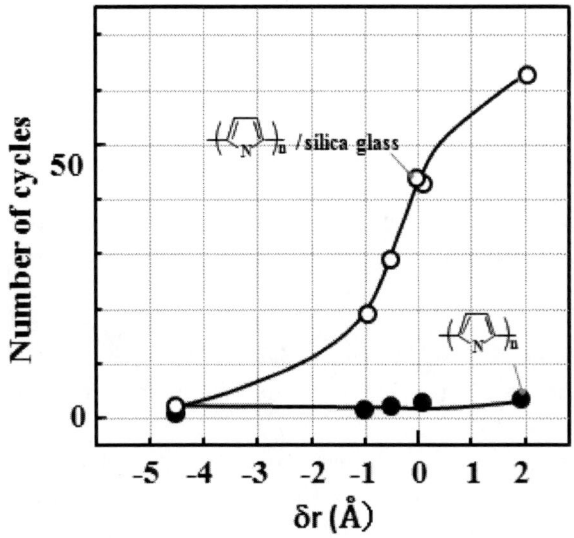

Figure 44. Relation between number of cycles of the redox process and δr in PPy-silica glass composite film. Here, δr shows (ionic radius of BF_4^- dopant) – (ionic radius of electrolyte anion).

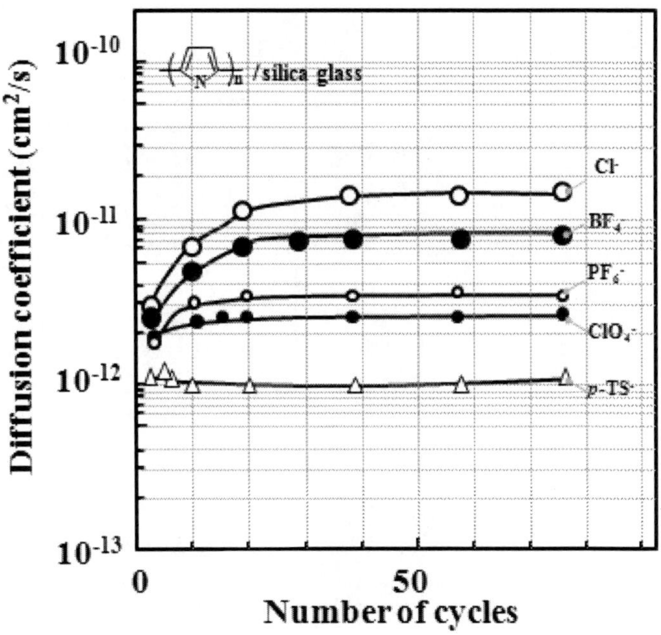

Figure 45. Relation between diffusion coefficient and number of cycles of the redox process in PPy-silica glass composite film.

However, it is assumed that the electrolyte anions in the acetonitrile solutions gradually participate in the redox reaction with repeated potential sweeps. It is then expected that the space which is occupied by dopant anions in PPy-silica glass composite film and the size of the electrolyte anions have direct effects on the diffusion rate. That is, it is concluded that the activation behavior as shown in Figure 43 occurs because the inactive dopant anions BF_4^- in the PPy-silica glass composite film are gradually exchanged with the active electrolyte anions in the acetonitrile solutions as the potential sweep is repeated.

Furthermore, as shown in Figure 46, there was a marked change in the surface morphology of the PPy-silica glass composite films before and after repeated potential sweeps, which might be due to the electrolyte anion exchange process occurring in PPy-silica glass composite film doped BF_4^- in acetonitrile solution.

In contrast, some dopant in the PPy films could not be exchanged easily with electrolyte anions in the acetonitrile solution by repeated potential sweeps because PPy films shrink in acetonitrile solutions.

5.2.3.4. Functional Application of PPy-Silica Glass Composite

The application of conductive polymers to electronic devices has been extensively studied [1, 2].

Rechargeable batteries and capacitors have been developed for practical use. In particular, the electrochemical properties of PPy-silica glass composite films clarified in this study may be interpreted as a learning effect and can be applied in principle to an ionic sensor, since the number of cycles to saturation of the oxidation-reduction current depends on the size of the electrolyte anions in the acetonitrile solutions.

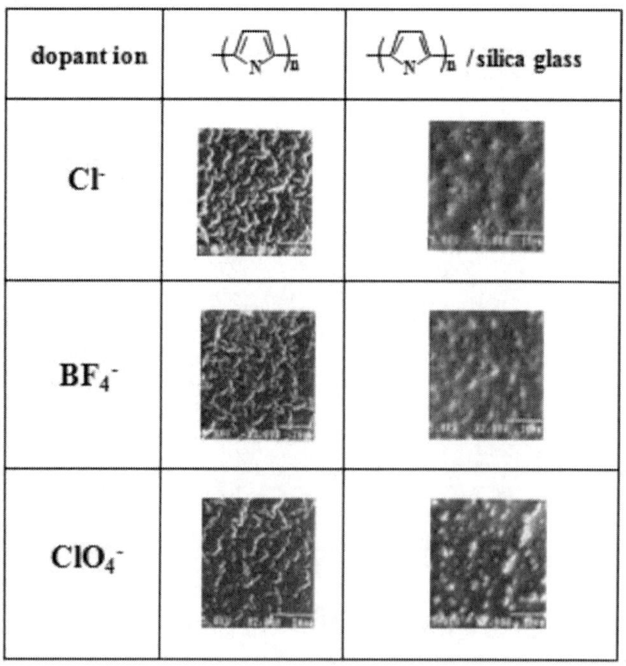

Figure 46. SEM photos of PPy film and PPy-silica glass composite film after repetition of redox process.

To verify this aspect, we examined the electrochemical behavior of PPy-silica glass composite films doped with BF4- in acetonitrile solutions containing electrolyte anions with different sizes (Cl^-, PF_6^-, and ClO_4^-) as shown in Figure 47.

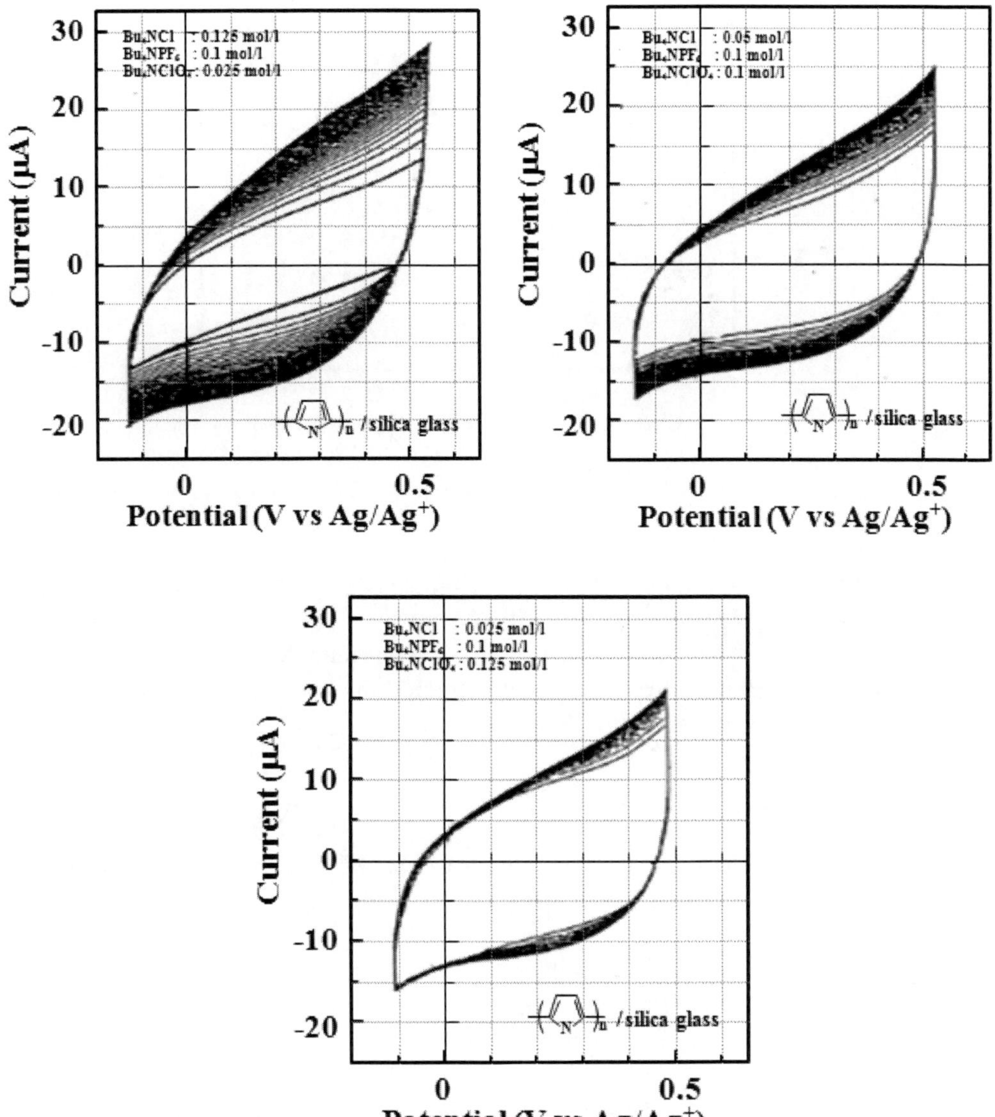

Figure 47. Cyclic voltammograms of PPy-silica glass composite film in acetonitrile solutions containing three linds of electrolyte.

Figure 48 shows the relationship between the number of cycles to saturation of the oxidation-reduction current and concentration of Bu4NCl in acetonitrile solution, which was evaluated from Figure 47.

The number of cycles to saturation of the oxidation-reduction current was 60 to 65, 28 to 32, and 19 to 20 for Cl^-, PF_6^-, and ClO_4^- anions, respectively. The number of cycles increased drastically in a stepwise manner with increasing concentration of Bu_4NCl.

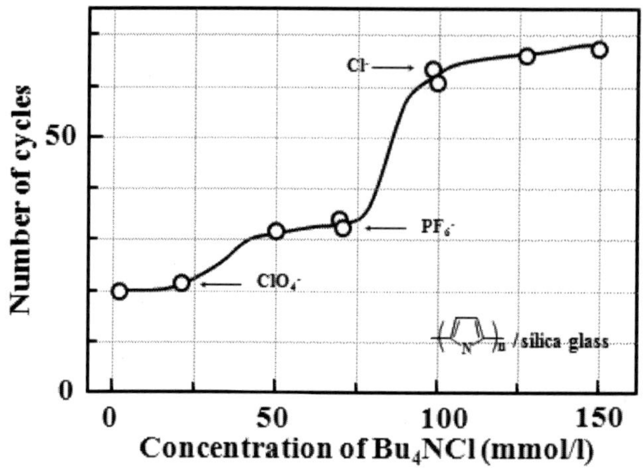

Figure 48. Relation between number of cycles of redox process and concentration of Bu_4NCl in PPy-silica glass composite film.

That is, at Bu_4NCl concentrations lower than 0.025 mol/l, the details (sensitivity, lifetime, the effect of solvent, and so on) are under way composite films recognize ClO_4^- anions, whereas the Cl^- anions are recognized at high Bu_4NCl concentrations above 0.1 mol/l. The number of cycles in the 0.05 mol/l to 0.075 mol/l region is due to the presence of PF_6^- anions. Further investigations aimed at understanding these details (sensitivity, lifetime, the effect of solvent, and so on) are under way.

5.2.4. Summary

The present experimental results on *in situ* measurements of PPy-silica glass composite films during electrochemical doping are summarized as follows.

1. In the cyclic voltammogram of PPy-silica glass composite films, the redox peak due to *p*-type doping and undoping appeared clearly and the redox process was activated by repeated electrochemical doping-undoping cycles.
2. PPy-silica glass composite film doped with some dopant could be exchanged relatively easily with electrolyte anions in the acetonitrile solution by repeated potential sweeps. The number of cycles to saturation of the oxidation-reduction current increased as the size of electrolyte anions decreased.
3. Application of this new type of PPy-silica glass composite film to an ionic sensor which responds to specific ions was also proposed.

5.3. Preparation and Functions of Conductive Polymer/Insulative Polymer Composite Films Using Molecular Self-Assembly

5.3.1. Introduction

Conductive polymers with an extensive π-electron conjugated system in their main chains are considered to be organic semiconductors with relatively small bandgaps and are being intensively studied as new materials in the field of physical chemistry.

To date, various applications of conductive polymers, such as in semiconductive devices and optoelectronic devices, have been proposed.

Unfortunately, except for rechargeable batteries and capacitors, these applications have not been developed to the stage of practical use [1, 2].

Chemical etching of polymer materials can drastically modify the physical and chemical properties of the polymer surface. Etching of semicrystalline polymers such as polyethylene has also been used as one way to elucidate their microstructure [68, 69].

In particular, in the case of the introduction of ionic groups such as sulfonic acid and sulfonate, highly interesting applications of polymers such as ion-exchange resins, polyelectrolytes, and biomaterials can be expected.

From the above mentioned point of view, we have tried to prepare sulfonated low-density polyethylene (SPE) films by introducing sulfonic groups on the surface of low-density polyethylene and investigated the sulfonation process. Furthermore, we have prepared conductive polymer/insulative polymer composite films based on conductive polymers grown on the surface of obtained SPE using the molecular self-assembly process [54, 55] and studied their physical properties and functional applications.

In the present work, we analyze, in detail, the characteristics of sulfonation layers in order to gain clear understanding of the physical properties of composite films and in the case of using PPy as a conductive polymer, report the growth process of PPy on the surface of SPE and the physical properties and functional applications of PPy/SPE composite films.

5.3.2. Experimental Procedure

5.3.2.1. Sulfonation of Low-Density Polyethylene

The samples used in this experiment are low-density polyethylene (PE, Sholex, density 0.923 g/cm^3) of thickness 15 - 150 μm, normally free from additives such as antistatic agents and antioxidants.

For sulfonation of PE, a 1000 ml glass container was evacuated to approximately 1 Pa and then 5 ml of fuming sulfuric acid (30 wt% of SO_3) was added in it. PE films were placed horizontally on the liquid 2 cm from the surface, and exposed to an atmosphere of fuming sulfuric acid for suitable periods of time. The samples were adequately rinsed in distilled water and dried in vacuum overnight. The directing SPE was obtained in the manner described above.

The degree of sulfonation was determined as the ratio of the weight increase [ΔW=W'-W (g)] to area for each film [S (m^2)], ΔW/S (g/cm^2), where W and W' denote PE film weight before and after sulfonation, respectively. For the infrared study, a Nihon Bunko A-100 spectrometer was used.

5.3.2.2. Preparation of PPy Electrolyte

Most syntheses have focused on the use of an organic solvent system, while, relatively less attention has been paid to aqueous systems. However, it was recently revealed that PPy can be synthesized chemically using ferric chloride ($FeCl_3$) as oxidant [1]. The dipping solution of the *in situ*-polymerized, *p*-type doped conductive polymer consisted of 0.005 mol/l $FeCl_3$ in deionized water, 0.02 mol/l sodium *p*-toluenesulfonate (*p*-TSNa), and 0.02 mol/l pyrrole monomer. The acidity of the solution was then adjusted with HCl to give a measured pH of 1. The dipping solution was aged for 30 min and filtered through a 0.5 μm filter prior to use.

5.3.2.3. Preparation of PPy Ultrathin Films by Molecular Self-Assembly

The molecular self-assembly process involves dipping a substrate into a dilute aqueous solution of a positively charged polymer (polycation) and then dipping it into a dilute solution containing a negatively charged polymer (polyanion) and is one of the fabrication techniques of monolayer and multilayer thin films. That is, the layer-by-layer molecular self-assembly technique is based on the spontaneous adsorption of a charged polymer onto a surface from a dilute polymer solution. This process is an extremely simple and effective way to produce thin film structures that are controllable at the molecular level. Each deposited layer depends on parameters such as solution concentration, pH, and ionic strength. The preparation of high-quality thin films with excellent uniformity and adhesion can be accomplished with a variety of different substrates, including substrates that are hydrophobic, hydrophilic as well as positively and negatively charged [56].On the other hand, it is well known that PAn and PPy thin films can be formed on the inner wall of the reaction vessel by means of a chemical polymerization process of aniline and pyrrole (*in situ* polymerization).In these thin films, the delocalized positive charge (polarons and/or bipolarons) of *in situ*-polymerized conductive polymer such as PPy and PAn can be used to fabricate thin films using a molecular self-assembly process. That is, the basis of this technique is the electrostatic attraction developed between the ionized functional groups of a polyanion and the delocalized positive charges along the backbone of a *p*-type doped conductive polymer.

Accordingly, since SPE possesses a negatively charged surface due to the introduction of sulfonic groups, as described above, PPy was deposited onto the SPE as a substrate by a single-dip, free-growth process in which the SPE was simply immersed in PPy dipping solution for a specific period of time, rinsed, and dried.

Figure 49 shows a schematic diagram of the PPy monolayer *in situ* polymerized onto the SPE substrate, i.e., a PPy/SPE composite film. The deposition of PPy was monitored using a UV-visible absorption spectrophotometer and the thickness of PPy film was evaluated using a multiple interference microscope and the absorbance-thickness calibration curve, which was prepared using a Hitachi U-3410 spectrophotometer. The electrical conductivity of PPy was measured using a standard four-probe technique in an evacuated chamber at room temperature.

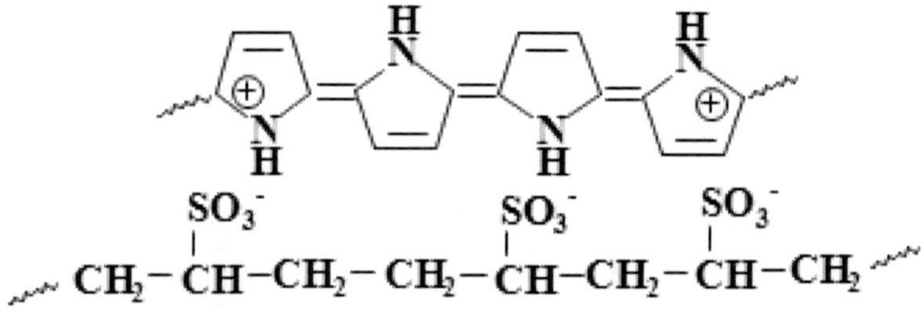

Figure 49. Schematic illustration of PPy thin film fabricated via the molecular self-assembly process on a sulfonated low-density polyethylene (SPE) substrate.

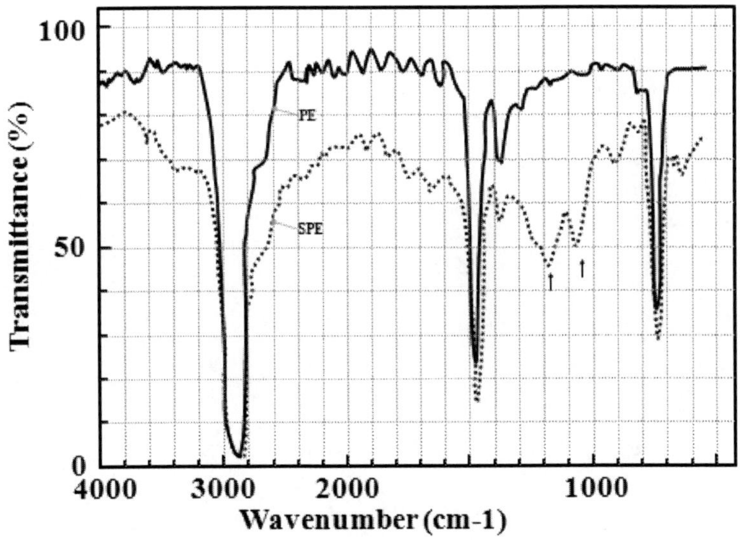

Figure 50. Infrared spectra of low-density polyethylene (PE) and SPE.

5.3.3. Results and Discussion

5.3.3.1. Properties of SPE

Figure 50 shows the infrared spectra of a PE film of thickness 40 μm before and after sulfonation. The sulfonation time T_s is 1 h. As is obvious from this figure, besides the dominant bands from PE (3000 - 2800, 1490 - 1300, and 730 - 720 cm^{-1}), the SPE show corresponding bands, which are the expected ones for the sulfonic group at about 1200 and 1040 cm^{-1} assigned, respectively, to the asymmetrical and symmetrical stretching mode of the $–SO_3H$ (SO_3^-).The absorption around 600 cm^{-1} is related to the C-S stretching vibration mode and water-sensitive bands appear at about 3400, 1710 - 1600, and 900 cm^{-1}. Simply by exposing PE to an atmosphere of fuming sulfuric acid, SO_3^- groups were introduced to the PE and the negatively charged substrate (SPE) can be formed easily.

Figure 51 shows the relative weight variation of SPE films with varying thickness (40, 70, and 150 mm) as a function of their reciprocal thickness values for different T_s values (0.5, 1.0, and 2.0 h). Good linear relationships for all values of T_s were obtained.

The slopes become steeper as T_s is increased and are related to the mean sulfonation degree. That is, if T_s is constant, the sulfonation degree is unified without regard to PE film thickness and the good linearly obtained shows the result of careful operation under similar experimental conditions, such as the same height for the films and the same reagent volume, among others, during the sulfonation periods.

The observed mean ΔW/S values for the SPE of 40 μm thickness are plotted against T_s in Figure 52. In the case of the shorter T_s, a linear increase of ΔW/S with T_s was observed, the slope being about 0.4 g/m^2h. In the time zone of T_s=2 - 3 h, ΔW/S shifted by about 2 g/m^2 upwards and has a discontinuous form.

In the case of longer T_s, the ΔW/S values also fit on a line with similar slope (0.4 g/m^2h) and a linear increase of ΔW/S is observed. At this time, we do not have a satisfactory explanation for this discontinuous plot but it is considered that a percolation threshold for the sulfonation degree could be present in the sulfonation process of PE films. That is, we speculate that the sulfonation of PE with T_s progresses linearly and after the surface of PE was sulfonated, the progress of sulfonation into PE bulk occurs from around the discontinuity point.

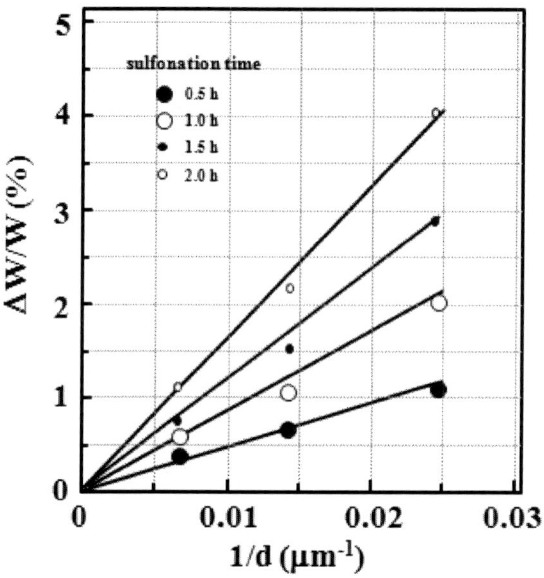

Figure 51. Relative weight variation of the samples after sulfonation for a given time Ts versus the reciprocal value of the thickness of the PE films.

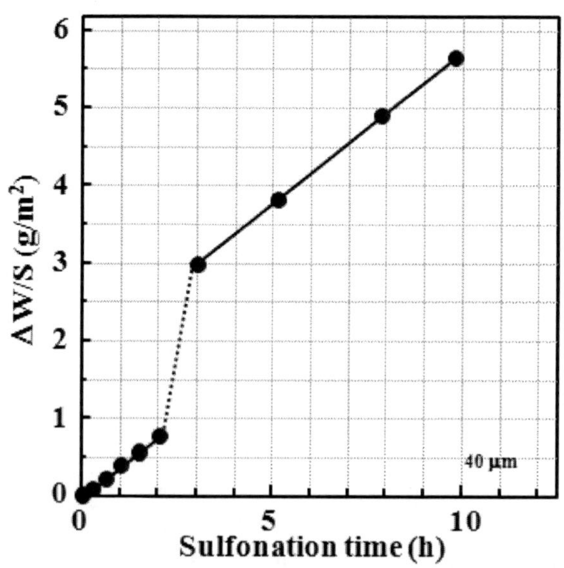

Figure 52. Degree of surface sulfonation of PE films as a function of sulfonation time Ts.

5.3.3.2. Preparation of PPy/SPE Composite Films

We have already reported that the fabrication of conductive polymer films using the molecular self-assembly process is considerably affected by the nature of the substrate. That is, homogeneous conductive polymer thin films with an excellent adhesion can be deposited and there is a linear correspondence between dipping time and the amount of PPy deposited onto the negatively charged substrate.

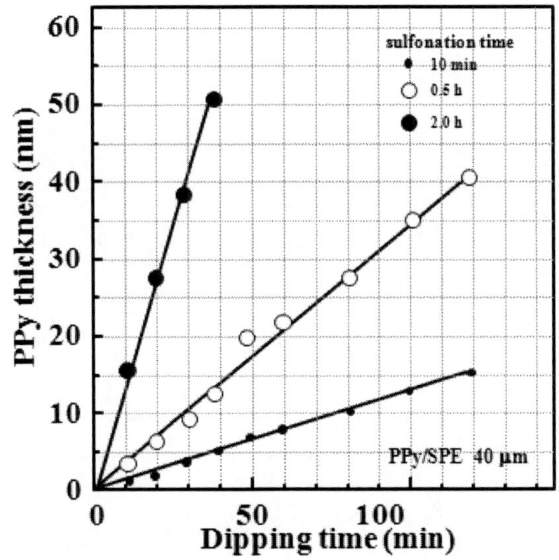

Figure 53. Relation between PPy film thickness and dipping time of SPE with various Ts.

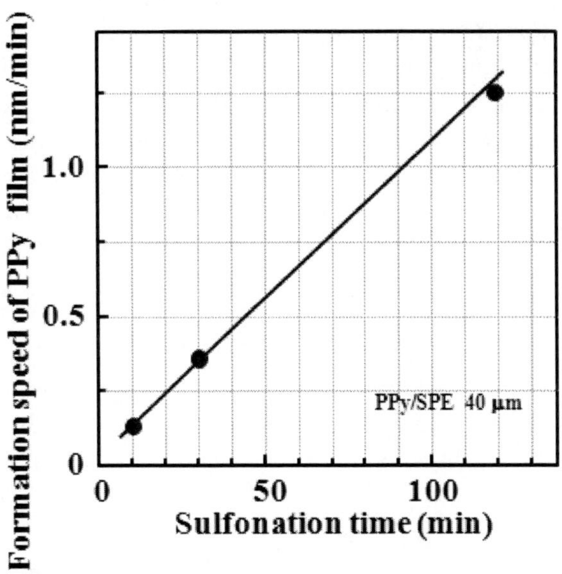

Figure 54. Relation between the formation speed of PPy film and Ts.

Figure 53 shows the relationship between PPy film thickness and dipping time of SPE (40 μm thickness) for various T_s values. PPy thickness increases linearly with the dipping time for each T_s value and as T_s is increased, i.e., as the amount of sulfonationm per unit area increases, the formation speed of PPy quickens and PPy thickness also increases at the same dipping time.Figure 54 shows the relationship between formation speed of the PPy film, evaluated from the slope of the straight line in Figure53, and T_s. It is suggested that the amount of PPy deposited per unit time onto the SPE is constant for the same T_s, because the formation speed of PPy increases almost in proportion to T_s. Figure 55 shows the relationship between PPy film thickness and T_s of SPE for constant dipping time.

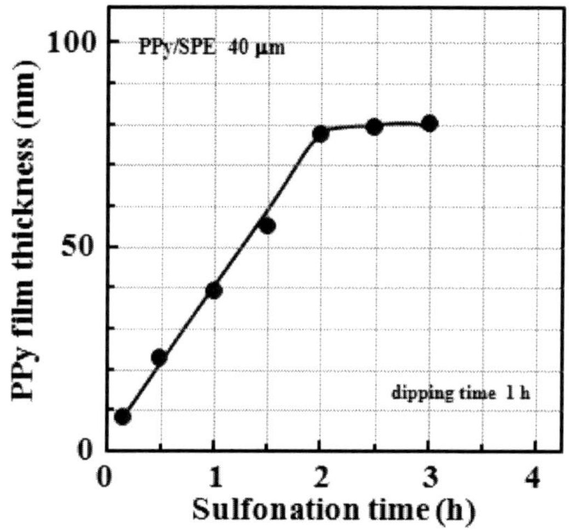

Figure 55. Relation between PPy film thickness and Ts of SPE with constant dipping time.

PPy film thickness exhibits an almost linear increase with increasing T_s, but for T_s beyond 2 h the thickness is saturated at about 80 nm and is subsequently independent of T_s.As shown in Figure 52, in the case of relatively short T_s, since the weight increase per unit area (ΔW/S) based on sulfonation is constant (0.4 g/m^2), the sulfonation degree also increases with increasing T_s and the SO_3^- groups are introduced uniformly onto the PE surface. As a result it is satisfactory to consider that the polymerization of PPy proceeds faster with increasing T_s and that the PPy film thickness also increases linearly. For T_s greater than 2 h, it is considered that sulfonation advances into the PE bulk, the discontinuous point was observed as shown in Figure 52.

Figure 56 shows the relationship between the conductivity of PPy and dipping time of SPE with T_s=0.5 h. Although the surface conductivity of SPE is about 3×10^{-15} S, the conductivity of the PPy film increases drastically as the dipping time proceeds and from a dipping time of about 10 min, shows a constant value of about 20 S/cm. This conductivity drastically changes with the composition of the dipping solution. That is, the conductivity of PPy increases with the increase in the composite concentration in the dipping solution, such as the concentration of $FeCl_3$, *p*-TSNa, pyrrole, and so on. For example, as shown in Figure56, the conductivity of the PPy film fabricated from a dipping solution with 0.05 mol/l $FeCl_3$ is about 150 S/cm during a relatively short dipping time. From these experimental results, it is suggested that the practical application of PPy/SPE composite films will find practical use as semitransparent electrodes in all kinds of organic thin-film devices.

As mentioned above, it is possible to successfully fabricate PPy monolayer films on the surface of SPE using the molecular self-assembly process and PPy/SPE composite films with high conductivity. When we use these composite films practically, it is necessary to clearly understand the forces of adhesion between SPE and PPy.

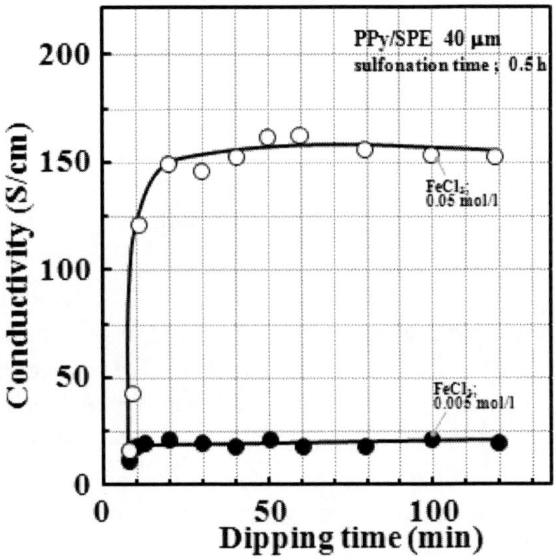

Figure 56. Relation between the conductivity of PPy and dipping time of SPE with Ts=0.5 h.

The contact technique and the direct method have been proposed for measuring adhesion condition, but these are not yet well-established techniques. In this experiment, we used the contact technique as a way to obtain qualitative data related to the adhesion of PPy/SPE composite films.

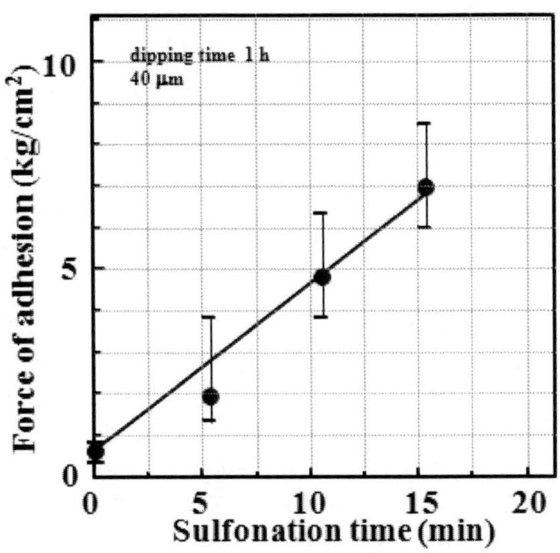

Figure 57. Relation between the forces of adhesion of PPy film and Ts.

Figure 57 shows the relationship between the forces of adhesion of PPy film and T_s. The thickness of the SPE film is 40 μm and the dipping time used to fabricate the PPy film on the surface of SPE is 2 h. The force of adhesion of PPy increases linearly with increasing Ts and is excellent in comparison with that of PPy fabricated on the surface of nonsulfonated PE. That is, PE is a nonpolar polymeric material and lacks forces of adhesion, but the fabrication technique of thin films using the molecular self-assembly process based on the electrostatic attraction between negatively charged groups and the PPy layer is highly useful for the formation of multilayer heterostructure composite films.

5.3.3.3. Functional Applications of PPy/SPE Composite Films

Since PPy/SPE composite films have relatively high electrical conductivity and excellent forces of adhesion between PPy film and SPE film, various applications such as antistatic, interferential materials, and transparent electrode materials have been suggested.

On the other hand, the doping effect is one of the most important and characteristic properties in conductive polymers and especially, in the case of one-dimensional conductive polymers such as PPy, this effect is marked and results in drastic changes of the nature of the material [1]. The doping-induced changes in conductive polymers include changes of electron states, optical, and electrical properties. They also include geometrical and mechanical parameters, as the dimensions of the polymer change as doping occurs [70]. These effects are due partly to doping-induced changes of polymer chain geometry, but also to steric effects from dopant insertion and extraction.

Both sides of PE were sulfonated and PPy/SPE composite films with PPy films were fabricated on both the surfaces of SPE. The composite films were formed into a strip of fancy paper shape (length: 1 cm, width: 1 mm, thickness of SPE: 30 mm, thickness of PPy layer: 80 mm×2) and a voltage (0.5 V) was applied between the PPy films on both the surfaces of SPE. As a result, we observed bending of the top of the composite film slightly bent to the negative polarity side. This is equivalent to volume changes of PPy layers based on electrochemical doping and dedoping as described above and the bending phenomena of composite film are due to expansion and/or shrinking of PPy layers. Accordingly, the use of these devices as smart materials such as a bimorph-type actuators is suggested. However, it is considered that because the PPy layer is extremely thin compared with the SPE layer, volume changes of PPy layers were not reflected sufficiently and the displacement became smaller.

As mentioned in Figure 55, in the fabrication of PPy thin films using the molecular self-assembly process, the thickness of the PPy films is saturated at about 80 nm, even if the dipping time is increased. In order to investigate the motion mechanism of actuators using PPy/SPE composite films, it is necessary to fabricate thick PPy layers. So, PPy films were prepared electrochemically on the PPy layers of composite films and actuator devices were fabricated. The electrolysis polymerization was carried out in a one-component cell using an HA-303 potential/galvanostat equipped with a Hokuto Denko Ltd. HF-201 coulombmeter. A PPy/SPE composite-film working electrode, a platinum plate counter electrode, and an Ag reference electrode were used in a three-electrode cell system. Deionized water containing 1 mol/l *p*-TSNa and 1 mol/l pyrrole was used as an electrolyte in this study. This electrochemical cell was set up in a constant temperature bath and a constant current of 0.14 mA was applied between the working electrode and the counter electrode. The thickness of the PPy layer fabricated electrochemically on the surface of the PPy/SPE composite film was about 40 μm.

Figure 58 shows photographs of the motion of an actuator using a PPy/SPE composite film observed in an aqueous solution containing 1 mol/l *p*-TSNa under ±0.5 V DC voltage. The PPy/SPE composite film bends to the negative side (reduction side) as a fulcrum the surface of electrolyte and the bending direction can be changed due to the polarity reversal of the applied voltage. If the applied voltage was removed under some condition, the state is stored and the PPy/SPE composite film possesses memory effects. The driving cycles of this device was reversible for about 100 cycles without any indication of degradation. The displacement and response speed of the end of the device considerably depends on the SPE thickness. For example, in the case of the device fabricated from 15-μm-thick SPE, the response speed and maximum displacement of the end of the device were about 0.03 mm/s and about 10 mm, respectively.

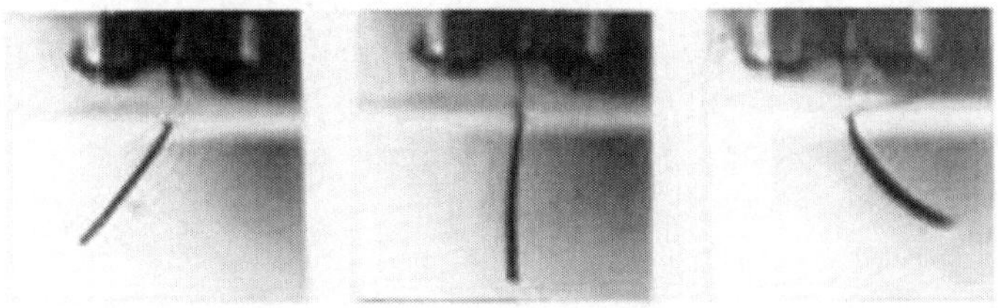

Figure 58. Photographs of bend of the PPy/SPE composite films.

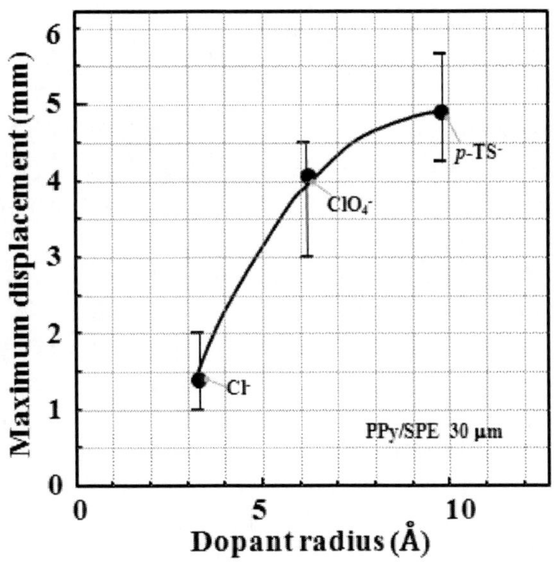

Figure 59. Relation between maximum displacement of PPy/SPE composite films and the size of the dopant.

The mechanisms of electrochemical swelling and shrinking of PPy/SPE composite film cannot be satisfactorily explained. However, as a tentative explanations, we suggest, for example, (1) doping and dedoping of bulky ions into the polymer network, (2) electrostatic repulsion between polycations induced by oxidation for the swelling, and (3) conformational change of polymer structure by injection and ejection of electrons (bonding angle, bonding distance, and so on) [70, 71].

Figure 59 shows the relationship between maximum displacement of the end of PPy/SPE composite films and the size of the dopant. The ionic radii of Cl^-, ClO_4^-, and *p*-TS^- are about 3.3, 6.2, and 9.8 Å, respectively.

Since the displacement becomes larger with increasing ionic radius, it is concluded that the primary mechanism of the electrochemical swelling and shrinking is (1) mentioned above.

However, in conductive polymer polycations are formed upon doping, it is believed that the mechanisms described in (2) and (3) are responsible for electrochemical swelling and shrinkage [71]. To clarify the mechanisms, more detailed experiments including those for studying the effects of solvents, are necessary.

5.3.4. Summary

We have investigated the sulfonation process of low-density polyethylene. Furthermore, we have considered polypyrrole (PPy) as an example and have reported on the fabrication of conductive polymer/insulative polymer composite films using the molecular self-assembly process and its physical properties. We have fabricated bimorph-type actuators on an experimental basis as an example of applications of these composite films and have investigated their motion properties. It is considered that composite films considered in this section are effective for fabricating not only actuators but also smart structures. We anticipate various applications of composite films in the near future.

5.4. Conductive Polymer/Insulative Polymer Composite Films

5.4.1. Introduction

To date, various conductive polymers have been synthesized, but PPy, the molecular structure of which is shown in Figure 1, is the most stable after doping and is one of the most attractive materials, together with PAn, because of its good environmental stability and relatively high electrical conductivity.

Recently, when the electroclysis polymerization was performed using pyrrole as a working electrode of the ITO substrate coated by the insulative polymer, in that case, the polymerization current that might be from the insulating working electrode is observed, and the conductive polymer/insulative polymer composite film was formed to be uniformly generated on the working electrode.

In the present work, we elucidate the detailed characteristics of fabrication processes in order to gain a good understanding of the physical properties of composite films, and also, the functional applications of the obtained composite films were examined.

5.4.2. Experimental Procedure

As shown in Figure 60, polyvinyl alcohol (PVA; saponification level: 78-82 mol%, the mean degree of polymerization: about 2000) was dissolved in distilled water. It was adjusted to the concentration of 20 % and then spin-coated to a suitable thickness (about 300-900 nm) onto the ITO-coated glass substrate. The electrical conductivity of the obtained PVA film was about 10^{-12} S/cm at room temperature.

The electrolysis polymerization was performed using an ITO-coated glass substrate with a working electrode of PVA a Ni plate as a counter electrode and an Ag wire as a reference electrode of the electrochemical cell shown in Figure 23. The composition of the polymerization liquid used for the electrolysis polymerization is 0.1 mol/l pyrrole (Tokyo Kasei Inc., purity 98 %) as the monomer, acetonitrile (Wako Pure Chemical Industries, Ltd., purity 99.5 %) as the solvent and four-fluorinated boric acid tetrabutylammonium tetrafluoroborate ($TBABF_4$ and Tokyo Chemical Industry Co., Ltd., 0.1 mol/l) as the supporting electrolyte.

The electrolysis polymerization was performed using a potentiostat (Hokuto Denko Inc., HSV-100), and the amount of passing electric charge at the polymerization potential of 1.5 V vs Ag/Ag^+ was measured. Moreover, the thickness of the composite film was measured by the multiple interference method using a multiple interference microscope (Olympus Corp, USPM-RU).

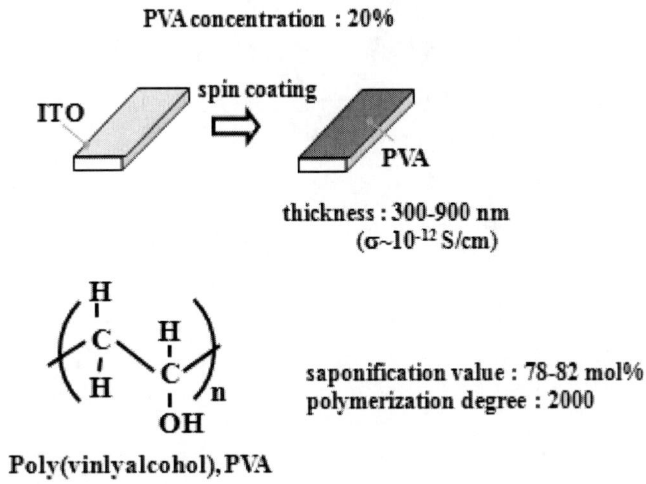

Figure 60. Working electrode used in this study.

5.4.3. Results and Discussion

5.4.3.1. PPy/PVA Composite Films

The electrolysis polymerization reaction does not occur because the working electrode used for the electrolysis polymerization is covered with insulating PVA (refer to Figure 60 for the molecular structure).

However, if the polymerization current does flow in this experiment, the polymerization reaction of pyrrole is induced, and PPy/PVA composite films are formed.

That is, when the ITO glass substrate coated with PVA, that had been left in air, as shown in Figure 61 (in air), is used as the working electrode, the current rises after about 10 s from the application of potential, and the polymerization current of several mA is observed. The working electrode is changed from transparent to back, and the formation of the PPy/PVa composite film is observed during this period. On the other hand, as also shown in Figure 61 (in vacuum), when the electrolysis polymerization is performed with the ITO glass substrate coated with PVA that had been deaerated for 1 h in a vacuum of about 13.3 Pa used as the working electrode, the polymerization current rises within a few seconds and a comparatively uniform composite film that is obtained. At this time, the observed polymerization currents are somewhat higher than that in the case with the used ITO glass substrate coated with PVA left in air. In this experiment, because distilled water is used as a solvent and the PVA film is formed by the spin-coated method, the influence of the absorbed water is considered. That is, because the absorbed water of PVA does not necessarily consistently influence the infiltration of the acetonitrile electrolyte into the PVA film in the thickness direction, it is possible to cause the light and shade pattern of the polymer. Therefore, it is possible that the existence of the adsorbed water influences the seepage velocity of the solvent. In the electrochemical polymerization reaction, it is considered that the electrolyte infiltrates the PVA film and the electrolysis polymerization is induced at the ITO electrode/electrolyte interface. As a result, the PPy/PVA composite films are obtained.

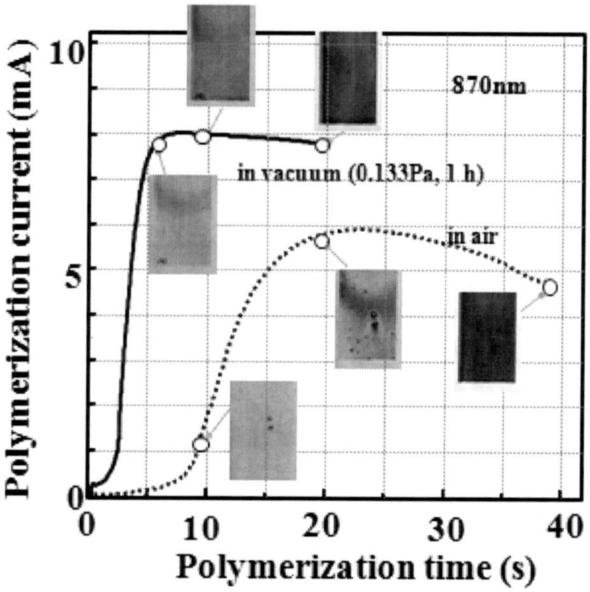

Figure 61. Relationship between polymerization current and polymerization time.

The interfacial phenomena become very complex because the interface in such an electrochemistry is a coexistence system of the electrode and the electrolyte, the movement of the charge owing to the potential gradient through this interface and the change, adsorption, and movement of the chemical species occur, the reaction involves a nonuniform system in which the solvent takes part, and the chemical changes of interface itself also are expected.

It is possible that the permeability of electrolyte in the PVA film depends on the basic physical properties, such as the hydrophobic (mixes well with water) or hydrophilic (does not to be easily mix with water) nature of the solvent. The composition of electrolyte was left as is, and a similar polymerization was attempted using propylene carbonate, chlorobenzene, ethyl lactate, and chloroform as solvents, in addition to acetonitrile. Consequently, it was confirmed that the PPy/PVA composite films can be obtained in the hydrophilic solvents such as the acetonitrile and ethyl lactate, but not in hydrophobic solvents such as chlorobenzene, propylene carbonate, and chloroform. Figure 62 shows a photograph of the working electrode obtained using electrolyte with (a) acetonitrile and (b) propylene carbonate as the solvent.

To confirm the difference in the permeability of the solvent into the PVA film, PVA film of 870 nm thickness is formed on the surface of a glass substrate on which a line had been drawn with an oil-based felt-tipped marker and which had been soaked in the propylene carbonate or acetonitrile at room temperature for 5 min.

The light and shade pattern is formed in the line drawn on the glass substrate soaked in acetonitrile, but there is no change in the glass substrate soaked in propylene carbonate, as shown in Figure 63. As the oil-based ink of the felt-tipped marker dissolves in these organic solvents, it is understood that the solvent infiltrated the ITO/PVA interface. From these results, it is considered that electrolyte infiltrates through the free volume of PVA.

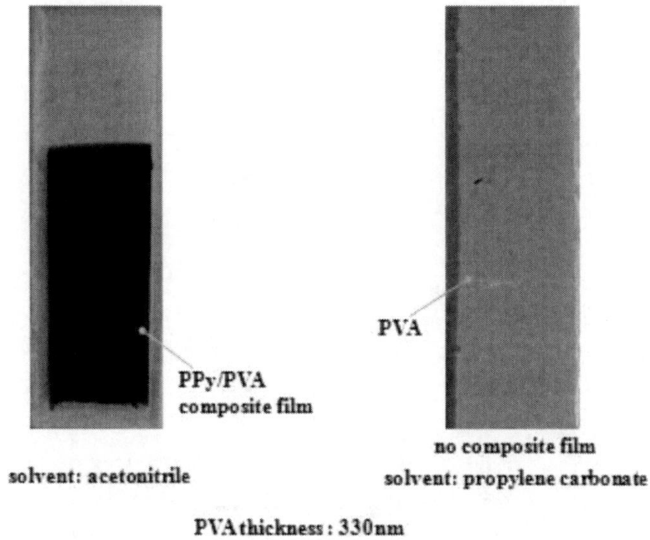

Figure 62. Photographs of composite film.

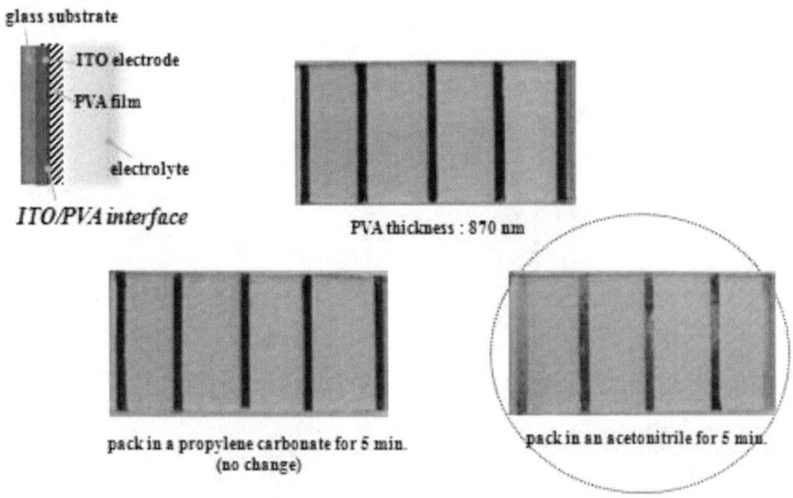

Figure 63. Difference in infiltration of two kinds of solvents.

On the other hand, the relationship between the polymerization current and the polymerization time observed when the PPy/PVA composite film is fabricated by using the substrate with the PVA coating on the ITO glass substrate after vacuum drying (0.133 Pa, 1 h) left for an arbitrary time in air is shown in Figure 64. To understand the appearance of the fabricated composite film, the photographs on the surface of the composite film, taken at an arbitrary time, for the cases of being left in air for 5 min and 30 min are also shown in this figure. When the air-exposure time of the substrate becomes long, the time needed for the polymerization current to rise lengthens gradually, the current value becomes small overall, and the inclination of the rising current is also gradual. It is understood that the composite film fabricated using the substrate with a short air-exposure time and processed in vacuum is comparatively uniform. From these experiments, in order to obtain uniform PPy/PVA composite films, it is important to remove the influence of the absorbed water as much as possible.

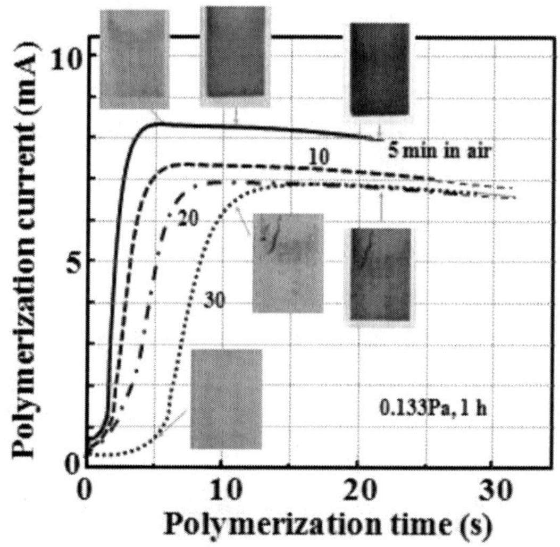

Figure 64. Relationship between polymerization current and polymerization time.

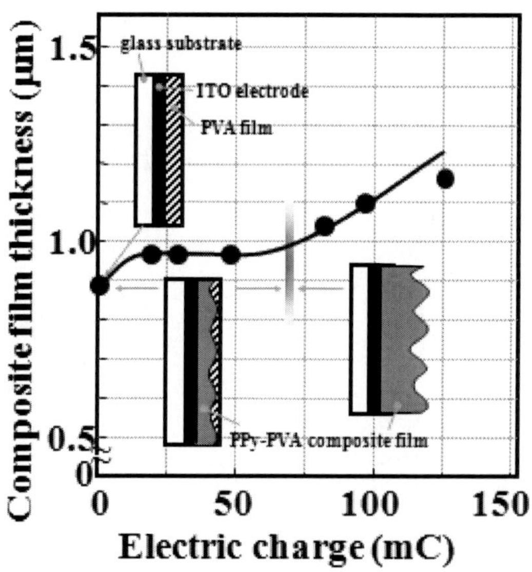

Figure 65. Relationship between composite film thickness and passing electric charge.

Figure 65 shows the relationship between the thickness of the PPy/PVA composite film and the quantity of passed electric charge. The thickness of PVA film is 870 nm. Immediately after the polymerization begins, because the polymerization reaction occurs on the ITO/PVA interface, PPy begins to grow, and it polymerizes within the PVA, the thickness of the

PPy/PVa composite film increases slightly (about 70-100 nm). The PVA bulk will polymerize when the polymerization of PPy is continued and the film thickness is constant at about 970 nm. In addition, if the polymerization reaction is induced, the overall thickness will increase because PPy penetrates into the PVA bulk and grows thicker than the PVA film.

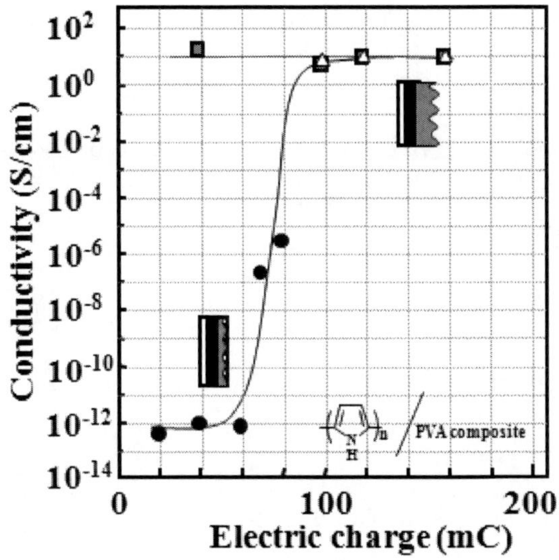

Figure 66. Relationship between electrical conductivity of composite film and passing electric charge.

Figure 66 shows the relationship between the electric conductivity of the PPy/PVA composite film and the amount of polymerization charge. In this figure, ■ sign shows the result of measuring the electric conductivity on the composite film side in contact with the ITO glass substrate side, by the four-probe method, and • and Δshow the average electrical conductivity of the composite film measured by the two- and four-probe methods, respectively. In the first stage of polymerization, because PPy grows in the PVA bulk, the electric conductivity of the composite film is comparatively low, about 10^{-12} S/cm, which is almost the same as the electric conductivity of a PVA single film. However, when the amount of polymerization charge becomes about 60-70 mC or more, PPy begins to grow thicker than PVA (870 nm) and the electric conductivity of the PPy/PVA composite film increases rapidly. When the polymerization charge becomes about 100 mC or more, the electric conductivity is around 10 S/cm.

The electrical conductivity of the conductive polymer/insulative polymer composite film described here is lower than that of a single conductive polymer and will be somewhat sacrificed. In addition, the electrically conductive pattern that consists of the polymer can be fabricated by using our new technique of composite film fabrication.

Figure 67 shows the procedure of the patterning.

a. A patterned conductive electrode is fabricated by a suitable method such as lithography.
b. The surface of the conductive electrode patterned with on insulative polymer film such as PVA (by the spin-coating method).
c. The electrolysis polymerization is carried out in this substrate as a working electrode.

d. After polymerization, the electrical conductive composite film that consists of conductive polymer/insulative polymer can be peeled off from the substrate.

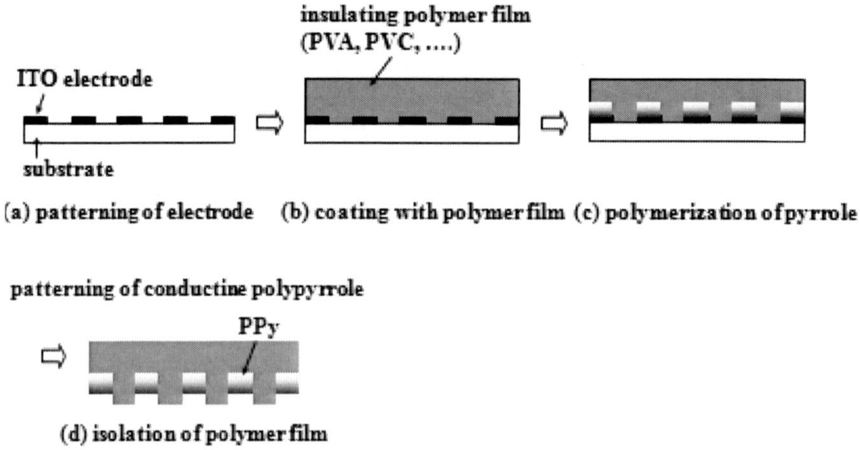

Figure 67. Schematic of procedure for patterning.

Figure 68 shows one example of the PPy/PVA composite film pattern fabricated by the procedure shown in Figure 11. The conductive pattern with an approximately 25 μm width shows the electric conductivity of about 10 S/cm.

5.4.3.2. PEDOT/PVA Composite Films

Jonas and Schrader found that poly(3,4-ethylenedioxythiophene), PEDOT, with the molecular structure shown in Table 1, is extremely transparent and its stability is excellent under a typical environment [72].

Polythiophene, PT and its derivatives have excellent environmental stability and a comparatively steady electrical conductivity. When the conductive polymer is actually applied, the environmental stability becomes extremely important.

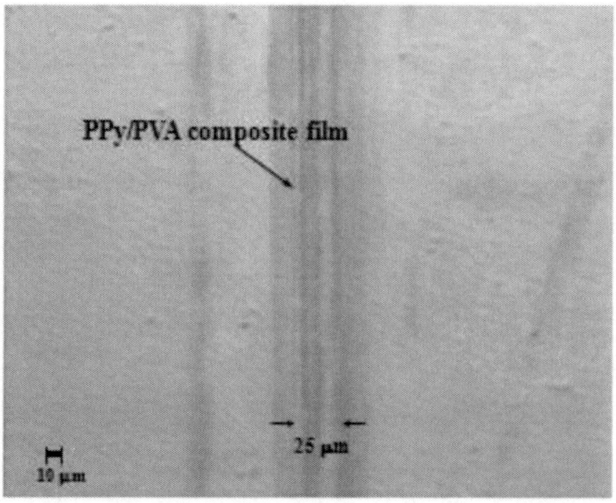

Figure 68. Example of patterning of composite film.

It is considered that doped PPy and PT, have excellent stability compared with that of positively charged polyacetylene, because of the stabilization effect of the positive charge with the nitrogen atom and the sulfur atom. Molecular structure of PPy and PEDOT and their main properties are summarized in Table 1.

The electrochemical polymerized PEDOT is rigid and fragile, therefore its use as a single film is considerably limited.

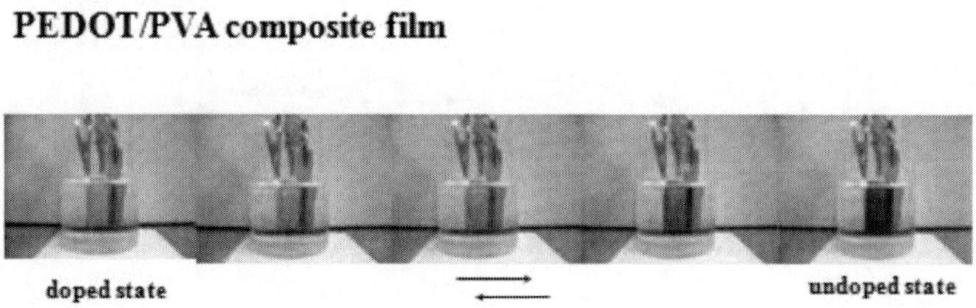

Figure 69. Color change of composite film owing on redox reaction.

Therefore, if the fabrication method of the composite film described in previous section (§5.4.3.1) can be applied also to the polymerization of PEDOT, and the conductive polymer characteristic similar to that of the PEDOT film is realized, composite films with excellent mechanical properties can be obtained. The electrochemical and optical properties of PEDOT/PVA composite films were examined from such a viewpoint.

Table 1. Molecular structures of PPy and PEDOT and their general properties

	Polypyrrole, PPy	poly(3,4-ethylenedioxythiophene), PEDOT
Chemical formulae	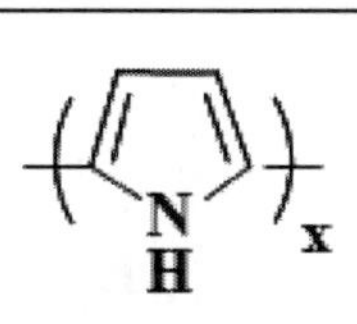	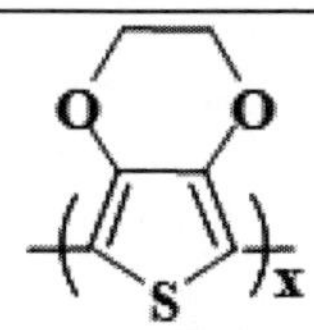
Properties	Black solid (Specific gravity1.4)	Black solid (Specific gravity 1.5)
conductivity	10~100 (S/cm)	20~500 (S/cm)
Pyrolysis temperature	~270 (°C)	~400 (°C)

Figure 69 shows the color change based on the redox reaction of PEDOT/PVA composite films were fabricated by the technique described in section 5.4.3.1. The composite film is light blue or almost transparent in the doped state, and deep blue in the undoped state. The color change is reversible in the PEDOT/PVA composite films. Such a color change of the composite film is a similar behavior to that of the redox reaction of a PEDOT single film. That is, it can be understood that the composite film sufficiently has the properties of a conductive polymer.

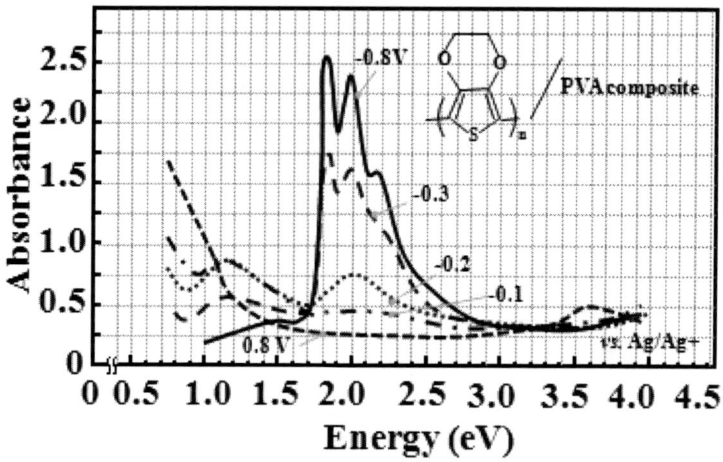

Figure 70. Optical absorption spectral change of PEDOT/PVA composite film taken during electrochemical doping.

The optical, electrical and electrochemical properties of the obtained composite films change suddenly owing to the reversible doping. For instance, Figure 70 shows the change in the optical absorption spectrum of the PEDOT PVA composite film with the doping potential. The composite films in the neutral and reduced states show strong absorption in the visible region and are deep blue in color. With increasing applied potential, the optical absorption in the visible region decreases and the composite film becomes transparent in the oxidized state from light gray.

In contrast, clear peaks that show the electrochemical properties of the PEDOT/PVA composite film are observed in the oxidation reduction curves as shown in Figure 71.That is, the peaks of the oxidation and reduction curves appear at around 0.75 V vs Ag/Ag^+ and about 0.5 and 0.8 vs Ag/Ag^+, respectively, and reversible doping is possible.

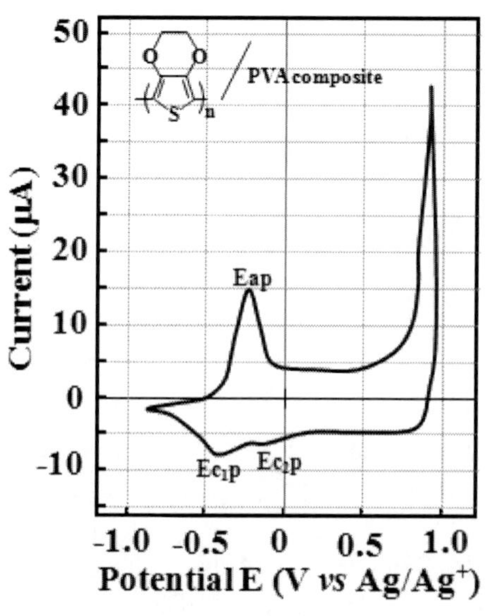

Figure 71. Cyclic voltammogram of PEDOT/PVA composite film.

5.4.3.3. Conductive Polymers and Biocompatibility

By applying this technique, we might be able to develop electrodes for electrical stimulation of the nervous system using conductive polymer, PPy. In order to establish this idea, we have to make preparations for PPy and/or PEDOT micropatterns, and evaluate biocompatibility in neuron cultures. Cultured cells on PPy and/or PEDOT extended their neuritis and survived over 7 days. For example, Figure 72 shows appearance of fibroblast L929 cultured on PEDOT doped with $TBABF_4$. Cultures were carried out in a humidified incubator with 5 % CO2/95 % air at 37 °C for 7 days. The fibroblast L929 exhibited good adhesion on PEDOT doped with $TBABF_4$-coated ITO surfaces.

After 24 h, it is understood that the cell is bonded on conductive polymer films and proliferates. 96 h later, the cell proliferated to all aspects of the conductive polymer film, PEDOT.

The compatibility of PPy was also similar to that of PEDOT. Accordingly, the PEDOT- and/or PPy-coated electrode may be useful to culture the cells on. In other words, when dealing with excitable cells such as neurons, it seems to be advantageous for a certain interest to culture the cells directly on an electrode or electr-conductive materials with which one can either stimulate the cells or defect the cell resposes.

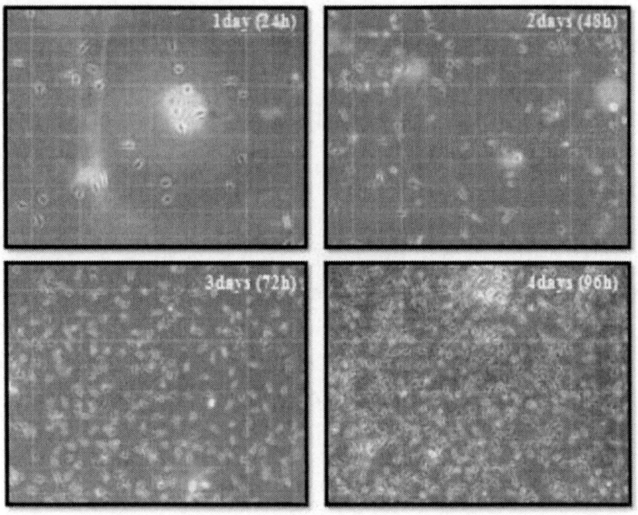

Figure 72. Appearance of fibroblast L929 culture on PEDOT doped with $TBABF_4$.

5.4.4. Summary

It has been revealed that when electrolysis polymerization was performed with a working electrode comprising an ITO substrate coated with an insulative polymer, a polymerization current, which normally does not flow, is observed, and a conductive polymer/insulative polymer composite film can be easily fabricated.

Because the electrolysis polymerization occurs between the electrode and the insulative polymer interface, the permeability of the electrolyte into the bulk of the insulative polymer is considered. The composite films are obtained in an electrolyte consisting of the hydrophilic solvent, but not in an electrolyte consisting of a hydrophobic solvent. It was proven that a conductive pattern can be easily fabricated by using this technique.

By using the new technique of composite film fabrication described here, patterning like that on a microelectrode is also possible. If the compatibility of the composite film and the neuron cells is excellent, the application, for example, to the formation of a biocompatible electrode can be considered, and technological advance to realize bioelectronics will be possible.

5.5. Conductive Polymer/Fullerene (C_{60}) Composite Film Prepared by the Electrophoretic Deposition

5.5.1. Introduction

A lot of effort has been devoted to the development of electronic devices that use semiconducting polymer materials such as conductive polymers. In contrast to traditional inorganic semiconductors, which require high synthesis temperatures, conductive polymers are synthesized near room temperature, and their excellent solubility enables the deposition of large films through cheap coating technologies such as spin-coating and printing. Due to these advantages, many researchers have worked on the development of, for example, polymer photovoltaic devices and polymer light-emitting devices.

It is well-known that the performance of photovoltaic devices made from conductive polymers is improved by using donor-acceptor composites, in which photoinduced charge transfer takes place [73, 74]. Since most conductive polymers have a donor-type or hole-transporting nature, possibilities for complementary acceptor materials have been extensively studied.

A few years after the theoretical proposal of this strategy [75], the (buckminster)fullerene C_{60} was the first suitable acceptor material found for conductive polymers [76-79], leading to an almost complete quenching of photoluminescence as well as a dramatic enhancement of the photon-electron conversion efficiency (more than two orders of magnitude above that of the polymer itself upon doping with a few mol-%) [76, 77], In this case, the electrons in the conductive polymer were transferred to the C_{60}, and the remaining holes became photocarriers migrating along the polymer chains. Because of the low C_{60} concentration in the composite, the C_{60} electrons do not play a role as photocarriers. Studies using field-effect transistors, which enable the study of the transport of carriers with either a negative or a positive charge, have clearly indicated that electron transport in a donor-acceptor composite is dominated by the percolation process [80, 81]. Therefore, an interpenetrating network or a bulk heterojunction of donor and acceptor materials that shows bipolar conduction is needed to harvest photocarriers of both signs.

However, the poor solubility of C_{60} in the common organic solvents [82] that dissolve conductive polymers has made it difficult to obtain an interpenetrating network of conductive polymer and C_{60}. The solution containing comparable amounts of conductive polymer and C_{60} has to be dilute, and the films obtained by the casting technique from such a dilute solution results in the growth of C_{60} crystals during slow solvent evaporation. It has been reported that the morphology of the film prepared by casting a toluene solution containing equivalent weights of poly(2-methoxy-5-(2'-ethylhexyloxy)-1,4-phenylene vinylene), MEH-PPV and C_{60} is badly affected by the precipitation of C_{60} [79, 83].

This fact has simulated extensive research by synthetic chemists into the development of acceptor materials with high solubilities, including a variety of chemically modified C_{60}s [83], perylene derivatives [80, 81] and cyano-substituted poly(*p*-phenylene vinylene) [84-86]. Although these materials are highly soluble in organic solvents and yield photocells with excellent photoconversion efficiencies, they require complicated chemistry.

Recently, we have shown that stable colloidal suspensions of a conductive polymer, poly(3-octadecylthiophene), can be easily prepared by a simple injection method without any additives, which enables quick film deposition using an electrophoretic process [14].

Although the films deposited by electrophoretic deposition are nanoporous, the sandwich devices made from them by vacuum deposition of the cathode show relatively stable light emission, indicating that a film without pinholes can be made by this process [15, 87]. A similar film-deposition technique utilizing a miniemulsion of conductive polymers has been reported by another group [18]. The preparation of light-emitting devices [88] and polymer composites [89] through the miniemulsion process has also been reported. Although the surfactants used in this process enable the regulation of colloidal size, our simple injection method does not employ any additives that could potentially disturb the purity of the films. The distinctive difference between these emerging film-deposition process employing colloidal particles and the traditional solution-based techniques is the separation of the film-deposition stage from the material solidification stage, Thanks to this unique feature, a comparable weight of unmodified C_{60} can be loaded into a conjugated polymer, as reported here.

5.5.2. Experimental Procedure

A gel-like fraction that did not dissolve in toluene was removed from MEH-PPV (Sigma-Aldrich Co. LLC.) by filtration prior to use. C_{60} (Materials Technologies Research Ltd.), toluene, and acetonitrile (Wako Pure Chemical Industries, Ltd.) were used as received. The colloidal suspensions of MEH-PPV were prepared as follows: 0.33 mg of MEH-PPV was dissolved in 1ml of toluene, and then the solution was injected into 4 ml of acetonitrile, which is not a MEH-PPV solvent, to obtain a 0.066 g/l colloidal suspension of MEH-PPV. The suspensions of C_{60} and the mixture solution of MEH-PPV and C_{60} were obtained in the same manner. The ratio of acetonitrile and toluene was kept at 4:1 throughout this study.

The method used for the electrophoretic deposition can be found in a previous publication [14]. About 2 ml of suspension was charged in a glass cell 1 cm wide, 1 cm long, and 3 cm high, and 250 V dc was applied between two indium tin oxide (ITO)-coated glass plates (10 Ω/sq.) spaced 5 mm apart soaked in the suspension. The films were dried naturally in air.

The optical absorption spectra were measured using a Hitachi 3410 spectrophotometer. A glass cell with a 1 mm optical path was used for the liquid samples. The photoluminescence spectra were taken with a Hitachi F-2500 photoluminescence spectrophotometer. An excitation wavelength of 450 nm was used throughout the work. A glass cell with a 1 mm optical path was used for the liquid samples, and the emission from the excited face was measured in order to avoid modulation due to self-absorption. Atomic force microscope (AFM) images were collected using a Burleigh Metris-2000 system operated in non-contact mode in air.

5.5.3. Results and Discussion

5.5.3.1. Colloidal Suspensions of MEH-PPV and C_{60}

We chose the conductive polymer MEH-PPV as the C_{60} host. Suspensions were prepared by the method described in our previous paper. [14] One unit of toluene solution with MEH-

PPV was injected into four units of acetonitrile, to obtain a suspension of MEH-PPV in acetonitrile/toluene (4/1). The C_{60} suspensions [90] and the mixture of MEH-PPV and C_{60} were obtained by the same method.

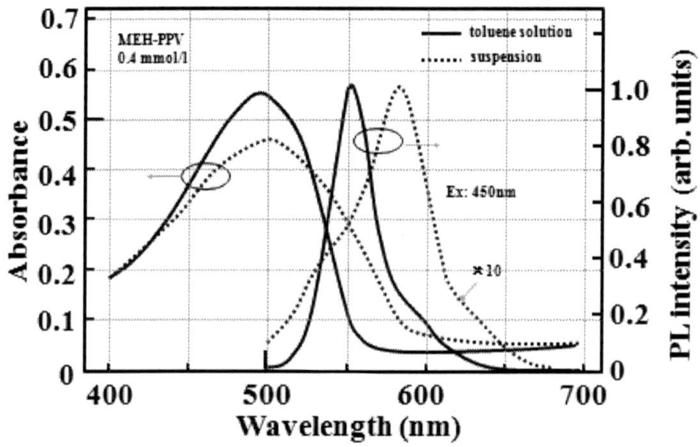

(a)

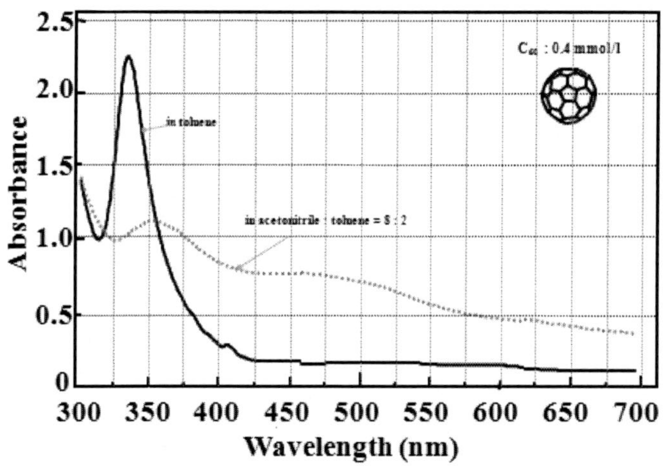

(b)

Figure 73. (a) Optical absorption and photoluminescence spectra of MEH-PPV in toluene and in an acetonitrile/toluene (4 : 1) mixture. (b) Optical absorption spectra of C_{60} in toluene and in an acetonitrile/toluene (4 : 1) mixture.The concentrations are 0.066 g/l and 0.29 g/l for MEH-PPV and C60 respectively, corresponding to 0.4 mmol/l for both.

Figure 73 (a) shows indicates the optical absorption and photoluminescence spectra of the MEH-PPV suspension. Those of a toluene solution of the same concentration of MEH-PPV are also shown for comparison. Both spectra of the suspension show a remarkable red-shift compared to those of the toluene solution. It is also noted that the photoluminescence from the MEH-PPV in the suspension is much weaker than that in the solution, by a factor of 10 %.These results clearly indicate the solidification of MEH-PPV in the acetonitrile/toluene mixture.The optical absorption spectra of the suspension and the toluene solution of C_{60} are shown in Figure 73 (b). A toluene solution of C_{60} is a transparent violet liquid, whereas the suspension is an opaque brown liquid. The absorption peak at around 340 nm in the solution broadens and red-shifts to around 360 nm in the suspension, clearly indicating the formation of C_{60} colloids in the suspension. The long tail stretching above 800 nm in wavelength must originate mainly from scattering suggesting a relatively large colloidal size.

Two methods for obtaining suspensions containing MEH-PPV and C_{60} have been proposed, as illustrated in Figure 74.

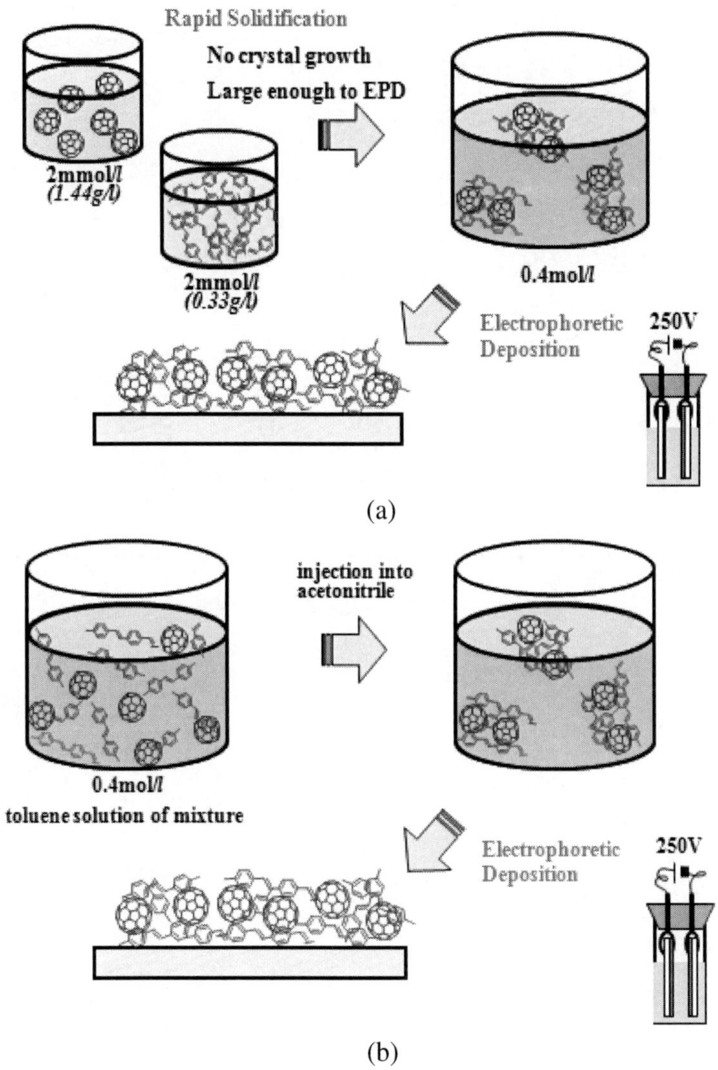

[a]mixture of suspensions.
[b]suspention of mixture.

Figure 74. The two strategies for preparing mixture suspensions of MEH-PPV and C_{60}proposed in this study: (a) mixing separate suspensions MEH-PPV and C_{60}; (b) mixing a suspension of toluene solution containing MEH-PPV and C_{60}.

One method is to mix separate suspensions of MEH-PPV and C_{60}; the other is to mix toluene solutions of them prior to injection into acetonitrile. The former method is expected to yield suspensions containing isolated MEH-PPV and C_{60} colloids, whereas a colloidal particle made by the latter method may consist of both components. Since our goal in this study was to load a large amount of pure C_{60} into MEH-PPV, suspensions heavily doped with 10, 20, 30, and 50 mol-% of C_{60}, corresponding to suspensions containing one C_{60} molecule per 9, 4, and 1 repeating units of MEH-PPV respectively, were prepared.

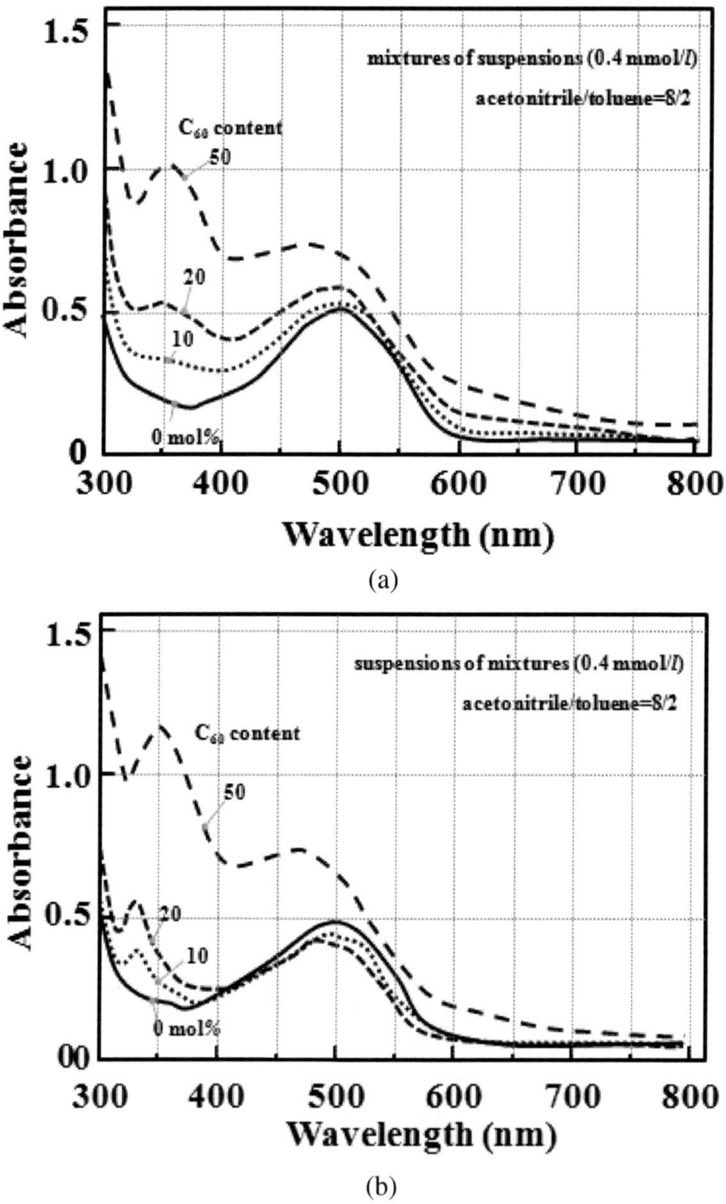

Figure 75. Optical absorption spectra of (a) mixtures of separate suspensions of MEH-PPV and C_{60}; (b) suspensions of the mixtures of MEH-PPV and C_{60} in toluene solution.

The mixtures of the suspensions of MEH-PPV and C_{60} gave the optical absorption spectra shown in Figure 75 (a), which were found to be a superposition of the relevant spectra shown in Figure 1. On the other hand, the suspensions of the mixtures of MEH-PPV and C_{60} show a sharp absorption peak at around 340 nm as shown in Figure 75 (b), up to 20 mol-%.

This clearly indicates that the C_{60} in the suspension of the mixture is molecularly dispersed, in contrast to the mixture of suspensions. However, the absorption spectrum of the suspension made from 50 mol-% solution shows the aggregation of C_{60}. The photoluminescence spectra depict the difference between the dispersion conditions of C_{60} in these suspensions. Although the photoluminescence from the MEH-PPV colloids was weakened by the introduction of the C_{60} colloids, as shown in Figure 76(a), the emission from MEH-PPV was only weakened by a factor of 10 %, even when the suspension contained 50 mol-% of C_{60}.

As reported by Yoshino et al., [77] C_{60} doping of 5 mol-% is enough to almost completely quench the photoluminescence from this type of poly(*p*-phenylene vinylene) derivative. Thus, it can be concluded that the C_{60} in these suspensions is a less effective luminescence quencher.

In contrast, the photoluminescence peak from MEH-PPV at around 580 nm was effectively quenched in the suspensions made from mixture solutions, as shown in Figure 76 (b). This confirms the molecular dispersion of C_{60} in these colloids that was suggested by the optical absorption spectra shown in Figure 75 (b). The shoulder found in the spectrum of the suspension of pure MEH-PPV at around 550 nm, which remains uncharged upon introduction of C_{60}, seems to originate from a fraction of MEH-PPV dissolved in the acetonitrile-toluene mixture, because this emission, whose peak wavelength coincides with that of the toluene solution, is not found in solid MEH-PPV films, as shown later.

It is worth nothing that effective photoluminescence quenching was absent in the parent solutions of the later suspensions, as shown in Figure 76 (c).

That is, since the MEH-PPV main chains and the C_{60} molecules were molecularly dispersed in a toluene solution, the distance between them was not close enough to allow photoinduced charge transfer between them. Therefore, the quenching of photoluminescence found in Figure 4(c) suggests the secondary aggregation of MEH-PPV and C_{60} colloids upon mixing.

5.5.3.2. Composite Films of MEH-PPV and C_{60}

The electrophoretic deposition of films from all suspensions mentioned above was easily achieved by the method reported in our previous paper [14].

Unlike the poly(3-octadecylthiophene) reported in that paper, both MEH-PPV and C_{60} films were found to be deposited on the positively biased electrodes, indicating that both colloids are negatively charged in the suspensions. Figure 77 shows the optical absorption spectra of the films. As shown in Figure 77 (a), all of the spectra coincide with those of the parent suspensions, suggesting the same deposition speed for both components in the mixture of suspensions.

The sharpness of the absorption peak from C_{60} indicated in Figure 77 (b) proves that the C_{60} molecules are molecularly dispersed in the films obtained by electrophoretic deposition, in suspensions of the mixture solution up to 20 mol-%.

The photoluminescence spectra of the electrophoretically deposited films also reflect the parent suspensions, as shown in Figure 78. That is, the photoluminescence from MEH-PPV is almost completely quenched by 10 mol-% of fullerene in the films deposited from the suspension of mixtures, whereas those deposited from the mixture of isolated suspensions show notable emission, even when the film contains 50 mol-% of C_{60}.

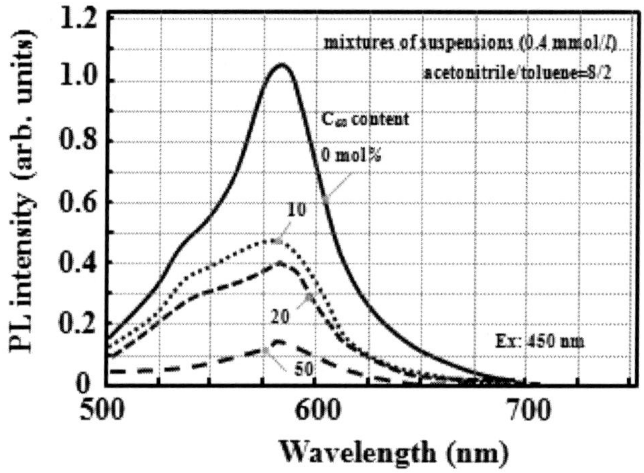

(a)

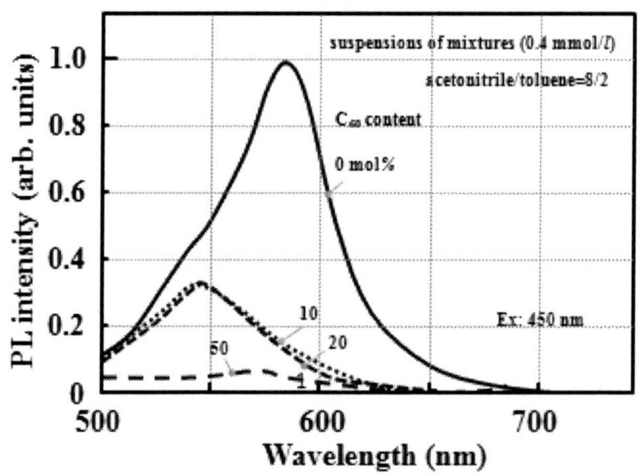

(b)

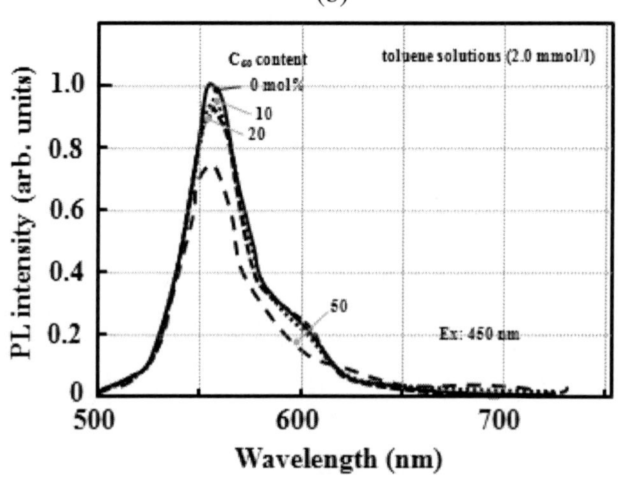

(c)

Figure 76. Photoluminescence spectra of (a) the mixtures of separate suspensions of MEH-PPV and C_{60}; (b) the suspensions of the mixture of MEH-PPV and $^{\circ}C_{60}$ in toluene solution; (c) the parent toluene solution of b).

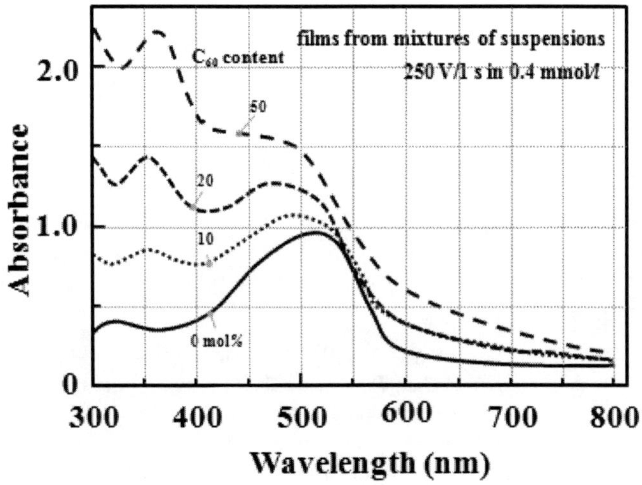

(a)

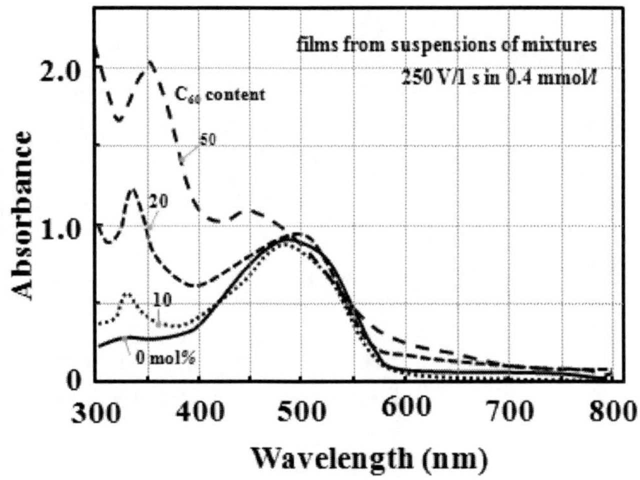

(b)

Figure 77. Optical absorption spectra of the composite films of MEH-PPV and C_{60} deposited in (a) the mixtures of separate suspensions; (b) the suspensions of the toluene solution mixtures. The deposition time was 1 s for all samples.

The emission peak at around 550 nm found in the suspensions is absent in the films, suggesting that the fraction this emission originates from is not collected by electrophoretic deposition. Figure 79 shows the atomic force microscopy (AFM) images of the composite films made from the mixture of the separate suspensions of MEH-PPV and C_{60}.

The films used here were naturally dried in the atmosphere, because this reveals the differences arising from the preparation conditions clearer than when washed with acetonitrile. The crater-like pores found in the films of pure MEH-PPV shown in Figure 79 (a) may originate from the residual toluene. Since acetonitrile evaporates much faster than toluene, toluene concentrated on the film during drying and started to dissolve the film entirely during the early stages, and locally later on. Such residual toluene removed the nanoporosity from the films. The formation of a pore by local dissolution is used to install the

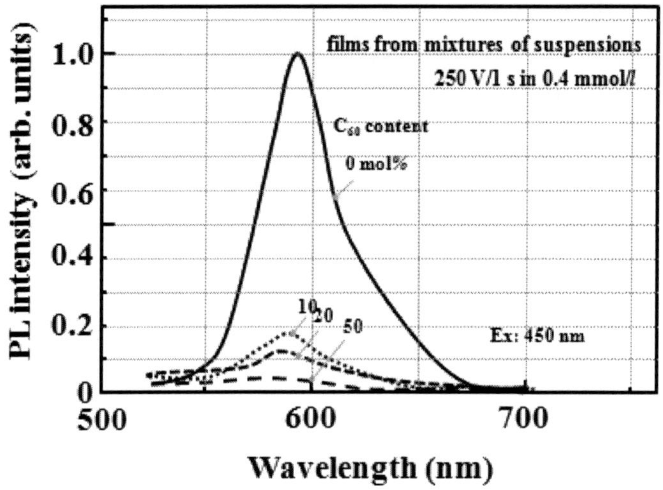

(a)

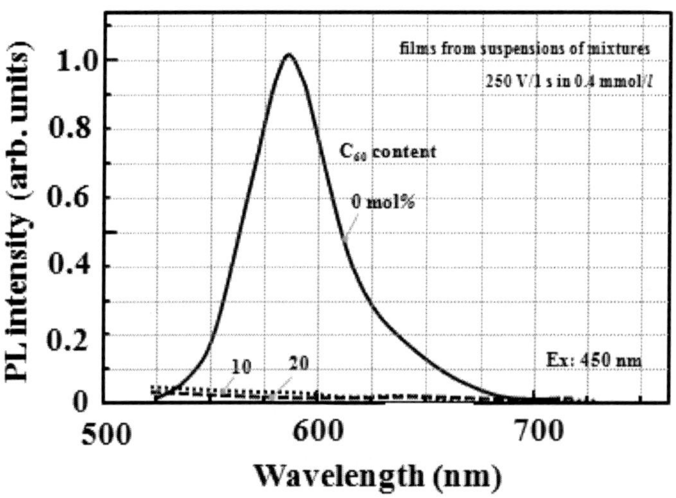

(b)

Figure 78. Photoluminescence spectra of the composite films of MEH-PPV and C_{60} deposited from (a) mixtures of separate suspensions; (b) the suspensions of the toluene solution mixtures. The deposition time was 3 s for all samples.

via-hole in polymer integrated circuits [12]. With increasing C_{60} fraction, the number of particles 0.1-0.2 μm in diameter increased. The particles are C_{60} as C_{60} solubility in toluene is low.On the other hand, the films deposited in the suspension of mixture solutions containing 10 mol-% of C_{60} showed a surface morphology similar to that of the pure MEH-PPV film, as shown in Figure 80 (a). That is, the film does not contain C_{60} particles, once again indicating the molecular dispersion of C_{60} in the film. With increasing C_{60} concentration, the films show nanoporosity, reflecting the colloidal particles in the parent suspension, as shown in Figure 80 (b), (c). This observation suggests that the solubility of composites of MEH-PPV and C_{60} decreases as the C_{60} concentration increases. The particle size notably increased with increasing C_{60} concentration, and the diameters of the particles found in Figure 80 (b), (c) are approximately 0.1 and 0.2 μm, respectively. Although the composite films of MEH-PPV and

C_{60} developed in this study exhibit notable nanoporosity, the fact that the nanoporous films of poly(3-octadecylthiophene) obtained through the same technique can be used as the emission layer of polymer light-emitting devices seems to prove the potential of the film for use in photocells. This potential is also supported by the report of successful photocell demonstration through the miniemulsion process. Since the performance of a photocell of donor-acceptor composites is highly influenced by the morphology [91], the composite of MEH-PPV and C_{60} with unique morphology as reported here should be expected to be useful in photocells. The application of the composite film to photocells is now under investigation.

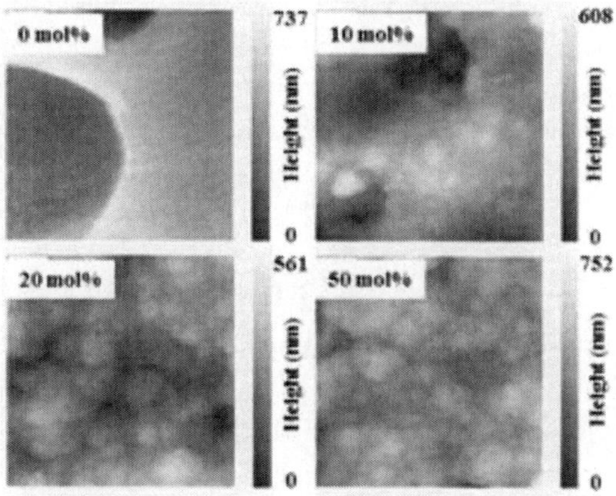

Figure 79. Atomic force microscopy images of the composite films of MEH-PPV and C_{60} deposited from the mixtures of separate suspensions. Films contain (a) 0 mol-%, (b) 10 mol-%, c) 20 mol-%, and d) 50 mol-% of C_{60}. The observation area is 2.7 μm (X) × 3.1 μm (Y). The deposition time was 3 s for all samples. The thickness of the films can be evaluated from the depth of a hole reaching the substrate as approximately 0.7 μm.

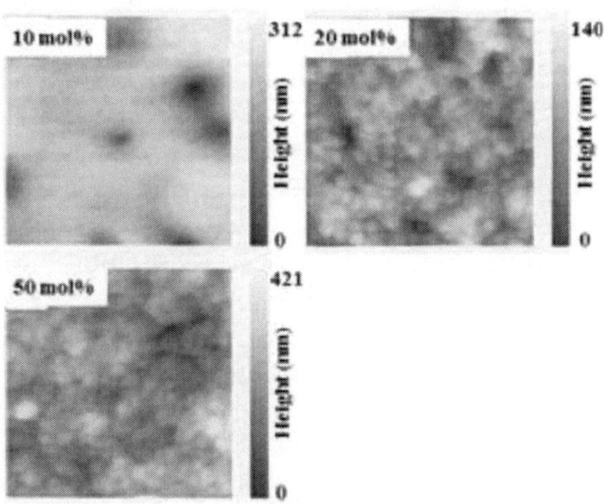

Figure 80. Atomic force microscope images of the composite films of MEH-PPV and C_{60} deposited from the suspensions of the mixtures in toluene solution. Films contain a) 10 mol-%, b) 20 mol-%, and c) 50 mol-% of C_{60}. The observation area is 2.7 mm (X) × 3.1 mm (Y). The deposition time was 3 s for all samples.

5.5.4. Summary

In this study, a novel way to load a large amount of unmodified C_{60} into a conductive polymer was reported. The principle is to separate the film-deposition stage from the material-solidification stage. That is, materials are quickly solidified from dilute solution into colloidal suspension, and are collected by electrophoretic deposition. The method reported by us for poly(3-octadecylthiophene) [14] has been found to be useful in systems consisting of MEH-PPV and C_{60}, indicating the versatility of the method.

The deposition from the mixture of the separate suspensions of both materials resulted in films consisting of isolated particles. On the other hand, by using a suspension made of a dilute solution of MEH-PPV and C_{60}, an MEH-PPV film with 20 mol-% of molecularly dispersed C_{60}, corresponding to a nearly 1 : 1 ratio by weight, could easily be obtained. Although we have focused solely on the case of a nanocomposite consisting of MEH-PPV and C_{60} in this paper, we believe that the simple technique proposed here is useful for making other nanocomposite systems containing poorly soluble components as well.

CONCLUSION OF NEW FABRICATION TECHNIQUES OF POLYMER-COMPOSITE FILMS

The world of the electronics which developed mainly on a silicon semiconductor is going to change drastically. A paradigm shift which is called organic electronics or printed electronics is finally more likely.

The Nobel prize for chemistry in the 20th century had been awarded to Alan J. Heeger, Alan G, MacDiarmid (February 7, 2007 demise), and Hideki Shirakawa for their pioneering works on the discovery and development of conductive polymers [40-42].

The interest in undoped organic semiconductors revived in the 1980s due to the demonstration of an efficient photovoltaic cell incorporating an organic hetero-junction p- and n-conducting materials [92] as well as the first successful fabrication of thin film transistors from conductive polymers and oligomers [8, 43, 93]. The main impetus, however, came from the demonstration of high-performance electroluminescent diodes from conductive polymers [4, 44]. Owing to the large efforts of both academic and industrial research laboratories during last 22 years, organic light-emitting devices (OLEDs) have progressed rapidly and meanwhile lead to first commercial products incorporating OLED displays. Other applications of organic semiconductors e.g., as logic circuits with organic field-effect transistors (OFETs) or organic photovoltaic cells (OPVCs) are expected to follow in the near future.

In Section 5, the following headings for fabricating composite films are intended to give an up-to date overview of the different facets of the subject of this book.

In the first heading, molecular self-assembly techniques are described. That is, multilayer heterostructure thin films of electroactive polymers such as poly(p-phenylene vinylene), PPV, -precursor and sulfonated polyaniline, SPAn, have been successfully fabricated via process based on the spontaneous self-assembly of conductive polyoion on a substrate. We have prepared SPAn/PPV multilayer heterostructures light-emitting diodes (LEDs) using this electroactive self-assembly technique and discussed the electroluminescent (EL) properties. The SPAn/PPV monolayer LED emits greenish-yellow light, but SPAn/PPV multilayer LED emits blueish-green light and there is a slight shift between the two spectra of these devices. These results can be interpreted in terms of the confinement effect of carriers in the superlattice structure consisting of SPAn/PPV multilayer system.

In the next heading, fabrication techniques of composite films based on a sol-gel method are introduced. Polypyrrole (PPy)-silica glass composite films were prepared by carrying out the electrochemical polymerization of PPy simultaneously with the synthesis of silica glass via a sol-gel process, and its electrochemical doping properties were investigated by cyclic voltammetry. These films could be easily doped electrochemically; the undoped and doped states were yellow and blue, respectively. The redox process in PPy-silica glass was gradually activated by repeating the electrochemical doping-undoping cycles. It was found that dopant anions in the PPy-silica glass composite film could be exchanged relatively easily with electrolyte anions in acetonitrile solution by potential sweeping. However, dopant anions in the PPy film could not be easily exchanged in acetonitrile solution. Finally, application of this new type of PPy-silica glass composite film in an ionic sensor which responds to specific ions has also been proposed.

As the third heading, composite films based on *in situ*polymerization technique are fabricated as follows. By exposing low-density polyethylene to an atmosphere of fuming sulfuric acid, sulfonated low-density polyethylenes (SPEs) were prepared and the degree of surface sulfonation of the treated film was determined from the weight increase per unit area. Polypyrrole (PPy) films on the surface of SPE were grown by molecular self-assembly and the properties of PPy/SPE composite films were investigated. Functional applications of PPy/SPE composite films are proposed, with particular attention to the movement properties of actuators using PPy/SPE composite films.And then, insulative/conductive composite films are prepared by using electrolysis polymerization technique. That is, poly(vinyl alcohol), PVA, produces a flexible composite polymer film with electrical, optical and electrochemical properties very similar to those of polypyrrole (PPy).

The rate of electrochemical depends on the diffusion rate of the electrolyte across the PVA film to the indium tin oxide (ITO) electrode. In particular, a solvent with a hydrophilic nature easily penetrates into the PVA film. By applying this new process, we demonstrate a unique method of forming an electrically conductive pattern in PVA film. It will be possible to develop electrodes for electrical stimulation of the nervous system using the conductive polymer, PPy. Then, by applying a similar technique, we fabricated poly(3,4-ethylenedioxy-thiophene), PEDOT/PVA, composite films and investigated their basic electrochemical properties.

The last heading of this Section is deal with fabrication method of composite films using electrophoretic deposition techniques. Conductive polymers doped with C_{60}s show excellent photoconversion efficiencies. However, due to the poor solubility of pure C_{60} in common organic solvents, it was thought that some chemical modifications to C_{60} were required in order to load conductive polymers with comparable weights of C_{60}. Here, we report a novel way to load large amounts of unmodified C_{60} into the conductive polymer, poly(2-methoxy-5-(2'-ethylhexyloxy)-1,4-phenylene vinylene), MEH-PPV. The principle of our approach is to separate the film-deposition stage from the material-solidification stage. That is, materials are quickly solidified from dilute solution into colloidal particles, which are large enough to be collected by electrophoretic deposition. Deposition from a mixture of separate suspensions of MEH-PPV and C_{60} resulted in films consisting of isolated particles. In contrast, by using a suspension made of a dilute solution of MEH-PPV and C_{60}, which was so dilute that traditional techniques such as spin-coating were not applicable, an MEH-PPV film doped with 20 mol-% C_{60}, corresponding to an almost 1 : 1 ratio by weight of molecularly dispersed C_{60}, was easily obtained. To the best of our knowledge, this is the first report on the successful loading of an equivalent weight ratio of unmodified C_{60} molecules into MEH-PPV.

REFERENCES

[1] Yoshino, K. and Onoda, M. *Polymer electronics –Conducting polymer and its electronically and optically functional devices-*, Tokyo: Corona Publishing Co., Ltd., 1996.

[2] Onoda, M., (Supervisor) *Physical properties and evaluation of conductive polymers for high performance organic electronic devices*, Tokyo-Osaka, CMC Publishing Co., Ltd., 2012.

[3] Nalwa, H. S.and Lohwer, L.S., (Eds.) *Handbook of Luminescence, Display Materials and Devices*, California: American Science Publishers; 2003.

[4] Burroughes, J.H., Bradley, D.D.C., Brown, A.R., Marks, R.N., Mackay, K., Friend, R.H., Burns, P.L., and Holmes, A.B. (1990). Light-emitting diodes based on conjugated polymers, *Nature*, 347 (6293) pp.539-541.

[5] Tada, K.and Onoda, M. (2006). Red emission from poly(9,9-dioctylfluorene) doped with phosphorescent dye through maskless dye-diffusion technique, *Applied Physics Letters*, 89 (4), 043508 (3 pages).

[6] Yu, G., Pakbaz, K. and Heeger, A.J. (1994). Semiconducting polymer diodes: Large size, low cost photodetectors with excellent visible-ultraviolet sensitivity, *Applied Physics Letters*, 64 (25) pp.3422-3424.

[7] Tada, K., Onoda, M., Zakhidov, A.A., and Yoshino, K. (1997). Characteristics of poly(p-pyridyl vinylene)/poly(3-alkylthiophene) heterojunction photocell, *Japanese Journal of Applied Physics*, 36 (3A) pp.L306-L309.

[8] Tsumura, A., Koezuka, H. and Ando, T. (1986). Macromolecular electronic device: Field-effect transistor with a polythiophene thin film, *Applied Physics Letters*, 49 (18) pp.1210-1212.

[9] Wenek, G.E. (1986). Electrically conductive polymer composites, In: T.A. Skothem (Editor), *Handbook of conducting polymers* (edition, pp.205-212) New York and Basel: Marcel Dekker Inc.

[10] Onoda, M., Tada, K. and Nakayama, H. (1999). Conducting polymer / insulating polymer composite films prepared by the molecular self-assembly process, *Japanese Journal of Applied Physics*, 38 (6A) pp.3736-3741.

[11] Bharathan, J.and Yang, Y. (1998). Polymer electroluminescent devices processed by inkjet printing: I. Polymer light-emitting logo, *Applied Physics Letters*, 72 (21) pp.2660-2662.

[12] Sirringhaus, H., Kawase, T., Friend, R.H., Shimoda, T., Inbasekaran, M., Wu, W., and Woo, E.P. (2000). High-resolution inkjet printing of all-polymer transistor circuits, *Science*, 290 (5499) pp.2123-2126.

[13] Van der Biest, O. O.and Vandeperre, L.J. (1999).Electrophoretic deposition of materials, *Annual Review of Materials Science*, 29 pp.327-352.

[14] Tada, K. and Onoda, M. (2002). Nanostructured conjugated polymer films by electrophoretic deposition, *Advanced Functional Materials*, 12 (6-7) pp.420-424.

[15] Tada, K. and Onoda, M. (2003). Polymer light-emitting devices for artificial fingerprints, *Japanese Journal of Applied Physics*, 42 (9A/B) pp.L1093-L1095.
[16] Tada, K. and Onoda, M. (2004). Loading fullerene into conjugated polymer without chemical modification, Advanced Functional Materials, 14 (2) pp.139-144.
[17] Tada, K. and Onoda, M. (2008). Uniform film of conjugated polymer for light-emitting device by electrophoretic deposition. *Journal of Physics D: Applied Physics*, 41 (3) 032001 (4 pages).
[18] Landfester, K., Montenegro, R., Scherf, U., Günter, R, Asawapirom, U., Patil, S., Neher, D., and Kietzke, T. (2002). Semiconducting polymer nanospheres in aqueous dispersion prepared by a miniemulsion process, *Advanced Materials*, 14 (9) pp.651-655.
[19] Onods, M., Manda, Y., Yokoyama, M., Sugimoto, R., and Yoshino, K. (1989). A photoelectron emission study of polythiophene derivatives, *Journal of Physics: Condensed Matter*, 1 (24) pp.3859-3864.
[20] Kaneto, K., Yoshino, K.and Inuishi, Y. (1977). Temperature Dependence of Reflectance Spectrum of $(SN)_x$Single Crystal, *Journal of the Physical Society of Japan*, 43 (3) pp.1013-1015.
[21] Onoda, M., Manda, Y., Iwasa, T., Nakayama, H., Amakawa, K., and Yoshino, K. (1990). Electrical, optical, and magnetic properties of poly(2,5-dimethoxy-p-phenylene vinulene), *Physical Review B Condensed Matter*, 42 (18) pp.11826-11832.
[22] Fukuda, M., Sawada, K. and Yoshino, K. (1989). Fusible conducting poly(9-alkylfluorene) and poly(9,9-dialkylfluorene) and Their Characteristics, *Japanese Journal of Applied Physics*, 28 (8) pp.L1433-L1435.
[23] Ohsawa, T., Kabata, T., Kimura, O., Onoda, M., and Yoshino, K. (1989). In situconductivity, electron spin resonance and absorption spectra measurements in polyaniline during electrochemical doping, *Japanese Journal of Applied Physics*, 28 (6) pp.996-999.
[24] Yoshino, K., Onoda, M., Manda, Y., Sawada, K., Sugimoto, R., and Inoue, S. (1989). Fusible poly(3-alkylselenophene) and its properties, *Japanese Journal of Applied Physics*, 28 (1) pp.L138-L140.
[25] Yoshino, K., Park, D.H., Park, B.K., Onoda, M., and Sugimoto, R. (1988). Large change of electrical conductivity and absorption spectrum of poly(3-alkylthiophene) at the solid-liquid phase transition, *Japanese Journal of Applied Physics*, 27 (9) pp.L1612-L1615.
[26] Yoshino, K. Love, P., Onoda, and Sugimoto, R. (1988). Dependence of absorption spectra and solubility of poly(3-alkylthiophene) on molecular structure of solvent, *Japanese Journal of Applied Physics*, 27 (12) pp.L2388-L2391.
[27] Yoshino, K., Kuwabara, T., Manda, Y., Makajima, S., and Kawai, T. (1990). Photoinduced Solubilization of Conducting Polymer and Its Application to Etching and Recording, *Japanese Journal of Applied Physics*, 29 (9) pp.L1716-L1719.
[28] Yoshino, K., Kuwabara, T., Iwasa, T., Kawai, T., and Onoda, M. (1990). Optical recording using conducting polymers, poly(p-phenylene vinylene) and its derivatives, *Japanese Journal of Applied Physics*, 29 (8) pp.L1716-L1719.
[29] Onoda, M., Nakayama, H., Amakawa, K., and Yoshino, K. (1992). Electronic properties of polyarylene-vinylene conducting polymers, *IEEE Transactions of Electrical Insulation*, 27 (3) pp.636-646.

[30] Yoshino, K., Nakao, K., Onoda, M., and Sugimoto, R. (1989). Instability of conducting polymer, poly(3-alkylthiophene) gel with solvent, temperature and doping and effect of alkyl chain length, *Solid State Communications*, 70 (6) pp.601-603.

[31] Yoshino, K., Nakao, K., Onoda, M., and Sugimoto, R. (1989). Novel Doping Effect of Conducting Polymer Gel, *Japanese Journal of Applied Physics*, 28 (4) pp.L682-L685.

[32] Yoshino, K., Nakao, K., Onoda, M., and Sugimoto, R. (1989). Photoluminescence of poly(3-alkylthiophene) under hydrostatic pressure, *Journal of Physics: Condensed Matter*, 1 (5) pp.1009-1012.

[33] Ohmori, Y., Manda, Y., Takahashi, H., Kawai, T., and Yoshino, K. (1990). Novel Temperature-Dependent Junction Characteristics of Poly(3-alkylthiophene) Schottky Diodes, *Japanese Journal of Applied Physics*, 29 (5) pp.L837-L839.

[34] Garito, A.F., Teng, C.C., Wong, K.Y., and Zammani'khamiri, O. (1984). Molecular optics: Nonlinear optical processes in organic and polymer crystals, *Molecular Crystals and Liquid Crystals*, 106 (3-4) pp.219-258.

[35] Sinclair, M., Moses, D., Akagi, K., and Heeger, A.J. (1988). Anisotropy of the third-order nonlinear-optical susceptibility in a degenerate-ground-state conjugated polymer:trans-$(CH)_x$, *Physical Review B Condensed Matter and Materials Physics*, 38 (15) pp.10724-10733.

[36] Stamm, U., Taiji, M., Yoshizawa, M., Yoshino, K., and Kobayashi, T. (1990). Femtosecond spectroscopy poly(3-mehylthiophene) films, *Molecular Crystals and Liquid Crystals*, 182 (1) pp.147-156.

[37] Yoshino, K., Kuwabara, T.and Kawai, T. (1990). Learning effect in the doping process of conducting polymer, *Japanese Journal of Applied Physics*, 29 (6) pp.L995-L997.

[38] Diaz, A. F., Kanazawa, K.and Gardini, G.P. (1979). Electrochemical polymerization of pyrrole, *Journal of the Chemical Society, Chemical Communications*, (14) pp.635-636.

[39] Decher, G., Hong, J. D.and Schmitt, J. (1992). Buildup of ultrathin multilayer films by a self-assembly process: III. Consecutively alternating adsorption of anionic and cationic polyelectrolytes on charged surfaces, Thin Solid Films, 210-211 (2) pp.831-835.

[40] Heeger, A.J. (2001). Nobel Lecture: Semiconducting and metallic polymers: The fourth generation of polymeric materials, *Reviews of Modern Physics*, 73 (3) pp.681-700.

[41] MacDiarmid, A.G. (2001). Nobel Lecture: "Synthetic metals": A novel role for organic polymers, *Reviews of Modern Physics*, 73 (3) pp.701-712.

[42] Shirakawa, H. (2001). Nobel Lecture: The discovery of polyacetylene film—the dawning of an era of conducting polymers, *Reviews of Modern Physics*, 73 (3) pp.713-718.

[43] Burroughes, J.H., Jones, C. A.and Friend, R.H. (1988). New semiconductor device physics in polymer diodes and transistors, *Nature*, 335 (6186) pp.137-141.

[44] Braun, D.and Heeger, A.J. (1991). Visible light emission from semiconducting polymer diodes, *Applied Physics Letters*, 58 (18) pp.1982-1984.

[45] Burn, P.L., Holmes, A.B., Kraft, A., Bradley, D.D.C., Brown, A.R., Friend, R.H., and Gymer, R.W. (1992). Chemical tuning of electroluminescent copolymers to improve emission efficiencies and allow patterning, *Nature*, 356 (6364) pp.47-49.

[46] Gustafsson, G., Cao, Y., Treacy, G.M., Klavetter, F, Colaneri, N., and Heeger, A.J. (1992). Flexible light-emitting diodes made from soluble conducting polymers, *Nature*, 357 (6378) pp.477-479.

[47] Doi, S., Kuwabara, T., Noguchi, T., and Ohnishi, T. (1993). Organic electroluminescent devices having poly(dialkoxy-p-phenylene vinylenes) as a light emitting material, *Synthetic Metals*, 57 (1) pp.4174-4179.

[48] Onoda, M., Uchida, M., Ohmori, Y., and Yoshino, K. (1993). Organic electroluminescence devices using poly(arylene vinylene) conducting polymers, *Japanese Journal of Applied Physics*, 32 (9A) pp.3895-3899.

[49] Ohmori, Y., Uchida, Muro., K., and Yoshino, K. (1991). Visible-light electroluminescent diodes utilizing poly(3-alkylthiophene), *Japanese Journal of Applied Physics*, 30 (11B) pp.1938-1940.

[50] Ohmori, Y., Muro, K., Onoda, M., and Yoshino, K. (1992). Fabrication and characteristics of Schottky gated field effect transistors utilizing poly(1,4-naphthalene vinylene) and poly(p-phenylene Vinylene), *Japanese Journal of Applied Physics*, 31 (5B) pp.L646-L648.

[51] Grem, G., Leditzky, G., Ullrich, B., and Leising, G. (1992). Realization of a blue-light-emitting device using poly(p-phenylene), *Advanced Materials*, 4 (1) pp.36-37.

[52] Hamaguchi, M.and Yoshino, K. (1995). Rubbing-induced molecular orientation and polarized electroluminescence in conjugated polymers, *Japanese Journal of Applied Physics*, 34 (6A) pp.L712-L715.

[53] Wegner, G. (1991) Ultrathin layers: innovation through research, *Advanced Materials*, 3 (1) pp.8-9.

[54] Royappa, A. T.and Rubner, M.F. (1991). Novel Langmuir-Blodgett films of conducting polymers. 1. Polyion complexes and their multilayer heterostructures, *Langmuir*, 8 (12) pp.3168-3177.

[55] Ferreira, M., Cheung, J. H.and Rubner, M.F. (1994). Molecular self-assembly of conjugated polyions: a new process for fabricating multilayer thin film heterostructures, *Thin Solid Films*, 244 (1-2) pp.806-809.

[56] Onoda, M. and Yoshino, K. (1995). Fabrication of ultrathin films and multi-layer heterostructures of conductive polymers and their functionality based on molecular self-assembly technique, *Hyomen*, 33 (4) pp.260-270.

[57] Murase, I., Ohnishi, T., Noguchi, T., and Hirooka, M. (1984). Highly conducting poly(p-phenylene vinylene) prepared from a sulphonium salt, *Polymer Communications*, 25 (11) pp.327-329.

[58] Yue, J.and Epstein, A.J. (1990). Synthesis of self-doped conducting polyaniline, *Journal of the American Chemical Society*, 112 (7) pp.2800-2801.

[59] Yue, J., Epstein, A. J.and MacDiarmid, A.G. (1990). Sulfonic acid ring-substituted polyaniline: A self-doped conducting polymer, *Molecular Crystals and Liquid Crystals*, 189 (1) pp.255-261.

[60] Yoshino, K., Nakajima, S., Fujii, M., and Sugimoto, R. (1987). Conducting polymer fiber prepared by melt-spinning method from fusible polythiophene derivatives, *Polymer Communications*, 28 (3) pp.309-310.

[61] Dubitsky, Y. A.and Zhubanov, B.A. (1993). Polypyrrole-poly(vinyl chloride) and polypyrrole-cellulose acetate conducting composite films by opposite-diffusion polymerization, *Synthetic Metals*, 53 (3) pp.303-307.

[62] Sakka, S. Science of sol-gel process: *Low temperature composition of functional glass and ceramics*, Agne Shofusha Co., Ltd., 1993.

[63] Ohsawa, T., Onoda, M., Morita, S., and Yoshino, K. (1991). Preparation of glass-conducting polymer composite by sol-gel method and its property, *Japanese Journal of Applied Physics*, 30 (11B) pp.L1953-L1954.
[64] Troung, V.T. (1992). Thermal degradation of polypyrrole: effect of temperature and film thickness, *Synthetic Metals*, 52 (1) pp.33-44.
[65] Pascual, R., Sayer, M., Vasant Kumar, C.V.R., and Zou, L. (1991). Rapid thermal processing of zirconia thin films produced by the sol-gel method, J*ournal of Applied Physics*, 70 (4) pp.2348-2352.
[66] Osaka, T., Naoi, K.and Ogano, S. (1988). Effect of polymerization anion on electrochemical properties of polypyrrole and on Li / $LiClO_4$ / polypyrrole battery performance, *Journal of The Electrochemical Society*, 135 (5) pp.1071-1077.
[67] Fujishima, A., Aizawa, M.and Inoue, T. *Method for measuring electrochemistry* (Vol.1), Gihodo Shupan Co., Ltd., 1989.
[68] Keller, A.and Sawada, S. (1964). On the interior morphology of bulk polyethylene, *Die Makromolekulare Chemie*, 74 (1) pp.190-221.
[69] Illers, Von K.H. (1968). Abbau von aus der schmelze kristallisiertem polyäthylen durch salpetersäure, *Die Makromolekulare Chemie*, 118 (1) pp.88-127.
[70] Otero, T.F., Rodriguez, J., Angulo, E., and Santamaria, C. (1993). Artificial muscles from bilayer structures, *Synthetic Metals*, 57 (1) pp.3713-3717.
[71] Osada, Y., (Supervisor) *The cutting edge of the soft actuator development*, Tokyo, NTS Publishing Co., Ltd., 1996.
[72] Jonas, F.and Schrader, L. (1991).Conductive modifications of polymers with polypyrroles and polythiophenes, *Synthetic Metals*, 41 (3) pp.831-836.
[73] Yoshino, K., Tada, K., Fujii, A., Conwell, E.M., and Zakhidov, A.A. (1997). Novel photovoltaic devices based on donor-acceptor molecular and conducting polymer systems, *IEEE Transactions on Electron Devices*, 44 (8) pp.1315-1324.
[74] Brabec, C.J., Sariciftci, N.S., Hummelen, J.C. (2001). Plastic solar cells, *Advanced Functional Materials*, 11 (1) pp.15-26.
[75] Zakhidov, A.A. (1988). Polarons and solitons in composite q-1-d systems: Vectorial separation of charges. Interchain molecular photocell, *Synthetic Metals*, 27 (1-2) pp.A51-A60.
[76] Morita, S., Zakhidov, A.A. and Yoshino, K. (1992). Doping effect of buckminsterfullerene in conducting polymer: Change of absorption spectrum and quenching of luminescene, *Solid State Communications*, 82 (4) pp.249-252.
[77] Yoshino, K., Yin, X.H., Muro, K., Kiyomatsu, S., Morita, S., Zkhidov, A.A., Noguchi, T., and Ohnishi, T. (1993). Marked enhancement of photoconductivity and quenching of luminescence in poly(2,5-dialkoxy-p-phenylene vinylene) upon C_{60}doping, *Japanese Journal of Applied Physics*, 32 (3A) pp.L357-L360.
[78] Sariciftci, N.S., Smilowitz, L., Heeger, A.J., and Wudl, F. (1992). Photoinduced electron transfer from a conducting polymer to buckminsterfullerene, *Science*, 258 (5087) pp.1474-1476.
[79] Smilowitz, L., Sariciftci, N.S., Wu, R., Gettinger, C., Heeger, A.J., and Widl, F. (1993). Photoexcitation spectroscopy of conducting-polymer–C_{60} composites: photoinduced electron transfer, Physical Review B, *Condensed Matter and Materials Physics*, 47 (20) 13835-13942.

[80] Tada, K., Harada, H., Onoda, M., Nayayama, H., and Yoshino, K. (1999). Percolation in carrier transport in FET with dye doped conducting polymers, *Synthetic Metals*, 102 (1-3) pp.981.

[81] Tada, K., Onoda, M., Nayayama, H., and Yoshino, K. (1999). Photocell with heterojunction of donor /acceptor polymers, Synthetic Metals, 102 (1-3) pp.982-983.

[82] Ruoff, R.S., Tse, D.S., Malhotra, R., and Lorents, D.C. (1993). Solubility of fullerene (C60) in a variety of solvents, *The Journal of Physical Chemistry*, 97 (13) pp.3379-3383.

[83] Yu, G., Gao, J., Hummelen, J.C., Wudl, F., and Heeger, A.J. (1995). Polymer photovoltaic cells: enhanced efficiencies via a network of internal donor-acceptor heterojunctions, *Science*, 270 (5243) pp.1789-1791.

[84] Halls, J.J.M., Walsh, C.A., Greenham, N.C., Marseglia, E.A., Friend, R.H., Moratti, S.C., and Holmes, A.B. (1995). Efficient photodiodes from interpenetrating polymer networks, *Nature*, 376 (6540) pp.498-500.

[85] Yu, G.and Heeger, A.J. (1995). Charge separation and photovoltaic conversion in polymer composites with internal donor/acceptor heterojunctions, *Journal of Applied Physics*, 78 (7) pp.4510-4515.

[86] Tada, K., Hosoda, K., Hirohata, M., Hidayat, R., Kawai, T., Onoda, M., Teraguchi, M., Masuda, T., Zakhidov, A.A., and Yoshino, K. (1997). Donor polymer (PAT6) — acceptor polymer (CNPPV) fractal network, *Synthetic Metals*, 85 (1-3) pp.1305-1306.

[87] Tada, K., and Onoda, M. (2003). Preparation and application of nanostructured conjugated polymer film by electrophoretic deposition, *Thin Solid Films*, 438-439 pp.365-368.

[88] Piok, T., Gamerith, S., Gadermaier, C., Plank, H., Wenzl, F.P., Patil, S., Montenegro, R., Kietzke, T., Neher, D., Scherf, U., Landfester, K., and List, E.J.W. (2003). Organic light-emitting devices fabricated from semiconducting nanospheres, *Advanced Materials*, 15 (10) pp.800-804.

[89] Kietzke, T., Neher, D., Landfester, K., Montenegro, R., Günter, R., and Scherf, U. Novel approaches to polymer blends based on polymer nanoparticles, (2003). *Nature Materials*, 2 (6) pp.408-412.

[90] Barazzouk, S., Hatchandai, S.and Kamat, P.V. (2001). Nanostructured fullerene films, *Advanced Materials*, 13 (21) pp.1614-1617.

[91] Liu, J., Shi, Y. and Yang, Y. (2001). Solvation-induced morphology effects on the performance of polymer-based photovoltaic devices, *Advanced Functional Materials*, 11 (6) pp.420-424.

[92] Tang, C.W., (1986). Two-layer organic photovoltaic cell, *Applied Physics Letters*, 48 (2) pp.183-185.

[93] Horowitz, G., Fichou, D., Peng, X., Xu, Z., and Garnier, F. (1989). A field-effect transistor based on conjugated alpha-sexithienyl, *Solid State Communications*, 72 (4) pp.381-384.

In: New Developments in Polymer Composites Research ISBN: 978-1-62948-340-5
Editors: S. Laske and A. Witschnigg

Chapter 4

NANOCOMPOSITES BASED ON BLOCK SULFONATED POLYARYLENETHIOETHERSULFONE COPOLYMERS AND IN SITU-GENERATED NANOPARTICLES OF ZIRCONIUM PHOSPHATE

***Zongwu Bai*[1,2,*], *Marlene D. Houtz*[1,4] *and Barney E. Taylor*[3,4]**
[1]University of Dayton Research Institute, University of Dayton, Dayton, Ohio, US
[2]AFRL/RXCC, Materials & Manufacturing Directorate, Wright-Patterson Air Force Base, Dayton, Ohio, US
[3]Universal Technology Corporation (UTC), Dayton, Ohio, US
[4]AFRL/RXAS, Materials & Manufacturing Directorate, Wright-Patterson Air Force Base, Dayton, Ohio, US

ABSTRACT

Novel nanocomposite membranes were successfully fabricated by an in-situ precipitation method. In this method, the swollen membranes of the acid form block sulfonated polythioethersulfone copolymer (B-SPTES) were immersed in aqueous solutions of zirconium dichloride oxide, followed by treatment in phosphoric acid solution. During this process, the zirconium phosphate (ZrP) nanoparticles, in situ, were precipitated into the swollen pores of the acid form B-SPTES membranes. Zirconium phosphate nanoparticles, which are in situ incorporated up to 20 wt%, were introduced into the B-SPTES copolymer matrix to improve proton conduction at high temperature and low relative humidity. ZrP/B-SPTES nanocomposite membranes showed enhanced proton conductivity up to 36 mS/cm at 120°C, 50% relative humidity. Transmission electron microscopy (TEM) analysis revealed that the ZrP nanoparticles were finely dispersed in B-SPTES copolymer domains with less than 10 nm in particle size.

Keywords: Nanocomposites, block sulfonated polymer, zirconium phosphate, proton exchange membrane, proton conductivity

* E-mail address: zongwu.bai@wpafb.af.mil (Corresponding author)

INTRODUCTION

Polymer electrolyte membrane fuel cells (PEMFCs) function by permitting electrochemical combination of a fuel and an oxidant. These feeds are supplied to the anode (fuel) and cathode (oxidant) in a manner such that power can be continuously drawn from the fuel cell. The feeds can't be allowed to combine and are separated from each other by an electrolyte whose principal function is to permit the transfer of ions through the membrane while prohibiting the direct contact of fuel and oxidant. In recent years, PEMFCs have been identified as promising power sources for vehicular transportation and for other applications requiring clean, quiet, and portable power. The polymer electrolyte membrane is one of the most critical components in polymer electrolyte membrane fuel cells. Polymer electrolyte membrane fuel cells based upon perfluorinated membranes have typically been operated in a temperature range between approximately 50 and 90°C [1]. This temperature range is a compromise between competing factors. Increasing the operating temperature above room temperature will improve the electrode kinetics of the oxygen reduction reaction [2, 3]. The upper limit of temperature results from the difficulty in maintaining membrane water content at temperatures at or above 100°C. In addition, temperatures above the polymer glass transition temperature (~110°C for protonated Nafion) can cause polymer chain rearrangements, which can lead to structural changes in the membrane and lower membrane stability, performance, and lifetime [4, 5]. Additionally, conventional membrane electrolytes are often based on costly per fluorinated polymers making commercialization economically challenging.

In last decades, many researchers have been focused on the synthesis of various sulfonated aromatic polymers that have high thermal, chemical and oxidative stability, good mechanical properties and low cost. Usually they are prepared either by post-sulfonation of commercial polymers or by direct synthesis of sulfonated polymer via copolymerization of sulfonated monomers. Hydrocarbon polymers containing polar groups that retain high amounts of water over a wide temperature range are particularly attractive and significantly less expensive to synthesize than perfluorinated polymers. Sulfonated high performance polymers based on arylene ether [6], sulfone [7, 8] and thioether [9, 10] linkages in the backbone have been synthesized and described in literature. These polymer systems are potential candidates for fuel cell applications as they exhibit good thermal and chemical stability and high proton conductivity. The block sulfonated polyarylenethioethersulfone (B-SPTES) copolymers [11], containing hydrophilic disulfonated and hydrophobic unsulfonated blocks, exhibit exceptional promise as an electrolyte separator material for fuel cell applications. These polymers have been observed to exhibit extremely high conductivity based upon both conventional four probe impedance tests and single cell fuel cell analysis. Morphological studies of the B-SPTES copolymers show a very well defined phase separation and a distinct morphological architecture (~50 nm) compared to random SPTES copolymers, which suggested that the hydrophilic groups aggregate as isolated domains with some local connection of hydrophilic domains as evidenced by the good conductivity of the B-SPTES copolymers. Additionally, these polymers are strong, chemically inert and thermally stable making them ideal for this application.

Optimized proton and water transport properties of the polymer membrane and proper water management are crucial for efficient fuel cell operation. While dehydration of the membrane reduces proton conductivity, an excess of water can lead to flooding of the electrodes. Both conditions result in poor fuel cell performance. The proton conductivity of the ionomer membranes relies on the presence of water. At temperatures above 100°C there is a drastic decrease in conductivity of the membrane because of the high evaporation rate of water. Thus, a modification of the polymer membranes for operation at higher temperatures would be an alternative approach [12]. So by combining optimized water management and proton conductivities into a composite polymer system, it becomes possible to improve the performance of the PEMs by addressing both of the conductivity issues at the same time. This results in a good balance of properties in the final polymer, and brings a system with potentially superior performance to currently available technologies. Previous research [13, 14], that focused on the composite approach to prepare the membranes based on proton conducting materials such as sulfonated polyaryleneetheretherketone and SPTES membranes showed excessive swelling, which rendered the membranes brittle upon drying.

The addition of an inorganic material into a polymer membrane can alter and improve physical and chemical polymer properties of interest (such as elastic modulus, proton conductivity, solvent permeation rate, tensile strength, hydrophilicity, and glass transition temperature) while retaining other important polymer properties to enable operation in a fuel cell at high temperatures. The hydration properties of the membranes are key characteristics that influence fuel cell performance. There have been several studies in literature to evaluate the proton conductivity of ZrP after its incorporation within nonconductive materials such as polymers. Examples are composite membranes prepared by the precipitation of zirconium phosphate in situ in polytetraflouroethylene (PTFE) [15], in Nafion [16], and in the composites prepared by mixing the PTFE emulsion and the ZrP powder followed by drying and pressing [17].Our effort is to develop a nanocomposite by introducing zirconium phosphate (ZrP) nanoparticles into the B-SPTES copolymer matrix to potentially improve the proton conductivity and decrease the water uptake. The obtained ZrP/B-SPTES nanocomposite was characterized with current techniques to reveal the microstructures, water uptake, and morphology. Proton conductivity of B-SPTES membranes was compared as functions of temperature (30–120°C) and water activity to elucidate the role of zirconium phosphate in altering the membrane performance in PEM fuel cells.

Experimental Methods

1. Preparation of ZrP/B-SPTES Nanocomposite Membrane

Materials: Zirconium dichloride oxide octahydrate ($ZrOCl_2 \cdot 8H_2O$, 98%, Sigma-Aldrich Co.), Phosphoric acid (85%, ACS reagent, Sigma-Aldrich Co.), and N-dimethylacetamide (DMAc, Anhydride, 99%, Sigma-Aldrich Co.), were used as received. Block sulfonated polyarylenethioethersulfone (B-SPTES) copolymer was prepared following the procedures, disclosed elsewhere [12]. Other reagents and solvents were obtained commercially and used without further purification. The ZrP/B-SPTES membranes were typically prepared as follows: To obtain uniform high-purity B-SPTES membranes as the starting material, the membrane was cleaned by rinsing with distilled water, followed by 2 M sulfuric acid (H_2SO_4) for 1 hour to remove any metallic/ionic impurities, and followed by another rinsing cycle in distilled water to remove any excess acid. The treated B-SPTES membrane, that is acid form, was then immersed in a 2 M aqueous solution of zirconium dichloride oxide octahydrate ($ZrOCl_2 \cdot 8H_2O$) for 4 hours at 80°C. After rinsing in cold water, the membrane was finally placed in a 2M phosphoric acid (H_3PO_4) solution overnight at 80°C for the formation of zirconium phosphate ($Zr(HPO_4)_2$), which was entrapped in the nano-pores of the B-SPTES membrane. The nanocomposite membrane was then repeatedly rinsed in distilled water to remove excess acid. The overall chemical reaction of this processing is shown in Scheme 1.

$$\mathbf{ZrOCl_2 + 2H_3PO_4 \longrightarrow Zr(HPO_4)_2 + H_2O + 2HCl}$$

Scheme 1. Chemical reaction of zirconium phosphate (ZrP).

2. Measurement of Water Uptake and Proton Conductivity

Water uptake of the nanocomposite membranes was determined from the difference in weight (wt) between the dry and the swollen membranes. The membrane, after vacuum drying at 100°C for 24 hours, was weighed and then soaked in water for 24 hours, and wiped with blotting paper (soaking paper) and weighed again. The percentage of water absorbed was calculated with reference to the weight of the dry specimen from the following equation (1):

$$Wateruptake(wt\%) = \frac{Wwet - Wdry}{Wdry} \times 100 \tag{1}$$

Proton conductivity was measured using AC impedance spectroscopy and utilized a standard 4-electrode measurement setup to eliminate electrode and interfacial effects. The Teflon sample fixture was placed inside a temperature and humidity controlled oven and the sample was exposed to the humidified air within the chamber. The two outer electrodes were made of platinum foil and these acted to source the current in the sample. Two inner platinum wire electrodes (spaced 1 cm apart) were then used to measure the voltage drop across a known distance. By measuring the impedance of the material as a function of frequency at a set temperature and humidity, the conductivity of the membrane is obtained using the magnitude of the impedance in a region where the phase angle is effectively zero. The conductivities were calculated by using equation (2):

$$\sigma = \frac{l}{R \times A} \tag{2}$$

Where σ is the proton conductivity, l = 1cm, R is the resistance, and A is the cross section area; $A = T \times W$, T is the thickness of the sample, W is the width of sample.

3. Thermogravimetry Analysis (TGA) and Thermomechanical Analysis (TMA)

Thermogravimetry analysis (TGA) was conducted using an Auto TGA 2950HR V5.4A thermogravimetric analyzer to study the thermal stability of nanocomposite membranes. The membranes were dried *in vacuo* for at least 12 hours for the removal of absorbed water before the experiments. The samples were pieces of thin films and had a weight total of 5–10 mg. The samples were evaluated over the range of 30–900°C at a heating rate of 10°C/min in N_2 and air. The in-plane coefficient of thermal expansion (CTE) of the nanocomposites was measured as a function of temperature via thermomechanical analysis (TMA). The technique utilized a TA Instrument TMA 2940 in tensile film mode with a heating rate of 4°C/min and typical sample dimensions of 100 mm x 15 mm x 0.25 mm. The CTE was determined from the slope of the resultant expansion temperature plots.

4. X-Ray Diffraction (XRD)

The wide angle x-ray scattering regime was collected using an evacuated Statton camera (Warhus, DE, USA) with a point-collimated x-ray beam achieved via two pinholes (0.05 mm), separated by ~15 cm at Cu Kα wavelength 1.5418 Å. In order to examine scattering from dry membranes, the ZrP/B-SPTES nanocomposite samples were placed in a vacuum oven at 80°C for 3 days, then quickly placed in a quartz capillary and sealed.

5. Scanning Electron Microscopy (SEM) and Transmissiom Electron Micrographs (TEM)

Surface morphology of the cryo-fractured dried and swollen membranes was examined by a Serion filed emission SEM (FEI, USA). The instrument was operated at 5Kv in the secondary electron imaging mode. High Resolution TEM analyses were conducted using a Philips CM200 Field Emission Gun (FEG) instrument operating at 200kV. A CCD camera was used for focusing at high magnification. Images were collected on an SO-163 film and digitized using a Minolta DiMAGE scanner at 2400 dpi resolution. The ZrP/B-SPTES nanocomposite film was dried, embedded in epoxy and ultra-microtomed. Ultra-microtoming was performed using a RMC PowerTome XL to the thickness of ~50 nm at room temperature using a 45° diamond knife with a 6° clearance angle.

Results and Discussion

1. The Reaction and Mechanism in the Fabrication of ZrP/B-SPTES Nanocomposites

Zirconium phosphate ($Zr(HPO_4)_2 \cdot H_2O$) was the desired additive to sulfonated polymer-based membranes because of its attributes including: (i) it has moderate proton conductivity when humidified (~10^{-3} S/cm) [18, 19]; (ii) it is a Brønsted acid with the ability to donate protons; (iii) it is thermally stable to temperatures above 180°C; (iv) it is hygroscopic and hydrophilic; and (v) it is easily synthesized in a manner that is compatible with the chemical and physical limits of the sulfonated polymer membrane [20]. In practice the zirconium phosphate phase may vary from amorphous to crystalline; the proton conductivity of zirconium phosphate can vary from 10^{-7} to 10^{-3} S/cm depending on the phase composition, structure and hydration state.

Zirconium phosphate was incorporated into B-SPTES copolymer membranes by using the procedure described in above. There is only the over-all chemical reaction for the preparation of zirconium phosphate nanocomposites as shown in Scheme 1. The detailed synthesis reactions of ZrP and ZrP/B-SPTES nanocomposite should include the following steps:

(1) Zr Ions Exchange with Protons of Sulfonated Group on B-SPTES

The membranes were dipped into a 2M solution of zirconium dichloride oxide ($ZrOCl_2$) for several hours at 80°C. The $ZrOCl_2$ solution diffuses into the membrane and the excess of Zr^{4+}ions within the membrane leads to an exchange with sulfonic acid protons in the membrane. The chemical reaction is shown in Scheme 2.

$$\mathbf{ZrOCl_2 + 4\ \text{-}SO_3H \longrightarrow Zr(SO_3)_4 + H_2O + 2HCl}$$

Scheme 2. Chemical reaction of zirconium sulphate (ZrS).

To accomplish this synthesis within the membrane, the procedure takes advantage of the fact that B-SPTES copolymers are ion-exchange membranes, in which the protons of B-SPTES copolymers are exchanged with zirconium ions in a solution of Zr^{4+} ions to form Zr-B-SPTES salt. The free Zr^{4+} ions are bonded with four of $-SO_3H$ groups to have zirconium sulphate (ZrS), and stay on the B-SPTES copolymer backbone. The proposed mechanism and structure is shown in Figure 1. From this structure, we understood that the chemical reaction of Zr^{4+} and $-SO_3H$.

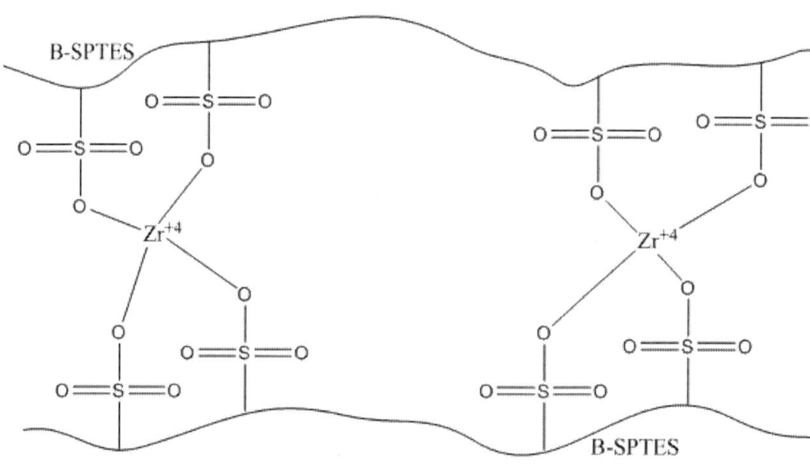

Figure 1. Chemical reaction structures of zirconium ions and B-SPTES copolymers.

(2) The Zirconium Ions are Subsequently Reacted in Place with Phosphoric Acid

The membranes were then immersed in 2 M phosphoric acid (H_3PO_4) overnight at 80°C. The chemical reaction is shown in Scheme 3. The phosphoric acid has two purposes: (i) it reacts with the Zr^{4+}ions to precipitate insoluble zirconium hydrogen phosphate in situ, and (ii) the acidic solution can re-protonate the sulfonate anions to regenerate the acidity of the membrane.

$$Zr(SO_3)_4 + 2\ H_3PO_4 \longrightarrow Zr(HPO_4)_2 + 4\ \text{-}SO_3H$$

Scheme 3. Chemical reaction of zirconium phosphate (ZrP).

Phosphoric acid (H_3PO_4) leading to the precipitation of the insoluble zirconium phosphate in the B-SPTES copolymer membranes (see Figure 2), thus, *in-situ*, form the zirconium phosphate nanocomposites. The ZrP/B-SPTES nanocomposite membranes were then repeatedly rinsed for several times in distilled water to remove any excess acid and $ZrOCl_2$ solution. After drying, the weight of nanocomposite membrane was increased about 20% as compared with the unmodified membrane.

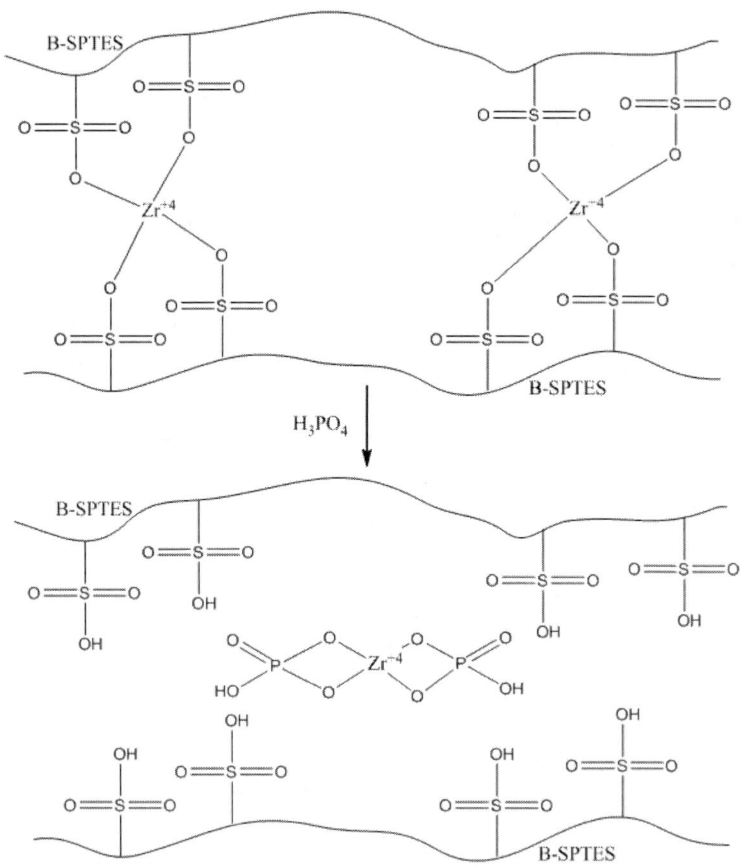

Figure 2. Chemical reaction scheme of zirconium phosphate in-situ formed in B-SPTES copolymer membranes.

The density of hydrophilic functional groups in the nanocomposite is responsible for the proton conductivity and its transport across the given polymeric membranes. Incorporation of ZrP in the sulfonated polymer matrix may lead chelation between –OH and S=O of ZrP and –SO_3H [21]. A schematic structure for proposed hydrogen bonding between ZrP and –SO_3H groups of sulfonated polymer matrix is presented in Figure 3. Due to chelation, proton-donating ability of –SO_3H reduced because of its bonding. Thus incorporation of ZrP leads reduction in proton conductivity values at lower temperature for prepared ZrP/B-SPTES nanocomposite membranes because of low membrane surface charge density.

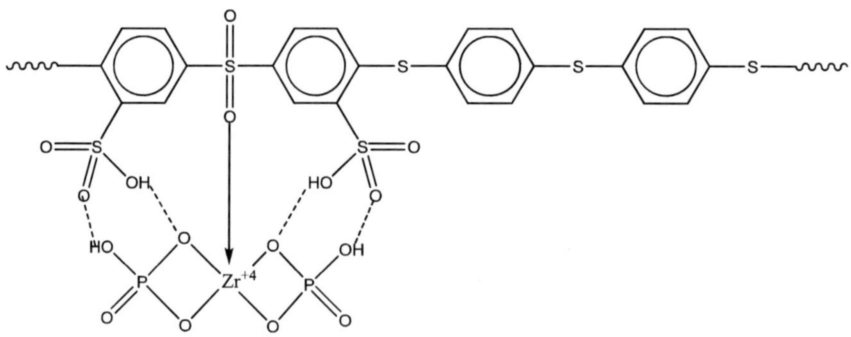

Figure 3. Possible crosslink mechanism of zirconium phosphate (ZrP) formed in the B-SPTES copolymer membranes.

2. Structures of Zirconium Phosphate (ZrP) and ZrP/B-SPTES Nanocomposite

Zirconium phosphate or Zirconium hydrogen phosphate (ZrP, $Zr(HPO_4)_2 \cdot H_2O$), was first synthesized by reflux [22]. Extensive investigations on synthesis methods and structural properties of ZrP have led to its use as a proton conductor [23]. There are two predominate structures of zirconium phosphate *alpha* and *gamma*. The structure of *alpha*-ZrP (see Figure 4) lends itself best to proton transport because it has a pendant -OH group which extends into the interlayer region and forms hydrogen bonded network with water.

The structural features of the zirconium phosphate and ZrP/B-SPTES nanocomposite were investigated by X-ray diffraction. XRD patterns were obtained for the ZrP/B-SPTES nanocomposite and B-SPTES copolymer membranes in the state of a thoroughly dried membrane (see Figure 5). The well-dried ZrP/B-SPTES nanocomposite membrane shows several sharp, well-defined peaks (see Figure 5 curve (b)), the peak at $2\theta = 11.7^o$ (d-spacing = 7.59 Å) is characteristic of the stacking of several *alpha* zirconium phosphate layers. This interlayer distance slightly differs from the one found by Clearfield [22] (see Figure 4, ca. 7.60 Å), which is probably due to an effect of confinement on the stacking formation inside the B-SPTES membrane. The broad peak at $2\theta = 18.6^o$ (d-spacing = 4.76 Å) are attributed to crystalline stacking in the B-SPTES-rich microphases, which is observed in the B-SPTES copolymer (see Figure 5 curve (a)) with peak at $2\theta = 19.0^o$ (d-spacing = 4.65 Å). The other peaks ($2\theta = 19.9^o$, d-spacing = 4.46 Å; $2\theta = 24.6^o$, d-spacing = 3.62 Å; and $2\theta = 34.2^o$, d-spacing = 2.62 Å) in Figure 5 curve (b) that are attributed to the presence of zirconium phosphate phases. These zirconium phosphate peaks match up with typical peaks found in *alpha* zirconium phosphate ($2\theta = 19.9^o$, d-spacing = 4.46 Å and $2\theta = 25.1^o$, d-spacing =3.54 Å) [24]. It is notable that diffraction maxima $2\theta = 19.9^o$ (d-spacing = 4.46 Å) for the ZrP/B-SPTES nanocomposite membrane may be attributed to the stacking of several *alpha* zirconium phosphate layers. This peak is pretty large compared to the other crystalline peaks suggesting that the packing along this direction does not extend over large distances.

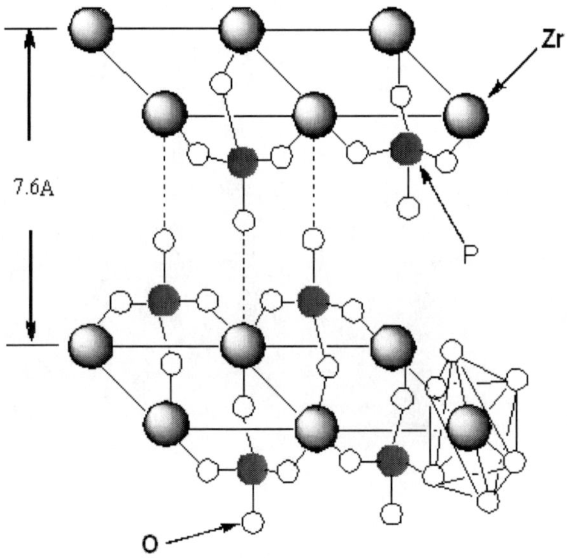

Figure 4. Sketch of *alpha*-zirconium phosphate (ZrP) crystal structure.

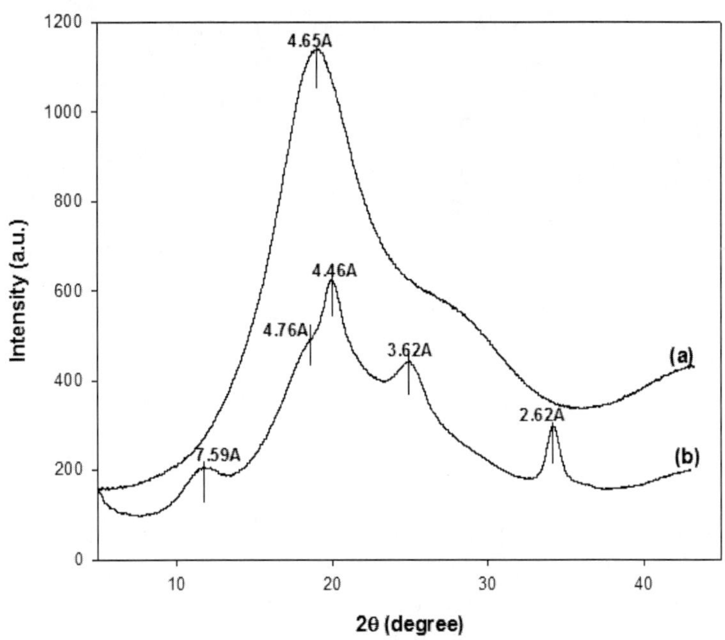

Figure 5. X-ray diffraction curves of (a) B-SPTES reference, (b) ZrP/B-SPTES nanocomposite membrane.

The width of the diffraction peak was used to estimate the zirconium phosphate particle size using the Debye-Sherrer formula with the Warren correction for instrumental effects. In the fully dried state, zirconium phosphate particles are estimated to be 11 ± 1 nm in size [25]. The peak narrowing reflects the growth of the zirconium phosphate particles, which can be seen from the TEM images in Figure 6. The zirconium phosphate particles have a size around 10 nm.

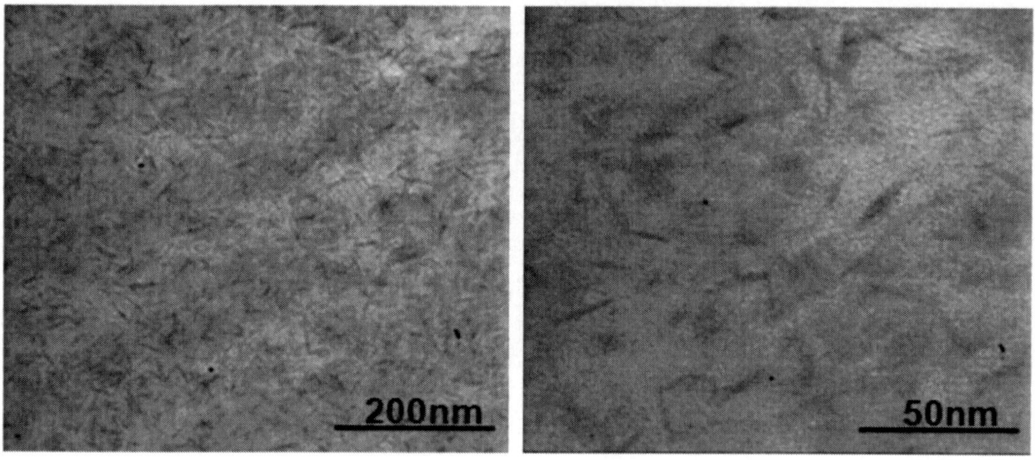

Figure 6. TEM micrographs of the ZrP/B-SPTES nanocomposite membranes.

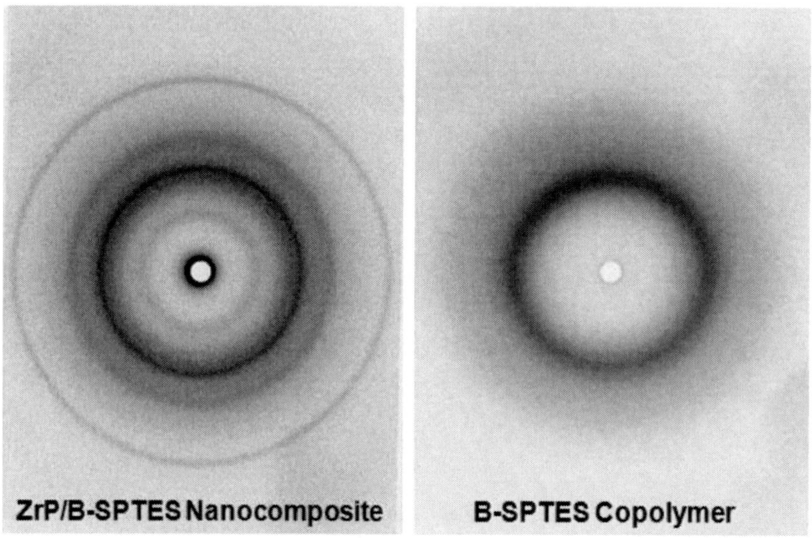

Figure 7. WAXS patterns for (a) ZrP/B-SPTES nanocomposite and (b) B-SPTES membranes.

The X-ray patterns are not very sharp (see Figure 7), and water uptake by the membrane disrupts the crystalline microstructure of B-SPTES copolymer. It is likely that the membrane swelling due to water uptake influences the crystalline packing of polymer chains as well as the structure of the zirconium phosphate phases.

3. Proton Conductivity of ZrP/B-SPTES Nanocomposites

For the PEMs applications in fuel cells, the proton conductivity of the membrane is particularly important since it has significant role in their fuel cell performance. The introduction of zirconium phosphate into the B-SPTES copolymers formed the nanocomposite. The zirconium phosphate in the nanocomposite may increase the proton conductivity as well as balance ratio of hydrophilicity to hydrophobicity which should greatly reduce dimensional changes during operation at higher temperatures and low relative humidity. Four probe proton conductivity measurements were performed at different temperatures between 35 and 85°C and relative humidities between 55 and 85%.

The conductivity data of nanocomposite membranes for different relative humidities are presented as a function of inverse temperature in Figure 8. It can be seen that conductivity values increased with increasing temperature for a given relative humidity. At a constant temperature, the conductivity increased with higher values of relative humidity. At higher temperature (>65°C), the proton nanocomposite membrane conductivity was comparable to the B-SPTES membrane for all values relative humidity measured. Lower proton conductivity at low temperature followed by a significant increase at high temperatures, suggests that the incorporation of ZrP has stabilized dimensional swelling and reduced water dilution at high temperature.

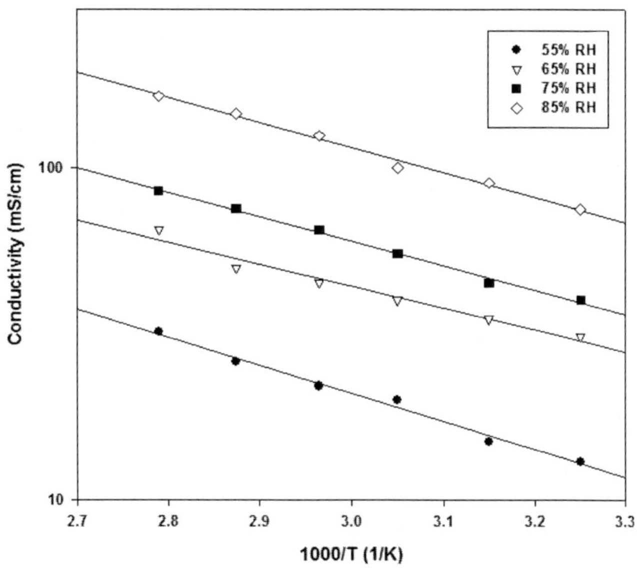

Figure 8. Temperature and relative humidity dependence of proton conductivity of ZrP/B-SPTES nanocomposite membranes.

The proton transport mechanism in ZrP at room temperature is dominated by surface transport that is four orders of magnitude greater than the bulk transport [26]. However the crystallinity also plays an important role in charge transport. In addition, the conductivity of ZrP is highly dependent upon the hydration, varying by two orders of magnitude as the relative humidity is increased [27]. Recent research has confirmed the dominance of the surface transport and demonstrated enhancements that can be made through modification of the P-OH groups [28, 29].

We expected the ZrP/B-SPTES nanocomposite membranes to exhibit an increased conductivity compared to B-SPTES copolymer membranes, based on the water content was higher in the ZrP/B-SPTES nanocomposite membrane and the zirconium phosphate may contribute extra protons to increase the charge carrier concentration. Figure 9 compares the proton conductivity of the ZrP/B-SPTES nanocomposite membrane with B-SPTES copolymer membrane. The ZrP/B-SPTES nanocomposite membranes had lower proton conductivity than B-SPTES copolymer membranes over the entire range of water activities and temperatures measured. The major proton conductivity change for the ZrP/B-SPTES nanocomposite membrane is associated with the water activity. The proton conductivity of the ZrP/B-SPTES nanocomposite membranes increased with temperature, but the variation of the ZrP/B-SPTES nanocomposite membrane conductivity with temperature was less than that observed for the B-SPTES copolymer membrane.

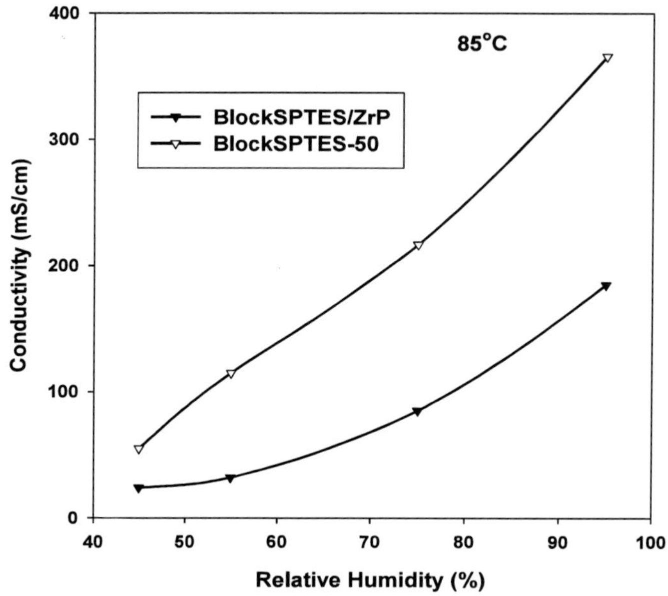

Figure 9. The proton conductivity of ZrP/B-SPTES nanocomposites as a function of the inverse temperature comparing with B-SPTES copolymer.

Table 1. Proton conductivity comparison of ZrP/B-SPTES nanocomposite and B-SPTES copolymer membrane

Sample	Proton Conductivity 85°C, 85%RH (mS/cm)	Proton Conductivity 120°C, 50%RH (mS/cm)
ZrP/B-SPTES nanocomposite	142	36
Block SPTES-50 copolymer	265	5

From Figure 9, we can see that the proton conductivity of ZrP/B-SPTES nanocomposite membrane (142 mS/cm) was reduced almost 50% compared to the B-SPTES copolymer membrane (265 mS/cm) at 85°C, 85% relative humidity. The ZrP/B-SPTES nanocompositecan be used to 120°C or higher. Table 1 shows the preliminary test results for the ZrP/B-SPTES nanocomposite and B-SPTES copolymer membrane. At 120°C, 50% relative humidity, the proton conductivity of the ZrP/B-SPTES nanocomposite is six times higher than that of the B-SPTES copolymer membrane.

4. Thermal Properties: Thermal Stability

Figure 10 shows the TGA curves of B-SPTES copolymer (in air) and ZrP/B-SPTES nanocomposite (in air and nitrogen). The B-SPTES copolymer exhibited two steps decompositions. The first weight loss was observed around 260°C and is attributed to the splitting-off of sulfonate acid group. The second decomposition step around 440°C corresponds to main chain decomposition. The 5% weight loss temperature Td = 292°C (onset ca.) for B-SPTES copolymer. The ZrP/B-SPTES nanocomposite showed a two-step degradation pattern in air, leaving behind some inorganic residue (12 wt. %). The first weight loss was observed around 270°C. It is attributed to the splitting-off of sulfonic acid group from the matrix of B-SPTES copolymer in nanocomposite. The second degradation step around 480°C corresponds to the main chain decomposition of B-SPTES copolymer. The 5% weight loss temperature Td = 315°C (onset ca.).

It is clear that the zirconium phosphate nanoparticles should have the influence on the thermal stability of ZrP/B-SPTES nanocomposite (480°C) comparing with B-SPTES copolymer (440°C). And also, it is notice that 5% weight loss temperature of ZrP/B-SPTES nanocomposite raised about 25°C, which indicated that the zirconium phosphate improved thermal stability of B-SPTES copolymer. The increased thermal stability makes ZrP/B-SPTES nanocomposite material suitable for high temperature PEM use.

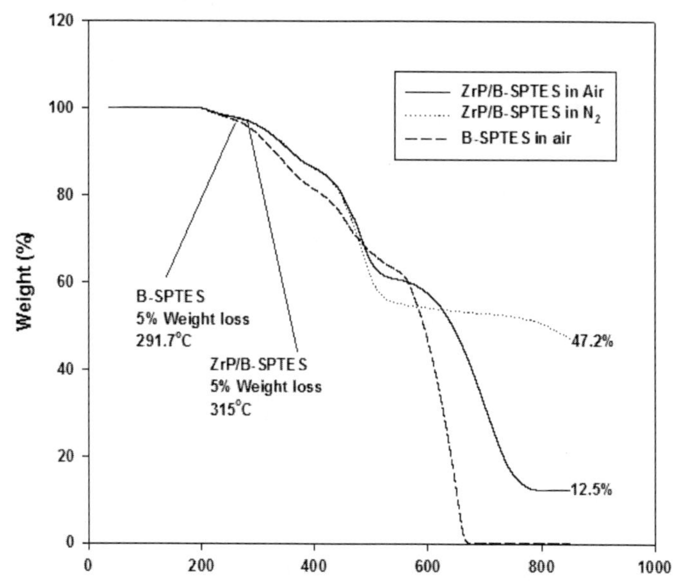

Figure 10. TGA of ZrP/B-SPTES nanocomposite and B-SPTES copolymer membranes.

5. Thermomechanical Properties

Dimensional stability of PEM materials is important for the operation of fuel cells at high temperatures and humidities. The temperatures, mechanical properties, and humidities are known to cause dimensional changes of polymer electrolyte materials.

Thermomechanical behavior of the ZrP/B-SPTES nanocomposite membranes was examined by TMA to determine the in-plane coefficient of thermal expansion (CTE). The ZrP/B-SPTES nanocomposite membranes were loaded uniaxially with a tensile stress of 0.5 MPa, and the growth in the sample dimension was monitored with increasing temperature. Below Tg, the applied stress is significantly below the yield stress of the polymers. However, above Tg the applied stress yielded a significant creep in the samples, allowing the accurate determination of Tgitself, albeit with an inaccurate measurement of the expansion coefficient above Tg. Thermal mechanical test of ZrP/B-SPTES nanocomposite membranes was used to evaluate the nanocomposite PEM's dimensional stability at high temperature for fuel cell applications.

The results for the ZrP/B-SPTES nanocomposites are shown in Figure 11. At low temperatures, we observed a broad linear regime with a small upward curvature (around 20°C, CTE = 3.8 x 10^{-3} mm/°C), indicating a nearly constant coefficient of thermal expansion (CTE Avg. ~8.0 x 10^{-4} mm/°C) along with increasing temperature. At temperatures near Tg, the applied stress can yield a significant creep in the samples. The dimensional change in tensile model shows a large jump in linear expansion (~170°C, CTE = 0.28 mm/°C) for ZrP/B-SPTES nanocomposites.

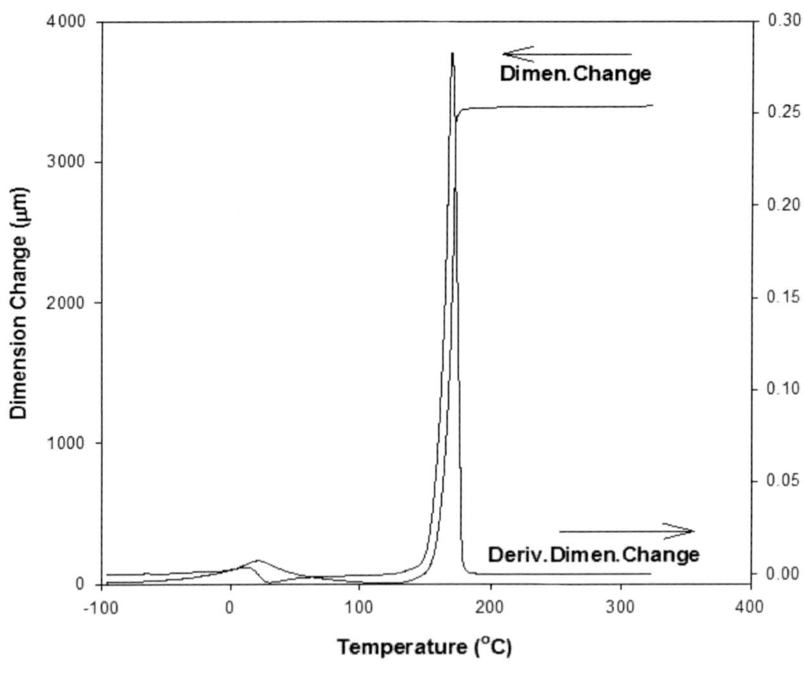

Figure 11. TMA result of ZrP/B-SPTES nanocomposite membranes.

6. Morphology and Microstructure

SEM observations have been conducted on the ZrP/B-SPTES nanocomposite membrane which has been previously cryo-fractured in order to visualize the structure along the membrane thickness. Figure 12 shows SEM micrographs of ZrP/B-SPTES nanocomposite membrane after drying. The dark region represents the B-SPTES-copolymer matrix, the white spots represents inorganic ZrP particles. The scale of the pictures from left to right is 3.0μm, 1.0μm, and 300nm respectively. The inorganic ZrP particles dispersed in the B-SPTES copolymer matrix appear as needle-like platelets without any preferential orientation and widely spaced (see Figure 12, left, center, and right). Some fluffy domains appears on the pictures which can be attributed either to some ZrP species highly dispersed in the B-SPTES copolymer matrix or more likely to ZrP particles embedded below the surface. This observation is in agreement with the results deduced from the WAXS region analysis (see Figure 7).The inorganic ZrP particles are homogeneously distributed along the membrane cross section. They appear as ~50 nm wide platelets with a thickness of ~10 nm corresponding to the stacking of about 12-15 layers of ZrP. These large dimensions suggest that the synthesis does not occur within the ionic domains but in the inter bundle areas where the spatial arrangement of the B-SPTES should be less compact. As a consequence, these particles present some ion conducting properties but probably do not promote the proton conduction on a local scale. However, they could help for water retention at elevated temperature by acting as nanotanks dispersed in the B-SPTES copolymer matrix with an inter distance of ~50 nm [11].

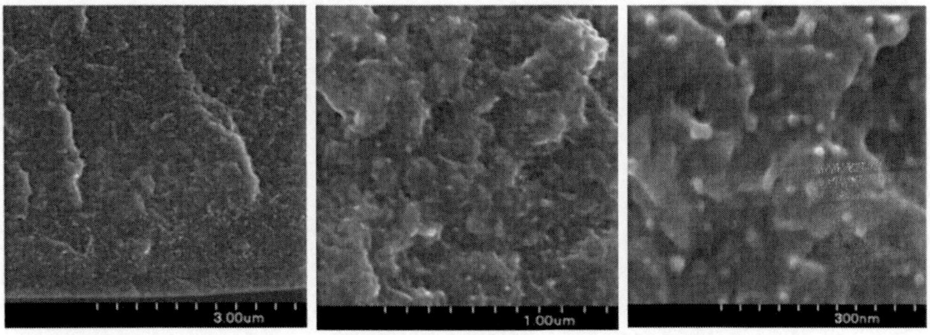

Figure 12. SEM images of ZrP/B-SPTES nanocomposite membrane, Dry film observation.

Conclusion

We have demonstrated the advantages of phase segregated hydrophilic domains in multiblock sulfonated polyarylenethioethersulfone (B-SPTES) copolymer PEMs that has resulted in high proton conductivities. Preliminary findings also revealed that nanostructuring the block copolymer PEMs with hydrophilic zirconium phosphate (ZrP) provided reasonable proton conductivities even at temperatures of 120°C. The nano-dispersant was preferentially sequestered in the hydrophilic domains of the block copolymer matrix and thus formed a part of the ionic clusters responsible for proton transfer. The inherent hydrophilicity of the nanoparticle enhanced the hydration capability/moisture retention of the PEM thus preserving high proton conductivity at the higher temperatures of the fuel cell's operation. The incorporation of the ZrP nanoparticles can also constrict the pores in the block copolymer matrix and hence create greater resistance to water molecules as a consequence of better dimensional stability.

REFERENCES

[1] Srinivasan, S. et.al. Mugerwa, M.N. (Eds.), (1993) Fuel Cell Systems, *Plenum Press*, New York, p234-246.

[2] Mukerjee, S., Srinivasan, S., (1993) Enhanced electrocatalysis of oxygen reduction on platinum alloys in proton exchange membrane fuel cells, *J. Electroanal. Chem.*, 357, 201-224.

[3] Li, Q.F., He, R.H., Jensen, J.O., Bjerrum, N.J., (2003) Approaches and recent development of polymer electrolyte membranes for fuel cells operating above 100°C, *Chem. Mater.*, 15, 4896-4915.

[4] Hinatsu, J.T., Mizuhata, M., Takenaka, H., (1994) Water uptake of perfluorosulfonic acid membranes from liquid water and water vapor, *J. Electrochem. Soc.*, 141, 1493-1498.

[5] Zook, L.A., Leddy, J., (1996) Density and solubility of Nafion: recast, annealed, and commercial films, *Anal. Chem.*, 68, 3793-3796.

[6] Ueda, M., Toyota, H., Teramoto, T., (1993) Synthesis and characterization of aromatic poly (ether sulfone)s containing pendant sodium sulfonate groups, *J. Polym. Sci. Polym. Chem. Ed.*, 31, 853-858.

[7] Allam, C., Liu, K.J., Mohanty, D.K., (1999) Preparation and properties of novel aromatic poly (thioethers) derived from 4, 4'-thiobisbenzenethiol, Macromol. *Chem. Phy.*, 200, 1854-1862.

[8] Nolte, R., Ledjeff, K., Bauer, M., Mulhaupt, R., (1993) Partiallysulfonated poly (arylene ether sulfone) - A versatile proton conducting membrane material for modern energy conversion technologies, *J. Membr. Sci.*, 83, 211-220.

[9] Bai, Z., Dang, T.D., Durstock, M.F., (2006) Proton conductivity and properties of sulfonated polyarylenethioethersulfones as proton exchange membranes in fuel cells, *J. Membr. Sci.*, 281, 508–516.

[10] Wiles, K.B., Wang, F., McGrath, J.E., (2005) Directly copolymerized poly (arylene sulfide sulfone) disulfonated copolymers for PEM-based fuel cell systems. I. Synthesis and characterization, *J. Polym. Sci. Polym Chem. Ed.*, 43, 2964-2976.

[11] Bai, Z., Yoonessi, M. Juhl, S.B., Drummy, L.F., Durstock, M.F., Dang, T.D., (2008) Synthesis and characterization of multiblock sulfonated poly (arylenethioethersulfone) copolymers for proton exchange membranes, *Macromolecules,* 41, 9483-9486.

[12] Javaid Zaidi, S.M., (2005) Preparation and characterization of composite membranes using blends of SPEEK/PBI with boron phosphate, *Electrochimica Acta,* 50, 4771–4777.

[13] Kerres, J.A., Ullrich, A., Meier, F., Haring,Th., (1999) Synthesis and characterization of novel acid–base polymer blends for application in membrane fuel cells, *Solid State Ionics,* 125, 243–249.

[14] Bai, Z., Price, G.E., Yoonessi, M., Juhl, S.B., Durstock, M.F., Dang, T.D., (2007) Proton exchange membranes based on sulfonated polyarylenethioethersulfone and sulfonated polybenzimidazole for fuel cell applications, *J. Membr. Sci.* 305, 69-76.

[15] Savadogo, O., (2004) Emerging membranes for electrochemical systems: II. High temperature composite membranes for polymer electrolyte fuel cell (PEFC) applications, *J. Power Sources,* 127, 135-161.

[16] Si, Y., Jiang, R., Lin, J.C., Kunz, H.R., Fenton, J.M., (2004) CO tolerance of carbon-supported platinum-ruthenium catalyst at elevated temperature and atmospheric pressure in a PEM fuel cell, *J. Electrochem. Soc.*, 151, A1820-A1824.

[17] Park, Y.I., Kim, J.D., Nagai, M., (2000) Increase of proton conductivity in amorphous phosphate–Nafion membranes, *J. Mater. Sci. Lett.*, 19, 1621-1623.

[18] Alberti, G., Casciola, M., (2001) Solid state protonic conductors, present main applications and future prospects, *Solid State Ionics,* 145, 3-16.

[19] Glipa, X., Lelout, J.M., Jones, D.J., Roziere, J., (1997) Enhancement of the protonic conductivity of a-zirconium phosphate by composite formation with alumina or silica, *Solid State Ionics,* 97, 227-232.

[20] Nakajima, H., Nomura, S., Sugimoto, T., Nishikawa, S., and Honma, I., (2002) High temperature proton conducting organic/inorganic nanohybrids for polymer electrolyte membrane, *J. Electrochem. Soc.*, 149, A953-A959.

[21] Song, X., Sayari, A., (1996) Sulfated zirconia-based strong solid-acid catalysts: Recent progress, *Catal. Rev. Sci. Eng.*, 38, 329-341.

[22] Clearfield, A., Pack, S.P., (1975) Factors determining ion exchange selectivity-I high temperature phases formed by α-zirconium phosphate and its sodium and potassium exchanged forms, *J. Inorg. Nucl. Chem.*, 37(5), 1283–1290.

[23] Alberti, G., Casciola, M., Constantino, U., Vivani, R., (1996) Layered and Pillared Metal (IV) Phosphates and Phosphonates, *Adv. Mater.*, 8, 291-303.

[24] Wang, Z., Ni, H., Zhao, C., Li, X., Zhang, G., Shao, K., Na, H., (2006) Influence of the hydroquinone with different pendant groups on physical and electrochemical behaviors of directly polymerized sulfonated poly(ether ethersulfone) copolymers for proton exchange membranes, *J. Membr. Sci.,* 285, 239-248.

[25] Nagarale, K., Gohil, G.S., Shahi, V.K., Rangarajan, R., (2005) Preparation of organic-inorganic composite anion-exchange membranes via aqueous dispersion polymerization and their characterization, *J. Colloid Interface Sci.*, 287, 198-206.

[26] Al-Othman, A. Tremblay, A.Y. Pell, W. Letaief, S. Ternan, M., (2010) Zirconium phosphate as the proton conducting material in direct hydrocarbon polymer electrolyte membrane fuel cells operating above the boiling point of water, *J. Power Sources,* 195, 2520-2525.

[27] Casciola, M., Bianchi, D., (1985) Frequency response of polycrystalline samples of $\alpha{-}Zr(HPO_4)_2{\cdot}H_2O$ at different relative humidities, *Solid State Ionics,* 17, 287-293.

[28] Carriere, D., Moreau, M., Lhalil, K., Barboux, P., Boilot, J.P., (2003) Proton conductivity of colloidal nanometric zirconium phosphates, *Solid State Ionics,* 162-163, 185-190.

[29] Sang, S.B., Wu, Q.M., Huang, K.L., (2007)Preparation of zirconium phosphate (ZrP)/Nafion1135 composite membrane and H^+/VO^{2+} transfer property investigation, *J. Membr. Sci.,* 305, 118-124.

In: New Developments in Polymer Composites Research ISBN: 978-1-62948-340-5
Editors: S. Laske and A. Witschnigg

Chapter 5

CONDUCTING POLYMER COMPOSITE FILMS BASED ON POLYANILINE: DIELECTRIC AND ELECTRICAL PROPERTIES

***Kateryna Fatyeyeva*[*,1,2,3]*, Christele Vanga Bouanga*[4]*, Alexander Pud*[5]*, Jean-François Bardeau*[4] *and Mohamed Tabellout*[4]**

[1]Normandie Univ, France
[2]Université de Rouen, PBS, Mont Saint Aignan cedex, France
[3]CNRS, Mont Saint Aignan cedex, France
[4]LUNAM Université, UMR 6283 CNRS, Institut des Molécules et Matériaux du Mans, département Physique de l'Etat Condensé, Le Mans cedex, France
[5]Institute of Bioorganic Chemistry and Petrochemistry, National Academy of Science of Ukraine, Kyiv, Ukraine

ABSTRACT

Composite polymer films based on different aliphatic polyamides (PA) (PA-6 and PA-12) and polyurethane (PU) (polymer matrix) with conducting polyaniline were prepared by an oxidative aniline polymerization inside a surface layer of the polymer host matrix. The detailed characteristics of the structure of the obtained composite films were studied by Raman spectrometry and optical microscopy. Dielectric and electrical properties of synthesized composite films were investigated thoroughly, and the obtained results reported. Dielectric measurements as a function of temperature and frequency have revealed the presence of relaxation processes influenced by the nature of the polymer matrix, i.e., one relaxation process for the polyurethane based composite film and two relaxation processes for polyamide based composite films. These relaxations are explained in terms of interfacial polarization caused by the polyaniline organisation. The real part of permittivity is found to be higher in the conducting composite films than in the virgin polymer matrix. Such behaviour is attributed to the interaction between the molecular chains of polymer and polyaniline and to the conductivity increase after the aniline polymerization.

* E-mail address: kateryna.fatyeyeva@univ-rouen.fr (Corresponding author)

Keywords: polyaniline, layered composite, polyamide, polyurethane, dielectric, conductivity

INTRODUCTION

Since intrinsically conducting polymers, especially π-electron conjugate polymers, including polythiophene, polyaniline (PANI), and polypyrrole, were found by MacDiarmid, Shirakawa and Heeger in 1977, they have been extensively studied for their promising applications [1-3]. Compared to many other conducting polymers, PANI is regarded as the most commercially promising one because of inexpensive monomer, which is easy to produce, its high yield, and thermal and chemical stabilities. Also, the remarkable capacity of PANI to switch from conducting (doped) to insulating (dedoped) states is the basis of its use in many applications (Fig. 1). This makes PANI a versatile potential material for applying in electronic and optical devices such as energy storage devices [4], electron field emitters [5], chemical and biological sensors and actuators [6] and so forth. PANI can be synthesized via numerous chemical [7, 8] or electrochemical polymerizations [9, 10]. By chemical doping and oxidation/reduction processes, electrical conductivity of PANI is commonly adjusted in a range of 10^{-8} to 10^{2} S/m [11]. However, like many other conducting polymers, PANI has poor mechanical properties and is insoluble in most of available organic solvents because of its stiff backbone and strong hydrogen bonding interaction.

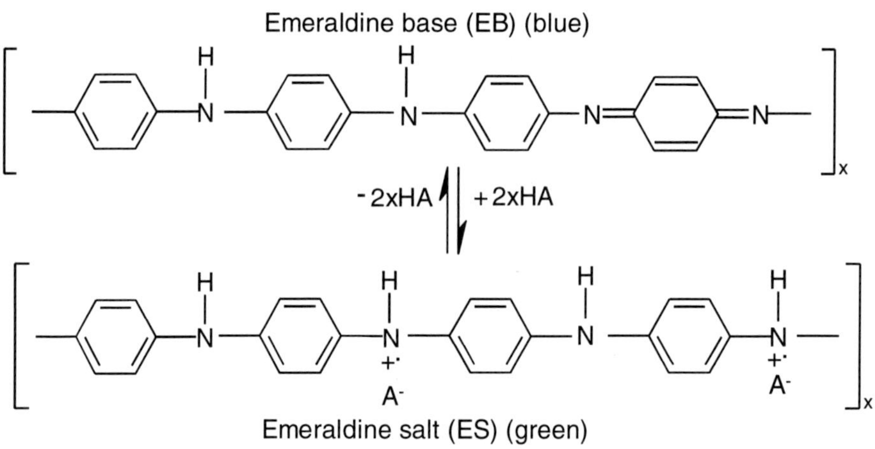

Figure 1. Transition between PANI base form and the protonated emeraldine form (one of the possible chemical representations). A^- represents an arbitrary anion.

Organic polymers find wide application in electronic products due to their remarkable merits such as light weight, flexibility, easy processing and resistance to corrosion. However, most organic polymers possess very low electrical conductivity that restricts their practical applications. Considerable research has been performed in order to overcome this problem. It was found that PANI combined with organic polymer matrix emerged as a new class of composite materials since the resulting material possesses unique synergism in properties [12-15]. PANI offers good compatibility with the polymer matrix. This compatibility not only insures the excellent mechanical characteristics but also gives rise to unique dielectric and electric properties [15-18].

Composites based on conducting polymers can be used for antistatic applications. The coatings for some of these applications require a high level of transparency, e.g., in the photographic industry. Besides, thin films of material are the most convenient form for the experimental studies as well as for the device fabrication. Therefore, the promising way of combining the electrical properties of PANI and the transparency of common polymers is the aniline polymerization inside the swelled polymer matrix by an oxidant solution in order to form surface layered composites [16, 19-22]. During the matrix polymerization, a conducting path of PANI through the transparent host matrix is formed. Several relaxation processes related to the film surface conductivity and influenced by the nature of the polymer matrix may be found in these composites.

The understanding of charge transport mechanism in the composite materials is very important from both fundamental and technological point of view. The dielectric relaxation spectroscopy (DRS) is one of the powerful techniques to characterize and realize the charge transport in complex materials. The studies of dielectric properties of the systems are useful for further possible technical applications. At present various theoretical models for the description of charge transport in conducting polymers have been developed [23, 24].

It should be noted that the dielectric behaviour of the composite materials and their conductivity vary depending on the structure of polymer chain, the doping level, the nature of acid-dopant, etc. [16, 19, 20, 25-29]. They are also greatly affected by the film-formation method and the selection of the dopant and the solvent [13, 19, 20, 26-29]. The method of DRS allows investigating the conductivity mechanism on the basis of measurement of dielectric parameters of conducting polymers [30].

The application of an electric field provokes the fluctuations of the electrical polarization and this leads to the electrical displacement. Such displacement depends on the material polarization. Dielectric polarization arises due to the existence of atomic and molecular forces, and appears whenever the charges in a material are somewhat displaced with respect to one another under the influence of an electric field. One can distinguish several types of polarization [31, 32], which depend on the applied frequency and temperature:

- electronic polarization – its effect involves distortion of the center of charge symmetry of the basic atom. Under the influence of applied field, the nucleus of an atom and the negative charge center of the electron shift, creating a dipole. Such polarization happens either at very high frequencies or at very low temperatures;
- atomic polarization happens at IR frequencies and corresponds to the atoms displacement in the case of the molecules deformations.

After these two kinds of polarization the real part of the dielectric complex permittivity ε' is designated as ε_∞ – it is a minimal value of ε' that is observed for the studied frequency region.

- dipoles orientation polarization – a phenomenon involving rotation of permanent dipoles under an applied field;
- the electrode polarization lies in the charge accumulation near electrodes and results from the ionic conductivity. These charges create an electric field with the direction opposite to that of the applied electric field. Such polarization appears at very low frequencies;

- Maxwell-Wagner-Sillars (MWS) effect – if the material consists of different phases, which possess different values of permittivity and conductivity, the charges can accumulate at the limits of these phases. Hence, the relaxation time of the MWS relaxation will depend on the conductivity and permittivity values of different components.

This chapter deals with the preparation of the layered composite films based on different aliphatic polyamides and polyurethane with PANI. The electrical and dielectric properties of the composite films have been studied and analyzed using the theoretical model.

METHODS

Materials

Aniline (Merck) was distilled under vacuum and stored under argon atmosphere in a refrigerator at 3 °C. Ammonium persulphate (APS) (Prolabo), hydrochloric acid (37 wt.%, Prolabo), $HClO_4$ (70 wt.%, Prolabo), *n*-hexane (Prolabo) and ammonia solution (25 wt.%, Prolabo) were of reagent grade and used as received. Films of PA-6 (25 μm) were purchased from Goodfellow Cambridge Ltd. Films of PA-12 (50 μm) were kindly donated by Arkema (France). The PU films were synthesized according to [33].

Synthesis of PANI

The synthesis of pure PANI was performed in order to compare its electrical and dielectric properties with the properties of the composite films. The emeraldine form of PANI (Fig. 1) was synthesized chemically by the oxidative polymerization according to [34]. The PANI-HCl powder was washed with the distilled water to remove any impurities. The obtained powder was dried under dynamic vacuum to a stable weight. The EB powder was prepared by the alkaline dedoping of the PANI-HCl emeraldine salt (ES) in a 5 % NH_4OH aqueous solution for 10 h. The powder was then washed and dried under vacuum and stored in a refrigerator to be used later. Thus obtained the dedoped PANI powder was redoped by stirring for 24 h in 1 M water solution of HCl, producing PANI-HCl complex. Finally, the doped powder was dried under dynamic vacuum to a constant weight.

Preparation of Polymer/PANI Surface Layered Composite Films

The surface layered conducting polymer composite films were prepared using the following procedure. The pristine polymer films were immersed into aniline at 20 ± 1 °C in argon atmosphere to be swelled during the definite time period as a function of the polymer matrix. The swelling degree α was calculated taking into account the changes in film weight:

$$\alpha = \frac{(m - m_0)}{m_0} \cdot 100\%, \tag{1}$$

where m_0 and m are the weight of the pristine and the swelled films, respectively.

The swelling process in the polymer matrix was performed till the achievement of the definite wt.% of aniline in the film (for example, for PA-12 during 2 h in order to reach 12 wt.% of aniline that corresponds to 7.5 wt.% of PANI [19]).

Swelled in that way polymer films were washed by *n*-hexane to remove the surplus of aniline from their surface. Then, one side of the swelled films was treated with the oxidizing solution of 0.1 M $(NH_4)_2S_2O_8$ in 1 M water solution of HCl in the home-made cell for running chemical aniline polymerization inside the surface layer of the polymer matrix in order to form the composite polymer/PANI film. After the completion of the polymerization process, the green transparent composite film was washed by distilled water and placed in Soxlet apparatus with *n*-hexane for 6 h to extract the non-polymerized aniline residua and by-products from the film matrix. The procedure was followed by drying the film at 60 °C under dynamic vacuum up to a stable weight.

Layered surface composite films obtained in that way were then treated with the 5 % ammonia aqueous solution for 24 h in order to convert (dedope) PANI inside the polymer matrix to the non-conducting emeraldine base (EB) state (Fig. 1). The dedoped composite films were then redoped in 1 M water solution of appropriate acid (HCl) to prepare the composite polymer/PANI doped films.

Dielectric Relaxation Spectroscopy

Dielectric permittivity measurements were performed using Novocontrol broadband dielectric spectrometer (Novocontrol GmbH, Germany) in wide frequency (0.1 Hz to 1 GHz) and temperature (173 K to 423 K) ranges. To cover the frequency domain from 0.1 Hz to 1 GHz, two experimental configurations were employed. From 0.1 Hz up to 10 MHz a Solarton SI 1260 analyzer combined with a broadband dielectric converter allowed taking impedance and dielectric measurements. The high frequency impedance analyzer (10 MHz to 1 GHz) was used with a precision coaxial line [32].

For both configurations, the sample cell was composed of two round golden parallel electrodes filled with the material to form a capacitor. In the case of the high frequency range the sample capacitor was used as a termination of a golden coaxial line and the impedance was then calculated from the complex reflection factor at the end of the line. The polymer film was placed between two electrodes. In the case of the pure PANI, the powder was compressed into disc of 10 mm in diameter and with the thickness of about 1 mm under the pressure of 3 t at room temperature. In the case of the layered surface composite films, the surfaces containing PANI were faced each other in order to respect the sample symmetry.

The sample temperature was varied by using a gas stream of nitrogen and the measured temperature close to the sample was controlled within the accuracy of ±0.2 K. The measurements determined the impedance of the sample capacitor Z

$$Z = Z' + iZ'' = \frac{U_s}{I_s}, \tag{2}$$

which was connected with the dielectric function of the sample by the equation:

$$\varepsilon = \varepsilon' - i\varepsilon'' = \frac{-i}{2\pi f Z_s C_0}, \tag{3}$$

where f denotes frequency and C_0 is the vacuum capacity of the sample capacitor.

The method of dielectric analysis was chosen to calculate dielectric parameters of the process (dielectric constant, imaginary and real permittivity, conductivity, relaxation time, etc.). The values of imaginary and real parts of dielectric permittivity $\varepsilon^*(\omega) = \varepsilon' - i\varepsilon''$ were analyzed according to the empirical Havriliak-Negami (HN) function [31]:

$$\varepsilon^*(\omega) = \varepsilon_\infty - i\left(\frac{\sigma_0}{\varepsilon_0 \omega^n}\right) + \frac{\Delta\varepsilon}{\left(1 + (i\omega\tau)^\alpha\right)^\beta}, \tag{4}$$

where $\omega = 2\pi f$; ε_0 denotes the vacuum permittivity; σ_0 is the dc-conductivity; n is the exponent factor, in most cases equal to 1; $\Delta\varepsilon$ is the difference between low and high frequency limits of ε' over the relaxation to which the HN function is applied. $\Delta\varepsilon$ is also proportional to the area below the curve of the ε'' relaxation peak; ε_∞ denotes the high frequency limit of the permittivity and is the unrelaxed value of permittivity; α and β are shape parameters; τ is the relaxation time.

The processing of the experimental dielectric spectra $\varepsilon^*(\omega)$ was made by means of WinFit 2.4 software of Novocontrol GmbH (Germany).

The characteristic relaxation time τ was taken at the position of the maximum of dielectric loss for each relaxation process and fitted according to the equation (4). The temperature dependence of that relaxation time was analyzed using Arrhenius equation [equation (5)] and activation energies of relaxation processes were obtained:

$$\tau = \tau_0 \exp\left(\frac{E_a}{k_B T}\right), \tag{5}$$

where τ_0 is the relaxation time at very high temperature; E_a is the activation energy; k_B is the Boltzmann's constant ($8.616 \cdot 10^{-5}$ eV/K).

Electrical Conductivity Measurements

To measure the surface resistance R of the conducting composite films as a function of the temperature, the four-electrode technique was used. Four copper wires were attached in parallel on the sample surface by the conducting silver paint for better electrical contact. The resistance values were measured using Agilent E3634A electrometer. The measurements were performed by changing the temperature at the rate of 0.2 °C/min in the cryostat equipped with a temperature-controlled heater. The temperature control was realized with the help of Novocontrol broadband dielectric spectrometer. The resistance measurements were averaged out of the results of (i) three measurements for each film, (ii) two films for each sample.

Resonance Raman Spectrometry Measurements

The Raman experiments were carried out by multichannel Horiba Jobin-Yvon T64000 spectrometer connected to a liquid N_2 cooled CCD detector and equipped with a confocal microscope. The 514.5 nm line of an Argon-Krypton ion laser was used as an excitation source for the Raman spectra. This excitation line was chosen since it allowed studying PANI in ES as well as in EB oxidation states (Fig. 1). It was previously shown that both phenyl and quinone units of PANI could be simultaneously enhanced by 514.5 nm excitation wavelength whereas vibrations of those units were resonance enhanced in blue and red excitation wavelengths, respectively [11].

To avoid any local degradation and perturbation of the polymer samples, the laser beam power was limited to 0.04 mW on the sample surface and the spectra were collected during the integration time of 300 s. The laser beam was focused on the sample in a ~1 μm spot size with an ×100 objective. The scattered Raman signal was collected by the same objective lens and was focused onto a pinhole (100 μm opening aperture) to reject out-of-focus signals. The diverging light coming out of the pinhole was then refocused onto the entrance slit of the spectrograph. Under those conditions, the depth resolution was determined to be 1.5 μm.

Optical Measurements

The electron micrographs of the composite films were performed using a Leica DMRX polarizing microscope.

Results and Discussion

Structure of the Composite Polymer/PANI Films

The composite films were prepared by the aniline chemical polymerization inside the polymer host matrix. However, the chemical aniline polymerization may be realized in the polymer matrix only under the condition of the matrix film swelling in the monomer solution and in the reaction media. This means that the structure and the thickness of the conducting (PANI) layer will depend on the swelling degree of the matrix. Thus, the swelling kinetics of the PA and PU films in aniline was investigated in order to determine the time necessary to get certain value of swelling and to understand the influence of the polymer matrix structure on the mechanism and kinetics of that process.

It was established that the swelling process of the PA-12 film in aniline went faster than in the PA-6 film. The swelling saturation is reached in ~ 3.5 h for the PA-12 film and only after more than 15 h for the PA-6 film (Fig. 2a and b, respectively). But the maximum swelling degree is higher in the case of the PA-6 film - 16 wt.% in comparison with 13 wt.% for the PA-12 film. The swelling process of the PU film in aniline was also established (Fig. 2c). One can note that in this case the aniline swelling saturation was rather quick and the maximum swelling degree (25 wt.%) was reached in only 10 min.

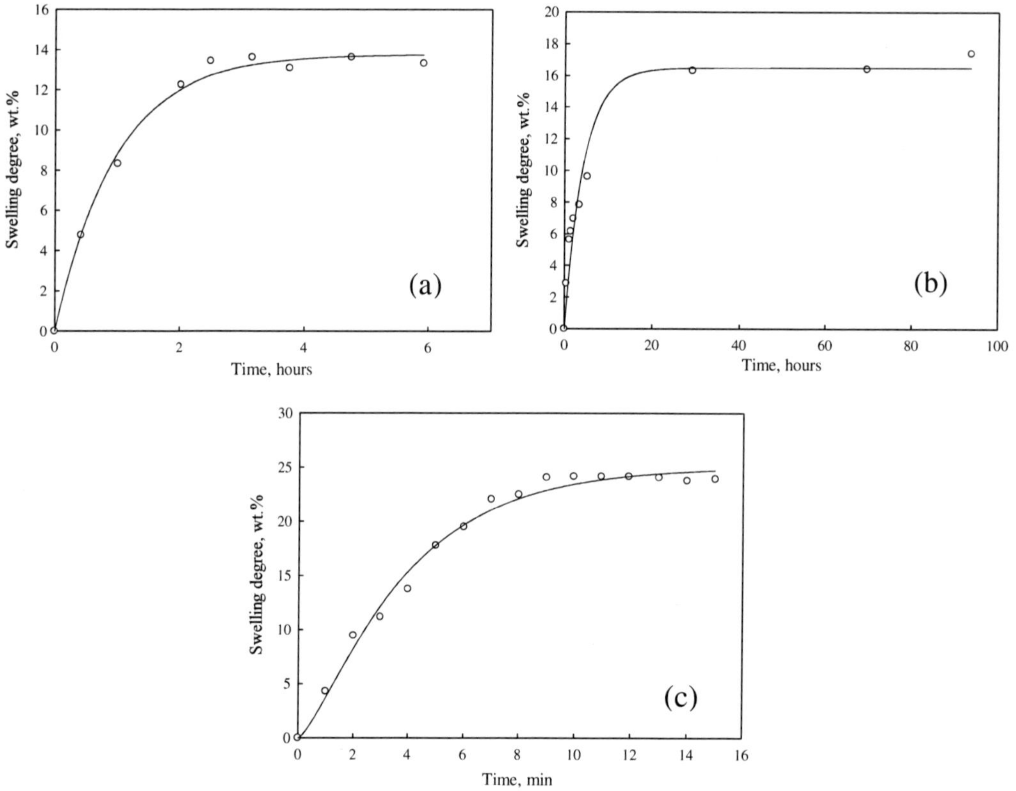

Figure 2. Swelling kinetic curves of the polymer matrix in aniline: *a* – the PA-12 film; *b* – the PA-6 film; *c* – the PU film.

This fact may be also explained by the different hydrophilicity of the polymer matrix due to the different ratio of polar/nonpolar groups. Indeed, the elementary chain of PA-12 consists of 11 aliphatic (CH_2) groups, whereas in the molecule of PA-6 there are only 5 such groups. On the other hand, different degree of crystallinity and, correspondingly, different free volume amount of the used polymer matrix films can be also considered as important factor. Really, it was found that the degree of crystallinity was higher in the case of the PA-6 film (38 %) as compared with the PA-12 (21 %) and PU (3 %) films [16, 20].

So, as the PA-6 matrix has higher crystallinity degree, it contains less volume of the amorphous phase than the PA-12 and PU films. Therefore, the swelling kinetics is much longer for the PA-6 matrix compared to the PA-12 and PU films (Fig. 2). The higher maximum swelling degree in the PU film may be also explained by the fact that aniline molecules can form hydrogen bonds with amide groups of PU, i.e., some physicochemical interaction of aniline with the polymer matrix takes place. And as the quantity of the polar amide groups is higher in the case of PU [33], the quantity of hydrogen bonds is higher too. Thus, the film of PU can hold more aniline molecules compared to the film of PA.

In order to obtain the composite polymer/PANI film, one surface of the polymer matrix, previously swelled in aniline, was treated by the oxidant solution (0.1 M $(NH_4)_2S_2O_8$ in 1 M water solution of HCl). The aniline polymerization process begins at the interface between the polymer matrix, containing aniline, and the acidified oxidant water solution. In this case both the acid and oxidant should diffuse from the solution bulk to the film surface. Though the aniline polymerization begins as the two-dimensional process with some fixed rate, a further development of the polymerization process in the film bulk has a much lower rate because it is strongly limited by a slow diffusion of the reagents into the polymer matrix [19]. The molecules of the monomer (aniline) should preferably be localized in a free volume of an amorphous part distributed among the polymer crystalline regions. Therefore, the aniline polymerization inside the polymer matrix should have different kinetics (i.e., different time spaces of induction period, different propagation step and termination stages) as compared to the aniline solution polymerization. And, really, the increase of the induction period of the aniline matrix polymerization was observed [19].

The other peculiarity of the aniline matrix polymerization stems from the solid phase nature of the polymer matrix and leads to an immobility of formed aniline dimers and other oligomers, which are actually trapped in the free volume of the matrix because of the large size and different intermolecular interactions. This fact influences the propagation step. Indeed, the chain propagation step can be realized only through the monomer (aniline) diffusion to these trapped oligomers. Besides, the oxidation reaction of aniline in the matrix bulk can be transmitted to some extent from the interface due to oxidation of oligoaniline chains, which will grow into the matrix through the interaction with the monomer molecules located there. It should be noted that in the case of the aniline matrix polymerization there are no problems with the PANI precipitation. All the forming aniline oligomers and PANI remain in the matrix. However, the presence of the overoxidized PANI may be observed under certain experimental conditions [19].

To determine the structure of the composite films thus obtained and the thickness of the conducting PANI containing layer, the Resonance Raman spectrometry was used. The spectra of virgin polymer films, pure doped and dedoped PANI powders and composite polymer/PANI films were measured under the same experimental conditions using the confocal micro-Raman spectrometry. As an example, the spectra of PA-6, doped PANI and composite PA-6/PANI doped film are presented in Fig. 3. As one can see, the spectrum of the composite film (Fig. 3, curve 3) does not contain any well-defined band characteristic of the PA-6 film (Fig. 3, curve 1). Due to the enhancement of the band intensity, originating from PANI, the signals from PA may be considered as negligible for the complete analysis. On the other hand, one can notice that the spectra of the composite film and the pure doped PANI powder are similar. Hence, taking into account the advantage of the confocal system and the enhancement of the PANI vibrational bands under the resonant conditions, the thickness of the conducting PANI containing layer may be determined. The dependence of the integral Raman intensity in the wavenumber region 1150-1670 cm^{-1} on a sample position is shown in Fig. 4.

Fig. 4 clearly shows that the distribution of PANI in the polymer matrix is identical for the PA and PU films. The obtained profile of the integrated Raman intensity is symmetric and, therefore, proves that the PANI containing layer can de considered as a homogeneous layer of a definite thickness. The experimental points were fitted by applying Gaussian's law. The distance *d* (Fig. 4) was obtained as a half of the width of the profile peak. Obtained results show that the used method of the aniline polymerization results in a double-layered conducting composite film. Correspondingly, one can suppose that the process occurs in an external layer of the composite film. This layer reduces the speed of diffusion of the oxidant solution into the interior and, accordingly, slows down the formation of PANI inside the polymer matrix. This fact is in good agreement with the results obtained by Byun and Im for the aniline polymerization in the Nylon (PA-6) film [35]. It should be mentioned that such a layered architecture is similar for the PA and PU films, since in both cases the same depth prtofile is obtained (Fig. 4).

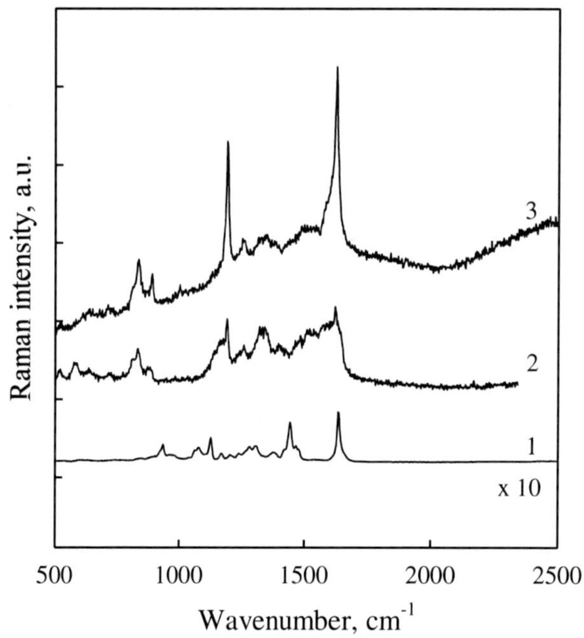

Figure 3. Raman spectra collected under the same experimental conditions (λ = 514.5 nm, t = 300 s, p = 0.04 mW). The intensity of the PA spectrum (*1*) was multiplied by the factor 10: *1* – the PA-6 film; *2* – the PANI powder in the form of ES; *3* – the composite PA-6/PANI doped film.

On the contrary, the thickness of the conducting layer strongly depends on the structure of the polymer matrix (Table 1). The thickness δ of the PANI containing layer was determined by taking into account the refractive index (n) of the polymer matrix and the full-width at half-maximum of the Raman profile (d):

$$\delta = d \cdot n \quad (6)$$

The thickness of the conducting layer in the case of the PA-6 film (4.7 μm) is found to be more than twice as much compared to the PA-12 film (2.0 μm). This feature can be explained by greater hydrophobicity of PA-12, which reduces the diffusion of the oxidant solution into the polymer host matrix. The highest thickness of the conducting PANI layer in the case of the PU matrix (7.8 μm) is in good agreement with crystallinity degree measurements. Indeed, the lowest degree of crystallinity was found for the PU film (3 %).

These results demonstrate the influence of the polymer matrix molecular structure on the properties of the obtained surface layered composite materials. In any case the composite film with the layered structure was obtained.

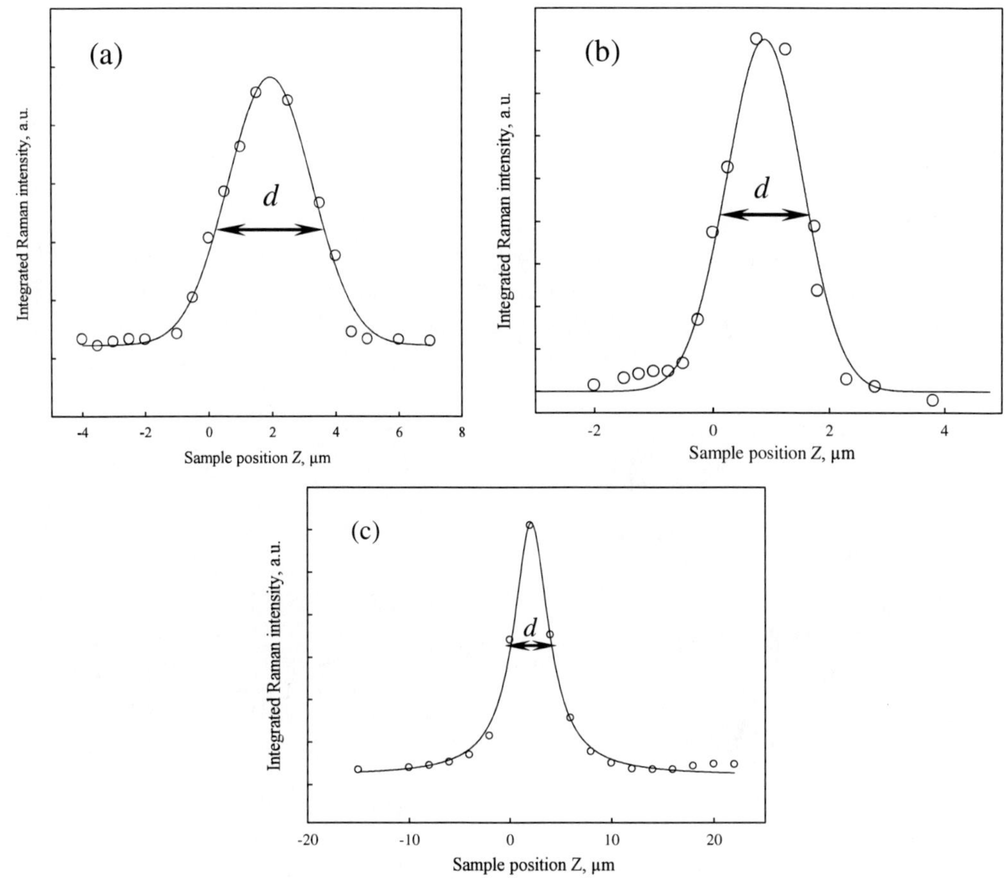

Figure 4. Evolution of the integrated Raman intensity of the PANI vibrational bands in the wavenumber region 1150-1670 cm^{-1}: *a* – the PA-6 matrix; *b* – the PA-12 matrix; *c* – the PU matrix.

Table 1. Results of the Raman depth analysis.

Matrix	**Distance *d* obtained from Raman depth analysis, μm**	**Refractive index (*n*)**	**Conducting layer thickness δ, μm**
PA-6	3.1	1.53	4.7
PA-12	1.3	1.54	2.0
PU	5.2	1.50	7.8

One can suppose that the structure of the polymer matrix may influence not only the thickness of the PANI containing layer, but also both the distribution of PANI inside the polymer matrix layer and the final composite structure. This feature is very important because the conductivity properties of the composites depend on the molecular distribution of the conducting clusters with respect to the polymer matrix [6, 16, 20]. The optical microphotographs for the HCl-doped PA-12/PANI composite film show that the pure PA-12 film has a uniform surface (Fig. 5a) in comparison with the doped composite film (Fig. 5b). Similar distribution of PANI was found by Zhang et al. for the PA-11/PANI composites with the help of TEM [36]. Also, the surface deformation of the composite films was clearly observed in the course of the AFM measurements [16, 20]. It can be concluded that the pure polymer matrix and the dedoped form of the composite polymer/PANI film exhibit comparatively flat surfaces, whereas doping (Fig. 1) leads to the appearance of the hilly features. Besides, not only doping (i.e., appearance of charge carriers on the PANI chains), but the acid-dopant also has influence on the surface morphology of the composite films [16].

It should be noted that aniline chemical polymerization inside the polymer matrix allows obtaining the conductive PANI containing layer in a thin subsurface layer of the dielectric matrix. Owing to a small thickness (2-8 μm) the PANI containing layers can quickly react on external environment changes, which allows applying them in sensors or optical devices.

One can suppose that the presence of the PANI conducting surface layer may result in the appearance of interfacial polarization effects, i.e., may influence the dielectric and electrical behaviour of the composite films.

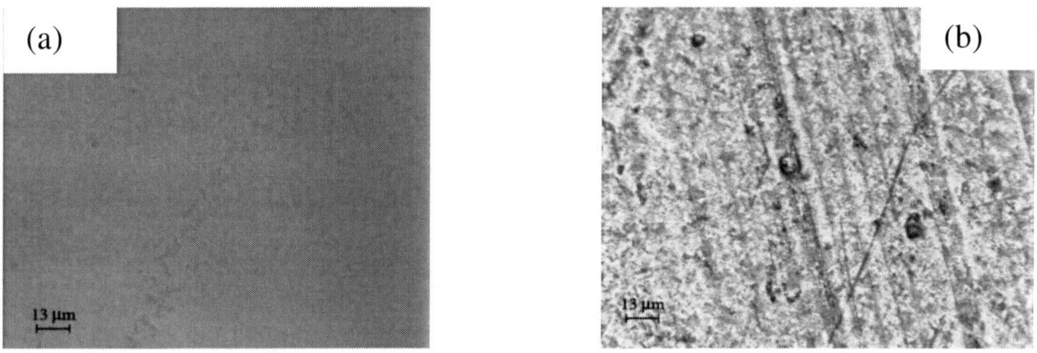

Figure 5. Optical micrographs of the pure PA-12 film (a) and of the composite PA-12/PANI-HCl film (b).

Dielectric Properties of Composite Polymer/PANI Films

In order to interpret the obtained data for the composite materials, the dielectric behaviour of the pure PANI samples in doped and dedoped forms (Fig. 1) was studied at first.

The frequency dependence of the measured real part ε' of the complex dielectric permittivity and the real part σ' of the conductivity for the dedoped PANI powder (pressed pellet) is presented at several temperatures in Fig. 6. The spectra in Fig. 6 are presented at temperatures higher than 273 K, though the measurements were performed beginning with 173 K. This is because the value of ε' below 273 K is almost temperature-independent having a weak frequency dispersion.

The real part of the permittivity exhibits a relaxation behaviour whose relaxation frequency is temperature-dependent. As one can see, ε' exhibits a low-frequency plateau corresponding to the static dielectric constant ε_s and then decreases with increasing frequency, forming a low plateau value ε_∞ in the high-frequency region (Fig. 6a).

The analysis of the dielectric permittivity using HN function (equation (4)) revealed a temperature dependence of the relaxation time τ (Fig. 7), which is well described by a simple Arrhenius law (see equation (5)). For the emeraldine base τ_0 appears to be $5.2 \cdot 10^{-11}$ s and the activation energy appears to be 0.54 eV. These results are in agreement with the calculations performed by Zuo et al. [37] for the PANI powders with different protonation levels. For example, for the EB they found $1.6 \cdot 10^{-10}$ s and 0.46 eV for τ_0 and the activation energy, respectively. At the same time, the obtained temperature dependent conductivity data (Fig. 7) can be fitted to Arrhenius equation of conductivity:

$$\sigma = \sigma_0 \exp(-\frac{E_a}{k_B T}), \tag{7}$$

where σ_0 is the conductivity at very high temperature; E_a is the activation energy, and k_B is the Boltzmann's constant ($8.616 \cdot 10^{-5}$ eV/K). The activation energy for the conductivity is found to be 0.59 eV, which is close to that of the relaxation time (0.54 eV).

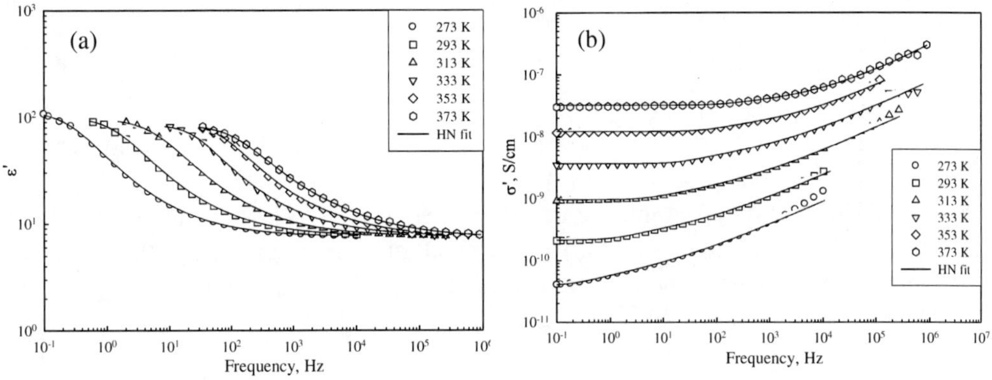

Figure 6. Frequency dependence of the real part of dielectric permittivity (a) and conductivity (b) for dedoped PANI at different temperatures. Solid lines are the best fit of HN function (equation (4)).

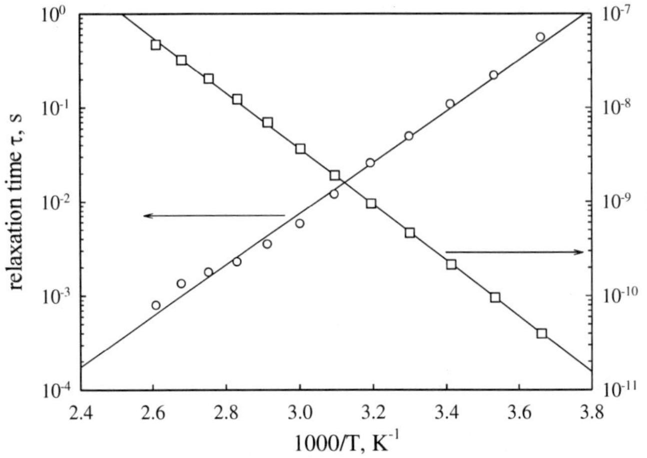

Figure 7. Arrhenius plot of the characteristic relaxation time and the conductivity for dedoped PANI. Solid line is the best fit to equations (5) and (7), respectively.

As the value of the activation energy is not very high, one can suppose that this relaxation process is a β-relaxation, as for the α-relaxation process the activation energy value should be much higher and its frequency relaxation appears at much higher temperatures [38]. In other words, the observed relaxation could be a β-relaxation process commonly observed in the traditional polymers since the activation energy is of the same order. On the other hand, as the value of the activation energy for the relaxation time is close to that of the conductivity (0.54 eV and 0.59 eV, respectively), one can surmise that the correlation between these processes exists and that both processes originate from the same transport mechanism. Therefore, the relaxation observed for dedoped PANI at low frequencies can be the well known conductivity relaxation [39]. In order to support this assumption the dependence of the dc-conductivity value versus the relaxation time was plotted in double logarithmic scale (Fig. 8). The linear behaviour of the dependence was obtained and its slope was found to be nearly equal to -1 with a linearity factor of 0.9951. Besides, it was established that the relaxation time τ and the dc-conductivity σ obey the following equation [39]:

$$\sigma(T) = \frac{\varepsilon_0 \cdot \varepsilon_s}{\tau(T)}, \tag{8}$$

where ε_0 is the permittivity of the free space ($8.85 \cdot 10^{-12}$ F/m). This relation is characteristic of the conductivity relaxation process.

Some deviation of the experimental and calculated according to equation (8) values was observed, which can be explained by the experimental error – for example, by the presence of the residual solvent (water) in the sample. Thus, not only the conductivity value of PANI, but the total conductivity of the sample was measured.

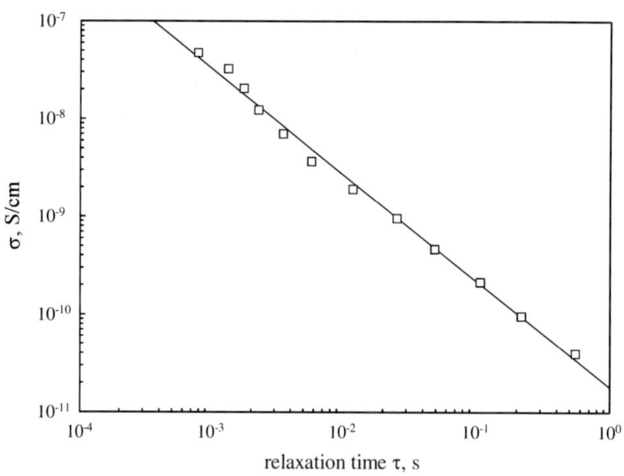

Figure 8. Dependence of the conductivity on the characteristic relaxation time for dedoped PANI.

It should be noted that in conducting polymers there is a strong charge trapping [37, 40] and its localized motion when applying an alternating electric field operates as an electric dipole. The dielectric relaxation in the presence of such an alternating electric field is a result of charge hopping among available localized sites [31]. So, for PANI the polarons and bipolarons are the relevant charge species involved in the relaxation and conductivity mechanisms [37, 41]. At low frequencies, such charge hopping could extend throughout PANI in the absence of strong pinning, which can lead to a continuous current, resulting in a dc-conductivity. It should be pointed out that the charge carriers exist even in the dedoped form of PANI as the sample reveals a dc-conductivity value although not very high (Fig. 6b).

The relaxation process represented in Fig. 6a arises from hopping and/or oscillations of these charges around fixed pinning centres [37].

The measured real ε' and imaginary ε" parts of the complex permittivity as a function of frequency for the doped PANI samples (pressed pellets) at 300 K are shown in Fig. 9. The doping process of PANI (Fig. 1) leads to a drastic increase of the permittivity. The dielectric constant ε' at low frequencies is found to be very high, but it drops very fast with increasing frequency (Fig. 9a). This may be due to the easy charge transfer through well ordered polymer chains in disordered regions as suggested by Joo et al. [42]. This result is characteristic of a conducting polymer and consistent with the literature [37, 43, 44]. Such behaviour can be also explained as one due to the increased contribution of the dc-conductivity [31].

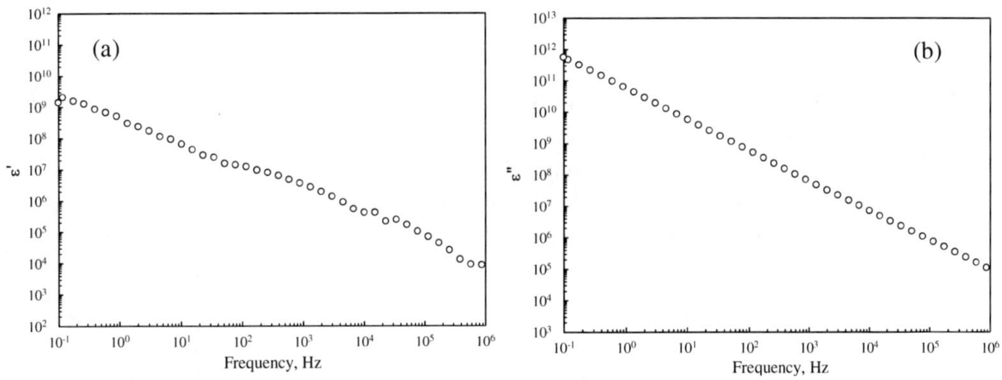

Figure 9. Frequency dependence of the real (a) and imaginary (b) parts of dielectric permittivity for PANI-HCl at 300 K.

Furthermore, the dependence of ε" on the frequency in the double logarithmic scale for the PANI doped sample is almost a straight line at low frequencies, with a slope equal to -1 (Fig. 9b). Such a dependence indicates that ε" is proportional to 1/f, and, correspondingly, that the conductivity is constant at low frequencies. This testifies to the presence of a dc-conductivity. Similar behaviour was also observed by other researchers, for example, Okutan et al. for polythiophenes [44].

As it was mentioned above, ε' increases with the frequency decrease for doped PANI and the very high polarization may hide the relaxation process if any in the studied low frequency range at room temperature. Nevertheless, one can say that the value of ε' decreases in the high-frequency region for HCl doped PANI, that looks like a relaxation process (Fig. 9a). Such behaviour is characteristic of the charge carriers systems [31]. The localized charge carriers when applying an alternating electric field can hop and jump to neighboring localized sites which form a continuous connected network, allowing the charges to travel through the sample and, thus, causing the electric conduction.

In order to study in detail the relaxation and conduction mechanisms in the PANI samples the measurements in the high frequency region (from 10^6 Hz till 10^9 Hz) were also performed (Fig. 10). As one can notice from Fig. 9a and 10, the relaxation process, which for dedoped PANI was observed in the low frequency region (Fig. 6a), shifts toward higher frequencies as a result of the doping process.

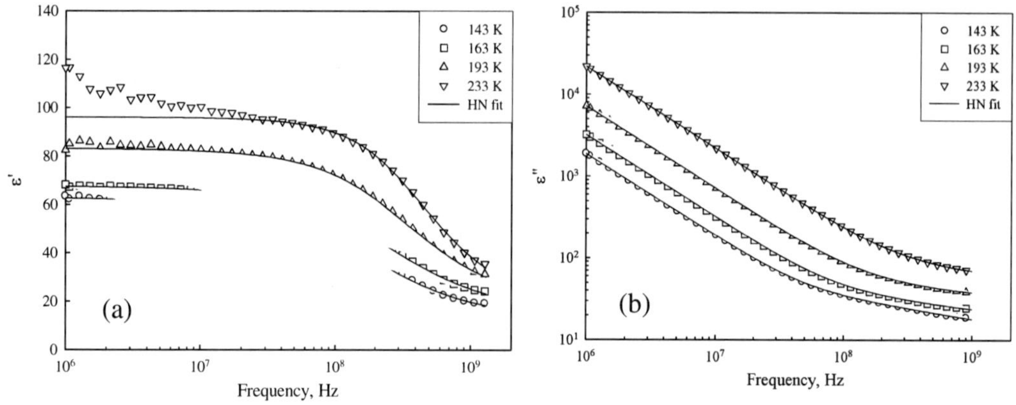

Figure 10. Frequency dependence of the real (a) and imaginary (b) parts of the dielectric permittivity for PANI-HCl. Solid lines are the best fit to HN equation ((4)).

Therefore, one can say that a relaxation process is present in the studied doped PANI sample and it is the interfacial polarization process. In order to study the relaxation behaviour with the temperature change and to estimate the activation energy of the process, dielectric measurements were performed for HCl-doped PANI in the high frequency range. The real and imaginary parts of dielectric permittivity and real part of the conductivity for the PANI-HCl sample as a function of the temperature are depicted in Fig. 10 and 11, respectively. As one can see, the real part of the dielectric permittivity exhibits a relaxation process of a rather high dielectric strength (Fig. 10a), as a result of the doping process. This relaxation process is associated with the charge carriers' mobility. However, for dedoped PANI form only a flat dielectric response was observed in this frequency region.

As in the case of dedoped PANI, the temperature dependencies of the relaxation time and dc-conductivity for doped PANI may be described according to the equations (5) and (7). It was found that the activation energy for PANI-HCl is 0.093 eV, i.e., much lower than that for the dedoped sample (0.54 eV). Also, the correlation between the relaxation time of the process and the conductivity values was found.

As it was shown previously, the surface conductive composite films have a layered film structure and a PANI clusters organisation. This can result in the appearance of interfacial polarization effects resulting in the occurence of the relaxation processes. Additional relaxation associated with the dipolar reorientation of chains segments or side groups in the polymer matrix may contribute the whole spectra of the composite film.

Fig. 12 presents the frequency dependence of ε" for the composite polymer/PANI films in the doped and dedoped states. The spectrum of the virgin polymer film is given for comparison. The solid lines in ε" are fits to the measured data using HN function (equation (4)). The results indicate that the aniline polymerization inside the polymer film and PANI presence in the film causes a great increase of ε" in the whole studied frequency region irrespective of the polymer matrix. This feature is observed for the composite films with PANI in the dedoped state as well as in the doped one. This may be due to the additional polarization which emerges from the presence of charge carriers in PANI even in the dedoped state. Besides, it was found that the value of ε' is rather high even in dedoped PANI (Fig. 6a).

Analyzing the obtained results we can conclude that the presence of PANI (~10 wt.%) even in thin subsurface layer leads to considerable changes in the dielectric spectra.

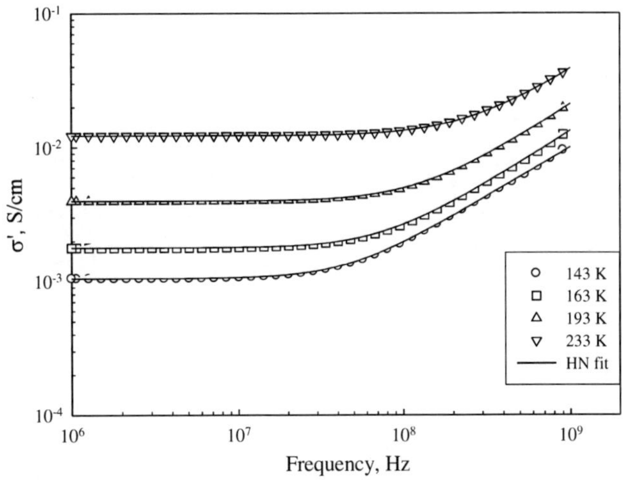

Figure 11. Real part of the conductivity for PANI-HCl as a function of temperature. Solid lines are the best fit to HN function (equation (4)).

The flat dielectric response for the virgin PU film indicates that no dipolar reorientation occurs at this temperature (Fig. 12a). The presence of dedoped PANI in the PU matrix does not modify the dielectric spectrum – within the whole range of studied frequencies the value of the dielectric loss ε" is practically constant. This indicates virtually no charge movement in the dedoped composite film in this frequency range. In the case of the PA films, two relaxation processes are observed in the virgin PA film and dedoped composite film (Fig. 13b and c). The first one - the high frequency process - is attributed to the γ-relaxation process associated with the local chain motion. The second process is most likely corresponding to a β-relaxation process characterised by a symmetric distribution of relaxation time ($\alpha \sim 0.5$, $\beta = 1$) and by a relatively high activation energy (Table 2).

However, a quite different behaviour is observed for the doped composite films (Fig. 12). In the case of PU based composite film one can observe a well-defined broad relaxation process labelled τ_1 (Fig. 12a). This relaxation process, characterized by shape parameters $\alpha = 0.6$ and $\beta = 1$, results from interfacial polarization effect caused by the accumulation of charge carriers at the PANI particles' interfaces. The found value of α is different from 1, which indicates the deviation from the ideal Debye relaxation process ($\alpha = \beta = 1$) [31, 32].

In the doped PA-12/PANI film the great increase in ε' can be explained by an additional polarization caused by a high concentration of charge carriers in conductive PANI particles dispersed within the insulating polymer matrix. As the conductivity of the dedoped PA-12/PANI film is closer to the virgin PA-12 film than that of the doped PA-12/PANI, such great variation of ε' and ε" was not observed in the former film, i.e., in the dedoped PA-12/PANI film. A similar behaviour was observed by Rocha et al. [45] in the case of the poly(vinylidene fluoride) (PVDF)/poly(*o*-methoxyaniline) (PVDF/POMA) blends, which were prepared by casting the blended solution of POMA and PVDF in dimethylacetamide.

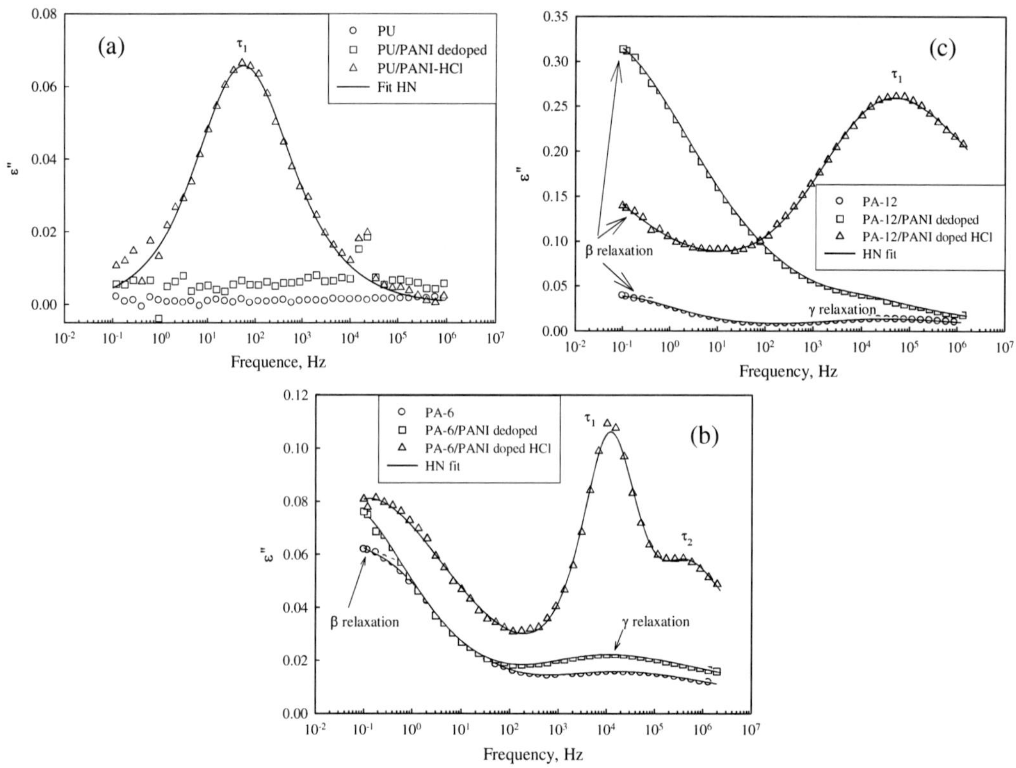

Figure 12. The variation of dielectric loss factor ε" as a function of the frequency: *a* – the composite PU/PANI films; *b* – the composite PA-6/PANI films; *c* – the composite PA-12/PANI films.

Table 2. The activation energies of polyamides.

Type of PA	Activation energy E_a, eV	
	γ	β
PA-6	0.326	0.525
PA-12	0.379	0.688

The doping process leads to the appearance in the studied frequency and temperature ranges of an additional relaxation process labelled τ_1. As it has been already mentioned, Resonance Raman spectrometry revealed a double-layered structure for the conducting composite films: the insulating and conductive layers. So, taking this fact into account, one can attribute the relaxation process found in the doped composite polymer/PANI film to the interfacial polarization phenomenon occurring between the conducting PANI containing layer and the insulating layer of pure polymer matrix. Especially as it was established that the additional relaxation process is present only when PANI is in its conducting doped state.

The performed dielectric relaxation measurements for the composite PA-12/PANI film doped by HCl as a function of the temperature (Fig. 13) showed that at high temperatures (more than 263 K) all relaxations are hidden or overlapped by the conductivity and α-relaxation process, associated with the chain motion in the polymer matrix, which dominate the spectra. As one can see in Fig. 13, the relaxation process τ_1 is temperature-dependent. The characteristic relaxation time τ_1 was taken at the position of the maximum of dielectric loss and its temperature dependence was analyzed according to equation (5). The obtained dependencies are shown in Fig. 14 and the obtained fit parameters can be found in Table 3.

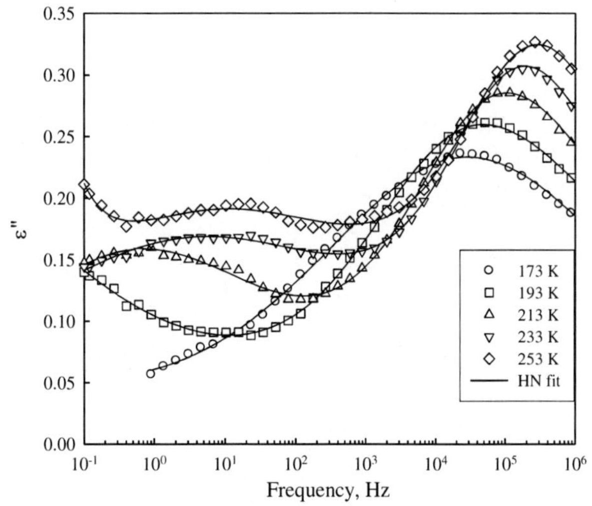

Figure 13. The frequency dependencies of ε" for the composite PA-12/PANI film doped by HCl.

As one can see from the obtained results, the activation energy of the relaxation process τ_1 has rather small values in comparison with β-relaxation. As it was established previously, the surface composite PA-12/PANI film consists of two layers - the first one being the conducting PANI containing layer and the second one being the pure polymer matrix. This interfacial polarisation leads to a Debye type relaxation process [31]. Therefore, it may be suggested that the charge carriers are responsible for this phenomenon.

Table 3. The fit parameters according to Arrhenius law [equation (5)] for the doped composite polymer/PANI films.

Polymer matrix / Parameter	PU		PA-6			PA-12	
	τ_1	τ_β	τ_1	τ_2	τ_β	τ_1	τ_β
τ_0, s	$9.5\cdot10^{-5}$	$5.9\cdot10^{-12}$	$1.3\cdot10^{-6}$	$1.2\cdot10^{-8}$	$5.8\cdot10^{-18}$	$2.0\cdot10^{-9}$	$7.1\cdot10^{-14}$
E_a, eV	0.15	0.54	0.04	0.05	0.67	0.12	0.54

It should be noted that the composite PA-6/PANI film (Fig. 12b) exhibits somewhat different behaviour. The doping by HCl leads to the appearance of two additional relaxation processes. These peaks are labelled τ_1 and τ_2. The temperature dependencies of these relaxation peaks for all composite films are presented in Fig. 14 and the fit parameters obtained by fit to Arrhenius law (equation (5)) for the doped films are shown in Table 3.

It should be mentioned that found relaxation processes were shifted to the high frequency region as compared to the PU film (Fig. 12). By analyzing these relaxations according to HN function (equation (4)) it was found that the relaxation processes were of a Debye type ($\alpha \approx 1$, $\beta \approx 1$). As we can see from the obtained results, the second peak (τ_2) observed in the composite film based on PA-6 has an activation energy of 0.05 eV. This value is found to be closer to the one found in pure PANI (0.093 eV) and, therefore, the second relaxation peak may be attributed to the interfacial polarization in PANI. On the other hand, analyzing the shape and the obtained values of the activation energy for τ_1 for PA-12 and PA-6 (Table 3) one can found that the relaxation peak τ_1 for the former is broader than for the latter. Thus, the first process in the case of the composite film based on PA-6 can be attributed to the interfacial polarization between the thin PANI containing layer and layer of pure PA-6 matrix, while the second one is due to the hopping conduction process inside the PANI conducting clusters as in the case of composite film based on PET matrix [23]. Such difference between the composite films can be explained by the PANI clusters organisation in the polymer matrix.

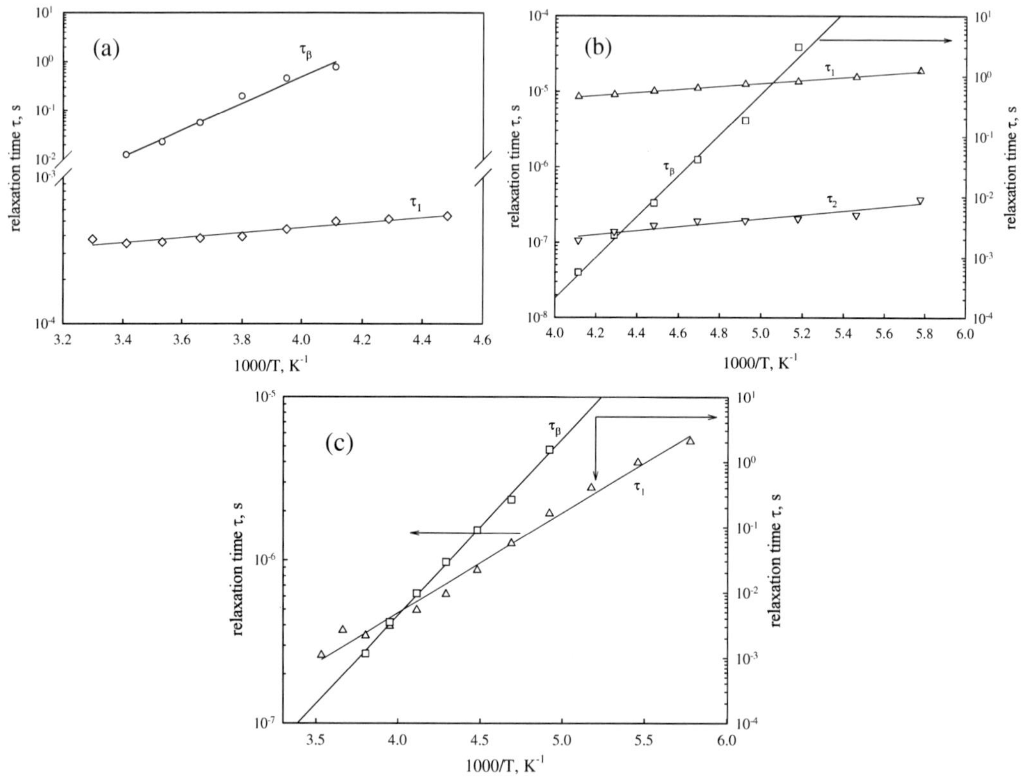

Figure 14. Arrhenius plots for the relaxation processes observed in the doped composite films: *a* – the PU/PANI film; *b* – the PA-6/PANI films; *c* – the PA-12/PANI film.

In the PA based composite films the relaxation process are found at higher frequencies compared with the composites based on the PET films [13, 23]. This fact may be explained by the higher conductivity values obtained for the PA films in comparison with those for the PET and PU films. It is known that the relaxation frequency is proportional to the conductivity. So, the increase of frequency by four decades signifies the increase of the conductivity value by four orders. Indeed, the resistance of the subsurface conductive layer of the PA/PANI film is four orders lower than that in the PET/PANI film case [23].

In order to study in detail the relaxation mechanism in the PU/PANI-HCl and PA/PANI-HCl composite films, the dielectric measurements as a function of temperature were performed. It was observed that the dielectric strength increased with the increase of temperature. It is known that for relaxations such as α and β, i.e., for relaxations associated with the molecular dynamics in polymers, the dielectric loss decreases with the temperature increase [31]. So, the revealed behaviour for composite films clearly indicates that found relaxation processes are of the interface polarisation type induced by the charge mobility. The relaxation frequency of the interfacial polarisation process is also found to increase with temperature (Fig. 14), i.e., it is a thermally activated process. The appearance of the peaks only in the spectrum of the doped film is, hence, due to the extended conjugated structure of doped PANI (Fig. 1), which causes delocalisation of the charge.

Electrical Properties of Composite Polymer/PANI Films

To confirm the assumption of the influence of the PANI conductivity and its organisation in the polymer matrix on the general conductivity of the composite materials, the electrical properties of the polymer composite materials has been studied. The information of conductivity behaviour for pure PANI is of great importance for a better understanding of the phenomena observed in the surface composite polymer/PANI materials. So, it has been discussed in the first place.

As expected, the doping process is seen to lead to an increase of the conductivity of pure doped PANI at low frequencies (not shown). Analyzing the obtained results one can say that at low frequencies the real part of the conductivity spectra of doped PANI is dominated by a dc-conductivity, whereas at high frequencies the spectra reveal an ac-conductivity increasing with the increase of frequency (Fig. 11).

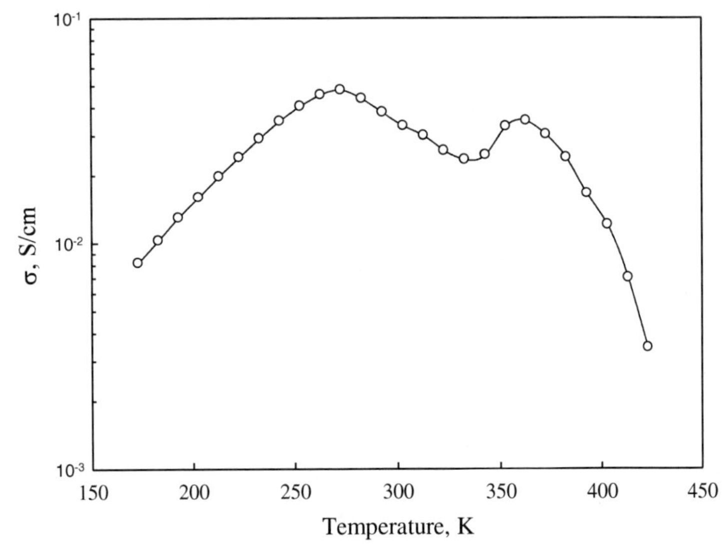

Figure 15. Temperature dependence of the electrical dc-conductivity for PANI-HCl.

The measurements of the temperature dependence of the electrical conductivity for the PANI sample doped by HCl were performed (Fig. 15). As it can be seen from this figure, the conductivity initially increases gradually with the rise in temperature from 173 K to ~300 K and then decreases. There are three potential causes for the change in the conductivity: (1) the loss of the adsorbed moisture; (2) the gradual deprotonation and (3) the oxidative degradation. The first process is likely to be responsible for the conductivity changes in the vicinity of 300 K. Other two processes lead to the conversion of conducting PANI into non-conducting one and a consequent loss of conductivity is anticipated and indeed observed. This decrease in conductivity is attributed to the separation of the dopant from the polymer backbones, which results in the dedoping process and, thus, a large portion of imine nitrogen is being in the non-conducting state (Fig. 1). In its turn, this leads to a decrease of the concentration of polarons. The assumption about the decreasing of the polarons concentration is also well confirmed by the EPR measurements performed by Kuo and Chen [46]. They performed conductivity studies for PANI doped with diphenyl phosphate and found that the temperature dependence of the spin density is consistent with that of the conductivity.

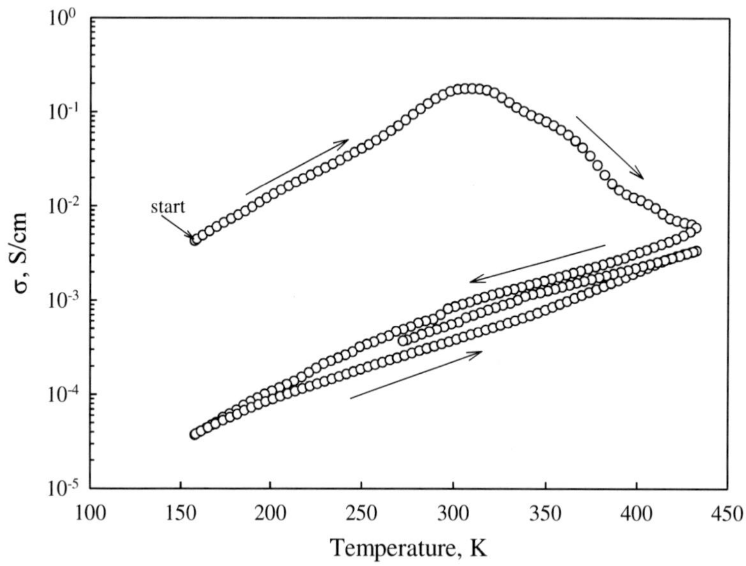

Figure 16. Temperature dependence of the electrical conductivity for the PANI-HCl pellet at 100 Hz.

In order to verify the proposed assumption about the deprotonation and loss of the moisture the additional conductivity measurements for the PANI-HCl sample as a function of temperature at a fixed frequency (100 Hz) were performed (Fig. 16). An observed difference between conductivity values in Fig. 15 and 16 is connected with the use of other way of measurement. In the first case we performed measurements of the conductivity dependence on the frequency and temperature simultaneously, i.e., the sample was being held at a certain temperature for some time (~20 min) until measurements at all frequencies were done. In the second case (Fig. 16) we measured the conductivity at a fixed frequency with temperature varying continuously with a temperature ramp of 2 °C/min. Therefore, the conductivity values obtained by the latter method are higher than those obtained by the former method. But, in spite of the slight difference in conductivity, the common behaviour is the same in both cases.

So, as one can see, the first scan is rather different from the next ones (Fig. 16). The conductivity maximum, which is observed at ~310 K is connected with the beginning of the moisture evaporation. The conductivity of PANI registered during the next scans is significantly less as compared with the first one, which means that not only the moisture was evaporated, but also the doping acid – HCl, which is a rather volatile acid indeed. But, nevertheless, some acid remains in the sample, since the conductivity value is still rather high in comparison with dedoped PANI.

The obtained conductivity measurements are also in agreement with the investigation of Prokes et al. [47], who studied the resistivity evolution of the PANI films and powders. They found that the resistivity value increased when exposed to high temperatures. Also, they performed resistivity measurements during the temperature cycling. The initial increase of the resistivity in the first scan reflects the loss of water, while the other ones are connected with the deprotonation and a potential degradation.

In order to calculate the activation energy value, the dependence of log σ on T^{-1} was plotted for the linear part of the temperature dependence of the conductivity for doped PANI (Fig. 17). This plot indicates an increase in conductivity with temperature that may be due to a hopping mechanism [38]. The linear variation of the conductivity versus temperature for doped PANI suggests a thermally activated process of an Arrhenius-type in the experimental temperature range.

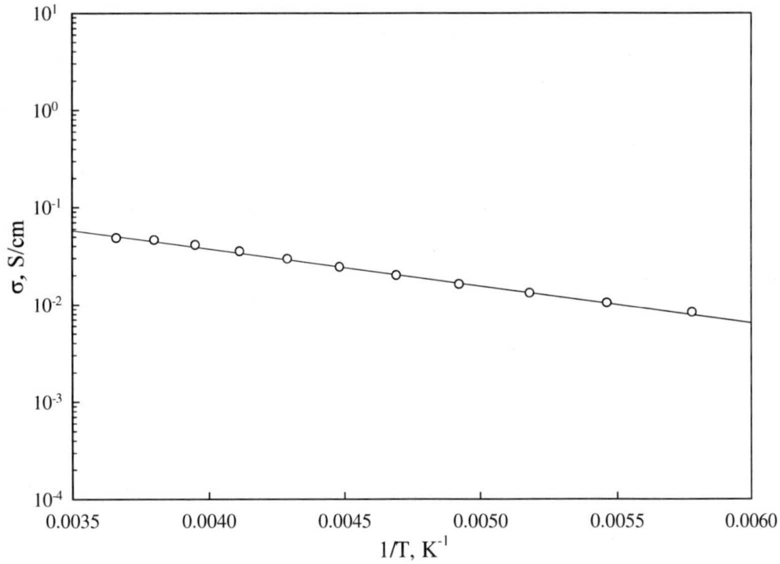

Figure 17. Arrhenius plot of the conductivity for the doped PANI pellet.

The value of the activation energy for the PANI-HCl pellet calculated from the slope (equation (7)) is equal to 0.076 eV. The obtained value is in good agreement with the activation energy value obtained for the PANI derivative – poly(*N*-methylaniline) doped with chloride ion (0.11 eV) [48].

The obtained value of the activation energy for the doped PANI sample (0.076 eV) is significantly lower than that for dedoped PANI (0.59 eV). Such difference can be explained by the presence of localized electrons in dedoped PANI which after the doping process are delocalized and, therefore, the conductivity process is facilitated.

To determine the nature of temperature dependence of electrical conductivity, the surface electrical resistance for the composite polymer/PANI films was measured as a function of temperature. It was found that the obtained dependences followed the Arrhenius equation (equation (7)) (not shown), indicating the semi-conducting behaviour of the composite films similar to the other semi-conductors [13, 45]. The activation energy of the conductivity as derived from the slope of Arrhenius plot was estimated to be 0.14 eV, 0.05 and 0.08 eV for the composite PU/PANI-HCl, PA-6/PANI-HCl and PA-12/PANI-HCl films, respectively. So, according to the variation of the activation energy depending on the type of the polymer matrix, the studied composites can be placed in the following row:

PA-6/PANI-HCl < PA-12/PANI-HCl < PU/PANI-HCl

The found values are close to that obtained by dielectric measurements (Table 3). This result indicates that both processes originate from the same electronic transport mechanism of the charge carriers (electrons) in the conducting PANI clusters. Therefore, in spite of the fact that the surface conductivity value of the composite surface films can not be measured using DRS, the conductive properties of these films can be evaluated through the study of the interfacial relaxation processes in the films.

CONCLUSION

Conducting polymer composite films based on PANI, PA and PU were prepared by the *in situ* aniline chemical polymerization. The electrical and dielectric properties of the obtained films were investigated in details. For comparison, pure PANI was also prepared and its electrical and dielectric properties in both doped and dedoped states were investigated to better understand the studied properties of the composite films.

It was found that the chemical aniline polymerization inside the polymer host matrix triggered off the interface of the oxidant solution and the polymer matrix swelled in aniline and, thereafter, occured deep in the polymer matrix. Besides, the polymerization process was limited by the monomer (aniline) diffusion into the reaction zone, and by the presence of the overoxidized PANI in the upper polymer matrix layer. All these features result in the layered structure of the composite film, which is confirmed by Raman spectrometry – the first layer being the conducting polymer/PANI layer and the second one being the pure polymer matrix. It has been established that the thickness of the formed PANI containing layer depends on the polymer matrix structure.

Dedoped PANI powder was characterized by the high permittivity value (~ 100) due to the presence of charge carriers. Also, a relaxation process was found at low frequencies for dedoped PANI and was attributed to the conductivity relaxation, since a correlation of the conductivity with the relaxation time was determined.

The electrical and dielectric properties for the composite polymer/PANI films differ from those observed for pure PANI because they are strongly dependent on the nature of the polymer matrix. In the case of composite films, besides the subglass relaxation processes (β and/or γ) occurring in the polymer matrix as well as in the dedoped composite film, additional relaxation processes are found in the doped composite related to its conductive properties. The found relaxation processes are thermally activated and are characterized by low activation energy. Interfacial polarization relaxations in the MHz region are attributed to the layered and clustered structure of the composite film. At higher frequency, conductivity relaxation appears to be connected with the interfacial polarization in the PANI clusters.

The temperature dependences of the surface electrical resistance of the composite films are well described by Arrhenius law. The corresponding activation energies are close to the values found by the dielectric relaxation spectroscopy, which proves the conducting nature of the PANI containing layer. Also, it was shown that the conductive properties of the layered conducting composite films can be evaluated through the study of the interfacil relaxation processes in the films.

REFERENCES

[1] Shirakawa, H., Louis, E.J., MacDiarmid, A.G., Chiang, C.K., Heeger, A. J. 1977. Synthesis of electrically conducting organic polymers: halogen derivatives of polyacetylene, $(CH)_x$. *J. Chem. Soc., Chem. Comm.* 474, 578-580.

[2] MacDiarmid, A.G. 2001. Nobel Lecture: "Synthetic metals": a novel role for organic polymers. *Rev. Mod. Phys.* 73, 701-712.

[3] Heeger, A.J. 2001. Nobel Lecture: Semiconducting and metallic polymers: the fourth generation of polymeric materials. *Rev. Mod. Phys.* 73, 681-700.

[4] Feng, Z., Nuli, Y., Yang, J. 2007. Conductive sulfur-containing material/polyaniline composite for cathode material of rechargeable magnesium batteries. *Acta Phys.-Chim. Sinica* 23, 327-331.

[5] Liu, Y., Zhang, Y., Ma, G., Wang, Z., Liu, K., Liu, H. 2013. Ethylene glycol reduced graphene oxide/polypyrrole composite for supercapacitor. *Electrochim. Acta* 88, 519-525.

[6] Arora, K., Sumana, G., Saxena V., Gupta, R.K., Gupta, S.K., Yakhmi, J.V., Pandey, M.K., Chand, S., Malhotra, B.D. 2007. Improved performance of polyaniline-irucase biosensor. *Anal. Chim. Acta* 594, 17-23.

[7] Cao, Y., Andreatta, A., Heeger, A.J., Smith, P. 1989. Influence of chemical polymerization conditions on the properties of polyaniline. *Polymer* 30, 2305-2311.

[8] Kulkarni, S.B., Joshi, S.S., Lokhande, C.D. 2011. Facile and efficient route for preparation of nanostructured polyaniline thin films: schematic model for simplest oxidative chemical polymerization. *Chem. Ing. J.* 166, 1179-1185.

[9] Prakash, R. 2002. Electrochemistry of polyaniline: study of pH effect and electrochromism. *J. Appl. Polym. Sci.* 83, 378-385.

[10] Okamoto, H., Kotaka, T. 1998. Structure and properties of polyaniline films prepared via electrochemical polymerization. Effect of pH in electrochemical polymerization media on th eprimery structure and acid dissociation constant of product polyaniline films. *Polymer* 39, 4349-4358.

[11] Bhadra, S., Khasrgir, D., Sungha, N.K., Lee, J.H. 2009. Progress in preparation, processing and applications of polyaniline. *Prog. Polym. Sci.* 34, 783-810.

[12] Huang, C., Zhang, Q. 2004. Enhanced dielectric and electromechanical responses in high dielectric constant all-polymer percolative composites. *Adv. Funct. Mater.* 14, 501-506.

[13] Pud, A., Ogurtsov, N., Korzhenko, A., Shapoval, S. 2003. Some aspects of preparation methods and properties of polyaniline blends and composites with organic polymers. *Prog. Polym. Sci.* 28, 1701-1753.

[14] Shang, S., Zeng, W., Tao X.-M. 2012. Fabrication of conducting polypyrrole/β-cyclodextrin nano- and microspheres using molecular templates. *RSC Adv.* 2, 4675-4682.

[15] Fang, F.F., Liu, Y.D., Lee, I.S., Choi, H.J. 2011. Well controlled core/shell type polymeric microspheres coated with conducting polyaniline: fabrication and electrorheology. *RSC Adv.* 1, 1026-1032.

[16] Fatyeyeva, K., Pud, A.A., Bardeau, J.-F., Tabellout, M. 2011. Structure-property relationship in aliphatic polyamide/polyaniline surface layered composites. *Mater. Chem. Phys.* 130, 760-768.

[17] Yuan, J.K., Dang, Z.M., Yao, S.H., Zha, J.-W., Zhou, T., Li, S.-T., Bai, J. 2010. Fabrication and dielectric properties of advanced high permittivity polyanilne/poly(vinylidene fluoride) nanohybrid films with high energy storage density. *J. Mater. Chem.* 20, 2441-2447.

[18] Neelgund, G.M., Bliznyuk, V.N., Pud, A.A., Fatyeyeva, K., Hrehorova, E., Joyce, M. 2010. Formation of nanostructured composites with environmentally-dependent electrical properties based on poly(vinylidene fluoride)-polyaniline core-shell latex system. *Polymer* 51, 2000-2006.

[19] Pud, A.A., Fatyeyeva, K.Yu., Bardeau, J.-F., Rogalsky, S.P., Tabellout, M., Shapoval, G.S. 2007. Polyamide-12/polyaniline layered composite films: specificity of the formation and Raman spectroscopy investigation. *J. Macromol. Sci., Part A: Pure Appl. Chem.* 44, 183-192.

[20] Vanga Bouanga, C., Fatyeyeva, K., Baillif, P.-Y., Bardeau, J.-F., Khaokong, C., Pilard, J.-F., Tabellout, M. 2010. Study of dielectric relaxation phenomena and electrical properties of conductive polyaniline based composite films. *J. Non-Crystalline Solids* 356, 611-615.

[21] Neoh, K.G., Tay, B.K., Kang, E.T. 2000. Oxidation and ion migration during síntesis and degradation of electroactive polymer-nylon 6 composite films. *Polymer* 41, 9-15.

[22] Hong, K.H., Oh, K.W., Kang, T.J. 2004. Polyaniline-nylon 6 composite fabric for ammonia gas sensor. *J. Appl. Pol. Sci.* 92, 37-42.

[23] Adohi, J.P., Vanga Bouanga, C., Fatyeyeva, K., Tabellout, M. 2009. Application of the Maxwell-Wagner-Hanai effective medium theory to the analysis of the interfacial polarization relaxations in conducting composite films. *J. Phys. D – Appl. Phys.* 42, P. 015302/1-015302/10

[24] Patil, R., Roy, A.S., Anikumar K.R., Jadhav, K.M., Ekhelikar S. 2012. Dielectric relaxation and ac conductivity of polyaniline-zinc ferrite composite. *Composites Part B: Eng.* 43, 3406-3411.

[25] Dhand, C., Arya, S.K., Singh, S.P., Singh, B.P., Datta, M., Malhotra, B.D. 2008. Preparation of polyaniline/multiwalled carbon nanotube composite by novel electrophoretic route. *Carbon* 46, 1727-1735.

[26] Dutta, P., Biswas, S., De, S.K. 2002. Dielectric relaxation in polyaniline-polyvinyl alcohol composites. *Mater. Research Bull.* 37, 193-200.

[27] Saravanan, S., Anantharaman, M.R., Venkatachalam, S. 2006. Structural and electrical studies on tetrameric cobalt phthalocyanine and polyaniline composites. *Mater. Sci. Eng. B* 135, 113-119.

[28] Tsotra, P., Friedrich, K. 2005. Electrical and dielectric properties of epoxy resin/polyaniline-DBSA blends. *J. Mater. Sci.* 40, 4415-4417.

[29] Lu, J., Moon, K.-S., Kim, B.-K., Wong, C.P. 2007. High dielectric constant polyaniline/epoxy composites via in situ polymerization for embedded capacitor applications. *Polymer* 48, 1510-1516.

[30] Pinto, N.J., Shah, P.D., Kahol, P.K., McCormick, B.J. 1996. Dielectric constant and ac conductivity in polyaniline derivatives. *Solid State Communications* 97, 1029-1031.

[31] Jonscher, A.K. Dielectric relaxation in solids. – Chelsea Dielectric Press. – 1996.

[32] Dielectric spectroscopy of polymeric materials: fundamentals and applications / Runt J.P., Fitzgerald J.J. – American Chemical Society, Washington DC. – 1997. – 485 p.

[33] Kebir, N., Campistron, I., Laguerre, A., Pilard, J.-F., Bunel, C., Couvercell, J.-P., Gondard, C. 2005. Use of hydroxytelechelic cis-1,4-polyisoprene (HTPI) in the synthesis of polyurethanes (PUs). Part 1. Influence of molecular weight and chemical modification of HTPI on the mechanical and thermal properties of Pus. *Polymer* 46, 6869-6877.

[34] MacDiarmid A.G., Xia Y.N., Wiesinger J.M. Methods for preparing conductive polyanilines // US patent 1995. - IC^6 C08F 6/00.- No.5403913.

[35] Byun, S.W., Im, S.S. 1998. Physical properties and doping characteristics of polyaniline – nylon 6 composite films. *Polymer* 39, 485-489.

[36] Zhang, Q., Jin, H., Wang, X., Jing, X. 2001. Morphology of conductive blends of polyaniline and polyamide-11. *Synthetic Metals* 123, 481-485.

[37] Zuo, F., Angelopoulos, M., MacDiarmid, A.G., Epstain, A.J. 1989. ac Conductivity of emeraldine polymer. *Phys. Rev. B* 39, 3570-3578.

[38] McCrum, N.G., Read, B.E., Williams, G. Anelastic and Dielectric Effects in Polymeric Solids. – Dover Publications, Inc., New York, 1991. – 616 p.

[39] Pelster, R., Nimtz, G., Wessling, B. 1994. Fully protonated polyaniline: hopping transport on a mesoscopic scale. *Phys. Rev. B* 49, 12718-12723.

[40] Papathanassiou, A.N., Grammatikakis, J., Sakkopoulos, S., Vitoratos, E., Dalas, E. 2002. Localized and long-distance charge hopping in fresh and thermally aged conductive copolymers of polypyrrole and polyaniline studied by combined TSDC and dc conductivity. *J. Phys. Chem. Solids* 63, 1771-1778.

[41] Bengoechea, M.R., Aliev, F.M., Pinto, N.J. 2002. Effects of confinement on the phase separation in emeraldine base polyaniline cast from 1-methyl-2-pyrrolidinone studied via dielectric spectroscopy. *J. Phys.: Condens. Matter.* 14, 1-10.

[42] Joo, J., Prigodin, V.N., Min, Y.G., MacDiarmid, A.G., Epstein, A.J. 1994. Phonon-induced nonmetal-metal transition of a doped polyaniline. *Phys. Rev. B* 50, 12226-12229.

[43] Jeon, S., Bae, S.C., Turner, J., Granick, S. 2002. Microviscosity of an ion-conducting polymer probed by fluorescence depolarization and dielectric spectroscopy. *Polymer* 43, 4651-4655.

[44] Okutan, M., Yeril, Y., San, S.E., Yilmaz, F., Gunaydin, O., Durak, M. 2007. Dieletric properties of thiophene based conducting polymers. *Synthetic Metals* 157, 368-373.

[45] Rocha, I.S., Mattoso, L.H.C., Malmonge, L.F., Gregorio, R. Jr. 1999. Effect of low contents of a polyaniline derivative on the crystallization and electrical properties of blends with PVDF. *J. Polym. Sci.: Part B: Polym. Phys.* 37, 1219-1224.
[46] Kuo, C.-T., Chen, C.-H. 1999. Characterization of polyaniline doped with diphenyl phosphate. *Synthetic Metals* 99, 163-167.
[47] Prokes, J., Krivka, I., Sulimenko, T., Stejska, 1 J. 2001. Conductivity of polyaniline films during temperature cycling. *Synthetic Metals* 119, 479-480.
[48] Abthagir, P.S., Saraswathi, R., Sivakolunthu, S. 2004. Aging and thermal degradation of poly(*N*-methylaniline). *Thermochimica Acta* 411, 109-123.

In: New Developments in Polymer Composites Research ISBN: 978-1-62948-340-5
Editors: S. Laske and A. Witschnigg

Chapter 6

THERMOSETTING BIOPOLYMERS REINFORCED WITH NATURAL FIBRES: A REVIEW

Carlo Santulli and Fabrizio Sarasini
Department of Chemical Engineering Materials Environment,
Sapienza-Università di Roma, Rome, Italy

ABSTRACT

The use of natural fibres for the reinforcement of polymers is a well-established practice: the biocomposites obtained with this procedure are increasingly used in sectors, such as automotive and construction industries. In particular, green composites represent a class of biocomposites increasingly emerging in polymer science: these are characterized by the use of a bio-based polymer matrix reinforced by natural fibres. This work discusses the replacement of oil-derived polymers by bio-based ones, reviewing the recent developments in the synthesis of thermosetting bio-based polymers and in their use for the production of green composites.

Keywords: Biodegradable composites, thermosetting, fatty oil acids, plant fibres

1. INTRODUCTION

The use of bio-based polymers as the matrix for composite materials is particularly desirable in terms of reduced environmental impact, resulting therefore in a lower carbon footprint during their whole lifecycle. The fossil fuels used throughout the petrochemical industry, and also for the synthesis of oil-based polymers are finite, so there is need of an additional source of durable, high performance raw materials. It is noteworthy that fossil fuels used for the manufacture of plastics account for approximately 7% of the worldwide use of oil and gas [1]. Renewable materials extracted from home-grown crops, trees, and agricultural wastes can provide most chemical building blocks for plastics, similarly to what petrochemicals can do. The drive towards manufacturing new materials and products, at the same time compatible with the environment and independent of fossil fuels, is testified, for instance, by the Technology Road Map for Plant/Crop-based Renewable Resources 2020, developed by the U.S. Department of Agriculture (USDA) and the U.S. Department of Energy (DOE). This set a target of 10% of basic chemical building blocks synthesised from plant-derived renewable substances by 2020, with the subsequent goal to increase these to 50% by 2050. One consequence of this trend is the replacement of oil-derived polymers with polymers from renewable resources (bio-based polymers) as the matrix in composite materials. The obtained materials are referred to as ''green composites'' [2,3], a term suggesting that the composite as a whole (both matrix and reinforcement) originates from renewable resources. On the other hand, the term biocomposites is used to indicate a class of composites made from natural fibres and petroleum-derived polymers such as polypropylene (PP), polyethylene (PE) and epoxies, even though some authors define as ''green composites'' all natural fibre composites [4], irrespective of the nature of the polymeric matrix (both bio and oil-based).

The most diffuse bio-based polymers, such as e.g., poly(lactic) acid or polycaprolactone, are thermoplastics. However, a number of industrial sectors, including e.g., boat building, aeronautical and furniture making, use mainly for their components thermosetting matrices, such as epoxies, unsaturated polyesters and phenolics. The recent availability of biodegradable thermosetting matrices offered possibilities to improve the environmental profile of these materials. This is particularly important, whenever green composites are produced using natural fibres, such as flax, hemp, sisal, coir, etc. The three major plant-based polymers are proteins, oils, and carbohydrates. Plant proteins are amino acid polymers derived mainly from oilseeds (i.e., soybeans) and grains (i.e., wheat and corn) and are usually produced as by-products of processing oils and starches.

Plant oils, such as soy oil, corn oil, and flax oil, can be derived from many crops: they are triglycerides and contain various fatty acids. Soybean, a major oil plant, contains about 20% oil [5]. Starch and cellulose, also called polysaccharides, are the main naturally occurring polymers in the large carbohydrate family. Agricultural fibre is also a member of the carbohydrate family. Natural fibres, such as flax, hemp, kenaf, jute and regenerated cellulose, consist mainly of cellulose, hemicellulose, and lignin, but are usually listed as a sole material when used as fibres in composites.

Recently also biodegradable resins are marketed, which are thermoplastics decomposed, through the action of microorganisms, into carbon dioxide and water in a short time. Biodegradable resins, like poly(lactic) acid, are mainly used in short lifetime applications, such as packaging. In structural composites, biodegradation might not represent a central issue. However, as concerns the end-of-life disposal of green composites, it is worth pointing out that a bio-based material is not by definition biodegradable: that is, bio-based and biodegradable are independent properties. The bio-based concept refers to the environmentally friendly origin of the material, while biodegradability is a final property defined by a number of international standards such as, for instance, EN 13432 that indicates the requisites that plastic products must possess to be biodegradable and compostable [6]. In this chapter, only bio-based thermosetting resins are dealt with. In this sense, functionalized (acrylated, epoxidized, acrylated/epoxidized, maleinated) plant oils (such as cardanol, soybean oil or castor oil) are becoming commercially available and are gradually modified to be included, in increasing quantities, into traditional oil-based thermosetting resins. The percentage of the bio-derived content in a thermosetting resin can vary greatly and only a few of them are nearly 100 % bio-based. The bio-based content varies, in general, from 12% to

90%, according to manufacturers' reports. The characterization of true bio-based content is more difficult to be performed: the relevant standard offering definitions and guidelines to this aim is ASTM D6866.

Increased interest in using natural fibres as the reinforcement of polymer composites is due to higher environmental awareness and the great potential of natural fibres. Natural fibres have many advantages to be used as reinforcements in polymer composites. They are considered as biodegradable and sustainable alternatives to other man-made and synthetic fibres. They are CO_2 neutral and leave no residues when they are incinerated. Their specific mechanical properties are good and they are neither abrasive nor toxic, having in addition low density and low price. However, they also possess disadvantages, like inhomogeneous quality and seasonal supply cycles. They have also limited water resistance and dimensional stability and tend to absorb moisture and lack good properties in toughness [4,7–9]. The fibres that are mainly used for reinforcements are bast fibres, such as flax and hemp, although other possible fibres of interest, in particular in the automotive industry, are being investigated, such as sisal (a leaf fibre). In particular, the investigation of flax fibres introduction in bio-based thermosetting led to the development of pre-pregs, which are particularly requested for application in the automotive, marine, sports and consumer goods sectors.

In this work, the potential of the different thermosetting bio-based resins is discussed, in particular for the production of composites with plant fibres. The difficulties in processing them with respect to traditional thermosetting resins are also discussed, once again in the perspective of the fabrication of a fully bio-based composite.

2. THERMOSETTING RESINS DERIVED FROM RENEWABLE RESOURCES

Thermosetting resins (or thermosets) are of vital importance in many industrial sectors due to their high flexibility, which leads to their high modulus, strength, durability, thermal and chemical resistance, as provided by high cross-linking density [10]. In non-food applications, the most widely applied renewable resources include plant oils, polysaccharides (mainly cellulose and starch), and proteins. In particular, plant oils have always found application as lubricants, varnishes or plasticizers. Over the past decades, there has been a growing attention to these oils for the substitution of fossil resources because of their availability, abundance and low price. Vegetable oils are a rich source of polymer precursors that can be modified to exhibit several types of functionalities, leading to new materials with a wide range of properties and are now being considered as the most important renewable raw materials for the production of bio-based polymers [11]. This interest is testified by the many available reviews, books and book chapters dedicated to the subject [1,5,11–15]: only a short summary on this subject is included in this work, which is mainly focused on the use of these bio-based polymers for the production of green composites. Plant oils are triglycerides (figure 1) (trimesters of glycerol with long-chain fatty acids) with varying composition of fatty acids depending on the plant, the crop, the season, and the growing conditions. The molecular parameters affecting the physical and chemical properties of plant oils are the length of fatty acids chains, the stereochemistry of their double bonds and the degree of unsaturation.

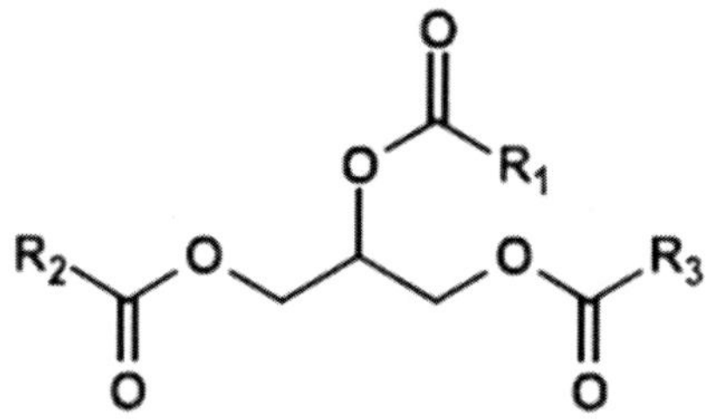

Figure 1. General triglyceride structure. R = fatty acid chain.

Table 1. Fatty acid distribution in various plant oils [5]

Fatty acid	#C:# DB*	Canola	Corn	Cottonseed	Linseed	Olive	Palm	Rapeseed	Soybean
Myristic	14:0	0.1	0.1	0.7	0.0	0.0	1.0	0.1	0.1
Myristoleic	14:1	0.0	0.0	0.0	0.0	0.0	0.0	0.0	0.0
Palmitic	16:0	4.1	10.9	21.6	5.5	13.7	44.4	3.0	11.0
Palmitoleic	16:1	0.3	0.2	0.6	0.0	1.2	0.2	0.2	0.1
Margaric	17:0	0.1	0.1	0.1	0.0	0.0	0.1	0.0	0.0
Margaroleic	17:1	0.0	0.0	0.1	3.5	0.0	0.0	0.0	0.0
Stearic	18:0	1.8	2.0	2.6	19.1	2.5	4.1	1.0	4.0
Oleic	18:1	60.9	25.4	18.6	15.3	71.1	39.3	13.2	23.4
Linoleic	18:2	21.0	59.6	54.4	56.6	10.0	10.0	13.2	53.2
Linolenic	18:3	8.8	1.2	0.7	0.0	0.6	0.4	9.0	7.8
Arachidic	20:0	0.7	0.4	0.3	0.0	0.9	0.3	0.5	0.3
Gadoleic	20:1	1.0	0.0	0.0	0.0	0.0	0.0	9.0	0.0
Eicosadienoic	20:2	0.0	0.0	0.0	0.0	0.0	0.0	0.7	0.0
Behenic	22:0	0.3	0.1	0.2	0.0	0.0	0.1	0.5	0.1
Erucic	22:1	0.7	0.0	0.0	0.0	0.0	0.0	49.2	0.0
Lignoceric	24:0	0.2	0.0	0.0	0.0	0.0	0.0	1.2	0.0
Average #DB/ triglyceride		3.9	4.5	3.9	6.6	2.8	1.8	3.8	4.6
Iodine value (mg/100g)		110-126	102-130	90-119	168-204	75-94	44-58	94-120	117-143

*Number of carbon atoms and C=C double bonds per fatty acid.

As expressed by the iodine value, plant oils can be classified into drying oils (iodine value higher than 130), semi-drying oils (iodine value between 90 and 130), and non-drying oils (iodine value lower than 90). The higher the iodine value, the more unsaturated the oil is. Other types of fatty acid functionalities (e.g., epoxy, hydroxyl, aromatic and aliphatic cyclic and furanoid groups) are available as well [5,14]. The multiple C=C bonds make these biological oils ideal monomers of natural origin for the preparation of biopolymers. Of the available vegetable oils, soybean oil is the most abundant, accounting for around 30% of the world's vegetable oil supply. Typically, the majority (~80%) of soybean oil produced each year is used for human food. A further 6% is used for animal feed, whereas the remainder (14%) finds non-food uses (soap, lubricants, coatings, paints, etc.). The non-food uses have mainly involved the polymerization of soybean oil derivatives, such as fatty acids and epoxidized oils [5]. Most common oils contain fatty acids that vary from 14 to 22 carbons in length, with 0 to 3 double bonds per fatty acid. The fatty acid distributions of several common oils are shown in table 1. The main structural features of these fatty acids are as follows [1]:

a) most fatty acids are straight-chain compounds with even numbers (8–24) of carbon atoms, b) the carbon–carbon double bonds in most unsaturated fatty acids exhibit a *cis* configuration, c) the unsaturated fatty acids with more than two carbon–carbon double bonds contain a bis-allylic group, and d) some special fatty acids contain other functional groups, such as hydroxyl or epoxy groups, in the vicinity of the double bonds. Generally, long-chain fatty acids (C18) are dominant in the oils derived from temperate crops, such as soybean, linseed, rapeseed, and sunflower oils, whereas the shorter fatty acids (C12–C14) dominate the oils from tropical crops, like coconut and palm oils [1]. A large variety of new polymeric materials have been prepared from vegetable oils and derivatives that exhibit industrially viable thermo-mechanical properties. One of the major efforts in this field has taken advantage of either the original C=C bonds of soybean oil [16–18], tung oil [19,20], corn oil [21], and fish oils [22,23] for cationic polymerization, or the C=C bonds of soybean oil derivatives for free radical polymerization [5] or olefin metathesis polymerization [24]. The other major process has involved the conversion of vegetable oils into polyols by epoxidizing the C=C bonds of the triglyceride oil, followed by oxirane ring opening of the epoxidized oil. Epoxidized vegetable oils can be cationically polymerized by latent thermal catalysts [25] or cured by amines [26] or anhydrides [27] to produce thermosetting resins, whereas the polyols can react with di-isocyanates to produce vegetable oil based polyurethane thermosets [28] or waterborne polyurethane dispersions [29,30].

Phenolic resins are thermosets widely used due to their superior mechanical strength, heat resistance and dimensional stability, as well as high resistance against various solvents, acids and water. Phenolic resins can be synthesized under both acidic and alkaline conditions, and acid-catalyzed resins (better known as novolacs) are prepared using a formaldehyde/phenol ratio between 0.75 and 0.85, while alkaline-catalyzed resins (better known as resoles) are prepared using a formaldehyde/phenol ratio higher than 1 [14]. In contrast to the resole resins, the novolac resins are only partially cross-linked, and therefore they must be cured in a subsequent step, in the presence of hexamethylenetetramine to produce a fully cross-linked material. Cashew nut shell liquid (CNSL), an agricultural by-product available in tropical countries, is one of the major and economical resources of naturally occurring phenols. Cashew nut shell liquid occurs as a reddish brown viscous fluid in the soft honeycomb structure of cashew nut shell obtained from the cashew tree, *Anacardium occidentale*. The cashew nut shell liquid contains four major components with an unsaturated C_{15}-chain, namely cardanol, cardol, anacardic acid and 2-methylcarbol [31,32]. Cardanol and its polymers have interesting structural features for chemical modification and polymerization into specialty polymers. The phenolic nature of the resulting cardanol makes it suitable for reaction with formaldehyde under a variety of conditions, to form novolac or resole resins [33]. Compared with traditional phenolic resins, cardanol–formaldehyde resins have improved flexibility (due to the plasticization effect of the long chain), leading to better processability, excellent resistance to the softening action of mineral oils and high resistance to acids and alkalis, while yielding a tensile strength lower than that of phenol–formaldehyde resins. Lignin, due to its similar structure with phenol, has also been considered as a natural substitute for phenolic-based resins [34]. Commercial lignins are available as co-products of various types of chemical pulping process in the paper industry, so that they are mainly considered as waste materials. Similarly to phenol–formaldehyde resins, lignins have an aromatic and cross-linked structure not yet fully elucidated, along with a great variety of functional groups, such as ether and hydroxyl groups, depending on the origin of lignin and the extraction technology. As a whole, lignin shows limited reactivity with formaldehyde due to steric hindrances, and its reactivity can be enhanced via chemical modification of lignin (hydrolysis, acetosolv, and organosolv processes) [35,36]. In a commonly used method (referred to as *phenolysis*), lignins are reacted with phenol in the presence of organic solvents such as ethanol or methanol, to fully substitute lignin with phenol [36]. Also tannin compounds, extracted from the wood and leaves of plants are considered as other raw renewable materials in the design of phenolic resins. Tannins are natural polyphenolic materials, composed mostly of flavan-3-ol repeating units, for which two phenolic rings are joined by a heterocyclic ring, together with smaller fractions of polysaccharides and sugars [14]. The reactivity of tannins is due to their phloroglucinolic or resorcinolic nuclei, determining the reactivity towards formaldehyde and consequently their potential to cross-link.

Epoxy resins represent one of the most versatile material and are traditionally obtained by the reaction between Bisphenol A (2,2-bis(4'-hydroxyphenyl)propane and epichlorohydrin, yielding diglycidyl ether of bisphenol A (DGEBA). In thermosetting materials, epoxies are currently combined with a large range of curing agents, such as amines and anhydrides. The resulting product is a three-dimensional network characterized by a broad spectrum of performances, depending on the nature of curing agent, the extent and density of cross-linking. For the production of sustainable epoxy resins, epoxidized plant oils and fatty acids have been largely utilized [37–39]. On industrial scale, the unsaturated fatty compounds can be converted into epoxidized plant oils by the *in-situ* performic acid procedure, but epoxidized triglycerides can be found in natural oils, such as vernonia, or can be synthesized from more common unsaturated oils, such as soybean oil or linseed oil, by an epoxidation reaction. The natural epoxy oil, vernonia oil, has an epoxy functionality of about 2.8 epoxy rings per triglyceride but, due to the low reactivity of epoxy groups and the tendency for intramolecular bonding, any epoxidized oil leads to poorly cross-linked materials with very limited thermal and mechanical properties. To alleviate these problems, attempts to chemically modify epoxidized oils have been carried out through a two-step procedure [40], which consists of carrying out transesterification reactions of soybean oil with allyl alcohol, followed by their epoxidation in the presence of benzoyl peroxide. In the presence of an appropriate anhydride-curing agent, the resulting networks can be more densely cross-linked with higher mechanical properties, when compared with commercially available epoxidized soybean oil. The modification of the double bonds of triglycerides can incorporate functionalities like maleates, hydroxyl or epoxy [41] groups making possible a further reaction via ring opening or polycondensation polymerization. After this reaction step, the product can be used without further modification in crosslinking reactions or it can be further modified through other chemical steps broadening the variety of functional molecules to be obtained. Appropriate reactions can allow the epoxidized triglycerides to attach vinyl functionalities. For example, acrylates have been incorporated by reaction of the epoxy groups with acrylic acid, so that acrylated epoxidized soybean oil (AESO) has been frequently reported in the preparation of bio-based polymers and composites [5]. The reaction of acrylic acid with ESO occurs by a standard substitution reaction and has first order dependence with respect to epoxy concentration and second-order dependence with respect to acrylic acid concentration. Although the reaction of ESO with acrylic acid is partially acid catalyzed by the acrylic acid, the use of additional catalysts is common. Tertiary amines such as N,N-dimethylaniline, triethylamine, and 1,4-diazobicyclo[2.2.2]-octane, are commonly used [5]. AESO can be blended with a reactive diluent, such as styrene, to improve its processability and control the polymer properties to reach values acceptable for structural applications. By varying the amount of styrene, it is possible to produce polymers with different moduli and glass transition temperatures (T_g). For instance, the pure AESO polymer was found to show a modulus of about 440 MPa: however, with a styrene content of 40 wt%, the modulus increased fourfold, up to 1.6 GPa [5]. The ultimate tensile strength of pure AESO was about 6 MPa, whereas the polymers with 40 wt% styrene showed much higher strengths, approximately equal 21 MPa. Other two triglyceride monomers that have been found to be promising candidates for use in the composites and engineering plastics fields are the maleinized soybean oil monoglyceride (SOMG/MA), and maleinized hydroxylated soybean oil (HSO/MA). The SOMG/MA is synthesized from the triglyceride oil in two steps. The first step is a standard glycerolysis reaction to convert the triglycerides into monoglycerides. The maleinization of the SOMG mixture at temperatures around 100 °C produces monoglyceride, diglyceride, and glycerol maleate half-esters. To increase the T_g and modulus of the SOMG/MA polymer, more rigid diols, such as e.g., neopentyl glycol (NPG) and bisphenol A (BPA), can be added during the maleinization reaction, which may increase the rigidity of the polymer network finally obtained. The maleinized hydroxylated oil (HO/MA) is synthesized with a similar approach to both the AESO and the SOMG/MA monomers. The double bonds of unsaturated oils are used to attach the polymerizable groups, similar to the method used for the AESO monomer, by converting the double bonds of the triglyceride to hydroxyl groups. The hydroxyls can subsequently be used to attach maleates, not differently from the situation described for SOMG/MA synthesis [5]. In this regard, by reacting the epoxidized triglyceride with an acid, the epoxies can be easily converted to hydroxyl groups. After hydroxylation, the oil can be reacted with maleic anhydride to functionalize the triglyceride with maleate half-esters. When the maleinization reaction is finished, the monomer resin can be blended with styrene, as previously described.

Another class of versatile materials with applications in several fields is represented by polyurethanes. In general, polyurethane polymers contain a significant number of urethane linkages (–NH-COO–), regardless of the specific nature of the macromolecule [14]. For making sustainable thermosetting materials, polyurethane are usually prepared starting from renewable polyols, while the second component, isocyanate, is mainly obtained from petroleum-resources [42]. As far as natural polyols are concerned, only biopitches, soy flours and castor oil have directly been used as raw polyols for the production of polyurethane thermosetting materials derived from renewable resources. To obtain vegetable oil-based polyols, the most used methods are based on the epoxidation of plant oils or air oxidation, followed by their soft-hydrolysis or maleinization carried out in one- or two-step procedures [43].

Recently, other renewable resources have been investigated for manufacturing thermoset resins. In addition to wheat gluten proteins and soy proteins, thermo-moulded egg-white proteins (albumin) and keratin (fibrous structural proteins) have been developed as new bioplastics [44], or chitosan, trehalose and glucose-related materials have received special attention as carbohydrate materials in the design of thermosetting materials [45,46]. A potentially 100% bio-based thermoset possessing thermo-gravimetric and thermo-mechanical properties comparable to commercial vinyl ester-based thermosets has been recently synthesized by Stanzione III *et al.*. [47]. For this purpose a lignin-derived chemical, vanillin, was used, which was then methacrylated in a two-step, one-pot synthesis to produce a vinyl ester resin (87 cP at 25 °C) with 1:1 mole ratio of the monofunctional monomer, methacrylated vanillin, to cross-linking agent, glycerol dimethacrylate. As a result, a hard, transparent thermoset with a broad glass transition, T_g = 155 °C, and a temperature of maximum decomposition rate, T_{max}, of 426 °C was obtained.

3. Composites with Biobased Thermoset Matrices and Biobased Reinforcements

Bio-based thermosetting polymers can be used as matrices in composites. Of course, this can happen also by introducing traditional reinforcement fibres e.g., glass fibres, in the matrix, as it has been often the case in literature. Some early examples, which are particularly suggestive for the development of a sufficiently strong interface, can be found in [47-48]. However, it is clear that the main point of the adoption of bio-based thermosetting matrices is the production of fully biodegradable materials, hence using biodegradable material also as filler/reinforcement. In this regard, a number of possibilities have been explored: these include the introduction of lignocellulosic fibres, in particular flax [49–53], hemp [51,54,55], jute [56,57], sisal [58–63], banana [64], coir [65,66], alfa [67]; other lignocellulosic material in different dimensions, such as wood flour [68–72], pine wood [55], nutshells [73] and rice hulls [74,75]; regenerated cellulose products, such as Lyocell [53], or micro-crystalline cellulose [76]. Among bio-based fillers, also protein materials have been considered, in particular keratin, obtained from pyrolised chicken feathers [77].

A second type of classification can be carried out, based on the type of bio-based thermosetting matrix used. This information, together with indication of the characterisation performed, has been included in table 2, for all the papers referenced above. One of the main aspects that need to be dealt with in the preparation of bio-based thermoset composites is the achievement of a higher level of compatibility with vegetable fibres. In [49], where a comparison between a conventional vinylester matrix and an AESO matrix is carried out, this does not appear to be the case, as far as the introduction of flax fibres is concerned. Moreover, the influence of the loading rate on the interfacial shear strength (IFSS) properties appeared to be somewhat higher when using AESO matrix. A significant consideration is nonetheless that fibre treatment is still required when dealing with biobased thermoset composites, whenever a better and sounder fibre adhesion is desired: in particular, some improvements were obtained by alkali treatment of banana fibres, introduced as the reinforcement of castor-oil based polyurethane matrix [65]. This type of matrix proved particularly promising in view of the production of composites with plant fibres. More specifically, castor oil demonstrated its suitability to provide successful grafting with maleic anhydride, offering additional possibilities of blending with other plant oils, such as tung oil. This yielded some benefits e.g., an improvement of the interface properties with jute fibres, resulting in considerably higher (by over 40%) impact strength, an effect that was ascribed to the presence of flexible fatty acid chains subsequent to blending [56]. However, this did not appear to be the case for flexural performance, as will be discussed later on.

Another significant characteristic of castor oil-based polyurethane matrices is the wide adaptability to fillers with different nature and density: a good example was their capability to offer some interfacial strength with particleboards fabricated using either Macadamia nut shells or wood flour. These materials possess properties sufficient for their prospected use in the construction industry, although their flexural performance is rather deceptive: this was attributed to the lower cellulose content of nut shells, which makes them possibly more suitable to a use in compression [73]. Other filler structures with limited directionality (more similar to spherical particles than to elongated fibres), such as rice hulls, were found to find a problematic adaptation and a consequent limited compatibility in matrices, such as linseed oil based ones. This was deemed to occur as an effect of higher hydrophilic character of the filler with respect to the matrix, further enhanced by the substantial presence of silica in rice hulls [73].

Table 2. Examples of green composites based on thermosetting polymers from renewable resources

Matrix	Filler (content)	Characterisation	Reference
Acrylated epoxidized soy oil (AESO) (>70% bio)	Flax	Interfacial shear strength	[49]
Bio-derived furan	Flax	Tensile tests, moisture absorption, combustion properties	[50]
Acrylated epoxidized soy oil (AESO)	Flax, hemp	Tensile and flexural tests, moisture absorption	[51]
Methacrylated soybean oil (MSO), methacrylic anhydride modified soybean oil (MMSO), acetic anhydride modified soybean oil (AMSO)	Flax	Charpy impact, flexural tests, thermogravimetry	[52]
Soy oil	Flax, Lyocell	Tensile, falling weight impact testing, interfacial stress	[53]
Castor oil polycaprolactone-based polyurethane	Hemp	IR spectroscopy, Differential scanning calorimetry, Dynamical mechanical thermal analysis	[54]
Castor oil polyurethane	Hemp, pine wood fibres	Flexural tests, Dynamical mechanical analysis	[55]
Bio-polyurethane	Jute	Tensile tests	[56]
Maleated castor oil (MACO)	Jute	Differential scanning calorimetry, Izod impact, flexural tests	[57]
Castor oil polyurethane	Sisal	Flexural tests	[58]
Tannin-phenolic	Sisal	Izod impact, Dynamical mechanical analysis, differential scanning calorimetry, thermogravimetry	[59]
Rubber seed oil[1] polyurethane	Sisal	FTIR, thermogravimetry, water absorption, tensile tests, flexural tests	[60]

[1]Rubber seed oil is a mixture of saturated (palmitic and stearic) and unsaturated (oleic, linoleic, linolenic) fat acids.

As a matter of fact, fillers with reduced hydrophilic character, owing to the presence of a larger amount of lignin in them, may provide better interfacial properties with bio-based thermosetting matrices. In practice, short-time creep results on linseed oil/wood flour composites showed that the creep deformation of the polymer is significantly reduced with the incorporation of wood flour [68].

In principle, the evolution of flexural properties of the obtained composites with increasing amount of vegetable fibre reinforcement can represent a suitable benchmark to understand whether an effective fibre-matrix interface is formed. This happens in particular because vegetable fibres are hierarchised hollow structures, consisting of multi-layered cell walls, which means they are particularly adapted to work under flexural loading, in that the material is concentrated where needed to withstand tensile and compressive stress, while no material is present along the neutral axis, where the total stress is zero.

The influence of the fibre dimension is clearly indicated in some studies, for example in [55], where using pine wood fibres and increasing the average particle diameter from 840 microns to 2000 microns and then to 3360 microns, the average flexural modulus is decreased from 2684 MPa to 1635 MPa and to 1223 MPa, respectively. A similar effect was observed on flexural strength, although in this case the decrease measured when passing from the lowest particle diameter to the medium one is approximately 40%, while passing from the medium to the highest one does not produce any signification variation of the flexural strength of the composite foam. This was explained by the better fibre-matrix adhesion and the higher density of the foams obtained with lower diameter fibres. In [55], the effect of reinforcement orientation on the flexural properties of the composites was also observed for castor oil polyurethane/hemp composites. With this aim, hemp fibre-threads were added to the matrix in three different ways: unidirectional (0°), bi-directional (0/90) and in a quasi-isotropic stacking sequence (0, ±45 and 90°). The lowest properties corresponded to the bi-directional fibre composite, while the quasi-isotropic materials have intermediate properties, with unidirectional showing the best performance. This was attributed to the fact that the lower (tensile) face is more important in determining the observed trend during flexural loading, so that the composite with longitudinally aligned fibres can withstand higher stresses.

In [57], maleated castor oil (MACO)/jute composites showed inadequate flexural strength, about 60% lower than that measured on unsaturated polyester (UPE)/jute composites. This was attributed to the higher hydrophobicity of the MACO resin compared to that of UPE, which reduced the interfacial interaction with the hydrophilic jute fibres. A possible positive effect can be obtained by polymerizing and curing MACO in combination with tung oil in situ during the composite formation, hence facilitating the formation of highly entangled interpenetrating polymer network encompassing the jute fibres [56]. Similar concerns were confirmed when comparing the flexural performance of castor oil polyurethane/sisal composites with respect to synthetic phenolic/sisal ones: even if 44 wt.% fibres, as plain weave, were introduced in the former against only 33 wt.% in the latter, the flexural strength of the composite was largely lower (1.6 MPa against 10.7 MPa, respectively). Some improvement was obtained by oven-treating the fibres for 24 hours at 60°C, hence reducing their tenor of humidity: in this case, the flexural strength was 3.7 MPa for castor oil polyurethane composites against 11.2 MPa for phenolic composites [58].

Aging of linseed oil/wood flour composites, including moisture absorption and UV treatment, has also been studied in [69] with particular reference to its effect on flexural properties. Although crosslinking and degradation processes can possibly occur simultaneously when these materials are in contact with oxygen, the results indicate that the crosslinking reactions carry on for a considerable period after curing. The increasing crosslinking density leads to the initially dangling chains forming chemical bonds to the network, becoming therefore elastically active chains, and it also reduces the permeability to oxygen in the outer layers of the laminate. The molecular mobility is consequently reduced and the rigidity of the material increased, especially in the earlier periods of aging (up to a few weeks) [78]. This results in a substantial increase of flexural modulus with aging time, which is also revealed for laminates absorbing humidity, although obviously on lower values of stiffness than for the dry ones. UV irradiation has an additional effect on glass transition temperature, which in the specific case of linseed oil resins can be brought from around 115°C to around 130°C as an effect of 300 hours of exposure in the UV chamber, leading consequently to some possible increase of the service temperature for composites [78].

One of the limitations of plant fibre composites is the lower amount of fibres that can be possibly introduced in the composite, owing especially to the variable diameter of vegetable fibres and to some inherent characteristics of woven tissues, such as fibrillation and disruptions in the warp-weft geometry, which lead to their more difficult fibre impregnation by the polymer matrix. A study by dynamical mechanical analysis on tannin-phenolic matrix composites reinforced with sisal fibres suggested that up to 60 wt.% of fibres can be introduced in the composite with increase in their storage modulus. In contrast, Izod impact tests indicate that this high level of reinforcement leads to some decrease of impact properties, suggesting to limit this introduction to the already high level of 50 wt.% [59].

These general indications suggest the potential of bio-based thermosetting polymers, based on fatty acids, to provide a suitable reinforcement for biomass materials, such as vegetable fibres, although the field is still under constant development: some crucial issues are widely recognised, such as effective polymer blending to provide suitable fibre-matrix compatibility [79].

Conclusion and Future Perspective

The activity in the field of bio-based thermoset polymer composites is growing, due to the pressure of some fields that are particularly interested in the development and applications of these materials, such as shipbuilding, aeronautics and automotive industry. The recent availability of plant oils (castor, soy, rapeseed, linseed, tung, cashew), which can be functionalised in different ways (acrylated, epoxidized, acrylated/epoxidized, maleinated) and used for the production of different matrices, particularly phenolic and polyurethane, has been presented as a significant trend for improved materials for the next decade.

The next step is obviously the introduction of bio-based filler: with this aim, a number of methods have been developed to improve blending of the polymer, starting from different sources of fatty oil acids, also considering their possible origin as by-products of other processes, such as those involved in the production of bio-diesel or foodstuff. On the other side, a growing number of possible fibres are currently investigated for composites production and the selection may depend on a number of factors, not excluding economical ones, especially when fibres are by-products of large production systems. However, the application of those which have had already significant applications in the field of composites with bio-based thermoplastic or conventional thermoset matrices, such as e.g., sisal can be recommended and represents an obvious possibility for development. Another option, which has been widely investigated is the production of WPC (wood-plastic composites), using bio-based thermoset matrices: in this respect, the use of wood flour is regarded as one of the most explored routes.

A number of issues remain to be solved nonetheless: in particular, the selection of a few optimised bio-based polymer blends, possibly tailored on the characteristics of the biomass selected for reinforcement, would substantially improve compatibility, if necessary with the assistance of fit-for-use chemical treatments of the filler. Moreover, mechanical characterisation of obtained composites has been limited so far to few basic test modes: for example, information on toughness of these composites is virtually non-existent and also data on impact and fatigue performance are very scarce. These brief suggestions may indicate that in the next future activity on these materials will gradually improve and diversify.

REFERENCES

[1] Lu Y, Larock RC. Novel polymeric materials from vegetable oils and vinyl monomers: preparation, properties, and applications. *ChemSusChem* 2009; 2:136–47.

[2] John M, Thomas S. Biofibres and biocomposites. *Carbohydrate Polymers* 2008; 71:343–64.

[3] Mohanty AK, Misra M, Drzal LT, Selke SE, Harte BR, Hinrichsen G. *Natural Fibers, Biopolymers, and Biocomposites* 2005.

[4] La Mantia FP, Morreale M. Green composites: A brief review. *Composites Part A: Applied Science and Manufacturing* 2011; 42:579–88.

[5] Wool RP, Sun XS. Bio-Based Polymers and Composites. Elsevier Science & Technology Books; 2005.

[6] Zini E, Scandola M. Green composites: An overview. *Polymer Composites* 2011; 32:1905–15.

[7] Faruk O, Bledzki AK, Fink H-P, Sain M. Biocomposites reinforced with natural fibers: 2000–2010. *Progress in Polymer Science* 2012; 37:1552–96.

[8] Hughes M. Defects in natural fibres: their origin, characteristics and implications for natural fibre-reinforced composites. *Journal of Materials Science* 2011; 47:599–609.

[9] Fuqua M a., Huo S, Ulven C a. Natural Fiber Reinforced Composites. *Polymer Reviews* 2012; 52:259–320.

[10] Yousefi A, Lafleur PG, Gauvin R. Kinetic studies of thermoset cure reactions: A review. *Polymer Composites* 1997; 18:157–68.

[11] Montero de Espinosa L, Meier MAR. Plant oils: The perfect renewable resource for polymer science?! *European Polymer Journal* 2011; 47:837–52.

[12] Belgacem M, Gandini A, editors. *Monomers, Polymers and Composites from Renewable Resources.* Elsevier; 2008.

[13] Mosiewicki M a., Aranguren MI. A short review on novel biocomposites based on plant oil precursors. *European Polymer Journal* 2013; 49:1243–56.

[14] Raquez J-M, Deléglise M, Lacrampe M-F, Krawczak P. Thermosetting (bio)materials derived from renewable resources: A critical review. *Progress in Polymer Science* 2010; 35:487–509.

[15] Xia Y, Larock RC. Vegetable oil-based polymeric materials: synthesis, properties, and applications. *Green Chemistry* 2010; 12:1893.

[16] Li F, Hanson M., Larock R. Soybean oil–divinylbenzene thermosetting polymers: synthesis, structure, properties and their relationships. *Polymer* 2001; 42:1567–79.

[17] Li F, Larock R. Novel Polymeric Materials from Biological Oils. *Journal of Polymers and the Environment* 2002; 10:59–67.

[18] Li F, Larock RC. New soybean oil-styrene-divinylbenzene thermosetting copolymers. III. Tensile stress-strain behavior. *Journal of Polymer Science Part B: Polymer Physics* 2001; 39:60–77.

[19] Li F, Larock RC. Thermosetting polymers from cationic copolymerization of tung oil: Synthesis and characterization. *Journal of Applied Polymer Science* 2000; 78:1044–56.

[20] Li F, Larock RC. Synthesis, structure and properties of new tung oil-styrene-divinylbenzene copolymers prepared by thermal polymerization. *Biomacromolecules* 2003; 4:1018–25.

[21] Li F, Hasjim J, Larock RC. Synthesis, structure, and thermophysical and mechanical properties of new polymers prepared by the cationic copolymerization of corn oil, styrene, and divinylbenzene. *Journal of Applied Polymer Science* 2003; 90:1830–8.

[22] Li F, Marks D., Larock R., Otaigbe J. Fish oil thermosetting polymers: synthesis, structure, properties and their relationships. *Polymer* 2000; 41:7925–39.

[23] Marks DW, Li F, Pacha CM, Larock RC. Synthesis of thermoset plastics by Lewis acid initiated copolymerization of fish oil ethyl esters and alkenes. *Journal of Applied Polymer Science* 2001; 81:2001–12.

[24] Rybak A, Fokou PA, Meier MAR. Metathesis as a versatile tool in oleochemistry. *European Journal of Lipid Science and Technology* 2008; 110:797–804.

[25] Park S-J, Jin F-L, Lee J-R. Effect of Biodegradable Epoxidized Castor Oil on Physicochemical and Mechanical Properties of Epoxy Resins. *Macromolecular Chemistry and Physics* 2004; 205:2048–54.

[26] Miyagawa H, Mohanty AK, Misra M, Drzal LT. Thermo-Physical and Impact Properties of Epoxy Containing Epoxidized Linseed Oil, 2. *Macromolecular Materials and Engineering* 2004;289:636–41.

[27] Miyagawa H, Mohanty AK, Misra M, Drzal LT. Thermo-Physical and Impact Properties of Epoxy Containing Epoxidized Linseed Oil, 1. *Macromolecular Materials and Engineering* 2004; 289:629–35.

[28] Lligadas G, Ronda JC, Galià M, Cadiz V. Polyurethane networks from fatty-acid-based aromatic triols: synthesis and characterization. *Biomacromolecules* 2007; 8:1858–64.

[29] Lu Y, Tighzert L, Dole P, Erre D. Preparation and properties of starch thermoplastics modified with waterborne polyurethane from renewable resources. *Polymer* 2005; 46:9863–70.

[30] Lu Y, Tighzert L, Berzin F, Rondot S. Innovative plasticized starch films modified with waterborne polyurethane from renewable resources. *Carbohydrate Polymers* 2005; 61:174–82.

[31] Phani Kumar P, Paramashivappa R, Vithayathil PJ, Subba Rao P V., Srinivasa Rao A. Process for Isolation of Cardanol from Technical Cashew (Anacardium occidentale L.) Nut Shell Liquid. *Journal of Agricultural and Food Chemistry* 2002; 50:4705–8.

[32] Tyman JHP, Johnson RA, Muir M, Rokhgar R. The extraction of natural cashew nut-shell liquid from the cashew nut (Anacardium occidentale). *Journal of the American Oil Chemists' Society* 1989; 66:553–7.

[33] Yadav R, Srivastava D. Kinetics of the acid-catalyzed cardanol–formaldehyde reactions. *Materials Chemistry and Physics* 2007; 106:74–81.

[34] Turunen M, Alvila L, Pakkanen TT, Rainio J. Modification of phenol-formaldehyde resol resins by lignin, starch, and urea. *Journal of Applied Polymer Science* 2003; 88:582–8.

[35] Vázquez G, Rodríguez-Bona C, Freire S, González-Álvarez J, Antorrena G. Acetosolv pine lignin as copolymer in resins for manufacture of exterior grade plywoods. *Bioresource Technology* 1999; 70:209–14.

[36] Çetin NS, Özmen N. Use of organosolv lignin in phenol–formaldehyde resins for particleboard production. *International Journal of Adhesion and Adhesives* 2002; 22:477–80.
[37] Meier MAR, Metzger JO, Schubert US. Plant oil renewable resources as green alternatives in polymer science. *Chemical Society Reviews* 2007; 36:1788.
[38] Sharma V, Kundu PP. Addition polymers from natural oils—A review. *Progress in Polymer Science* 2006; 31:983–1008.
[39] Seniha Güner F, Yağcı Y, Tuncer Erciyes a. Polymers from triglyceride oils. *Progress in Polymer Science* 2006; 31:633–70.
[40] Zhu J, Chandrashekhara K, Flanigan V, Kapila S. Curing and mechanical characterization of a soy-based epoxy resin system. *Journal of Applied Polymer Science* 2004; 91:3513–8.
[41] Khot SN, Lascala JJ, Can E, Morye SS, Williams GI, Palmese GR, et al. Development and application of triglyceride-based polymers and composites. *Journal of Applied Polymer Science* 2001;82:703–23.
[42] Petrovic Z. Polyurethanes from Vegetable Oils. *Polymer Reviews* 2008; 48:109–55.
[43] Pechar TW, Sohn S, Wilkes GL, Ghosh S, Frazier CE, Fornof A, et al. Characterization and comparison of polyurethane networks prepared using soybean-based polyols with varying hydroxyl content and their blends with petroleum-based polyols. *Journal of Applied Polymer Science* 2006; 101:1432–43.
[44] Jerez A, Partal P, Martínez I, Gallegos C, Guerrero A. Egg white-based bioplastics developed by thermomechanical processing. *Journal of Food Engineering* 2007; 82:608–17.
[45] Nair LS, Starnes T, Ko J-WK, Laurencin CT. Development of injectable thermogelling chitosan-inorganic phosphate solutions for biomedical applications. *Biomacromolecules* 2007; 8:3779–85.
[46] Berger J, Reist M, Chenite A, Felt-Baeyens O, Mayer JM, Gurny R. Pseudo-thermosetting chitosan hydrogels for biomedical application. *International Journal of Pharmaceutics* 2005; 288:197–206.
[47] Stanzione III JF, Sadler JM, La Scala JJ, Reno KH, Wool RP. Vanillin-based resin for use in composite applications. *Green Chemistry* 2012; 14:2346.
[48] Henna PH, Kessler MR, Larock RC. Fabrication and Properties of Vegetable-Oil-Based Glass Fiber Composites by Ring-Opening Metathesis Polymerization. *Macromolecular Materials and Engineering* 2008; 293:979–90.
[49] Spārniņš E, Nyström B, Andersons J. Interfacial shear strength of flax fibers in thermoset resins evaluated via tensile tests of UD composites. *International Journal of Adhesion and Adhesives* 2012; 36:39–43.
[50] Pohl T, Bierer M, Natter E, Madsen B, Hoydonckx H SR. Properties of compression moulded new fully biobased thermoset composites with aligned flax fibre textiles. *Plastics Rubber And Composites* 2011; 40:294–9.
[51] Williams GI, Wool RP. *Composites from Natural Fibers and Soy Oil Resins* 2000: 421–32.
[52] Skrifvars M, Adekunle K, Dan A. Biobased Composites Prepared by Compression Molding with a Novel Thermoset Resin from Soybean Oil and a Natural-Fiber Reinforcement 2010; 116:1759–65.

[53] Akesson D, Skrifvars M, Hagstrom B, Walkenstrom P SJ. Processing of Structural Composites from Biobased Thermoset Resins and Natural Fibres by Compression Moulding, Journal of Biobased Materials and Bioenergy 3 (3), 2009, 215-225. *Journal of Biobased Materials and Bioenergy* 2009; 3:215–25.

[54] Lee N, Kwon OJ, Chun BC, Cho JW PJ. Characterization of castor oil/polycaprolactone polyurethane biocomposites reinforced with hemp fibers. *Fibers and Polymers* 2009; 10:154–60.

[55] Marcovich NE, Aranguren MI, Ra I, Zolta B. Microfoams Based on Castor Oil Polyurethanes and Vegetable Fibers. *Journal of Applied Polymer Science* 2007; 105:2791–800.

[56] Neto JRA. Influência da Adição de uma Carga Nanoparticulada no Desempenho de Compósitos Poliuretano / Fibra de Juta. *Polimeros-Ciencia e Tecnologia* 2007; 17:10–5.

[57] Mistri E, Routh S, Ray D, Sahoo S, Misra M. Green composites from maleated castor oil and jute fibres. *Industrial Crops and Products* 2011; 34:900–6.

[58] Milanese AC, Cioffi MOH, Voorwald HJC. Flexural behavior of Sisal/Castor oil-Based Polyurethane and Sisal/Phenolic Composites. *Materials Research* 2012; 15:191–7.

[59] Ramires EC, Frollini E. Tannin–phenolic resins: Synthesis, characterization, and application as matrix in biobased composites reinforced with sisal fibers. *Composites Part B: Engineering* 2012; 43:2851–60.

[60] Bakare IO, Okieimen FE, Pavithran C, Abdul Khalil HPS, Brahmakumar M. Mechanical and thermal properties of sisal fiber-reinforced rubber seed oil-based polyurethane composites. *Materials & Design* 2010; 31:4274–80.

[61] Silva RV, Ueki MM, Spinelli D, Bose WW TJ. Thermal, mechanical, and hygroscopic behavior of sisal fiber/polyurethane resin-based composites. *Journal of Reinforced Plastics and Composites* 2010; 29:1399–417.

[62] Wang HJ, Zhang CY, Rong MZ, Zhang MQ CT. Interfacial effects in short sisal fiber/maleated castor oil foam composites. *Composite Interfaces* 2008; 15:95–110.

[63] Wu SP, Qiu JF, Rong MZ, Zhang MQ, Zhang LY. Plant oil-based biofoam composites with balanced performance. *Polymer International* 2009; 58:403–11.

[64] Merlini C, Soldi V BG. Influence of fiber surface treatment and length on physico-chemical properties of short random banana fiber-reinforced castor oil polyurethane composites. *Polymer Testing* 2011; 30:833–40.

[65] Barbosa V, Jr, Ramires EC, Tanaka R, Ilce A FE. Biobased composites from tannin-phenolic polymers reinforced with coir fibers,. *Industrial Crops and Products* 2010; 32:305–12.

[66] Barreto ACH, Junior AEC, Freitas JEB, Rosa DS, Barcellos WM, Freire FNA, Fechine PBA MS. Biocomposites from dwarf-green Brazilian coconut impregnated with cashew nut shell liquid resin . *Journal of Composite Materials* 2012; 47:459–66.

[67] Maafi EM, Tighzert L MF. Elaboration and characterization of composites of castor oil-based polyurethane and fibers from alfa stems,. *Journal of Applied Polymer Science* 2010; 118:902–9.

[68] Mosiewicki MA, Marcovich NE, Aranguren MI. Creep Behavior of Wood Flour Composites Made from Linseed Oil-Based Polyester Thermosets. *Journal of Applied Polymer Science* 2011; 121:2626–33.

[69] Mosiewicki MA, Rojas O, Sibaja R, Borrajo J, Aranguren MI. Aging study of linseed oil resin / styrene thermosets and their composites with wood flour 2007; 881:875–81.

[70] Mosiewicki M a., Dell'Arciprete G a., Aranguren MI, Marcovich NE. Polyurethane Foams Obtained from Castor Oil-based Polyol and Filled with Wood Flour. *Journal of Composite Materials* 2009; 43:3057–72.

[71] Casado U, Marcovich NE, Aranguren MI, Mosiewicki MA. High-Strength Composites Based on Tung Oil Polyurethane and Wood Flour: Effect of the Filler Concentration on the Mechanical Properties. *Polymer Engineering & Science* 2009; 49:713–21.

[72] Aranguren MI, González JF, Mosiewicki M a. Biodegradation of a vegetable oil based polyurethane and wood flour composites. *Polymer Testing* 2012; 31:7–15.

[73] Wechsler A, Zaharia M, Crosky A, Jones H, Ramírez M, Ballerini A, Nuñez M S V. Macadamia (Macadamia integrifolia) shell and castor (Rícinos communis) oil based sustainable particleboard: A comparison of its properties with conventional wood based particleboard. *Materials & Design* 2013; 50:117–23.

[74] Quirino RL, Larock RC. Rice Hull Biocomposites . I . Preparation of a Linseed-Oil-Based Resin Reinforced with Rice Hulls. *Journal of Applied Polymer Science* 2011; 121:2039–2049.

[75] Quirino R, Larock R. Rice Hull Biocomposites , Part 2: Effect of the Resin Composition on the Properties of the Composite. *Journal of Applied Polymer Science* 2011; 232:2050–2059.

[76] Ramires EC, Megiatto JD, Gardrat C, Castellan A FE. Biobased Composites from Glyoxal-Phenol Matrices Reinforced with Microcrystalline Cellulose, 20 (2), 2010, 126-133. *Polimeros-Ciencia e Tecnologia* 2010; 20:126–33.

[77] Senoz E, Stanzione JF, Reno KH, Wool RP, Miller MEN. Pyrolyzed chicken feather fibers for biobased composite reinforcement. *Journal of Applied Polymer Science* 2013; 128:983–9.

[78] Mallégol J, Lemaire J, Gardette J-L. Drier influence on the curing of linseed oil. *Progress in Organic Coatings* 2000; 39:107–13.

[79] Karger-Kocsis J. Editorial corner – a personal view. Thermoset polymers containing bio-based renewable resources. *Express Polymer Letters* 2009; 3:676.

In: New Developments in Polymer Composites Research ISBN: 978-1-62948-340-5
Editors: S. Laske and A. Witschnigg

Chapter 7

FUNDAMENTAL INVESTIGATIONS OF REINFORCEMENT OF EXPANDED PERLITE IN POLYPROPYLENE

Hannelore Isabella Mattausch[1,*], Stephan Laske[1], Kristin Cirar[2], Helmut Flachberger[2] and Clemens Holzer[1]

[1]Chair of Polymer Processing, Department of Polymer Engineering and Science, Montanuniversitaet Leoben, Austria

[2]Chair of Mineral Processing, Department of Mineral Resources and Petroleum Engineering, Montanuniversitaet Leoben, Austria

ABSTRACT

The term perlite describes oversaturated, volcanic and glassy rocks, which have chemically bound water (2-5 wt%). Upon heating, perlite can be expanded up to 20 times of its original volume. Important applications of perlite are in the field of building industry as aggregates, in refrigeration engineering as insulating material or in the pharmaceutical industry as inert part in the production of pills. As mineral filler in polypropylene (PP), expanded perlite can increase the thermal conductivity, the viscosity and the mechnical properties of polypropylene composites. There is only limited research work available that gives attention to polymers filled with expanded perlite. This new filler could establish new possibilities in enhancing the properties of polymers. This chapter focuses on fundamental properties such as crystallinity, mechanical properties and viscosity of expanded perlite/PP compounds. The composites were compounded with a co-rotating twin screw extruder ZSK 25 (Coperion GmbH, Weingarten, Germany). For producing the closed cell expanded perlite a new technology, the bublon process, was used. Two different partical sizes (D50 of 0.525 mm and D50 of 0.125 mm) were chosen and the filler content was varied at 5, 10, 15 wt%. For the analysis of the effects of the screw geometry, two setups have been chosen for the processing of the materials.

Keywords: Polypropylene, expanded perlite, mechanical properties, viscosity, crystallinity

---

* E-mail address: hannelore.mattausch@unileoben.ac.at (Corresponding author)

INTRODUCTION

Polypropylene (PP) is widely used in many different applications such as automotive or building industry due to the good performance and favorable cost-benefit ratio. To overcome the disadvantages of PP such as low toughness or slight flame-retarding effect, mineral fillers such as layered silicates or expanded perlite can be employed. The major drawbacks of layered silicates use in PP are the hydrophilic nature and high cost compared to other mineral fillers. Compatibilizers such as polymer grafted with maleic acid anhydride can promote the adhesion between two materials with different polarities (e.g., polyolefins and layered silicates). By contrast alternative fillers offer new opportunities to modify the compound's property profile. Perlite is an oversaturated volcanic glasslike rock, that can be expanded up to 20 times of its original volume upon heating above 800 °C due the chemically bound water (2-5 wt%), which is known as expanded perlite [1].

This spherical shaped filler can be used without functionalization of the filler itself or of the polymer matrix. A new technology, the bublon process, was used to produce expanded perlite. Due to this new process the main advantages of expanded perlite are reduced water absorption, improved resistance to wear, excellent insulating properties, higher compaction resistance and a glazed as well sealed surface, which can give improvements in many different applications as building industry, in refigeration engineering or the pharmaceutical industry [2, 3].

To obtain the best material quality, homogeneity and interfacial adhesion between polymer and filler, particles have to be improved and tailor-made for the desired property profile. The positive examples of good performance of mineral fillers in polymer matrix are often due to the predominant interfacial interactions between filler and polymer matrix. This can be also be positively influenced by the adequate processing technique (e.g., screw geometry configuration, screw rotation speed, applied shear, residence time etc.). To achieve good performance, a homogenous dispersion of the filler in the polymer matrix is needed because the enhancement of the material correlates strongly with the morphology of the composites. The filler has also to be dispersed homogenously in the polymer matrix because poor distribution can lead to (re)agglomeration of the fillers in the matrix during processing [4, 5]. But the dispersion of inorganic particles in polymer matrix is often not easy possibly. So the combination of the adequate processing technique and adequate chemical modification method of the filler is important for reaching high quality polymer compounds with an enhanded property profile.

Due to the limited research work of polymer composites filled with expanded perlite, the influence of processing parameters on various properties of the resulting composite is not well investigated. This chapter deals with fundamental properties of PP/expanded perlite composites and the influence of processing on the materials' properties. Several investigations to the determine the basic property profile of PP/expanded perlite composites have been conducted at the Chair of Polymer Processing. There are still many challenging topics for further research to reach the full potential polymer/expanded perlite composite systems.

This topics could be the behavior of bimodal size distributed expanded perlite particles in polymer matrices or the surface modification of the expanded perlite particles to achieve certain functionalization of the resulting compound.

EXPERIMENTAL

Materials

The polypropylene BB412E with a melt flow rate (MFR) of 1.3 g/10 min; 230 °C/2.16 kg was supplied by Borealis (Linz, Austria). The mineral filler expanded perlite was provided by the Chair of Mineral Processing, Montanuniversitaet Leoben (Austria) and produced by the new Bublon process. Expanded perlite was produced with two different particle sizes (D_{50} of 0.525 mm and D_{50} of 0.125 mm). The mineral filler with D_{50} of 0.525 mm particle size is labeled as PC and with D_{50} of 0.125 mm as PX.

Bublon Process

The new Bublon process was developed in close cooperation with the Chair of Mineral Processing (Montanuniversitaet Leoben, Austria) and Binder and Co AG (Gleisdorf, Austria). The processing steps are the conditioning of the perlite run (crushing, screening, heating), expanding in a tower oven and packaging. Expanded perlite can be produced with a glazed and sealed surface and the expansion is without producing flue gar for CO_2 emissions.

Processing of PP/Expanded Perlite Composite

The composites were prepared with a co-rotating twin screw extruder ZSK 25 (Coperion GmbH, Weingarten, Germany) with a single string die. Two different screw geometries were chosen. The medium shearing screw geometry (MS) consists of 2 kneading and 3 conveying blocks, which can be seen in Figure 1a. While the low shearing screw geometry (LS) has one kneading segment before the side feeder and 2 conveying segments (see Figure 1b).

The minimum residence times were measured for each composites with adding colored masterbatch to the dosed polymer in the hopper and measuring visually the time until the first change in color of the materials appears at die opening. Compounds with various filler content (5, 10, 15 wt%) were prepared at a processing temperature of 200 °C and a screw speed of 150 rpm. The fillers were dosed gravimetrically through sidefeeder after the first kneading elements. For the various analysis methods, plates with a thickness of 4 mm (tensile tests) and 2 mm (viscosity) were produced using a hydraulic vacuum press Collin 200 PV (Dr. Collin, Ebersberg, Germany).

Rheological Properties

The rheological properties of the composites were investigated by using a rotational rheometer MCR501 (Anton Paar GmbH, Graz, Austria). Thereby a plate-plate measurement set-up was chosen. To ensure that the strain used is within the linear viscoelastic range, strain sweep tests were performed for each sample with a constant frequency of 1 rad/s. The specimens had a diameter of 25 mm and 2 mm thickness.

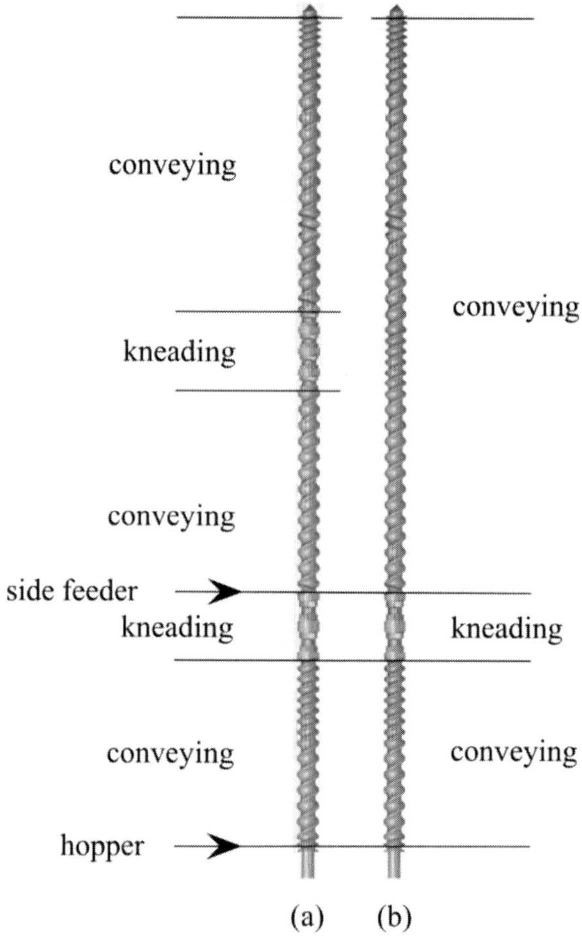

Figure 1. Medium shearing (a) and low shearing (b) screw geometry.

The materials were measured over a frequency range of 0.1 to 500 rad/s at 200 °C. All rheological measurements were conducted under nitrogen atmosphere. At least 3 specimens were measured for each produced composite.

Thermal Properties

The thermal properties of the produced composites were analyzed by a differential scanning calorimeter (DSC) DSC 1 (Mettler Toledo, Schwerzenbach, Switzerland). The experiments were conducted under a heating rate of 10 K/min, a cooling rate of 20 K/min and under flowing nitrogen atmosphere. The samples were heated from room temperature to 220 °C. About 9-10 mg of the material was placed in sealed aluminium cups with pierced lid. Then the samples were cyclically heated and cooled.

The degree of crystallinity was determined as the ratio between the melting enthalpy of the samples (first heating) and the melting enthalpy of the 100% crystalline polymer. The enthalpy value for 100% crystalline PP (209 J/g) grades was taken from [6]. At least 3 specimens of each composistion were analyzed.

Mechanical Properties

The tensile tests were carried out on a universal testing machine UTS Eurotest 250 (UTS Testsysteme, Ulm, Germany) according to ISO 527-1. For the measurement of the Young's modulus an extensometer was placed on the specimen and was removed immediately after obtaining the Young's modulus value. Afterwards the test was continued till the break of the sample. The specimens for the tensile tests were milled out of a pressed plate. At least 6 specimens were tested for each material.

Scanning Electron Microscope

The morphology of the PP/expanded perlite composites were analyzed by scanning electron microscope (SEM). The images were taken at the Materials Center Leoben Forschung GmbH (Leoben, Austria) on the typ EVO MA25 (Carl Zeiss GmbH, Vienna, Austria) at 20 kV. The two mm thick samples were broken in liquid nitrogen and afterwards sputtered with gold.

RESULTS AND DISCUSSION

The results of the conducted investigations for the application of expanded perlite as filler for polypropylene are summarized and accordingly interpreted in the following. Also the influence of particles size (D_{50} of 0.525 mm as PC and D_{50} of 0.125 mm as PX) and screw geometry (medium shearing MS and low shearing LS) on the composites' properties were analyzed.

As one major influence factor on the property profile of the final composite, the morphology of the samples were investigated by SEM images. In Figure 2 the cross sections of PP plates filled with 5 wt% PX (LS and MS) are depicted. The overview of the LS sample shows that the particles were not destroyed by processing, while in MS produced materials mostly destroyed particles could be found. The undestroyed spherical shaped filler in LS materials have an inner structure. This honeycomb like structure stabilizes the expanded perlite particles against forces applied during processing. If a particle surface breaks, the polymer can flow inside the particle, increases the specific surface and enhances thereby the adhesion between polymer and filler (see Figure 2c).

In higher filled composites (PP PX filled with 15 wt%) the morphology is quite similar to the PP PX filled with 5 wt%. With the low shearing screw the particles are almost intact (see Figure 3a) but with more shearing due to the sharper screw geometry the particles are broken (see Figure 3c). Furthermore no difference of expanded perlite distrubtion can be observed.

The composites produced with HS has a different morphology compared to the LS materials (see Figure 4). Almost all particles are destroyed but the fragments are dispersed quite homogenously. This effect is indepentent of the filler content.

The minimum residence time increases with higher filler content and with more induced shear during processing, which can be seen in Table 1. Due to the bigger particle slize an slight increase of residence time can be observed. With higher amount of filler the viscosity increases, which limits the conveying ability of the screw geometry and leads therefore to an increased minimum residence time. Due to the additional kneading block at medium shearing screw geometry, the minimum residence time and the residence time profile are extended. Thereby the kneading segments allow a higher black flow of the polymer melt compared to conveying elements.

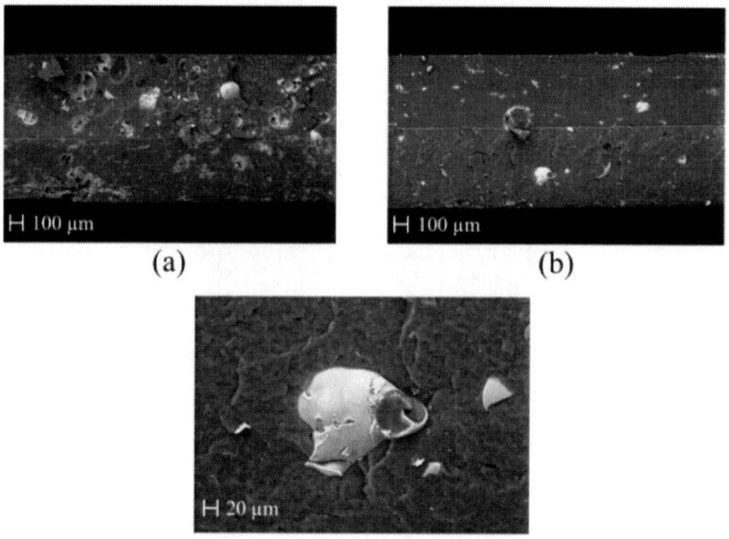

(a) (b) (c)

Figure 2. Morphologies of PP/PX (filled with 5 wt%) composites procuded with LS and MS screw geometry: a) cross section of PP PX LS; b) cross section of PP PX MS; c) single PX particles produced with MS screw geometry.

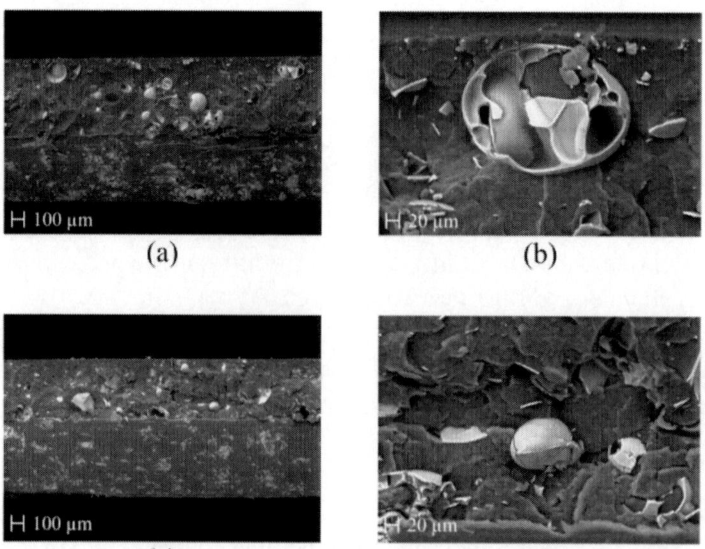

(a) (b) (c) (d)

Figure 3. Morphologies of PP/PX (filled with 15 wt%) composites procuded with LS and MS screw geometry: a) cross section of PP PX LS; b) single particle in PP PX LS; c) cross section of PP PX MS; c) single particle in PP PX MS.

(a)

(b)

Figure 4. Morphologies of PP/PX (filled with 15 wt%) composites procuded with LS (a) and MS (b) screw geometry.

Table 1. Minimum residence time of unfilled PP and PP/expanded perlite composites

sample	screw geometry	minimum residence time (s)
PP	MS	55
PP + 5 wt% PC	MS	58
PP + 10 wt% PC	MS	59
PP + 15 wt% PC	MS	62
PP + 5 wt% PX	MS	55
PP + 10 wt% PX	MS	58
PP + 15 wt% PX	MS	58
PP	LS	47
PP + 5 wt% PC	LS	49
PP + 10 wt% PC	LS	52
PP + 15 wt% PC	LS	55
PP + 5 wt% PX	LS	48
PP + 10 wt% PX	LS	50
PP + 15 wt% PX	LS	50

Rheological measurements can provide fundamental insight in the processing behavior. Especially fillers (particle size, particle shape, molecular weight, molecular weight distribution, degree of branches etc.) have a significant influence on the melt state of the material.

The complex viscosities of unfilled PP and PP/expanded perlite with medium shearing screw geometry can be seen in Figure 5for both particle sizes. The trend, which can be observed in this figure, is representative for composites with LS screw geometry.

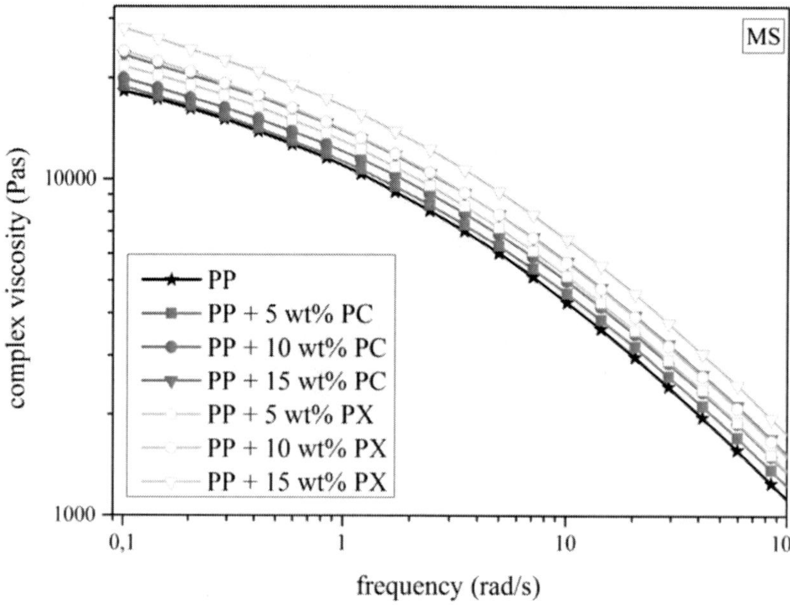

Figure 5. Complex viscosities of unfilled PP and PP/expanded perlite composites (MS screw geometries).

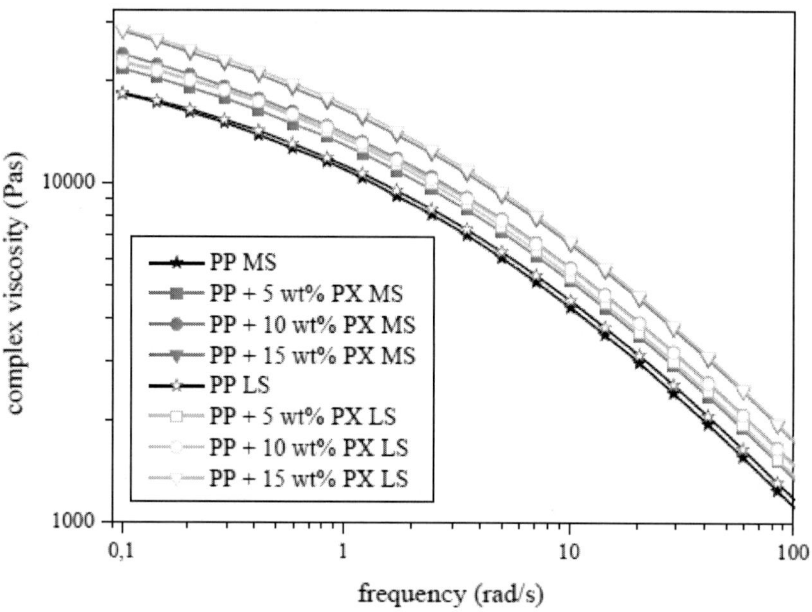

Figure 6. Comparison of complex viscosities of unfilled PP and PP/expanded perlite composites in the dependency of screw geometries (MS and LS).

With increasing filler content, the complex viscosity also increases. While the measured values for the various composites differ, the principle of the viscosity curves remained the same shape. The lowest complex viscosity exhibit the unfilled polymer matrix. The overall highest complex viscosity was found with 15 wt% PX LS. The general increasing of the complex viscosity with the addition of mineral fillers are due to the adsorption of the polymer chains on the filler particles [7]. The influence of screw geometry on the resulting viscosities is not significant (see Figure 6). The results of the rheological evaluation fits with the results of the minimum residence time measured during compounding.

The DSC is a very good characterization method to investigate the effect of processing conditions on the thermal behavior of polyolefin composites. Results obtained from DSC measurements for PP composites are presented in Table 2.

The crystallinities of the filled PP decreases with increasing amount of filler. The volume is occupied by filler acts as constraint to crystallization of the polymer chains [8-13]. It can also be observed that the crystallinities are slightly higher for the compounds processed with low shearing screw geometry. The size of the filler particles has no significant impact on the measured crystallinities.

Also the calcutated standard deviation remains below 1% except of two samples, which shows a good reproducability of the measurements.

For the characterization of the influence of the particle size, filler content and screw geometery on the mechanical properties the Young's modulus, the tensile stress at break, the elongation at break and the tensile strength were measured.

In Figure 7 the Young's modulus of PP/expanded perlite with two different particles sizes and screw geometries are depicted. The composites show an improvement in Young's modulus with the addition of expanded perlite. The higher the filler content of expanded perlite, the more the Young's modulusincreases.

The best improvement (about 44 %) could be achieved with the LS screw geometry and bigger particle sizes. In literature it was found that with increasing amount of perlite in polyethylene (PE) the Young's modulus improves because the flexing of the matrix is prevented by relatively high modulus filler particles [14]. The bigger particles were destroyed during processing, but the small fragements were dispersed quite homogenously as it can be seen in the SEM images. Therefore PP PX composite can achieve a higher increase in the Young's modulus.

Table 2. Crystallinities of unfilled PP and PP/expanded perlite composites

sample	screw geometry	crystallinity (%)	standard deviation (SD)
PP	MS	32,1	0,6
PP + 5 wt% PC	MS	30,4	0,8
PP + 10 wt% PC	MS	28,4	0,5
PP + 15 wt% PC	MS	27,8	0,2
PP + 5 wt% PX	MS	30,1	0,9
PP + 10 wt% PX	MS	28,6	0,8
PP + 15 wt% PX	MS	27,3	0,6
PP	LS	34,2	0,3
PP + 5 wt% PC	LS	33,1	0,7
PP + 10 wt% PC	LS	30,1	0,4
PP + 15 wt% PC	LS	28,2	0,5
PP + 5 wt% PX	LS	29,7	4,5
PP + 10 wt% PX	LS	30,5	0,6
PP + 15 wt% PX	LS	26,3	2,3

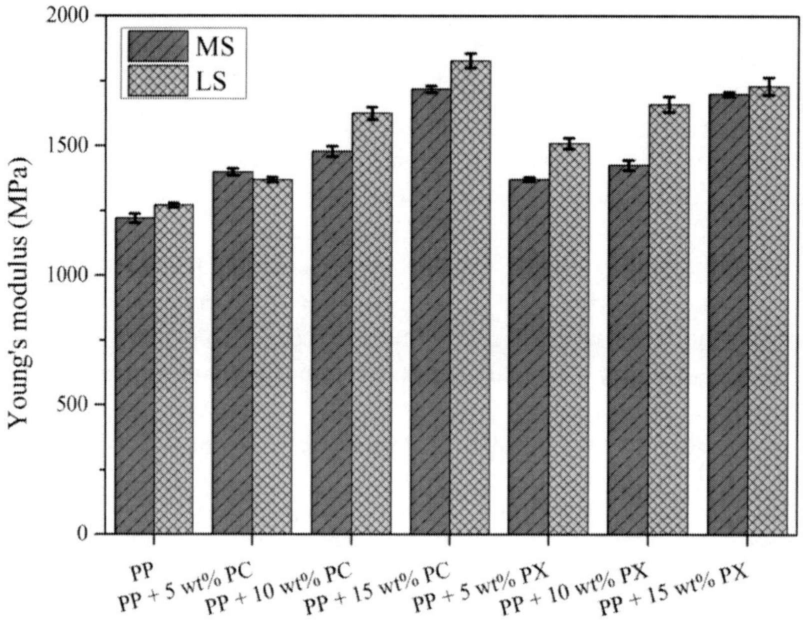

Figure 7. Young's modulus of unfilled PP and PP/expanded perlite composites (different screw geometries).

The tensile stress at break of unfilled PP and filled PP/expanded perlite composites can be observed in Figure 8. No significant change in this mechanical values could be analyzed regarding the influenced of filler content, particles size and applied shear during processing.

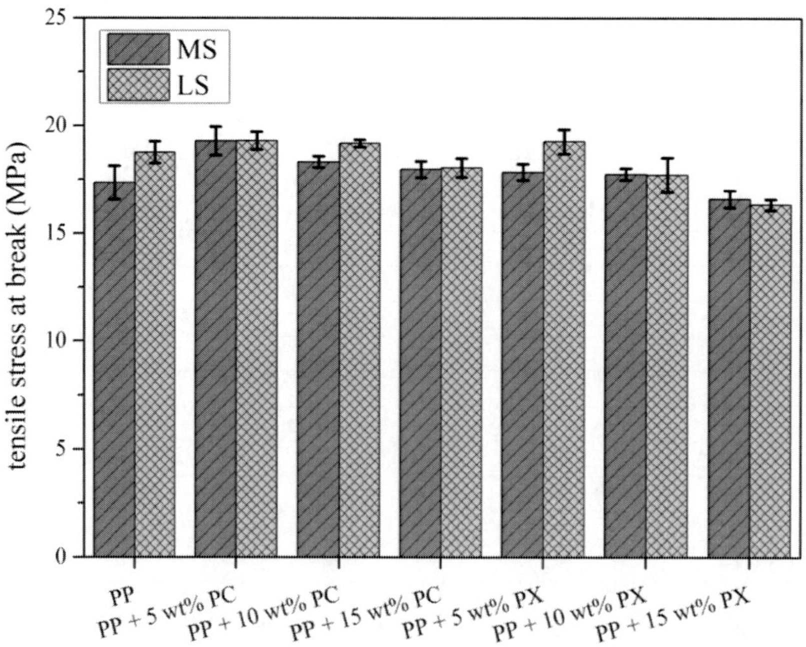

Figure 8. Tensile stress at break of unfilled PP and PP/expanded perlite composites (different screw geometries).

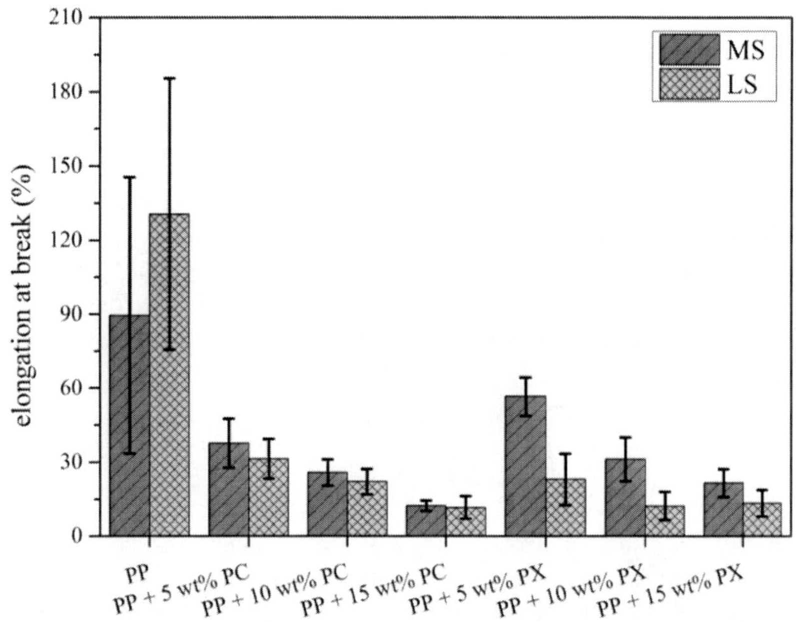

Figure 9. Elongation at break of unfilled PP and PP/expanded perlite composites (different screw geometries).

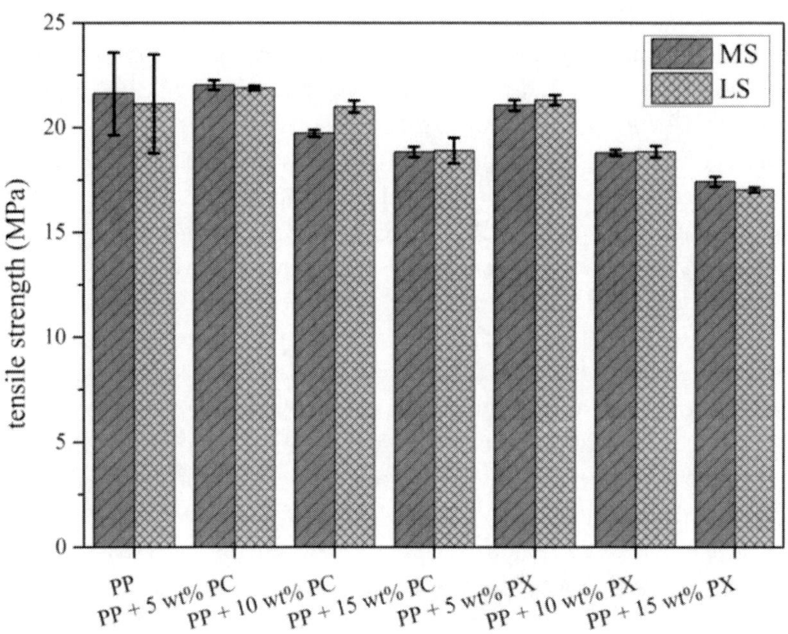

Figure 10. Tensile strength of unfilled PP and PP/expanded perlite composites (different screw geometries).

The values of elongation at break of the unfilled und filled PP composites are depicted in Figure 9. The elgonation at break is decreased with increasing filler content compared to the unfilled polymer matrix. The compounds become more brittle and expanded perlite forms stress concentration points. Due to that the samples are fractured more rapid when stress is applied [14, 15]. No significante influence of screw geometry on the elongation of break could be investigated for the PC type of expanded perlite. For the smaller particles of the PX type the compounds produced with the medium shearing screw achieved higher elongation at break values. This behavior becomes less expressed with higher filler loading. Thereby the achieved homogenity of the expanded perlite particles inside the polymer matrix with more induced shearing is presumably the cause for the observed difference.

The tensile strength decreases with smaller particles sizes and higher filler content (see Figure 10). The influence of the screw geometry is only marginal. Nevertheless a significant dependency could not be found due to the high mean variation of the measured values of the unfilled polymer matrix.

Conclusion

In this chapter fundamental investigations of the influence of screw geometries and different particles sizes and filler content on the morphology, rheological, thermal and mechanical properties of PP/expanded perlite composites are illustrated and summarized.

The homogenous dispersion of the expanded particles is one major influence factor on the property profile of the final composite. The low shearing screw geometry lets most of the filler particles undestroyed. Processing with the medium shearing screw produces lots of fragments with only a small number of intact expanded perlite particles, while the distribution of the single particles is more homogenous. The filler loading has no observable influence on this behavior. The expanded perlite particles show a honeycomb like inner structure which stabilizes them against forces applied in e.g., processing. In partly destroyed particles the polymer melt can flow inside this structure and increases the particles specific surface. This leads to an enhanced adhesion between expanded perlite and the polymer matrix.

During processing of the various PP/expanded perlite composites the minimum residence time was measured. Thereby the filler content was discovered to have the highest impact on the minimum residence time under constant processing conditions. As expected the change of some conveying elements to kneading blockshas increased the minimum residence time for all the produced compositions due to their low conveying ability compared to conveying elements.

Many fillers have a significant influence on the rheological properties of polymer melts, due to their size, shape, compatibility, material properties etc. The obtained rheological values show, that the addition of the expanded perlite to the polymer matrix, increases the viscosity values with increased amount of added filler. Thereby the curves of the filled polymer remained their shape compared to unfilled PP independent of the filler concentration. No significant influence of the different screw geometries could be observed.

The crystallization properties of the composites were obtained by DSC measurements. Thereby the measured crystallinities showed significant decrease through the present filler, which interferes with the crystallization of the polymer. The influence of the expanded perlite partcle size on the cristallization behaviour was marginal.

For the characterization of the mechanical properties of the produced composites the Young's modulus, the tensile stress at break, the elongation at break and the tensile strength were analyzed. All compositions show an increase of the Young's modulus with the addition of expanded perlite compared to unfilled PP. The material processed with a low shearing screw and bigger particle size showed the highest increase in Young's modulus (about 44 %).

The addition of expanded perlite leads to a significant decrease in elongation at break. The influence of filler, filler content and screw geometry on other mechanical values are only marginal.

Expanded perlite exhibit great reinforcement potential for polypropylene. Although this spherical shaped filler has also disadvantages such as limited elongational behavior like other mineral fillers. The closed shell structure of the particle offers wide possibilities for further developments.

ACKNOWLEDGMENTS

We want to thank Alexander Eder for the engineering drawings of the screw geometries. We acknowledge Christian Brendinger and Barabara Strohmayer for the assistance in material characterization.

REFERENCES

[1] Sodeyama, K., Sakka, Y., Kamino, Y., Seki, Y., (1999). Preparation of fine expanded perlite. *J. Mater. Sci.* 34, 2461-2468.

[2] Tekin, N., Kadinci, E., Demirbas, Ö., Alkan, M., Kara, A., Dogan, M. (2006). Surface properties of poly(vinylimidazole)-adsorbedexpanded perlite. *Micropor. Mesopor. Mat.* 93, 125-133.

[3] Akkaya R., (2013). Effects of pH, concentration and temperature on radionuclides sorption onto polyhydroxyethylmethacrylate-expanded perlite composite. *J. Radioanal. Nucl. Chem.* 295, 351-355.

[4] Vargas, A.F., Orozco, V.H. Rault, R., Giraud, S., Debauc, E., Lopez, B.L. (2010). Influence of fiber-likenanofillers on the rheological, mechanical, thermal and fireproperties of polypropylene: An application to multifilamentyarn. *Composites Part A* 41, 1797-1806.

[5] Sternstein, S.S., Zhu, A.J. (2002). Reinforcementmechanism of nanofilledpolymermelts as elucidaded by nonlinearviscoelasticbehavior. *Macromolecules* 35, 7262-7273.

[6] Marinelli, A.L., Bretas, R.E.S. (2003). Blends of polypropyleneresinswith a liquidcrystallinepolymer. I. Isothermalcrystallization. *J. Appl. Polym. Sci.* 87, 916-930.

[7] Lou, J., Harinath, V. (2004). Effects of mineral fillers on polystyrenemeltprocessing. *J. Mater. Process. Tech.* 152, 185-189.

[8] Akin-Öktem, C., Tincer, T. (1994). Preparation and characterization of perlite-filled high-densitypolyethylenes. II. Thermal and Flow properties. *J. Appl. Polym. Sci.* 54, 1115-1122.

[9] Chacko, V.P., Karasz, F.E., Farris. R.J. (1982). Morphology of $CaCO_3$-filled polyethylenes. *J. Polym. Sci. Polym. Phys.* 20, 2177-2195.

[10] Maiti, S.N., Mahapatro, P.K. (1990). Crystallization of i-PP/$CaCO_3$ composites and itscorrelationwithtensileproperties. *Int. J. Polym. Mater.* 14, 205-222.

[11] Chiang, W.Y., Yang, W.D. (1988). Polypropylene composites. I. Studies of the effect of grafting of acrylicacid and silane coupling agent on the performance of polyproylene mica composites. *J. Appl. Polym. Sci.* 35, 807-823.

[12] Ek, C.G., Kubat, J., Rigdahl, M. (1987). Stress relaxation, creep and internal stresses in high densitypolyethylenefilledwith calcium carbonate. *Rheol. Acta* 26, 55-63.

[13] Xavier, F., Sharma, Y. N. (1986). Structure-property relations in polypropylene mica composites. *Polym. Compos.* 7, 42-49.

[14] Akin-Öktem, C., Tincer, T. (1994). Preparation and characterization of perlite-filled high-densitypolyethylenes. I. Mechanicalproperties. *J. Appl. Polym. Sci.* 54, 1103-1114.

[15] Sheldon, R.P. (1982). Composites Polymeric Materials, *Applied Science*, London.

In: New Developments in Polymer Composites Research ISBN: 978-1-62948-340-5
Editors: S. Laske and A. Witschnigg

Chapter 8

NEW THERMALLY CONDUCTIVE LIQUID CRYSTALLINE POLYESTER AND ITS COMPOSITES

Shusuke Yoshihara *

Frontier Materials Development Laboratories, Kaneka Corporation, Japan

ABSTRACT

Thermally conductive thermoplastics have attracted attention as heat dissipative materials for a vast number of electronic devices. However, the thermal conductivity (TC) of polymers is very low (0.1-0.5 W m^{-1} K^{-1}) because of various defects. Even with large concentrations of thermally conductive filling material, it is difficult to enhance the TC of polymer composites. Moreover, a high filler content leads to undesirable weight gain and poor processability. One solution for this is to enhance the TC of the polymer matrix. The TC of polymer composites can significantly increase when polymer matrices serve as good heat conductors between filler particles. Here I show thermoplastics that exhibit high TC as a result of lamellar crystal alignment. Smectic liquid crystalline PB-*n* polyesters have been found that show a parallel orientation of lamellae by shear flow during injection molding. The polymer chains are aligned in the normal direction (ND) with respect to the molding surface, thus leading to high TC (1.2 W m^{-1} K^{-1}) in this direction. Furthermore, the composites containing thermally conductive filler particles exhibit a dramatic enhancement in the TC. This technique enables the production of polymer materials with high TCs without a high filler content, and hence lightweight composites with good processability. An injection molding process is suitable for producing these materials and no special equipment is required. New products based on this technique have been developed.

Keywords: Composites, injection molding, lamellar crystal, liquid-crystalline polymers, thermal conductivity

* E-mail address: Shusuke_Yoshihara@kn.kaneka.co.jp; Address: 5-1-1, Torikai-Nishi, Settsu, Osaka 566-0072 Japan.

INTRODUCTION

Thermally conductive thermoplastics have attracted attention as heat dissipative materials for a vast number of electronic devices such as automobile parts, communication equipment, and lighting apparatuses. A metal such as aluminum is currently used to dissipate the heat generated from a car part or a high-power LED, but the development of resin materials as an alternative to metals is desirable, particularly for the purpose of weight reduction. Polymers are excellent and easy-to-process electric insulators. However, their thermal conductivity (TC) is very low (0.1-0.5 W m^{-1} K^{-1}). It is technically difficult to increase a TC of the resin to the level of a metal, but there has recently been a report of the heat dissipation performance of polymer composites with TC values greater than 10 W m^{-1} K^{-1}, which is similar to that of metals[1]. Therefore, the value of 10 W m^{-1} K^{-1} is one of the targets of this study. The general technique to enhance the TC of polymer materials is an incorporation of a large amount of thermally conductive inorganic fillers. The filler forms a continuous thermal network at high filler loading, as a result of which composite TC increases. However, the high filler content leads to undesirable weight gain and poor processability. It is well known that enhancing the TC of polymer matrices is a solution for this[2].Here, I report a technique to enhance the TC of a thermoplastic cresin itself by injection molding [3] and introduce the TC enhancement mechanism and possible applications.

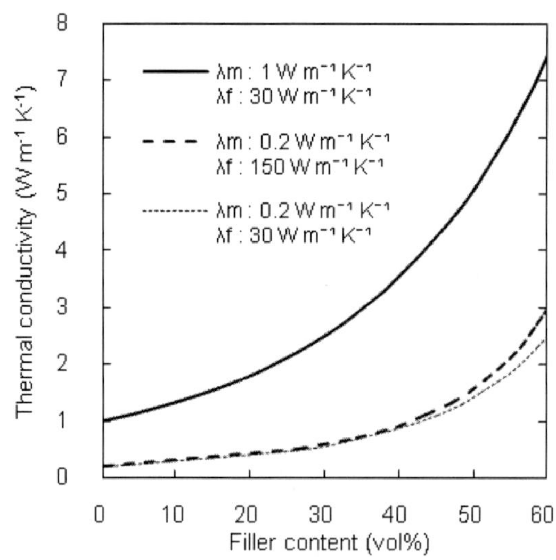

Figure 1. TC of polymer/filler composite calculated by Bruggeman's equation.

IMPORTANCE OF A POLYMER MATRIX WITH HIGH TC

Bruggeman's Theory

The equation of Bruggeman's theory [4] is given as

$$1-V=\frac{\lambda_c-\lambda_f}{\lambda_m-\lambda_f}\left(\frac{\lambda_m}{\lambda_c}\right)^{1/3} \qquad (1)$$

where V is the volume fraction of the filler, and λ_c, λ_f, and λ_m are the TCs (W m^{-1} K^{-1}) of the composite, filler, and polymer matrix, respectively. Figure 1 shows the relationship between the TC of the composite and the filler content at three different values of λ_m and λ_f: (λ_m,λ_f)= (0.2, 30); (0.2, 150); and (1, 30). To enhance the composite TC, Bruggeman's theory suggests that it is more effective to enhance the TC of the polymer matrix than to enhance that of the filler particles. In other words, the TC of the polymer composite can be significantly increased when the polymer matrix serves as a good heat conductor between the filler particles. Such a thermally conductive matrix enables the production of polymer materials with (1) significantly high TC at a high filler loading or (2) higher TC than that of common thermoplastic composites at comparable loading levels.

TCs of Polybutylene Terephthalate(PBT)/Hexagonal Boron Nitride (h-BN) Composites

Hexagonal boron nitride (h-BN)has plate-shaped particles with a structure similar to that of graphite (Figure 2), and exhibits a high TC of 400 W m^{-1} K^{-1} in the in-plane direction (ID) [6] while being an electrical insulator. Thus, h-BN is generally used to produce insulating polymer composites with high in-plane TC. First, PBT and h-BN with a mean particle size of 45μm (Momentive Performance Materials Inc.PT110) were mixed using a twin-screw extruder (TECHNOVEL Corp., KZW15TW). PBT and the composites were molded with a mini injection molder (DSM Xplore, Micro Injection Moulding Machine 5.5 ml) into plates with dimensions of 10×40×1 mm^3asshown in Figure 3a. The TC of molded PBT itself was isotropic and the value was 0.25 W m^{-1} K^{-1}. Figure 3b shows the TCs of PBT/h-BN composites with h-BN fractions ranging from 0 to 50 vol%. The TC values increased as the h-BN content increased, but a fraction of more than 50 vol% was required to attain a TCmore than 10 W m^{-1} K^{-1}.

TCs of Common Polymer Materials

Table 1 shows isotropic TC values of common polymer materials[5]. Although high-density polyethylene (HDPE) has the highest TC value of 0.44 W m^{-1} K^{-1} among common isotropic thermoplastics, the average value is only about 0.2 W m^{-1} K^{-1}.

PREVIOUS STUDIES TO INCREASE THE TC OF POLYMERS

It is known that a polymer shows high TC along the molecular chain direction. Choy and co-workersreported that HDPE extended to 25 times the orignal length exhibited a TC of more than 10 W m^{-1} K^{-1} along the fiber axis and only 0.2 W m^{-1} K^{-1} in the vertical direction[7]. These results are due to the stronger covalent bonds within the molecular chain compared to the van der Waals forces between the chains. Recently, there are reports that a more extended HDPE nanofiber showed a TC of 104W m^{-1} K^{-1} [8] and spider silk shows ananomalously high TC of 416 W m^{-1} K^{-1}, which is the same level as that of copper [9]. In addition, to get highly oriented polymers, some studies have focused on subjecting polymers in the molten state to a strong magnetic field [2,10,11] or by rubbing liquid crystal molecules[12].

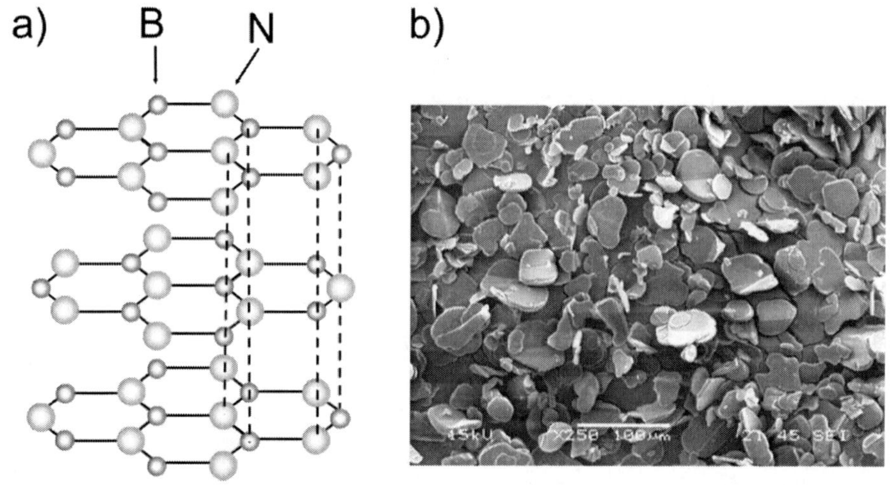

Figure 2. Hexagonal boron nitride (h-BN): a) Crystalline structure, b) SEM image, Scale bar, 100 μm.

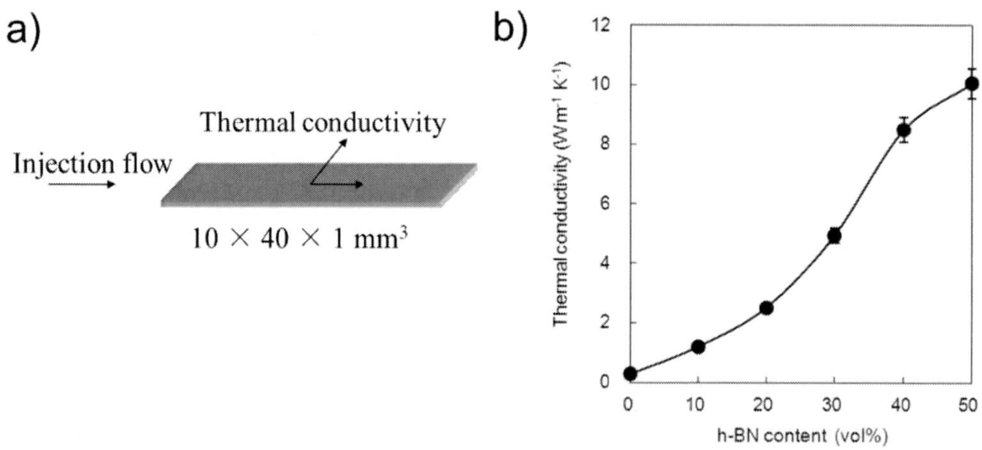

Figure 3. TC of injection-molded PBT/h-BN composites: a) The dimension of the molded plate, b) In-plane TC as a function of the volume fraction of h-BN.

Our Concepts for the Development of a Thermally Conductive Polymer Matrix

The methods using drawing, subjecting the polymer to a magnetic field, and rubbing are impractical for molding various complex shapes. Plastics for manufacturing end-use products are commonly formed by injection molding. Thus, the most suitable approach is to enhance the TC of polymers simultaneously with a simple injection molding process. Furthermore, polymer should function as a thermally conductive matrix even under the incorporation of filler particles. Hence, the following two points were development concepts:

1) Polymer composites can be molded by a simple injection molding process.
2) Dramatic enhancement of TC is observed by incorporation of thermally conductive filler particles.

Table 1. Thermal conductivity of selected polymers [5]

Material	Thermal Conductivity at 25°C [W m^{-1} K^{-1}]
Low density polyethylene (LDPE)	0.30
High density polyethylene (HDPE)	0.44
Polypropylene (PP)	0.11
Polystyrene (PS)	0.14
Polymethylmethacrylate (PMMA)	0.21
Nylon-6 (PA6)	0.25
Poly(ethylene terephthalate) (PET)	0.15
Poly(butylene terephthalate) (PBT)	0.29
Polycarbonate (PC)	0.20
Polyetheretherketone (PEEK)	0.25
Polyphenylene sulfide (PPS)	0.30
Polysulfone (PSU)	0.22
Polyvinyl chloride (PVC)	0.19
Polyvinylidenedifluoride (PVDF)	0.19
Polytetrafluoroethylene (PTFE)	0.27
Polyimide, Thermoplastic (PI)	0.11
Poly(dimethylsiloxane) (PDMS)	0.25
Epoxy resin	0.19

Strategy for TC Enhancement of a Polymer Matrix

Liquid Crystalline Polyester

The liquid crystalline (LC) phase is a state where the properties of the material are a mixture of those of a conventional liquid and those of a solid crystal. Here, nematic and smectic liquid crystalline polymers are briefly explained as representative examples. One of the most common LC phases is the nematic phase, where molecules have no positional order but align themselves to achieve long-range directional order with their long axes roughly parallel. Thus, although the molecules are free to flow with their centers of mass randomly distributed as in a liquid, they still maintain their long-range directional order (Figure 4a). The smectic phases form well-defined smectic layers (Figure 4b). To the best of our knowledge, all commercially available liquid crystalline thermoplastics are nematic liquid crystals. Since the polymer chains align parallel in the flow direction during injection molding, these materials have good fluidity. Almost all of anematic polymer is a fully aromatic polyester with high heat resistance.

Here, the LC polyester we developed is a smectic type and the properties are much different from other nematic LC polyesters. The polyester consists of rigid mesogens connected by flexible spacer groups as shown in Figure 4b, leading to a low melting point. So, smectic polymers tend to be inferior to fully aromatic LC polymers regarding their heat resistance. One of the polymers could form laterally large and well-stacked crystalline lamellae [13].

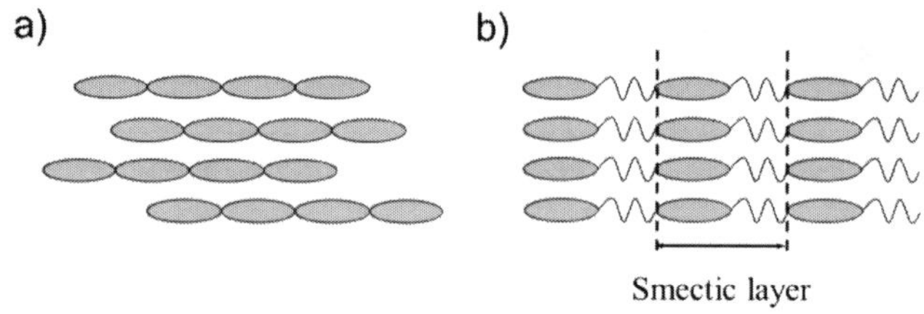

Figure 4. Liquid crystalline polymers: a) Nematic liquid crystal, b) Smectic liquid crystal.

TC Enhancement of Thermoplastics by Lamellar Crystal Alignment

We have developed a method for producing polymer materials that are not only lightweight and highly processable but can achieve dramatically high TCs afforded by the lamellar crystal alignment of polymer matrices using main-chain PB-*n* polyesters, as shown in Figure5a, where *n* is the even number of methylene units in the spacer moiety. For example, the PB-10 polyester consists of 1,10-dodecanedioic acid and 4,4′-biphenol and forms a smectic I (SmI) LC phase[14]. The polymer chains are highly extended, allowing them to be accommodated in the monoclinic lattices of SmI, and they form lamellae that are laterally large and well stacked, with a thickness of 47nm[14,15].The number average molecular weight (M_n) and polydispersity index (PDI) of the polymer used in this study were determined as 10.0 kg mol^{-1} and 1.9, respectively, by gel permeation chromatography (GPC) in a *p*-chlorophenol/toluene (3/8 volume ratio) solution with polystyrene standards (Viscotek HT-GPC with a RI detector). The crystal-smectic and smectic-isotropic phase transition temperatures of the polymer were204and 246°C, respectively, as determined by differential scanning calorimetry (DSC) at a heating rate of 10°C min^{-1} (PerkinElmer Pyris 1 DSC).

Injection molding in the LC state (240°C) induced lamellae alignment in the flow direction and oriented the polymer chains along the normal direction (ND) with respect to the molding surface as shown in Figure 5band 6 (for more details, see Reference [3]).Tokita and coworkers observed similar parallel orientations of lamellae using a main-chain polymer showing smectic CA phase[16].The molded PB-10 exhibited a high TC of 1.2 W m^{-1} K^{-1} in the ND (Figure 5c), which is 2-6 times higher than that of other isotropic polymers, compared to the modest value of 0.30 W m^{-1} K^{-1}observed in the ID. The TCs of PB-12 and PB-18 in the ND were also high, i.e., 1.1 and 0.82 W m^{-1} K^{-1}, respectively. Figure 7 shows an SEM image of a cross-sectional surface of injection-molded PB-18. The 29-nm-thick lamellar crystals were also well aligned parallel to the flow direction.

To achieve polymer chain orientation in the ND, a number of reports have suggested using a strong magnetic field [2,10,11].However, with our proposed technique, orientation is achieved using a simple injection molding process.

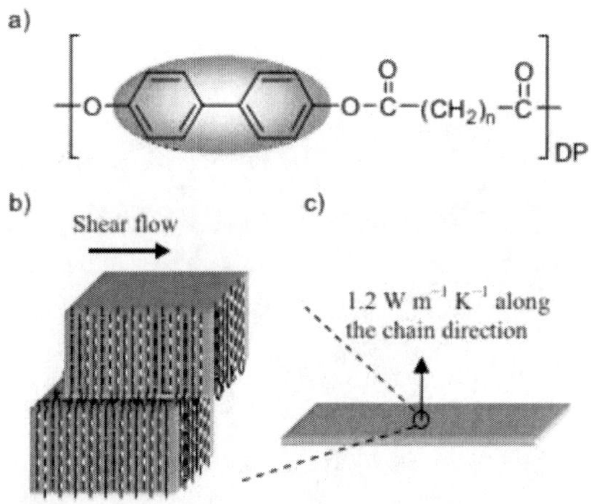

Figure 5. Concept for TC enhancement of a polymer: a) Chemical structure of PB-n, b) Parallel orientation of a lamellar crystal with respect to the molding surface induced by injection shear flow, c) TC of molded PB-10.

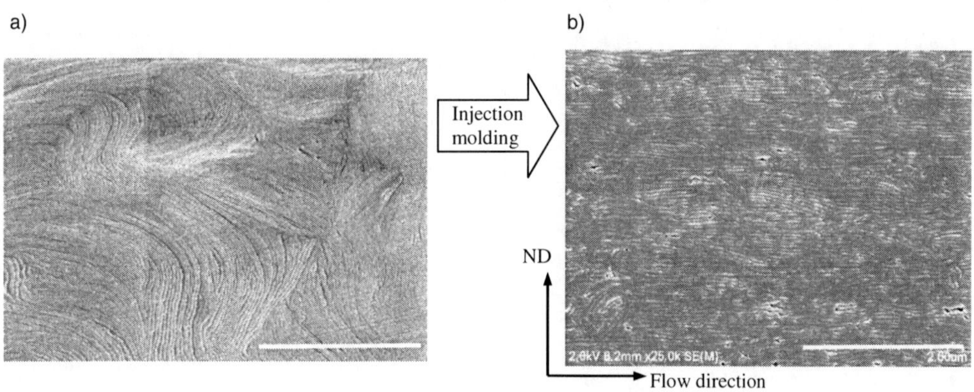

Figure 6. SEM image of crystalline lamellae of PB-10: a) Before injection molding. Scale bar, 5 μm, b) After injection molding. Scale bar, 2 μm.

Influence of the Molecular Weight of PB-10 on Lamellar Crystal Alignment

PB-10 with different molecular weights, PB-10-II and PB-10-III, were prepared by controlling the reaction period of melt condensation. The PB-10-II with a lower molecular weight (M_n: 8.00 kg mol^{-1}, PDI: 1.9) gave a similar type of lamellar crystal alignment and high TC (1.2 W m^{-1} K^{-1}) in the ND. In contrast, the PB-10-III having a higher molecular weight (M_n: 16.0 kg mol^{-1}, PDI: 2.4) showed a random orientation of lamellae even after injection molding, and the TC was isotropic with values around0.53 W m^{-1} K^{-1}. Alt and coworkers investigated the alignment behavior of a main-chain polymer showing smectic phase depending on various shear conditions[17]. However, the parallel orientation of lamellae was not generally observed, which they attributed to molecular entanglements or high interlayer connectivity that might inhibit the sliding of lamellae, which is necessary for parallel orientation. This reasoning is also applicable to the results for PB-10-III. We speculate that the flow behavior results from the greater interconnectivity of the lamellae due to the higher molecular weight.

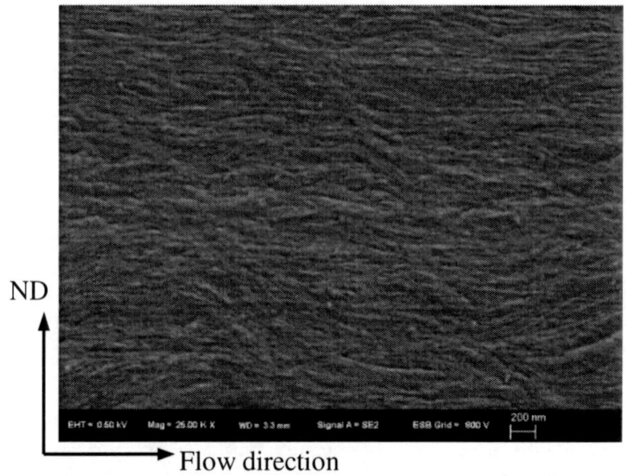

Figure 7. SEM image of crystalline lamellae of injection-molded PB-18. Scale bar, 200 nm.

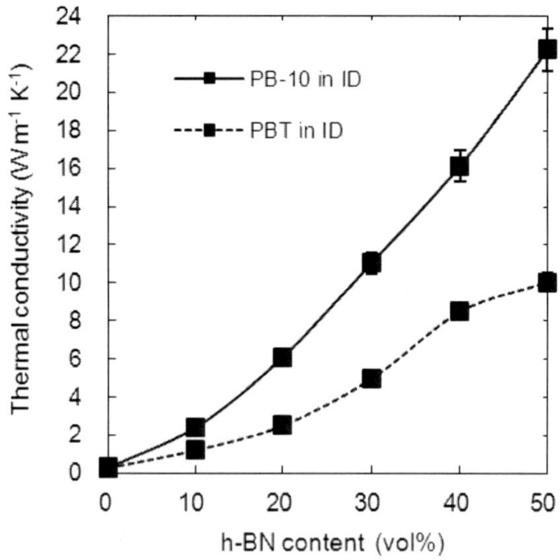

Figure 8. Effect of h-BN filler content on in-plane thermal conductivity of polymer composites.

TCs of PB-10/h-BN Composites

Next, PB-10 and h-BN were mixed using a twin-screw extruder and the composites were molded by injection molding. Two-dimensional wide-angle x-ray diffraction (2D-WAXD) measurements revealed that the molded composites had an interesting structure in which both polymer lamellae and the plate-shaped h-BN arranged their normal in the ND (see Reference [3]). Figure 8 shows in-plane TC values of the composites compared with PBT/h-BN composites. Surprisingly, a significant enhancement of TC was observed with the PB-10 matrix, although PB-10 is a poor conductor in this direction similar to PBT.

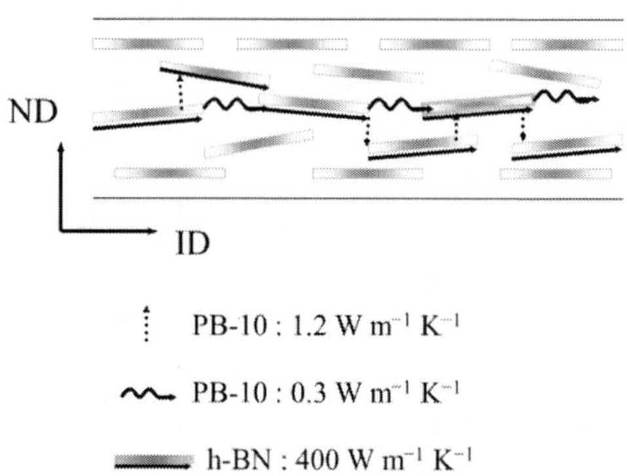

Figure 9. Heat conduction model in the in-plane direction for injection-molded PB-10/h-BN composites.

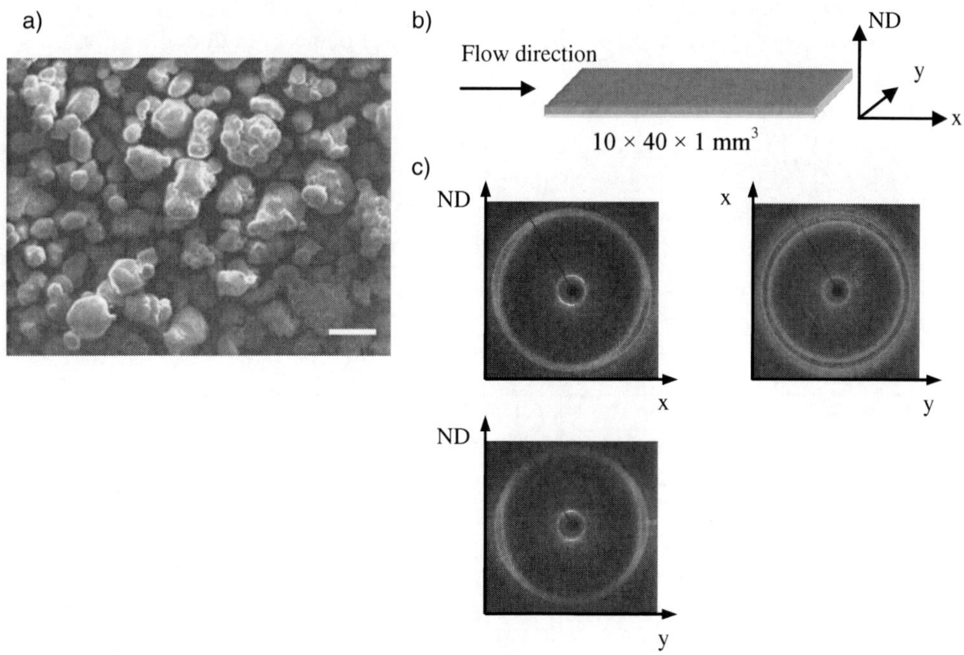

Figure 10. Characterization of injection-molded PB-10/MgO composites: a) SEM image of the MgO particles. Scale bar, 50 μm, b) Shear geometry in three characteristic directions, c) 2D-WAXS patterns along the x, y, and ND axes measured for the composites at 11vol% loading.

On the basis of these results, we propose a heat conduction model in which polymer chains aligned in the ND serve as effective heat paths between the h-BN filler particles, which conduct heat mainly in the in-plane direction (Figure9).

When the filler content was 50vol%, the TC reached 22 W m^{-1} K^{-1}, which to the best of our knowledge is a new record for the TC of polymer materials with injection moldability and electrical insulation properties. Moreover, to attain a TC of 10 W m^{-1} K^{-1}, the target value, PB-10 requires an amount of h-BN 20vol% lower than other common polymers. This leads to lighter composites with good processability.

TCs of PB-10/Magnesium Oxide (MgO) Composites

In addition, PB-10 composites with 11 vol% magnesium oxide (MgO) filler particles, roughly spherical in shape, were prepared to investigate the polymer chain orientation by injection molding and the resulting TC value. An MgO with a mean particle size of 50μm (Figure 10a) was purchased from Ube Material Industries, Ltd. (RF-50-SC). The particles have electrical insulation properties, a density of 3.65 g cm^{-3}, and a TC value of 42 W m^{-1} K^{-1}[18,19].The composite was prepared by mixing PB-10 with MgO particles at 220°C using a twin-screw extruder. The orientation of the polymer chains at the center of the injection-molded composite plate was observed using 2D-WAXD (Figure 10b). The 2D-WAXD pattern (11vol%) in Figure 10c shows the outer (4.24 Å) and inner (21.0 Å) reflections. In both (ND, x) and (ND, y) patterns, outer peaks derived from the inter-mesogen reflections appear on the equator. No specific orientation was observed in the (x, y) pattern. These profiles are similar as those observed in the pure injection-molded resin [3], indicating that the polymer chains were aligned perpendicular to the (x, y) plane even under incorporation of MgO particles. Therefore, the PB-10 composite exhibited dramatic TC enhancement in the ND. Figure 11 shows a TC value in the ND of 1.6 W m^{-1} K^{-1},which is 3-4 times higher than the value of 0.44 W m^{-1} K^{-1} of the PBT composite at 12vol% loading. To achieve a TC value of 1.6 W m^{-1} K^{-1}, the PBT composite requires a MgO fraction of about 50vol% according to Bruggeman's theory. These findings indicate that PB-10 has a great advantage over other common polymers in the use of not only plate-shaped filler particles such as h-BN but also spherical particles such as MgO.

Table 2. Properties of new thermally conductive thermoplastics, SP series

Item	Method	Condition	Unit	SP047 N	SP237 N
Thermal conductivity	Xe Flash(ASTM E1461)	out-of -plane	W m^{-1} K^{-1}	2.0	**3**
		in-plane		**14**	7.6
Density	ISO 1183	23°C	g cm^{-3}	1.82	2.08
DTUL	ASTM D648	1.82MPa	°C	181	167
Flammability	UL94	$0.8mm^t$	-	V-0 eq.	V-0 eq.
Spiral flow	220°C	$1mm^t$	mm	120	70
Dielectric breakdown	ASTM D149	23°C(AC50Hz)	kV mm^{-1}	30	22

*Values were obtained from our laboratory and are not guaranteed.

Development of New Products

Properties

Based on the current technique, a polymer structure was optimized to improve various properties, and other additives such as a reinforcing agent and a flame retardant were compounded, providing lightweight polymer materials with high TC, electrical insulation properties, and good processability. The representative properties of our new products named "SP series" are listed in Table 2. The main features are that the in-plane TC of SP047 Nis 14 W m^{-1} K^{-1} and the out-of-plane TC of SP237 N is 3 W m^{-1} K^{-1}. The densities are 1.8-2.1 gcm^{-3}, which are lower than those of metals. Therefore, the products are expected to be applied as alternative materials for metals. Meanwhile, the DTUL of each material is not so high (160-180°C).

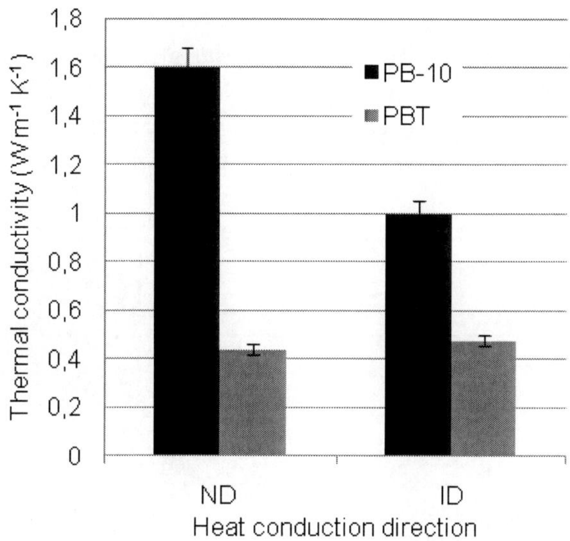

Figure 11. TC of composites for PB-10 and PBT with MgO fraction of 11 and 12vol%, respectively.

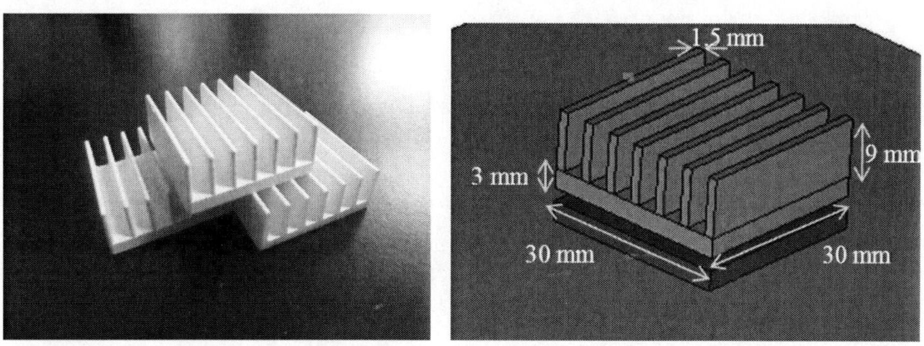

Figure 12. Heat sink used for evaluation of heat dissipation of various composites.

Heat Dissipation Performance of theComposites

To confirm the heat dissipation performances of the thermally conductive composites, samples having TCs values of 5-14 W m^{-1} K^{-1} were prepared and molded into heat sink components, as shown in Figure 12, by injection molding. As a comparison, the performance of an aluminum die-cast component with the same shape and a TC of 180 W m^{-1} $K^{-1}$was also evaluated, where the thermal emissivity was adjusted to 0.85 of resin by an alumite treatment.

The contact between the heat sink and heater was made using adhesive tape, and the temperature of the heater with a power of 1.2 W was measured with a thermocouple under room conditions. Figure 13 shows the relationship between the TC of the heat sink materials and the heater temperature under equilibrium. The temperature was 210°C without a heat sink. Using a heat sink, the temperature decreased as the TC of the heat sink material increased. When the SP047 N with the TC value of 14 W m^{-1} K^{-1} was used, the temperature was reduced to 65°C. On the other hand, the aluminum die-casting heat sink decreased the heater temperature to 60°C, indicating that SP047 N has a comparable heat dissipation performance to the die-cast aluminum. This result was attributed to the small size of the heat sink used in this study. When the TC of the material increased to some extent, the heat transfer to the end of the fin is saturated and the heat radiation to the air becomes rate-limiting. From these findings, SP047 N is expected to be an alternative material to metals such as aluminum for the purpose of reducing the weight of small electronic devices.

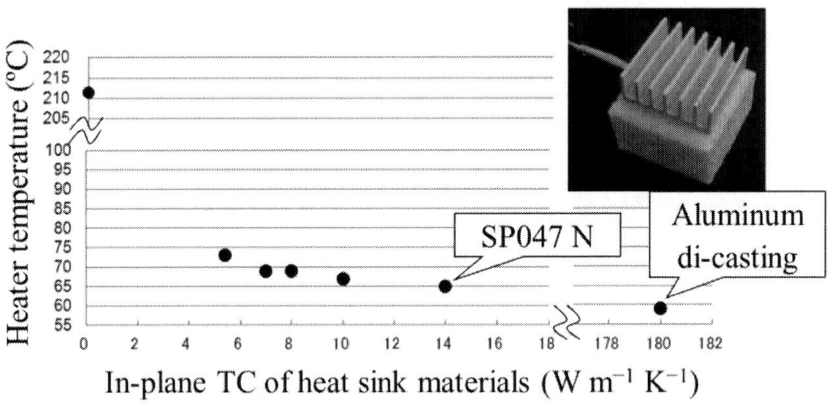

Figure 13. The relationship between a TC of heat sink materials and heater temperature.

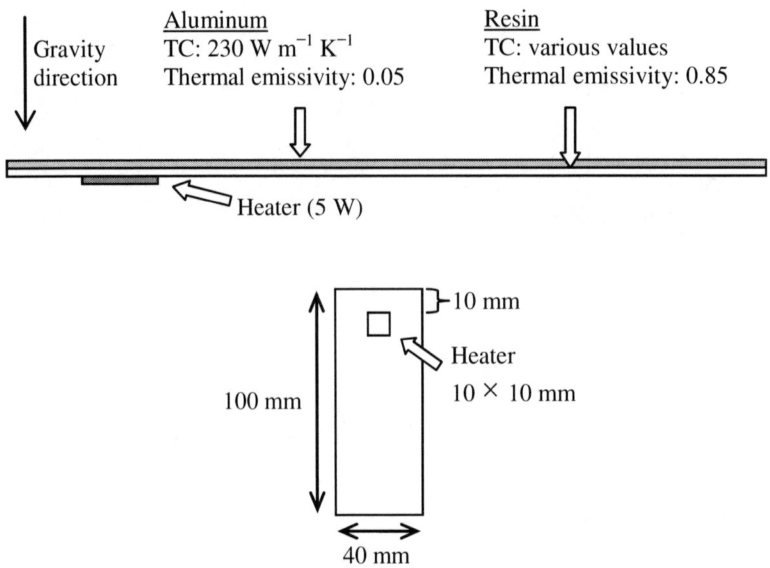

Figure 14. The conditions of the heat dissipation simulation with a resin/metal complex.

Application of SP237 N with Out-of-plane High TC

There are a lot of cases where a structural body is formed from a resin and metal complex in electronic equipment. In this case, the out-of-plane TC of the resin is important to effectively utilize the good heat transfer ability of the metal. Simulations were conducted to investigate the relationship between the TC of the resin and the heat dissipation performance of a resin/aluminum complex. Figure 14 shows the conditions of the simulation. A thermally conductive resin board was adhered to a heater with a power of 5 W. As shown in Table 3, the different shaped aluminum boards with a TC of 230 W m^{-1} K^{-1} were placed on various locations on the resin board, following which the heater temperature was measured. Here, the thermal resistance at each interface between the materials was ignored. Without the aluminum board, the heat dissipation performance was high as the in-plane TC of the resin increased. In contrast, with the aluminum board present, the heat dissipation performance became high as out-of-plane TC of the resin increased. This result indicates that the resin effectively conducted heat from the heater to the aluminum board and the aluminum distributed the heat well. From these results, in the case of a resin/metal complex, a composite with a high out-of-plane TC such as SP237 N is effective in terms of heat dissipation.

Table 3. Results of heat dissipation simulation for resin/aluminum composition

	TC of resin [W $m^{-1}K^{-1}$]		**Heater temperature (5 W)[°C]**			
	In-plane	**Out-of-plane**				
SP047 N	**14**	1.6	**182**	126	131	**169**
SP237 N	7.6	**3.0**	210	**119**	**129**	201
Components 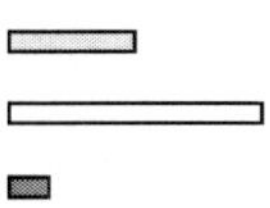	Aluminum Resin Heater			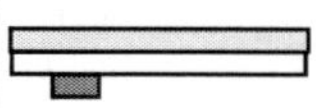	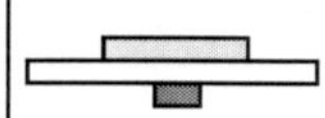	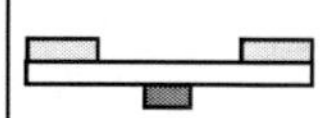

CONCLUSION

We have developed a method for enhancing the TC of various polymers. Shear flow in the LC state of PB-*n* during injection molding induces the parallel orientation of lamellae, in which polymer chains are aligned in the ND with respect to the molding surface, thus leading to a high TC in this direction. This flow behavior could be observed even when filler particles were incorporated. The composites exhibited dramatic enhancement in TC because the polymer chains aligned in the ND served as effective heat paths between filler particles. Based on this technique, we have developed new products named "SP series" whose TC values are 14 W m^{-1} K^{-1} in the ID or 3 W m^{-1} K^{-1} in the ND with electrical insulation properties. These products are suitable as heat dissipative materials for a vast number of electronic devices such as automobile parts, communication equipment, and lighting apparatuses.

REFERENCES

[1] Janssen, R. H. C., Douren, L., van Dijk ,H. K.,(2010),Thermally conductive plastics: balancing material properties with application needs, *LED professional Review*, 17, 38-41.

[2] Hasegawa, M., Suyama, N., Shimoyama, N., Aoki, H., Nunokawa, T., Kimura, T., (2011), Enhanced thermal conductivity of semi-aliphatic liquid crystalline polybenzoxazoles using magnetic orientation, *Polym. Int.*, 60, 1240-1247.

[3] Yoshihara, S., Ezaki, T., Nakamura, M., Watanabe, J., Matsumoto, K., (2012), Enhanced thermal conductivity of thermoplastics by lamellar crystal alignment, *Macromol. Chem. Phys.*, 213, 2213-2219.

[4] Bruggeman,D. A. G., (1935), Berechnung verschiedener physikalischer konstanten von heterogenen substanzen, *Ann. Phys.*, 24, 636-664.

[5] Han, Z., Fina, A., (2011), Thermal conductivity of carbon nanotubes and their polymer nanocomposites: A Review, *Prog. Polym. Sci.*, 36, 914-944.

[6] Huang, M. T., Ishida, H., (1999), Investigation of the boron nitride/polybenzoxazine interphase, *J. Polym. Sci. Part B: Polym. Phys.*, 37, 2360-2372.

[7] Choy,C. L., Luk, W. H., Chen, F. C., (1978),Thermal conductivity of highly oriented polyethylene, *Polymer*, 19, 155-162.

[8] Shen, S., Henry, A., Tong, J., Zheng, R., Chen, G., (2010), Polyethylene nanofibres with very high thermal conductivities, *Nat. Nanotechnol.*, 5, 251-255.

[9] Huang, X., Liu, G., Wang, X., (2012), New secrets of spider silk: exceptionally high thermal conductivity and its abnormal change under stretching, *Advanced Materials*, 24, 1482-1486.

[10] Harada, M., Ochi, M., Tobita, M., Kimura, T., Ishigaki, T., Aoki, W., (2003), Thermal-conductivity properties of liquid-crystalline epoxy resin cured under a magnetic field, *J. Polym. Sci. Polym. Phys.*, 41, 1739-1743.

[11] Kimura, T., (2008), Development of high performance polymers by magnetic field-induced orientation, *J. Net. Polym. Jpn.*, 29, 51-56.

[12] Geibel,K., Hammerchmidt,A.,Strohmer, F., (1993),*In situ* photo polymerized, oriented liquid-crystalline diacrylates with high thermal conductivities, *Adv. Mater.*, 5, 107-109.

[13] Tokita, M., Osada, K., Yamada, M., Watanabe, J., (1998), Chain-folded lamellar structure in the smectic H phase of a main-chain polyester, *Macromolecules*, 31, 8590-8594.

[14] Tokita, M., Okuda, S., Yoshihara, S., Takahashi, C., Kang, S., Sakajiri, K., Watanabe, J., (2012), Identifying smectic I phase of main-chain PB-10 polyester consisting of 4,4'-biphenol and 1,10-dodecanoic acid by fibre X-ray diffraction, *Polymer*, 53, 5596-5599.

[15] Krigbaum, W. R., Watanabe, J., Ishikawa, T., (1983), *Thermotropic polyesters*. 2. investigation of the mesophase properties of polymers based on 4,4'-dihydroxybiphenyl, *Macromolecules*, 16, 1271-1279.

[16] Tokita, M., Tokunaga, K., Funaoka, S., Osada, K., Watanabe, J., (2004), Parallel and perpendicular orientations observed in shear aligned SCA liquid crystal of main-chain polyester, *Macromolecules*, 37, 2527-2531.

[17] Alt, D. J., Hudson, S. D., Garay, R. O., Fujishiro, K., (1995), Oscillatory shear alignment of a liquid crystalline polymer, *Macromolecules*, 28, 1575-1579.

[18] Lee, J. H., Eun, J. H., Park, S. Y., Kim, S. G., Kim, H. J., (2003), Hydration of r.f. magnetron sputtered MgO thin films for a protective layer in AC plasma display panel, *Thin Solid Films*, 435, 95-101.

[19] Jeong, I., Kim, J. Lee, J. Hong, J., (2010), Electrical insulation properties of nanocomposites withSiO2 and MgO Filler, *Trans. Electr. Electron. Mater.* 11, 261-265.

In: New Developments in Polymer Composites Research ISBN: 978-1-62948-340-5
Editors: S. Laske and A. Witschnigg

Chapter 9

DETERMINATION OF BINODAL CURVES IN TERNARY POLYMER SYSTEMS: CO-/POLYMERS COMPATIBILITY

Francisco Torrens[1,*] and Gloria Castellano[2]
[1]Institut Universitari de Ciència Molecular,
Universitat de València, València, Spain
[2]Facultad de Veterinaria y Ciencias Experimentales,
Universidad Católica de Valencia *San Vicente Mártir*, València, Spain

ABSTRACT

A simple method is presented for the numerical evaluation of binodal curves, in ternary systems polymer (P)–liquid 1 (L1)–L2 and P1–P2–solvent, by exploiting the SIMPLEX method. The objective function (OF) is the weighted sum of the square differences of chemical potentials of the two phases of every component, obtained by evaluating first derivatives of Gibbs free energy of mixture with respect to moles of components. Our method: (1) is stable since it does not require OF derivatives and (2) is applied in many cases by changing the equation of state. Comparisons are reported with simplified iterative procedures from the literature, for mixtures of (1) two polymers with identical characteristics in solvent and (2) solvent–nonsolvent–polymer with solvent–polymer interaction parameter equal to zero. A miscibility step up is applied by optimizing intermolecular H-bonding between two polymers. Polystyrene (PS) allows compatibility in chloroform with H-acceptors poly(vinyl pyridine/pyrrolidone). In order to enhance miscibility, PS is co-polymerized with methacrylic acid (MAA)/vinyl phenol (VPh). Two PS co-polymers are characterized, bearing 8 wt.% of OHs. Miscibility gaps *via* ternary phase diagrams of binodals, interpolymer interaction parameter from viscometry and association equilibrium constant by infrared permit examining the spacing effect of the interacting moieties along PS chain. The polymer miscibility rises with content of MAA in the co-polymer chain; however, when the polar group is VPh, the trend is disrupted.

* Address: Universitat de València, Edifici d'Instituts de Paterna, P. O. Box 22085, E-46071 València, Spain. (Corresponding author)

Keywords: ternary system, phase diagram, thermodynamics, optimization, SIMPLEX, polymer compatibility, hydrogen bonding, association constant, styrene-co-methacrylic acid co-polymer

INTRODUCTION

The numerical determination of binodal curves was considered for binary polymer–polymer, and ternary polymer–liquid-1 L(1)–liquid-2 L(2) and polymer-1 P(1)–polymer-2 P(2)–solvent systems. Reports were presented in order to obtain analytic strategies [1,2] and refined numerical tools [3]. The base is chosen equation of state, *e.g.*, SAFT, for numerical evaluation of L–L phase equilibrium in the polymer systems. With the aim of testing the reliability of our algorithm, Flory–Huggins (FH) theory is used for the Gibbs free energy [4], despite case studies in the literature showed that its validity is limited and FH equation for liquids with volume of mixing needs corrections [5]. A general algorithm, efficient for reproducing the binodal curves in mixed polymers, results interesting, especially if it is able to guarantee stability when input parameters or governing equations are changed. In the present report, a method is applied for the evaluation of the binodal curves for general ternary systems polymer–L(1)–L(2) and P(1)–P(2)–solvent. Our approach assumes, for the first examples, a general expression of the Gibbs free energy of mixture derived from FH theory [6–11]; then, it is applied to real mixtures *via* the simplified PC-SAFT equation of state.

As a rule, differentiation of Gibbs free energy with respect to the number of moles of the i-th component gives the chemical potential μ_i. Binodal curves are obtained by imposing liquid–liquid phase equilibrium at a given temperature and pressure, *i.e.*, by equating the chemical potentials of every component between two phases, which yields to a nonlinear, nonconvex system of equations that is solved resorting to numerical methods. Our algorithm could be applied with any analytical expression of the Gibbs free energy, and FH theory is used in the first examples with the aim of comparing the results with consolidated solutions from the literature. Other numerical approaches are attained; nevertheless, commercial packages based on Newton–Raphson (NR) schemes [12]: (1) depend on the choice of the Gibbs free energy since information on the derivatives is required and (2) their convergence is affected by the starting point chosen at the first iteration. Furthermore, NR algorithms require special care in the multidimensional case, and the strict monotony of the first partial derivatives is recommended in order to guarantee success. Usually, the algorithms proposed, *e.g.*, by von Solmes *et al.*, are based on *alternating tangent method*, an NR modification. Contrary to the schemes based on first derivatives, our approach avoids the typical NR numerical instabilities since (1) the calculation of the second derivatives of the Gibbs free energy is not needed, (2) the method does not require the particular properties of the smoothness of the objective function (OF) and (3) it is particularly efficient and less time consuming. On the other hand, for the simple cases in which, *e.g.*, two components behave in the same way with respect to the third, robust iterative procedures or at-hand simplified solutions in which only one unknown variable must be determined were available for decades [13–15]. Nevertheless, the general problem remains open, and the algorithms are not able to guarantee convergence without limitations. Milani and Milani developed an approach based on a genetic algorithm that is based on trivial algebraic operators, directly applied on random initial populations [16]. The OF was represented by the weighted sum of the square differences of the chemical potentials of the phases of every component. Once that a concentration v_i of one phase of one component in the ternary system was fixed, *e.g.*, v_1, the minimization of OF was guessed on a three-variable domain, being v_2, v'_1 and v'_2 unknown concentrations.

In earlier publications, it was examined the poly[styrene (PS)-*co*-methacrylic acid]–*b*–poly(4-vinylpyridine) treatment in solution under liquid–liquid phase-separation conditions by viscometry [17], compatibility between PS (co-)polymers *via* H-bonding [18], interaction parameters, and phase behaviour of epoxy/PS and epoxy/co-polymer PS–*b*–poly(methyl methacrylate) [19]. A reliable evaluation of a solution point for a given concentration in one phase of one component is obtained with limited computational effort. The next section presents the materials and methods. Next, the thermodynamic analysis is described. In the results section, our numerical model is applied for several cases and compared with previous solutions to test its reliability. Next, two examples of bicomponent systems are treated in A Real-Case Binary Example subsection, by using the simplified PC-SAFT method for the calculation of the parameters of the governing equations. The compatibility between PS co-polymers *via* H-bonding is analyzed.

Materials and Methods

Chloroform Prolabo (France) of analytical grade was distilled in the presence of P_2O_5. Poly(*N*-vinyl-2-pyrrolidone) (PVP) Fluka (Buschs, Switzerland) was precipitated in benzene and vacuum-dried up to constant weight. Poly(4-vinylpyridine) (PVPy) was ultrafiltered from a commercial sample from Polysciences (Warrington, PA). Solutions at 2% (w/v g·dL^{-1}) of commercial PVPy in mixed water/methanol 30/70 (v/v) were ultrafiltered at 3 kg·cm^{-2} pressure. This fraction was precipitated with ether, reprecipitated after dissolution in $CHCl_3$ and vacuum-dried up to constant weight. Poly(styrene–methacrylic acid) (PSMAA) co-polymer synthesis was performed by free-radical polymerization of 40 g of styrene and methacrylic acid (MAA) in methanol 50 wt.%, at 65ºC under $N_{2(g)}$ with 0.3 g of α,α'-azobisisobutyronitrile (AIBN) initiator. The reaction yielded 10% (12 h) of random co-polymers. The solution was precipitated twice with mixed tetrahydrofuran (THF)/ethanol. Product was dried in a vacuum oven at 60ºC (12 h). Reference PS was obtained as above: 50 g of styrene in 150 mL of toluene at 60ºC with 0.2 g of AIBN for 48 h. Poly(styrene/vinylphenol) (PSVPh) was synthesized as above, by starting from 40 g of styrene and vinylphenol (VPh), from decarboxylation of *p*-hydroxycinnamic acid in toluene at 60ºC with 0.3 g of AIBN, yielding analogous conversion. Precipitation and drying were performed as PSMAA.

Sample characterization. PSMAA solutions in mixed benzene/ethanol 90/10 (v/v) were titrated with sodium ethoxide in binary solvent by potentiometry. PS, PSMAA and PVPy were characterized by liquid chromatography (LC) that consisted of solvent-delivery system model 590 and U6K universal injector from Waters (Milford, MA). The detection was done with a refractive index detector model 2410 (Waters). PS/PSMAA LC was equipped with three sets of TSK columns, pore-size name Gel H_{XL}. For PVPy, TSK columns (pore-size names Gel Super AW 2500, 3000, 4000) were used [20]. The mobile phases THF for PS/PSMAA and $CHCl_3$ for PVPy were adjusted to 1 mL·min^{-1}. The volume of the injection of samples was 100 μL. Eluents were filtered and degassed *via* regenerated cellulose 0.45 μm pore diameter filters from Micro Filtration Systems (Dublin, CA). Composition of PSVPh samples was analyzed by Fourier-transform infrared (FTIR) from solutions in $CHCl_3$, after calibration with VPh solutions. PSVPh samples were LC-characterized as PS/PSMAA. PVP K-30 Fluka was eluted *via* TSK Gel H_{XL} columns with dimethylformamide.

Viscometry. Intrinsic viscosity coefficients of every polymer/co-polymer, in binary $CHCl_3$/polymer solute or ternary $CHCl_3$/co-polymer/polymer, were determined by extrapolation to infinite dilution of co-polymer with binary $CHCl_3$/polymer. *Binodal isotherms in size-exclusion chromatography (SEC) phase diagrams.* Co-existence data were obtained by equilibrating mixtures located in the region of two phases and analyzing compositions by SEC [21–24]. Titrated phase diagram was isothermal at room temperature, and its determination was limited to homopolymer + co-polymer content <30 wt.% in the ternary system since, otherwise, mixture excessive viscosity prevented a clear phase separation.

Infrared (IR) Spectra

$CHCl_3$/PSVPh/PVPy, $CHCl_3$/PSVPh/PVP and $CHCl_3$/PSVPh/ethylpyrrolidone (EP) systems. A Perkin–Elmer 225 IR spectrometer was used with CaF_2 cells of 0.12 mm width, thermostated at 40°C. The band of OH in PSVPh at 3 600 cm^{-1} was analyzed. PSVPh solution in $CHCl_3$ with concentration $c_o = 1.2\times10^{-3}$ M was measured. Spectra performed in mixed PSVPh–PVPy/PVP/EP, at concentrations c_o (PSVPh) and c_1 (PVPy/PVP/EP), decayed in OH band intensity at 3 600 cm^{-1} as c_1 grew, because of increasing formation of H-bonds between OH of PSVPh and N/O=C of PVPy/PVP/EP. *$CHCl_3$/ethylphenol (EPh)/EP, $CHCl_3$/EPh/PVP and $CHCl_3$/EPh/ethylpyridine (EPy) systems.* Band of OH in EPh at 3 260 cm^{-1} was monitored. The procedure was as above: measuring solution of EPh in $CHCl_3$, at $c_o = 1.2\times 10^{-3}$ M, and in mixtures in solution with c_o (EPh) and c_1 (EP/PVP/EPy). *$CHCl_3$/PSMAA/PVPy and $CHCl_3$/PSMAA/PVP systems.* The spectrometer was a Nicolet FTIR analyzing OH…N H-bond band at 1 607 cm^{-1}. To prevent band deformation with that of MAA C=O at 1 750 cm^{-1}, the latter was discounted. Examination of $CHCl_3$/PSMAA/PVP was performed by interfering band of C=O of MAA simultaneously in PSMAA/PVP (1 750 cm^{-1}).

Thermodynamic Analysis

Theory of FH is the base to describe the phase behaviour of the ternary system, composed of L(1)–L(2)–polymer or P(1)–P(2)–solvent. Gibbs free energy of mixing ΔG_m for solution is [25]:

$$\Delta G_m = RT(n_1 \ln\phi_1 + n_2 \ln\phi_2 + n_3 \ln\phi_3 + \chi_{12}n_1\phi_2 + \chi_{13}n_1\phi_3 + \chi_{23}n_2\phi_3 + \chi_T n_1\phi_2\phi_3) \quad (1)$$

where R is the gas constant, T, the temperature, n_i and ϕ_i, the number of moles and volume fraction of component i (i = 1, 2: liquid; i = 3: polymer), χ_{ij}, the binary interaction parameter between components i and j, and χ_T, ternary interaction parameter [26]. Tompa calculated binodal curves for nonsolvent [L(1)]–solvent [L(2)]–polymer with $\chi_T = \chi_{23} = \chi_{12} = \chi_{13}$, *via* a recursive algorithm [27]. Scott [28] and Tompa applied FH theory of polymer solutions to mixtures with and without added solvent. Scott obtained the Gibbs free energy of mixing of a total volume V of two polymers as:

$$\Delta G_m = (RTV/V_r)[(\phi_A/x_A)\ln\phi_A + (\phi_B/x_B)\ln\phi_B + \chi_{AB}\phi_A\phi_B] \tag{2}$$

where V_r is reference volume that is taken close to molar volume of the smallest polymer repeating unit, ϕ_A and ϕ_B, the volume fractions of polymers A and B, respectively, x_A and x_B, the degrees of polymerization of polymers A and B in terms of the reference volume V_r, χ_{AB}, interaction parameter between A and B, and is related to the enthalpy of interaction of the polymer repeating unit, each of molar volume V_r. In the presence of solvent [P(1)–P(2)–solvent systems], Scott derived *via* Eq. (2):

$$\Delta G_m = (RTV/V_S)[\phi_S\ln\phi_S + (\phi_A/x_A)\ln\phi_A + (\phi_B/x_B)\ln\phi_B + \chi_{AB}\phi_A\phi_B + \chi_{AS}\phi_A\phi_S + \chi_{BS}\phi_B\phi_S] \tag{3}$$

where index S refers to solvent, V_S is the reference volume and equates the solvent molar volume, χ_{AS} and χ_{BS}, interaction parameters between polymers A and B, and solvent. Notice that x_A, χ_{AB}, χ_{AS} and χ_{BS} are considered in terms of the reference volume V_S. Scott calculated binodals for special cases, *e.g.*, by varying χ_{AB} and assuming $\chi_{AS} = \chi_{BS} = 0$ with $x = x_A = x_B$. In the general case, binodal curves are obtained by evaluating chemical potentials of each component.

Differentiation of Gibbs free energy [Eqs. (1) and (3)] with respect to the number of moles of component i-th gives the chemical potential μ_i. Phase equilibrium L–L at a specified temperature and pressure is imposed, *i.e.*, the chemical potentials of every component between the two phases are equated, which yields to recover point by point binodal curves *via* a nonlinear, nonconvex system of equations solved by resorting to numerical methods. By denoting with v_1, v_2, v_3 and v'_1, v'_2, v'_3 the concentrations of components 1, 2, 3 in phase one and phase two, the system of equations is written as:

$$\begin{cases} \mu_1 = \mu'_1 \Leftrightarrow F_1 = \mu_1 - \mu'_1 \\ \mu_1 - \mu_1^o = RT\left[\ln v_1 + (1-v_1) - v_2\dfrac{x_1}{x_2} - v_3\dfrac{x_1}{x_3} + (\chi_{12}v_2 + \chi_{13}v_3)(v_2+v_3) - \chi_{23}\dfrac{x_1}{x_2}v_2v_3\right] \\ \mu_2 = \mu'_2 \Leftrightarrow F_2 = \mu_2 - \mu'_2 \\ \mu_2 - \mu_2^o = RT\left[\ln v_2 + (1-v_2) - v_1\dfrac{x_2}{x_1} - v_3\dfrac{x_2}{x_3} + (\chi_{21}v_1 + \chi_{23}v_3)(v_1+v_3) - \chi_{13}\dfrac{x_2}{x_1}v_1v_3\right] \\ \mu_3 = \mu'_3 \Leftrightarrow F_3 = \mu_3 - \mu'_3 \\ \mu_3 - \mu_3^o = RT\left[\ln v_3 + (1-v_3) - v_1\dfrac{x_3}{x_1} - v_2\dfrac{x_3}{x_2} + (\chi_{31}v_1 + \chi_{32}v_3)(v_1+v_2) - \chi_{12}\dfrac{x_3}{x_1}v_1v_2\right] \end{cases} \tag{4}$$

where x_1, x_2 and x_3 are the number of segments in components 1, 2 and 3. Equations (4) are obtained taking as generalized expression for entropy and heat of mixing in multicomponent system, Eqs. (5):

$$\begin{cases} \Delta S_m^* = -k\sum n_i \ln v_i = -k\left(n_1 \ln v_1 + \sum n_i \ln v_i\right) \\ \Delta H_m = z\sum x_i n_i v_i \Delta w_{ij} \\ \Delta G_m = \Delta H_m - T\Delta S_m^* \end{cases} \tag{5}$$

where ΔS^*_{m} is entropy of mixing calculated from external configurational considerations neglecting first-neighbour interactions, ΔH_{m}, the heat of mixing, z, number of primary molecules combined in a given nonlinear molecule, k, Boltzmann's constant, x_1, number of segments in species i, n_i, number of polymer molecules of species i in solution, Δw_{ij}, change of interaction free energy associated with contact formation between the pair of segments of molecules i and j. The optimization method was applied in our program for theoretical simulation of molecular associations and chemical reactions (AMYR) [29] and empirical conformational energy program for peptides vs. 2 (ECEPP2) [30].

RESULTS AND DISCUSSION

Binary Solvent–Polymer Mixture

In order to test the reliability of our numerical model, consider a simple case of a system of two components, consisting of solvent v_1 and homogeneous polymer v_2, characterized by parameter x representing the ratio of molar volumes of polymer and solvent. At first attempt, for simplicity, one assumes as governing equation, *i.e.*, OF, FH expression for chemical potentials of both components:

$$\begin{cases} \Delta\mu'_1 = \ln v'_1 + (1 - 1/x)v'_2 + \chi v'^2_2 \\ \Delta\mu'_2 = (1/x)\ln v'_1 + (1 - 1/x)v'_1 + \chi v'^2_1 \end{cases} \tag{6}$$

By equating the chemical potentials for the two coexisting phases for both components, one obtains an optimization problem in which, *e.g.*, v_2 is fixed and, at different values of χ, v'_2 is the sole optimization variable. It is possible to plot the binodal curves for binary system in the plane v_2–χ. The problem can be solved *via* standard numerical tools based on NR and simplified expressions; nevertheless, the example is meaningful because it (1) allows testing the reliability of the results obtained *via* our optimization algorithm and (2) shows that our approach does not depend on the expression chosen as equation of state, since it is the only required value. In A Real-Case Binary Example subsection, comparisons are presented between our algorithm and von Solms *et al.* NR approach, for real mixtures of technical interest, *e.g.*, PS–methyl*cyclo*hexane (MCH) for different molecular weights of PS, when simplified PC-SAFT equation of state is used instead of FH theory.

Symmetrical Case, Polymer (1)–Polymer (2)–Solvent Solutions

A symmetrical case for a ternary system of two polymers with identical characteristics in a solvent is discussed. Polymers are assumed with the following parameters: $x_2 = x_3 = 1000$

and $\chi_{12} = \chi_{13} = 0$. For the solvent, one takes $x_1 = 1$, whereas for polymers, three different interaction parameters $\chi_{23} = 4$, 8 and 10, respectively. Scott solved the problem by exploiting the symmetry information and reducing it, for a given concentration v_1, to a simple nonlinear equation in one unknown. In the general case, a system of three nonlinear equations in three unknowns should be tackled.

Liquid (1)–Liquid (2)–Polymer Solutions

The most common ternary system containing polymers is that of two nonpolymeric liquids and one polymeric species. A simple case of a ternary system of nonsolvent NS or L(1), solvent S or L(2), and polymer P is taken into consideration, in which the interaction constants between one of the liquids and the polymer equals that between the two liquids ($\chi_{12} = \chi_{13}$). For this ternary system, an iterative procedure was proposed by Tompa in the special case of $\chi_{23} = 0$ and $x_1 = x_2 = 1$, which may be used as a comparison with the general algorithm and that is summarized:

$$\begin{cases} 2\chi(v_1 - v'_1) - \ln\dfrac{v_1}{v'_1} = -\left(1 - \dfrac{1}{x_3}\right)(v_3 - v'_3) + \chi_{12}\left(v_1^2 - v'^2_1\right) \\ 2\chi(v_1 - v'_1) - \ln\dfrac{v_1}{v'_1} = \ln\dfrac{v'_2}{v_2} \\ 2\chi(v_1 - v'_1) - \ln\dfrac{v_1}{v'_1} = \dfrac{1}{x_3}\ln\dfrac{v'_3}{v_3} \end{cases} \tag{7}$$

For an arbitrarily chosen v_1, a trial value of v'_1 is assumed. From the first part of Eqs. (7), one obtains $v'_3 - v_3$, from the third part, v'_3/v_3 and, hence, v'_3 and v_3. The condition $\Sigma v_i = 1$ gives v'_2 and v_2. Their ratio can be compared with that calculated from the second part of Eqs. (7). The divergence is used as a guide for a better choice of v'_1. The comparisons show that the general procedure is able to reproduce, with good accuracy, previously presented nongeneral procedures.

A Real-Case Binary Example

The comparison between our algorithm and the *alternating tangents approach* [31] was reported, for the determination of binodal solutions in the polyisobutylene–diisopropyl ketone and PS–MCH systems, at different polymer molecular weights. Following this notation, the governing equations for phase equilibrium is formulated in general, which applies to the evaluation of binary/ternary L–L equilibria. Denoting by NC the total number of components, one assumes as independent variables the logarithmic mole fractions in phases 1 and 2, and logarithm of factors **K** that are defined as:

$$K_i = \frac{v_i}{v'_i} \tag{8}$$

The variables form the full set of independent variables for the ternary case that gives ninc (3·NC) variables. For the binary case, the logarithm of temperature is an additional variable: seven (3·NC+1) variables are obtained. At equilibrium, the fugacities of each component in the two phases are equal:

$$v_i\hat{\Phi}_i = v'_i\,\hat{\Phi}'_i \Leftrightarrow K_i = v_i/v'_i = \hat{\Phi}'_i/\hat{\Phi}_i \tag{9}$$

where $\hat{\Phi}_i$ and $\hat{\Phi}'_i$ are the fugacity coefficients in phases (1) and (2), respectively. Equation (9) together with the mass balances provides Eqs. (10) that must be satisfied at the equilibrium:

$$\begin{cases} \ln K_i - \ln\hat{\Phi}'_i + \ln\hat{\Phi}_i = 0 \quad (i = 1,\ldots,\mathrm{NC}) \\ \sum v'_i = 1 \\ \sum v_i = 1 \end{cases} \tag{10}$$

One independent parameter is fixed in order to solve Eqs. (10). Lindvig *et al.* suggested to add a *specification equation* in the form $y_S = S$, where **y** is the vector collecting all variables, *s*, the index of variable to be specified and *S*, fixed value of input. The total number of equations and variables is seven/nine (binary/ternary cases). For binary systems with the simplified PC-SAFT, it is abridged:

$$f = \ln v_1 + \ln\hat{\Phi}_1 - \ln(1-v_1) - \ln\hat{\Phi}_2 - \left[\ln v'_1 + \ln\hat{\Phi}'_1 - \ln(1-v'_1) - \ln\hat{\Phi}'_2\right] \tag{11}$$

Numerical values are obtainable from the simplified PC-SAFT equation of state [32–34]. The PC-SAFT parameters are gotten. The experimental data are taken from Shultz and Flory [35].

Compatibility between Polystyrene Co-Polymers and Polymers in Solution via Hydrogen Bonding

The ternary diagrams for the $CHCl_3$/PSMAA/PVPy system (*cf.* Figure 1) show the construction of the binodals, which required c_B and c_C calculation for every phase in equilibrium. The compositions of phases in equilibrium, expressed *vs.* ternary $\omega_A{}^T$ and binaryweight fraction ω_C, were represented in the ternary diagram as the intersection of a perpendicular to axis $\omega_A{}^T$ and the straight line that joins co-ordinate ω_C with the triangle upper vertex. The PS, PMAA, PVPy, PVPh, PVP and $CHCl_3$ specific volumes were taken as 0.917, 0.412, 0.820 [23], 0.862, 0.775 [36,37] and 0.674 $mL\cdot g^{-1}$ at 25ºC. The region of two phases is located below the binodal, and all compositions above are homogeneous systems. The behaviour of the system *vs.* the amount of MAA donor group in PS-co-polymer is illustrated. The upper isotherm corresponds to a ternary system with unmodified PS. PVPy displays a high incompatibility with PS. The greater the content of MAA in PSMAA, the greater the compatibility between PSMAA–PVPy in $CHCl_3$; however, the behaviour cannot be extended beyond 7.2% of MAA, because of complex co-acervation at 8 wt.% of MAA.

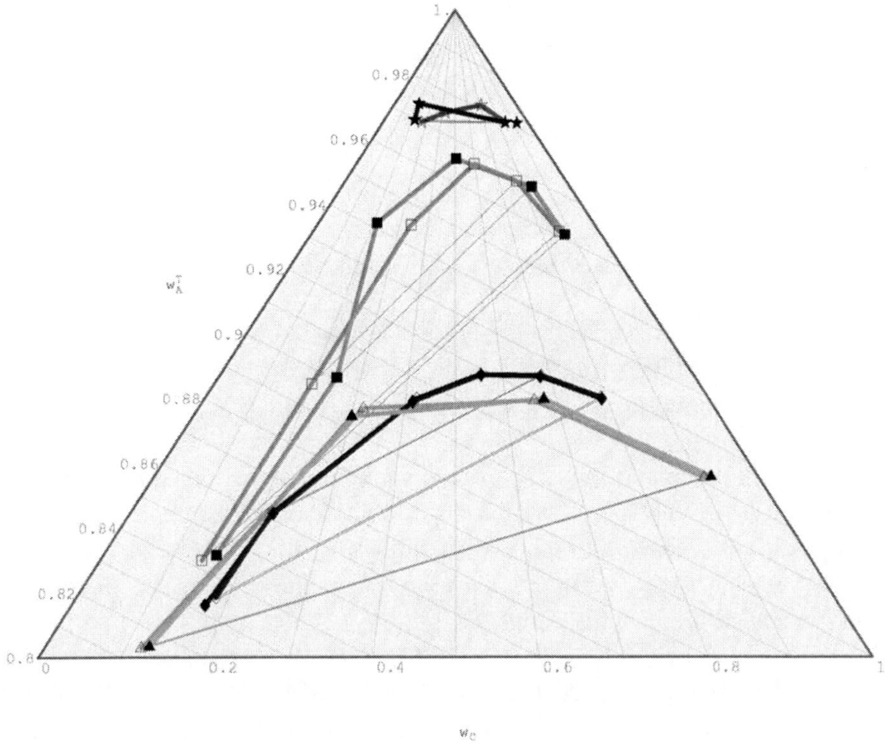

Figure 1. Comparison between binodal isotherms from SEC (empty symbols) and viscometry (filled symbols): $CHCl_3$/PS/PVPy (*), $CHCl_3$/PSMAA(3)/PVPy (□), $CHCl_3$/PSMAA(6)/PVPy (◇) and $CHCl_3$/PSMAA(7)/PVPy (△).

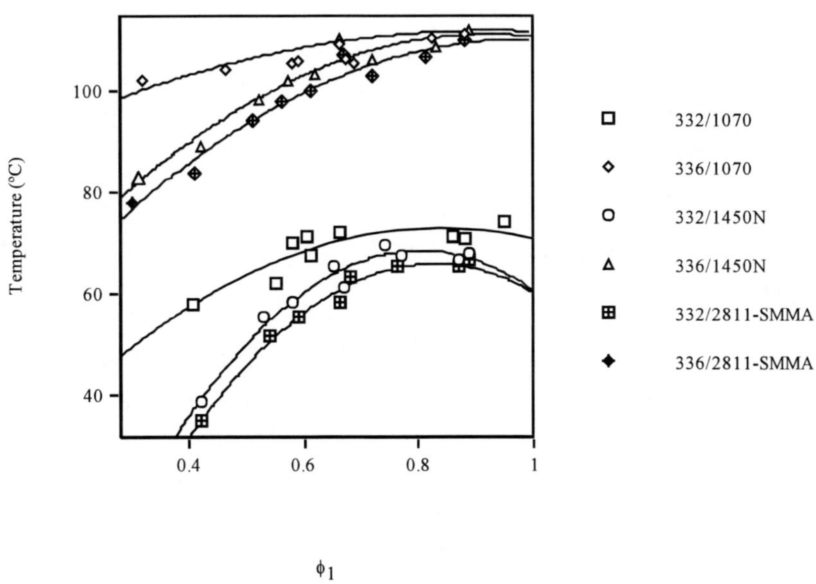

Figure 2. Phase diagrams for binary systems: (a) DER332/PS1070, (b) DER336/PS1070, (c) DER332/PS1450N, (d) DER336/PS1450N, (e) DER332/P2811–SMMA and (f) DER336/P2811–SMMA.

Parameter $\chi > 0$ of FH was lower for a given temperature, as temperature windows of phase separation *vs.* the volume fraction of compound-1, ϕ_1, in phase diagrams of systems (*cf.* Figure 2) were smaller. Values of $\chi_{c=0} > 0$ at a given temperature, in certain temperature ranges, were not lower as phase separation window in phase diagrams is smaller, which confirms that χ depends on T and ϕ_2.

INTERASSOCIATION CONSTANTS

Results were performed by FTIR on specific interactions, with H-bonds formation producing intermacromolecular complexes. In the $CHCl_3$/PSMAA/PVPy system, absorbance measurements of H-bond –COOH…N≤ band at frequency $\nu = 1\,607$ cm^{-1} were accomplished.

The measurements were performed by evaluating the fraction of complexed –OH, α, and the equilibrium constant. With spectra conforming to PSMAA–PVPy complex obtained in the $CHCl_3$/PSMAA/PVPy system with the concentrations c_o of PSMAA and c_1 of PVPy, absorbance was evaluated at 1 607 cm^{-1} as $d = c_o\alpha\varepsilon l$, being constants ε and l complex absorbance index and cell length, respectively.

With spectra corresponding to PSMAA–PVPy complex obtained in $CHCl_3$/PSMAA/PVPy system, with concentrations c_o of PSMAA and c_2 of PVPy, absorbance $d' = c_o\alpha'\varepsilon l$ was evaluated at 1 607 cm^{-1}. Ratio of absorbances of bands was $d/d' = \alpha/\alpha' = r$. Conditional constant of complex formation was:

$$K = \frac{c_o\alpha}{c_o(1-\alpha)(c_1 - c_o\alpha)} = \frac{c_o\alpha'}{c_o(1-\alpha')(c_2 - c_o\alpha')} \tag{12}$$

From $c_o = 1.2\times10^{-3}$, $c_1 = 1.0\times10^{-3}$, $c_2 = 1.2\times10^{-3}$ mol·L^{-1} and $r = 0.885$, $\alpha = 0.524$ was obtained. Bands of PSMAA (≈1 750 cm^{-1}) were not used because of proximity of complex band. Association constant $K = 2.96$ L·mmol^{-1}. For $CHCl_3$/PSVPh/PVPy (*cf.* Figure 3) or $CHCl_3$/PSVPh/PVP, the measurements of the intensity of the PSVPh –OH band were performed at 3 600 cm^{-1} in pure solvent or in the presence of PVPy/PVP. The measurements allowed the percentage of complexed –OH and association constant. The $CHCl_3$/PSVPh(4) spectra at concentration $c_o = 1.2\times10^{-3}$ mol·L^{-1}, and $CHCl_3$/PSVPh(4)/PVPy with equal PSVPh c_o and at PVPy concentration $c_1 = 1.0\times10^{-3}$ mol·L^{-1} were analyzed. The band absorbances were $d_o = c_o\varepsilon l$ and $d_1 = (c_o - c_o\alpha)\varepsilon l$, being α fraction of complexed –Ph–OH. Ratio of absorbances was $d_1/d_o = 1 - \alpha$. Conditional constant of complex formation was:

$$K = \frac{c_o\alpha}{c_o(1-\alpha)(c_1 - c_o\alpha)} \tag{13}$$

From $d_1/d_o \approx 0.568$, it resulted $\alpha \approx 0.432$ and $K \approx 1.58$ L·mmol^{-1}.

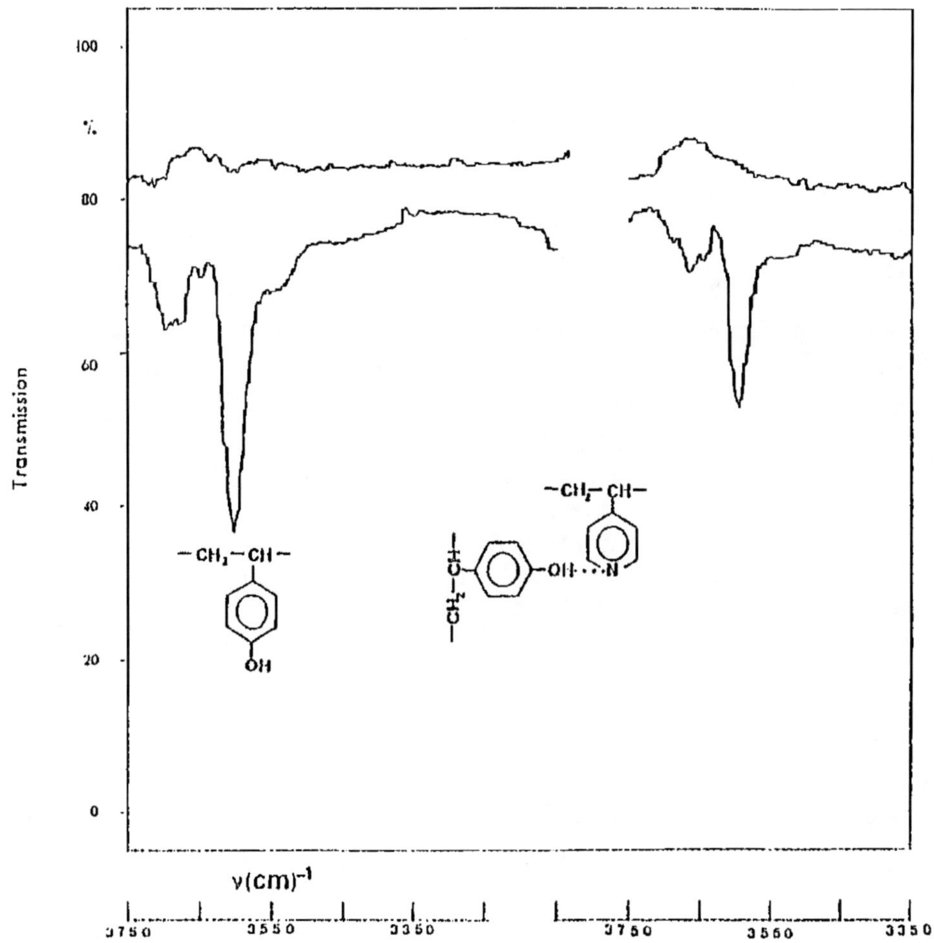

Figure 3. FTIR spectrum with the formation of H-bonds producing intermolecular complexes.

CONCLUSION

From the present results and discussion the following conclusions can be drawn.

1. The optimization algorithm for the numerical evaluation of binodal curves was presented in ternary systems: liquid-1 (L1)–L2–polymer (P) and P1–P2–solvent. It exploits SIMPLEX minimization. The objective function represented by weighted sum of square differences of chemical potentials of the two phases of every component, obtained by evaluating first derivatives of Gibbs free energy of mixture with respect to moles of components. Equilibrium conditions between the two phases of equal component were derived *via*Flory–Huggins theory and simplified PC-SAFT. Reliability of our approach was assessed *via* comparing with simple cases reported in the literature.
2. Viscometric changes intervened during formation of the interpolymer complex, based on PSVPh–PSMAA/PVPy–PVP chains. The changes correlated with the

experimental conditions. The interpolymer complex formation was confirmed *via* infrared and chromatography, highlighting the compatibility between homopolymers with the formation of intermolecular interactions, especially H-bonds. The infrared experiment results correlated well with the viscosity data.

3. Binodal isotherms allowed the degree of compatibility of mixed co–homopolymers *vs*. polar-group concentration. MAA incorporation in PS caused miscibility rise with PVPy. At higher concentrations, co-acervation produced phase separation in polymer-diluted zones. PSVPh/PVPy achieved maximum miscibility. A rise in polar-group concentration did not rise miscibility, because of the competition between the rise in specific interactions and the decay in PSVPh solubility. For PSMAA/PSVPh with PVP, the miscibility was greater than above. In viscometry, polymers interactions parameter varied in concordance with miscibility change in ternary diagrams, rising with compatibility. Infrared permitted conditional constant of complex formation in solution, assuming A_1:B_1; for PSMAA/PVPy, constant was twice than PSVPh/PVPy because MAA OH adopted more configurations than Ph–OH, facilitating H-bond; in PSVPh/PVP and EPh/EP for molecules, constant was greater than for polymers; in PSVPh/PVPy and EPh/EP for molecules, the constant was higher.

ACKNOWLEDGMENTS

The authors want to dedicate this manuscript to Prof. Dr. Agustín Campos, who was greatly interested in this research and would have loved to see its conclusion. F. T. thanks support from the Spanish Ministerio de Ciencia e Innovación (Project No. BFU2010–19118).

REFERENCES

[1] Sollich, P.; Warren, P. B.; Cates, M. E. *Adv. Chem. Phys*. 2001, 116, 265–336.
[2] Sollich, P. *J. Phys.: Condens. Matter* 2002, 14, R79–R117.
[3] Von Solms, N.; Kouskoumvekaki, I. A.; Lindvig, T.; Michelsen, M. L.; Kontogeorgis, G. M. *Fluid Phase Equilib*. 2004, 222/223, 87–93.
[4] Lindvig, T.; Michelsen, M. L.; Kontogeorgis, G. M. *Fluid Phase Equilib*. 2002, 203, 247–260.
[5] Favre, E.; Nguyen, Q. T.; Clement, R.; Neel, J. *Eur. Polym. J*. 1996, 32, 303–309.
[6] Flory, P. J. *Principles of Polymer Chemistry*; Cornell University Press: Ithaca, NY, 1953.
[7] Flory, P. J. *J. Chem. Phys*. 1941, 9, 660–661.
[8] Flory, P. J. *J. Chem. Phys*. 1942, 10, 51–61.
[9] Huggins, M. L. *J. Chem. Phys*. 1941, 9, 440–440.
[10] Huggins, M. L. *Ann. N.Y. Acad. Sci*. 1942, 43, 1–32.
[11] Huggins, M. L. *J. Am. Chem. Soc*. 1942, *64*, 1712–1719.
[12] Emilianenko, M.; Lin, Z.-K.; Du, Q. *Comput. Mater. Sci*. 2006, 35, 61–74.
[13] Tompa, H. *Trans. Faraday Soc*. 1949, 45, 1142–1152.
[14] Bramford, C. H.; Tompa, H. *Trans. Faraday Soc*. 1950, 46, 310–316.

[15] Tompa, H. *Polymer Solutions*; Butterworths Scientific Publications: London, 1956.
[16] Milani G.; Milani, F. *J. Comput. Chem.* 2007, 28, 2203–2215.
[17] Torrens, F.; Soria, V.; Monzó, I. S.; Abad, C.; Campos, A. *J. Appl. Polym. Sci.* 2006, 102, 5039–5049.
[18] Torrens, F.; Soria, V.; Codoñer, A.; Abad, C.; Campos, A. *Eur. Polym. J.* 2006, 42, 2807–2823.
[19] Torrens, F.; Monzó, I. S.; Gómez-Clarí, C. M.; Abad, C.; Campos, A. *Polym. Compos.* 2008, 29, 1337–1345.
[20] García-Lopera, R.; Figueruelo, J. E.; Porcar I.; Campos A.; Abad C. *J. Liq. Chromatogr. Relat. Technol.* 2005; 28:245–266.
[21] Gómez, C. M.; Verdejo, E.; Figueruelo, J. E.; Campos, A.; Soria, V. *Polymer* 1995, 36, 1487–1498.
[22] Campos, A.; Gómez, C. M.; García, R.; Figueruelo, J. E.; Soria, V. *Polymer* 1996, 37, 3361–3372.
[23] Gómez, C. M.; Figueruelo, J. E.; Campos, A. *Polymer* 1998, 39, 4023–4032.
[24] Campos, A.; Gavara, R.; Tejero, R.; Gómez, C.; Celda, B. *J. Polym. Sci., Part B: Polym. Phys.* 1989, 27, 1599–1610.
[25] Young, T.-H.; Chuang, W.-Y. *J. Membr. Sci.* 2002, 210, 349–359.
[26] Pouchly, J.; Zivny, A.; Solc, K. *J. Polym. Sci, Part C* 1968, 23, 245–256.
[27] Scott, R. L. *J. Chem. Phys.* 1949, 17, 268–279.
[28] Scott, R. L. *J. Chem. Phys.* 1949, 17, 279–284.
[29] Torrens, F.; Rubio, M.; Sánchez-Marín, J. *Comput. Phys. Commun.* 1998, 115, 87–89.
[30] Torrens, F.; Castellano, G. *J. Argent. Chem. Soc.* 2006, 94(1-3), 27–47.
[31] Lindvig, T.; Michelsen, M. L.; Kontogeorgis, G. M. *Ind. Eng. Chem. Res.* 2004, 43, 1125–1132.
[32] Gross, J.; Sadowski, G. *Ind. Eng. Chem. Res.* 2001, 40, 1244–1260.
[33] Gross, J.; Sadowski, G. *Ind. Eng. Chem. Res.* 2002, 41, 1084–1093.
[34] Gross, J.; Sadowski, G. *Ind. Eng. Chem. Res.* 2002, 41, 5510–5515.
[35] Shultz, A. R.; Flory, P. J. *J. Am. Chem. Soc.* 1952, 74, 4760–4767.
[36] Elias, V.H.-G. *Makromol. Chem.* 1961, 50, 1–19.
[37] Chaufer, B.; Sebille, B.; Quivoron, C. *Eur. Polym. J.* 1975, 11, 683–688.

In: New Developments in Polymer Composites Research
Editors: S. Laske and A. Witschnigg
ISBN: 978-1-62948-340-5

Chapter 10

INFLUENCE OF CHEMICAL STRESS ON SULFUR-POLYMER COMPOSITE STRUCTURE

Milica M. Vlahović[1,*], Predrag B. Jovanić[2], Sanja P. Martinović[1], Tamara Đ. Boljanac[1] and Tatjana D. Volkov Husović[3]

[1]Institute for Technology of Nuclear and Other Mineral Raw Materials, Belgrade, Serbia
[2]University of Belgrade, Institute for Multidisciplinary Research, Belgrade, Serbia
[3]University of Belgrade, Faculty of Technology and Metallurgy, Belgrade, Serbia

ABSTRACT

Sulfur- polymer composites are high performance thermoplastic materials made of mineral aggregate, filler, and modified sulfur binder, instead of cement and water as in ordinary Portland cement- based composites at temperatures above the hardening point of sulfur (120ºC). Using sulfur to obtain modified sulfur binder is based on its physico-chemical characteristics. According to our terminology, modified sulfur binder considers a mixture of elemental sulfur and modified sulfur- sulfur polymer. Contemporary experience all over the world shows that composite materials with modified sulfur binder instead of cement and water have significant chemical and physico- mechanical advantages comparing with Portland cement- based composites.

Modern trends of obtaining materials with desired properties are based on combining ingredients with different properties in different proportions, as well as on the application of various manufacturing procedures and additional material processing.

Having in mind that while in service use, all materials are exposed to divers external influences that provoke some type of response, the idea of this research was to change the quality of sulfur- polymer composite by imposed chemical stress. Since sulfur- polymer composites are relatively new building materials that can potentially replace conventional material made with Portland cement as a binder in many branches of construction, it was found plausible to choose Portland cement- based composite as a referent material. In order to quantify the changes in the material structure as its response to the specific imposed stimulus, the scanning electron microscopy (SEM) and *energy dispersive X-ray analysis (EDX) were applied.*

* E-mail address: milica.vlahovic@yahoo.com (Corresponding author)

According to the applied analyses on micro level, it can be concluded that the initial structure of sulfur- polymer composite was rearranged. As a macro level result material with better resistance to chemically imposed stress compared with the referent material was obtained.

Keywords: sulfur- polymer composite, Portland cement-based composite, scanning electron microscopy (SEM), energy dispersive X-ray analysis (EDX)

1. Introduction

The rapidly expanding world economies have resulted in a sharp increase in energy demand. Much of this energy demand is projected to come from oil, which is increasingly becoming more "sour" or greater in sulfur content. At the same time the world's regulatory agencies are enacting increasingly stringent environmental regulations. Not only that these regulations are applied to emissions from the refining complex itself, but on the products produced by the refinery as well. These regulatory and market dynamics have created a challenging environment for a refiner to operate in, especially with respect to recovery and disposition of sulfur (sulfur management).

Figure 1 illustrates the net result of this crude quality decline trend. As shown, the crude quality will increasingly become heavier and higher in sulfur content.

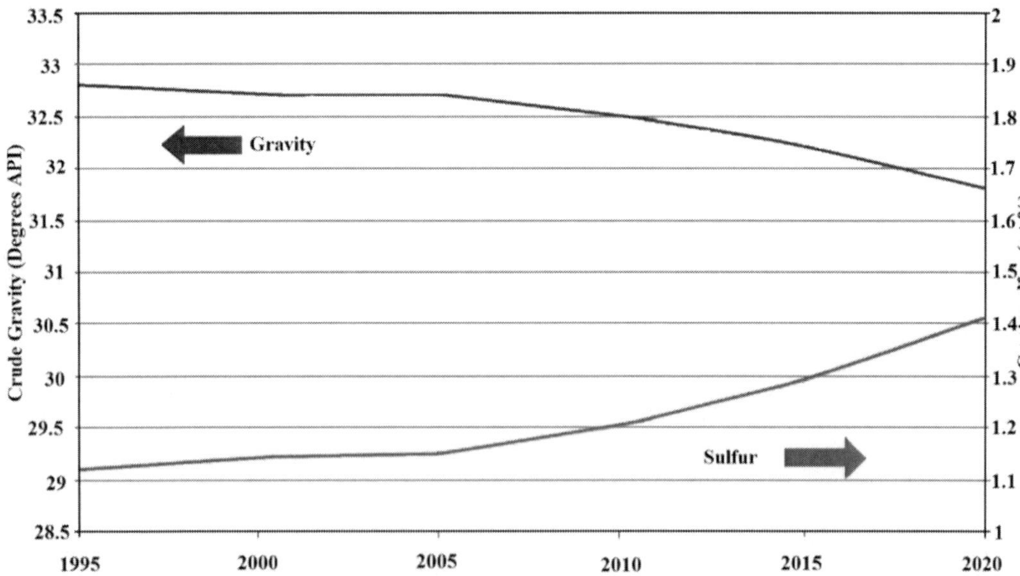

Figure 1. Gradual Global Trend toward Heavier and Higher Sulfur Crude [1].

Modern science and applied research in recent years have been facing with the expansion of new materials as a result of introducing nanotechnology into existing processes and materials. In general, by developing processing equipment, it is possible to modify structures of the starting materials on a very small scale (nano-, ato-) [2].

A wide range of materials that had a great impact in the development of practical applications of materials were polymers. Organic polymers, which played a key role as undeniably remarkable and useful materials, have left to the world waste called plastic. It is interesting that polymers have recently become far more important as carriers of catalysts, especially for waste treatment processes (catalytic depolymerization). Not less important in practice, but always left aside, were polymeric materials based on inorganic components. From the perspective of nowadays-used technologies and process solutions, these polymers can play an important role in various technological processes. The difficulty lies in the fact that their production on an industrial scale is still problematic. However, inorganic polymers, in combination with other materials can generate composite materials of very high quality and possibilities of applications [3].

Beside plenty of inorganic materials that can in some circumstances generate polymer chains, special attention should be paid to sulfur. Sulfur is an element that was being removed from the atmosphere by slow natural processes during eons. In these processes, sulfur was bound to metals giving ores. The other natural cycle of sulfur removal, by binding to organic materials through food chains, led to its storage in crude oil. The modern development of industrial society over the last 200 years, the so-called technological revolution, has in fact reversed those natural processes towards restoring sulfur to the environment by exploitation of minerals and crude oil. Now, in addition to the obvious problem of carbon dioxide, with all of the effects, we have a serious, but quite a neglected problem, with the treatment of produced quantities of the waste sulfur. Analyzing the global trends of environmental protection, the focus is still on the methodology of solving the problem of carbon dioxide and greenhouse effects. Different models have been proposed, but some of them, for example a model of CDM credits, are just passing the problem geographically around the world, without its efficient resolution at the source.

Sulfur is the 16th most abundant element in nature. It is found in the earth's crust, in the ocean and even in meteorites. Sulfur occurs naturally all over the world and it is the most prolific where sulfur-rich gas and oil is processed and refined. Canada is the biggest exporter and China is the biggest importer of sulfur [4].

The waste sulfur from the ore processing was converted into additional product. The best-known process of removing waste sulfur from the extraction of metals from ores, is converting gaseous sulfur oxides into sulfuric acid and its further treatment towards production of artificial fertilizers. Nevertheless, well known effect of sulfur rain on vegetation in areas of large systems for exploitation of minerals still exists. Modern industrial production still requires further development of technologies that would provide incorporating waste sulfur into useful products, thus removing it from the environment, like sulfur binding to metals in order to give ores in the beginning of the sulfur cycle.

Production of sulfur in all forms is schematically presented in Figure 2.

In the last decades the availability of sulfur has considerably grown in many countries. This is mainly due to the current environmental restrictions regarding the petroleum and gas refining processes, which limit the maximum quantity of sulfur present at combustibles. Unlike treatment of carbon dioxide, sulfur from oil processing was briquetted and stored leading to more or less controlled disposal. Extremely large quantities of sulfur are thus obtained as a by-product of these processes, together with coal processing and refining of copper in the mining sector.

Elemental sulfur production and demand is schematically presented in Figure 3.

Production and end use of sulfur is schematically presented in Figure 4.

The development of new applications for sulfur becomes fundamental.

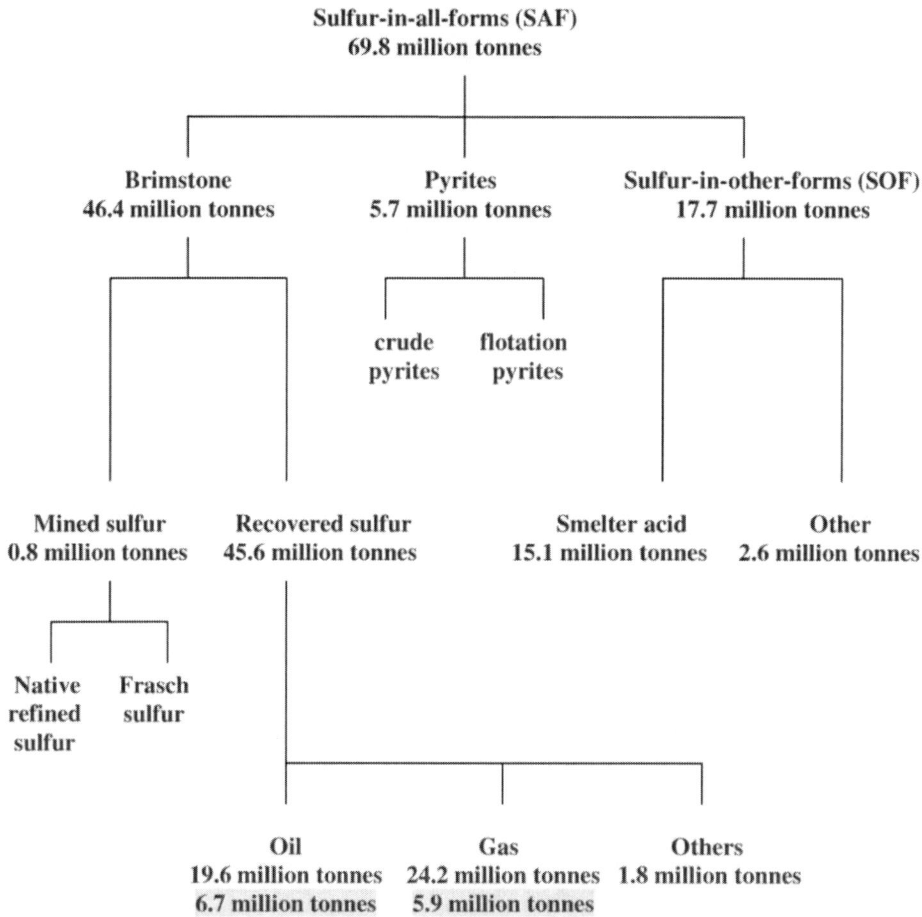

Figure 2. Sulfur-in-All-Forms Production.

Building materials certainly represent media that should be examined as potential acceptors for large quantities of wastes from various sources. The fact is that wide range of hazardous waste can be inerted by their incorporating into usable building materials.

During the 1960s, there was a remarkable investment in environmental protection against discharge of sulfur into the atmosphere, thus making sulfur a surplus commodity on the market, particularly in the United States and Canada. This was a crucial point, that made the interest for usage of sulfur as a structural binder to grow further, and initiated extensive research programs which became very active in the 1970s and focusing on various properties of the material, including durability [5].

On the other side, Portland cement may be partially or completely substituted with other binders such as modified heated sulfur to form a stable, hard concrete product. The baseline condition is defined as the production of concrete using traditional cementitious binders derived from limestone and clay that rely on the chemical bonds formed upon contact with water to bind together aggregate material (sand and rock). This binder ("clinker") is a key component of Portland cement. Sulfur substitutes avoid the process and combustion emissions associated with the manufacture of Portland cement. In the case of a modified sulfur alternative, the sulfur itself is generated as a by-product of natural gas processing and petroleum refining. Unlike concrete made from Portland cement (which can be cold mixed), concrete made with modified heated sulfur needs to be heated during production. Although modified sulfur, together with aggregates, needs to be mixed at 135°C when making sulfur concrete, this process uses less energy than conventional cold concrete mixing processes because it avoids the energy needed to heat limestone to 1450°C to manufacture Portland cement [4,6].

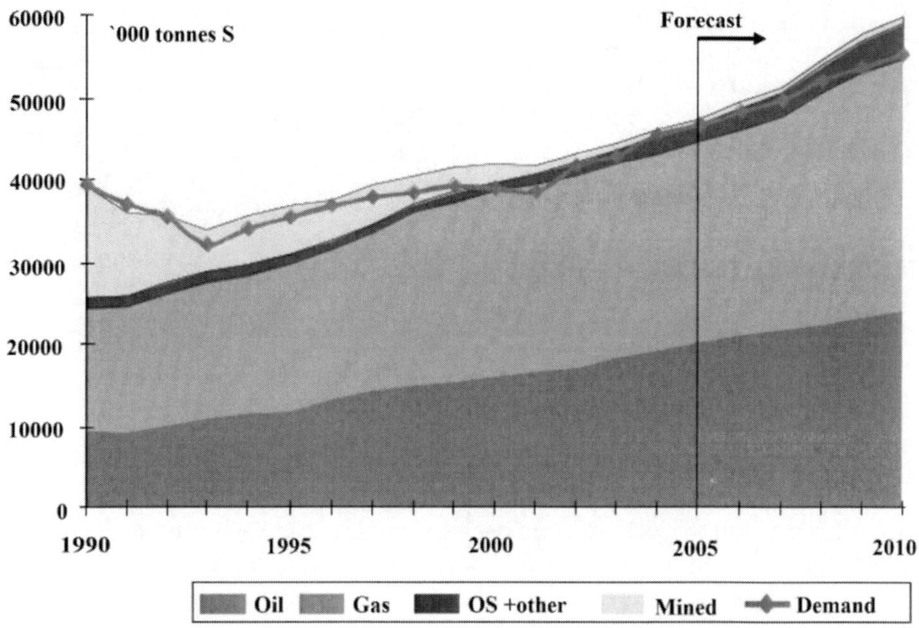

Figure 3. World Elemental Sulfur Production and Demand.

Despite the need to be hot mixed (with heat likely obtained from the combustion of fossil fuels), concrete and cement products made with modified heated sulfur releases far fewer GHG emissions than concrete made with Portland cement. Modified heated sulfur avoids the process emissions released in the calcination process of clinker production as well as the combustion emissions typically generated to supply heat to the calcination process [7].

Due to a relatively low melting point for sulfur, the maximum temperature in sulfur concrete product applications is generally 113°C. Sulfur concrete products maintain high strength properties, allowing their use in a wide variety of pre-cast applications such as traffic barriers, drainage tiles, paving stones, and marine defenses.

Based on literature review and other reliable sources, it is evident that there will be a continuous abundant supply of sulfur in the future due to strict global environmental regulations [5]. The report from the World Business Council for Sustainable Development report, Cement Industry Energy and CO_2 Performance "Getting the Numbers Right" [8], clearly indicates that global production of sulfur will continue to increase, thus assuring its continued availability.

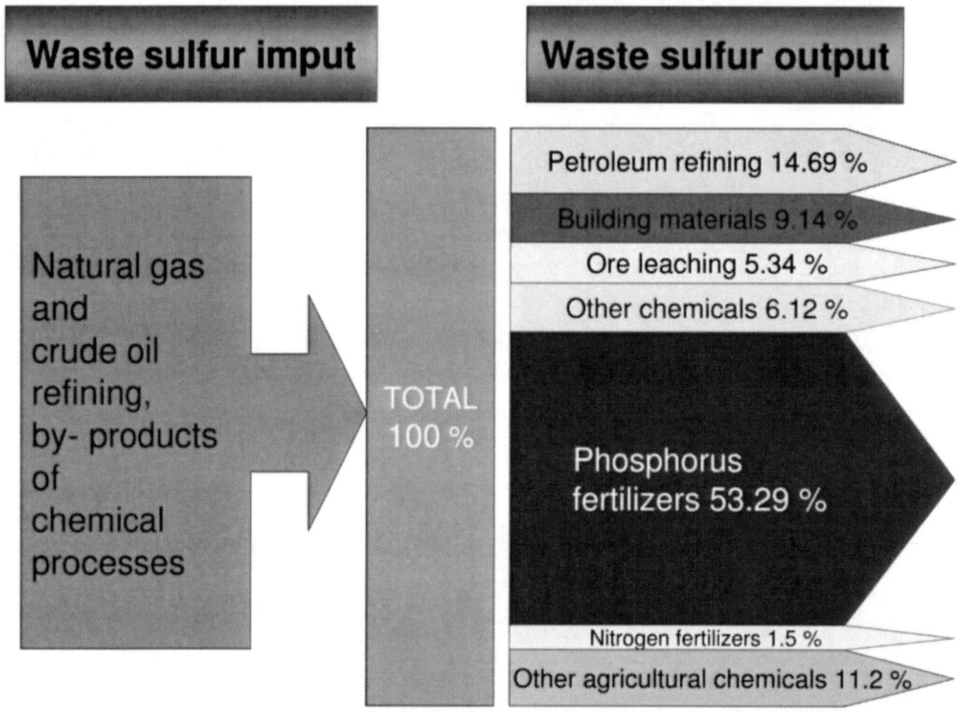

Figure 4. Production and end use of sulfur.

Construction materials such as sulfur concrete and sulfur modified asphalt continue to receive more attention since they are environmentally friendly and cost-effective. Beginning in the 1970s, successful projects in which sulfur concrete has been used as a construction material have been carried out in different levels. Most of these took place mainly in North America. Few and mainly research or test pilot projects have been carried out in Europe, with only Denmark conducting commercial or industrial activities on sulfur concrete.

It is, therefore, not surprising that such environmental demands and the surplus sulfur draw more attention for the use of sulfur concrete as a construction material. Research activities on sulfur concrete as a construction material are currently reported to be going on in Spain, Italy, and The Netherlands, while interest on medium scale industrial uses of sulfur concrete plans are underway in Poland. In South East Asia, preliminary reports are indicating a growing interest in the use of sulfur concrete as a result of the increase of surplus sulfur from refinery industry. Similar reports come from the Middle East (Saudi Arabia), South America (Chile) and Africa (South Africa), where attempts to use sulfur concrete were reported earlier, or are in progress [5].

The use of sulfur to produce sulfur- polymer composites- sulfur concrete, sulfur mortar and sulfur- modified asphalt is a relatively new technology that has to be proven in practice. The fact is that these materials are already in use, but refining their application quality is still certainly needed. Sulfur concrete has a relatively simple composition and manufacture, and very interesting characteristics and properties. The recent research programs on sulfur concrete have been continuing to study different properties related to the material performance, most of which are reported to be excellent.Its extremely high corrosion resistance, mechanical strength and fast hardening, make it a high performance material suitable for several applications, especially the ones in which other materials, fail [5]. Furthermore, sulfur- polymer composites can be used for solidification and encapsulation of different waste materials (fly ash, cement kiln dust, phosphogypsum, mercury) into a sulfur-polymer matrix, thus obtaining the sustainable development of construction and industrial sectors and for corrosion protection of reinforcing steel and concrete [9-13].Sulfur- polymer composite should also find widespread application in chemical and fertilizer replacement for materials in acid and salt environments as well as in metallurgical operations since they fail in acid and salt environments. Sulfur- polymer composite coated fertilizers were developed to provide better impact strength and controlled-release properties compared to sulfur- coated fertilizers.

2. Experimental

The project scenario is defined as the production of sulfur-polymer composite specimens using modified sulfur binder. The reference specimens were prepared using Portland cement binder. The main objective of this research was to develop a sulfur-polymer composite with simple and clear formulation, using modified sulfur binder, obtained from secondary sulfur from oil refining process, and having adequate physical, mechanical and environmental properties for its future utilization. Other interests were to understand the physico-chemical behavior of sulfur-polymer composite by analyzing its structure at microscale and to propose several applications for this material, especially those where it could offer a great potential or extend the current range of uses.

2.1. Materials

The initial components for technological procedure for sulfur- polymer composite obtaining in this research were: sulfur, modified sulfur, aggregate and filler. The same type of aggregate, cement and water were used for Portland cement composite production.

2.1.1. Sulfur and Modified Sulfur

At first, the elementary sulfur was used as a binder in production of sulfur- polymer composite materials. However, despite excellent mechanical properties after preparation, the samples exhibited low stability, so spalling and failure occurred after a short period [5]. The development of modified sulfur binder contributed to better endurance of sulfur- polymer composites, which led to its primary use for road construction and repairing, and as building materials [14-17].

Sulfur, the basic component for a modified sulfur binder, originated from the oil refining process by Claus's procedure in the Oil Refinery Pancevo, and its purity was 99.9%. Modification of sulfur was carried out by polymerization with cyclic hydrocarbon polymer compound, dicyclopentadiene (DCPD).The procedure performed according to the patent [18-21] consisted of mixing DCPD and melted sulfur at a temperature ranging from 120 to 140 °C and ambient pressure for 30 min, followed by rapid cooling and solidification of the obtained sulfur polymer.

Because of the importance of preparing the sulfur- polymer composite with modified sulfur, both elementary and the obtained modified sulfur were investigated by scanning electron microscope (SEM), type JEOL JSM-5800 with EDX, Figure 5, and their microstructures were analyzed according to the literature [22]. The results showed that the elementary sulfur was composed of dense orthorhombic crystals of alpha form (S_α), Figure 5 (a), while modified sulfur consisted of plate monoclinic crystals of beta form (S_β), partially polymerized in zigzag chains, Figure 5 (b). Therefore, it was proved that modification of sulfur was achieved.

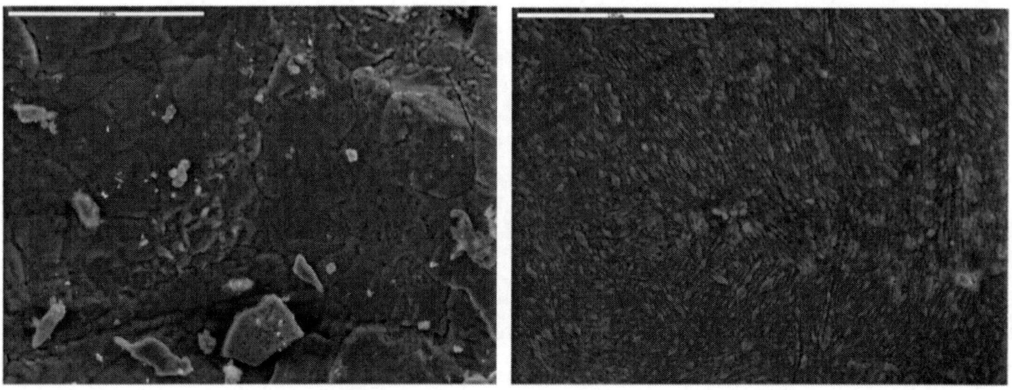

(a) (b)

Figure 5. Structure of sulfur: (a) elementary; (b) modified.

2.1.2. Aggregate

Sand with maximum grain size of 2 mm, obtained by sieving the locally available classical building mixture of sand and gravel, was used as an aggregate. Chemical analysis indicated that the aggregate mainly consisted of oxides of silica (89.98 wt.%), aluminium (3.61 wt.%), calcium (0.84 wt.%), iron (0.62 wt.%), potassium (0.59 wt.%), sodium (0.57 wt.%), and magnesium (0.19 wt.%).

2.1.3. Filler

The filler used in this production was alumina (Almatis, Germany).

The choice of filler is important because it forms with sulfur paste that coats and binds the aggregate particles. The main functions of filler are to [23]: control the viscosity of the sulfur- filler paste and to provide nucleation sites for crystal formation and growth, fill the voids in the aggregate and, hence, reduce the required amount of sulfur, reduce shrinkage during hardening, reinforce the matrix and therefore increase the strength.

The chemical composition and density of the applied filler is given in Table 1.

2.1.4. Cement

Ordinary Portland cement (65.5 wt.% CaO, 21.9 wt.% SiO_2, 5.4 wt.% Al_2O_3, 2.2 wt.% Fe_2O_3, 1.1 wt.% K_2O, 0.1 wt.% Na_2O, 1.3 wt.% MgO, and 3.9 wt.% SO_3) was used in this study.

Table 1. Chemical composition and density of alumina

Chemical composition (wt. %)										Density,
SiO_2	Al_2O_3	Fe_2O_3	CaO	MgO	Na_2O	K_2O	TiO_2	P_2O_5	SO_3	ρ (g/cm^3)
0.15	99.3	0.02	0.03	0.004	0.40	0.03	0.02	-	-	3.86

2.1.5. Samples Preparation and Characterization

The sulfur- polymer composite was prepared according to the manufacturing technological procedure described in the patent [24]. The process consisted of mixing both melted sulfur and modified sulfur into preheated and homogenized dry mixture of aggregate and filler at sulfur melting temperature, 132-141 °C. The content of filler was chosen in accordance with its influence on rheology, and in order to achieve the highest density of the mixture. The heated aggregate and filler were then properly mixed with the molten modified sulfur binder until a homogenous viscous mixture was obtained. The temperature was controlled because the viscosity of the mixture must allow its workability and handling while at the same time avoiding the thermal degradation of the thermoplastic material. After homogenization via 2 minutes of mixing, the concrete mixture was casted into 120 ºC preheated molds and compacted with a vibrating table for 10 seconds in order to settle the products in the molds.

After vibrating, any extra material was removed to get a well-finish at top surface and then allowed to harden inside the molds, at room temperature. A very important characteristic of sulfur- polymer composite is rapid hardening (from 15 minutes to several hours, depending on size and shape of the sample) that allows removal from the mold and curing in a relatively short period of time (only 24 h at room temperature). After 3 hours, samples were removed from the mold and cured at room temperature for 24 hours. The mechanical properties were measured after 75 hours.

The Portland cement composite mixtures, made with the same aggregate, Portland cement and water to cement ratio w/c of 0.5, were mixed in a Hobart mixer. The overall mixing time was about 6 min. The mixture was poured into molds and shaken on a vibration table. The specimens were demolded 24 h after casting and then cured in a moist room at a temperature of 20 ± 2 °C with 95–98% relative humidity for 27 additional days, before being subjected to the tests.

For investigating physico-mechanical characteristics of the composite materials and further testing, prism samples with dimensions (4 x 4 x 16cm) were prepared.

Mechanical strength (compressive and flexural) of the composite samples was conducted using the "Amsler" press with maximum load of 200 kN, and method for testing the strength of concrete according to the standard [25]. Each reported value is the average of three values obtained from three different samples, to ensure the reliability of the test results.

The composition and relevant starting mechanical properties of the prepared reference samples are shown in Table 2.

The proper size distribution of the aggregate and filler, and the right choice of the manufacturing technological process yielded sulfur- polymer composite particulate material with extremely low porosity, a tight microstructure and high compression and flexural strength.

Table 2. Chemical composition and mechanical properties of reference sulfur- polymer composite samples after 3 days of curing and Portland cement composite samples after 28 days of curing

Sample	Chemical composition (wt.%)				ρ_B (g/cm^3)	Mechanical strength (MPa)	
	Aggregate	Binder	Filler	Water		σ_c	σ_f
SPC	63	30	7	-	2.34	49.2	8.4
PCC	63	24	-	13	2.18	46.3	6.9

SPC = sulfur- polymer composite, PCC = Portland cement composite,
ρ_B = bulk density, σ_c = compressive strength, σ_f = flexural strength

2.2. Methods

2.2.1. Immersion Test

The durability of a composite structure is closely associated to the permeability of its surface layer, which is responsible for limiting the entry of substances that can initiate or propagate possible deleterious actions (acids, alkalis, salts, water, oxygen, etc.).

In this research, five series of sulfur- polymer composite samples were subjected to varying degrees of chemical stress by exposing all specimens to the same corrosion environment. The different degrees of chemical stress were induced by immersing the specimens with dimensions of (4 x 4 x 16cm) in 10% HCl solution for 7, 14, 21, 60 and 180 days.

At the same time, Portland cement composite samples with identical dimensions were subjected to the same aggressive environment. Since serious deterioration was noticed on those samples, immersion tests were stopped after 21 days.

2.2.2. Microstructure Analysis

In order to determine behavior of sulfur- polymer composite and sulfur as a binder, structural and chemical analyses using a scanning electron microscope were performed.

The results of previously reported studies indicated that the investigated materials reacted completely differently in the presence of a chemical agent [26-32]. In the case of the reference material, Portland cement composite, a total destruction after 21 days was noticed, and therefore it was not possible to continue with the treatment, unlike sulfur- polymer composite that developed specific "defense mechanism" to the influence of a chemical agent.

Bearing in mind that the testing of materials is related to the definition and quantification of the structure and homogeneity of the material, further investigation of the structure with increased resolution, in this case using a scanning electron microscope, was a logical step. Beside high resolution, the advantage of a scanning electron microscope is the possibility of using a so-called deep sharpness, which means that a specific specimen preparation is not required.

Generally speaking, images obtained by the SE detector are used to determine the morphological structure of the specimen. Image generation by the BSE detector can give information of the ingredients in the specimen. This kind of analysis needs additional sample preparation. The two principles of scanning electron microscopy are schematically presented in Figure 6.

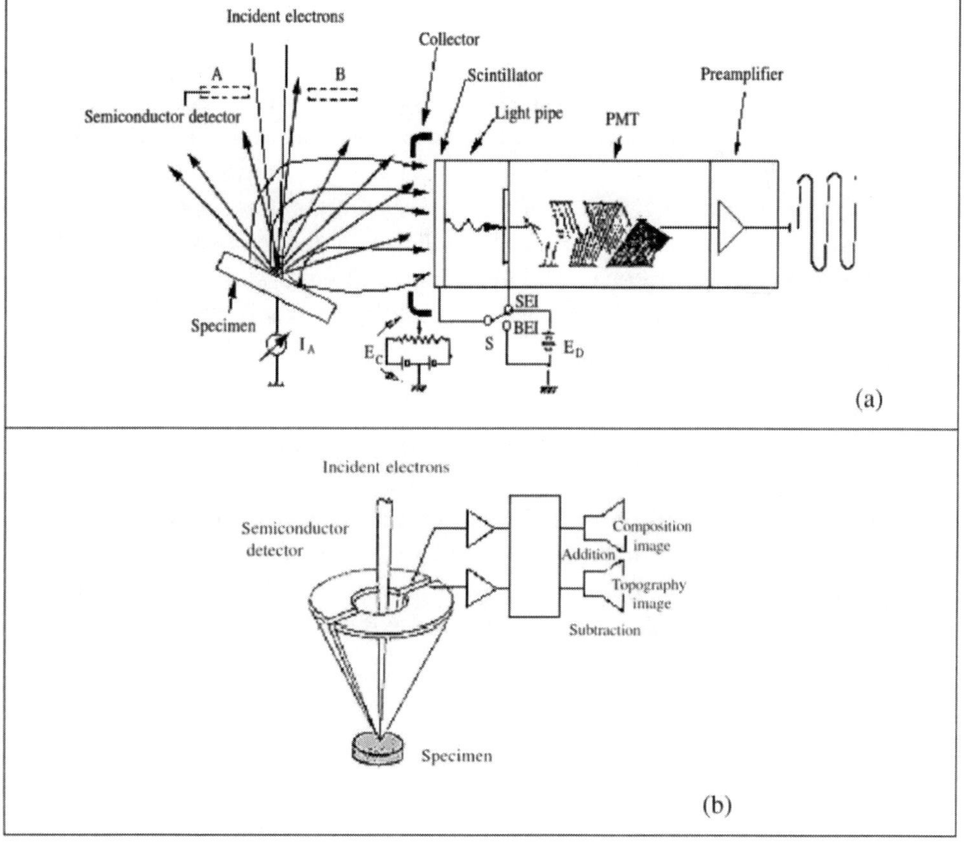

Figure 6. Information collection in scanning electron microscope by various detectors, (a) secondary electrons, SE; (b) back- scattered electrons, BSE.

The scanning electron microscope provides the measurement of characteristic X-rays generated by interactions of the primary electron beam and the investigated sample. Energy dispersive X-ray spectroscopy (EDX) is based on measuring the number of generated x-photons and their energies, to obtain so-called EDX spectrum. Based on this spectrum, it is possible to identify elements that are found in the examined structure. In practice, EDX analysis for determining elemental composition of individual points or small areas is rarely used unlike other options such as line analysis and especially mapping, which is more important for investigating homogeneity of materials. Mapping is a special feature of the scanning electron microscope that provides specimen observation only in an energy range that refers to the chosen element thus offering the surface distribution of that element.

The following possibilities provided by scanning electron microscope were used:

- Analysis of images generated by secondary electrons (SE), which gives the morphology of the specimen,

- Analysis of images generated by back-scattered electrons (BSE), which offers information on the morphological characteristics of structure components, and
- Energy dispersive X-ray analysis (EDX), which enables elemental chemical analysis of selected fields of the specimen.

The application of hydrochloric acid as an agent for imposed chemical stress was used as an external excitation of the material, whereby the surface structure changes as response to these impacts both of the investigated and reference materials were analyzed. The testing methodology considered monitoring the structure changes of sulfur- polymer composite before the treatment and up to 180 days of the influence of the agent for imposed chemical stress. These structures were compared with the structures of the reference material, Portland cement composite, before the treatment and treated up to 21 days. The reference material was completely degraded after 21 days, and therefore it was not possible to continue with the treatment.

3. Results and Discussion

Macroscopic observation of the sulfur- polymer composite samples after six months of immersion in the solution of hydrochloric acid confirmed the absence of damage or deterioration on the surface. There was also no significant mass loss or surface roughness.

The aim of the applied SEM analysis was to determine morphological changes during time in terms of the sulfur distribution.

3.1. SEM Analysis of Sulfur- Polymer Composite

First analyses were performed on unpolished samples. Images obtained by the SE detector were used for determination of the morphological changes in the specimen.

SEM- pictures of sulfur- polymer composite specimens are presented in Figure 7.

In the initial structure, Figure 7(a), distinct, clearly defined aggregate and sulfur binder that looks like a monolithic integral part of the system can be noticed. The structure shows further changes in the direction of homogenization and "drowning" of the aggregate in the binder as the time of chemical treatment passes, Figure 7 (b, c). Based on the analysis of the structural changes, a mechanism of secondary homogenization of the material under the influence of the chemical agent, with clearly visible secondary binding of the aggregate and primary binder phase, also expressed after 180 days, is evident, Figure 7 (e).

In order to quantify the structural changes, the analysis of the polished samples was done. Structural changes were monitored using back-scattered electrons (BSE) that indicated different structural elements in the samples.

BSE microphotographs of sulfur- polymer composite specimens are presented in Figure 8.

Clearly distinguished binder phase can be noticed in BSE microphotographs given in Figure 8. A capability of back-scattered electrons to provide obtaining images of clearly defined material structure during the interaction with the material is used and it is possible to differentiate structure ingredients with the appropriate resolution. In this particular case, it was necessary to deduce if there was a segregation of sulfur under the influence of the chemical agent and whether in that case sulfur became the binding phase allowing the improvement of material mechanical properties, as previously observed [26,27,30,32]. According to the obtained BSE microphotographs, it is evident that there is a segregation in the observed specimens, which confirms the assumption that the chemical agent initiates the secondary homogenization induced by the chemical activation of sulfur.

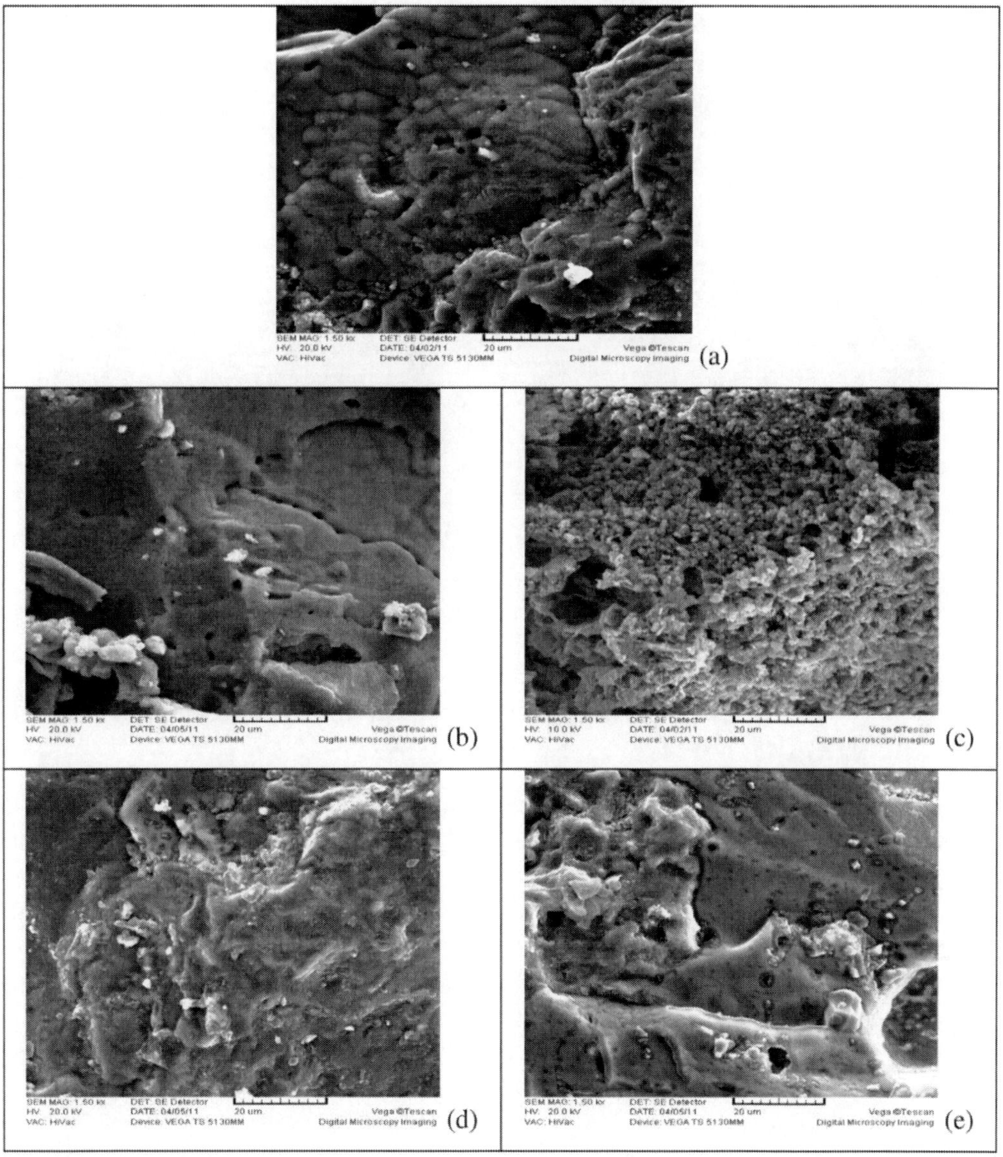

Figure 7. SEM- pictures of sulfur- polymer composite specimens: (a) untreated, (b) treated for 14 days, (c) treated for 21 days, (c) treated for 60 days, and (d) treated for 180 days.

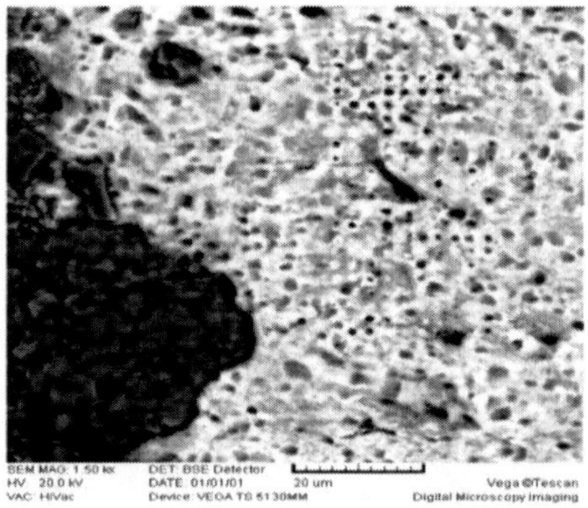

(a)

(b)

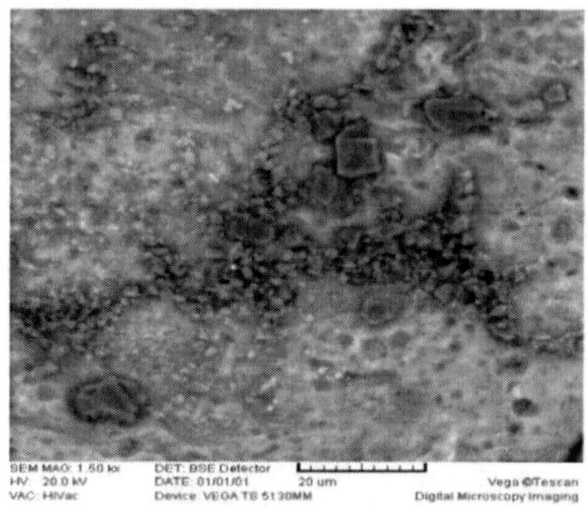

(c)

Figure 8. BSE microphotographs of sulfur- polymer composite specimens: (a) untreated, (b) treated for 21 days, and (c) treated for 180 days.

In order to provide a justification for the assumptions, one of the advantages of scanning electron microscopy- EDX analysis that is capable of detecting elements in the specimens, was used.

EDX analysis was performed in terms of sulfur and silicon distribution in the investigated specimens.

The initial analyses using *energy dispersive X-ray method were realized on the surface in order to monitor changes of the investigated elements fractions.* In this case, beside sulfur, silicon was chosen as the most dominant constituent of the aggregate and the second dominant element in sulfur- polymer composite (26,4%).

The obtained EDX spectra and corresponding BSE microphotographs of sulfur- polymer composite specimens during the exposure to chemical stress are presented in Figure 9. On the given EDX spectra, the sulfur domination in binder phase of the samples treated by the chemical agent can be observed.

BSE microphotograph	EDX spectrum
0 days	

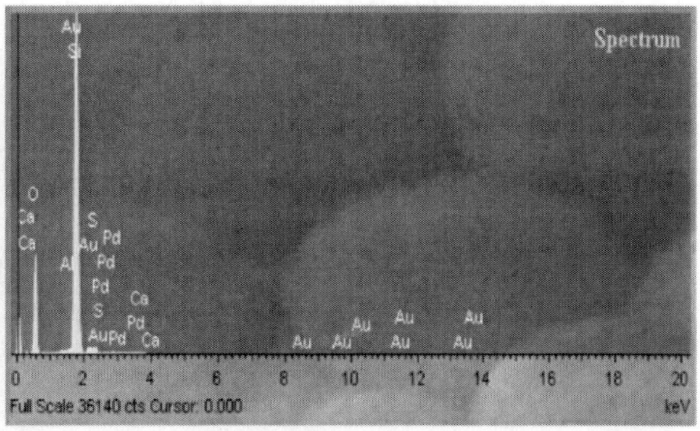

BSE microphotograph	EDX spectrum
21 days	

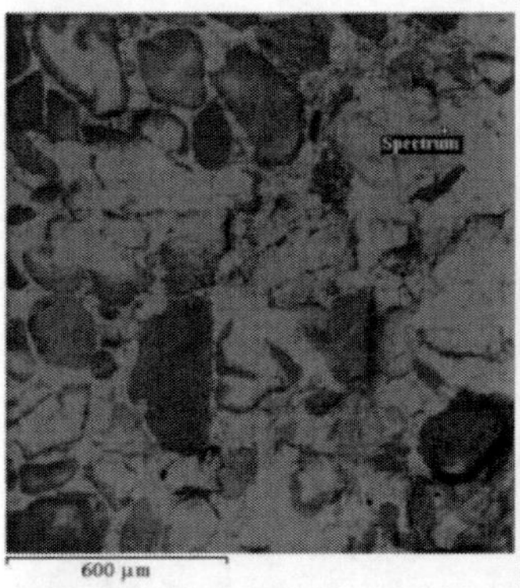

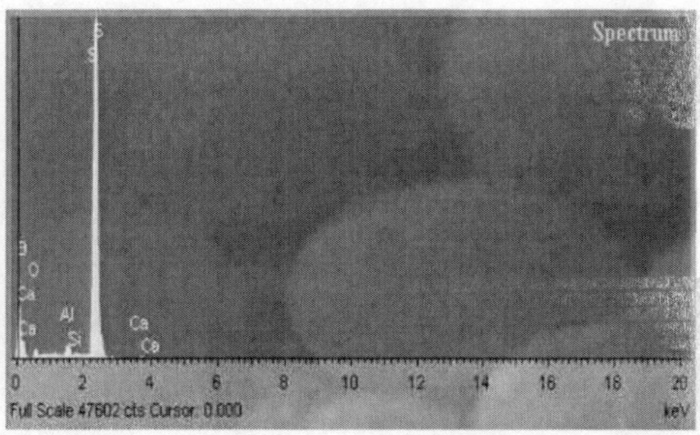

BSE microphotograph	EDX spectrum
180 days	

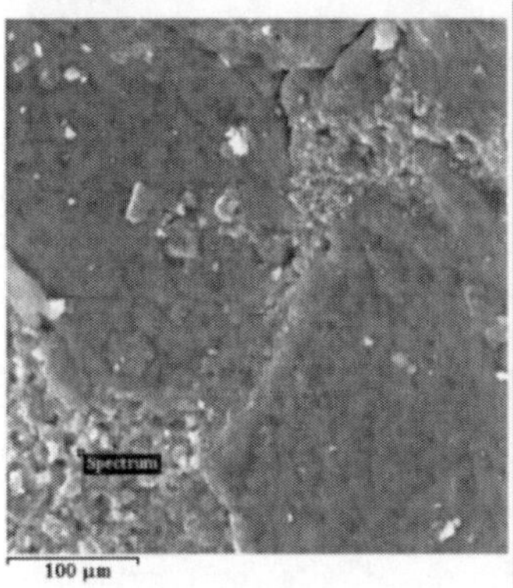

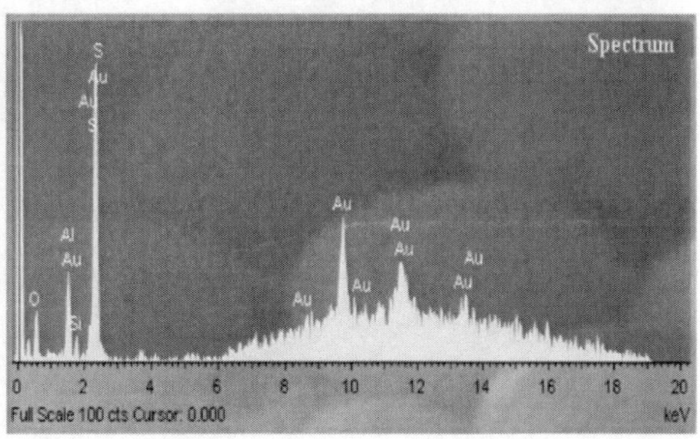

Figure 9. BSE microphotographs and EDX spectra of sulfur- polymer composite specimens during the treatment.

EDX spectrum of the initial, untreated specimen of sulfur- polymer composite showed that, except from the most dominant element sulfur, the other dominant element is silicon, followed by aluminum and calcium. The idea was to determine the effect of chemical agent by following changes in the ratio of sulfur and silicon. EDX spectrum of sulfur- polymer composite specimens treated for 21 days exhibited changes in sulfur to silicon ratio, which indicated the possibility of sulfur segregation under the influence of used chemical agent. Also, further change in sulfur to silicon ratio can be noticed on EDX spectrum of sulfur- polymer composite specimens treated for 180 days. It should be mentioned that the presence of precious metals on EDX spectra is a result of specimen preparation using steaming technique.

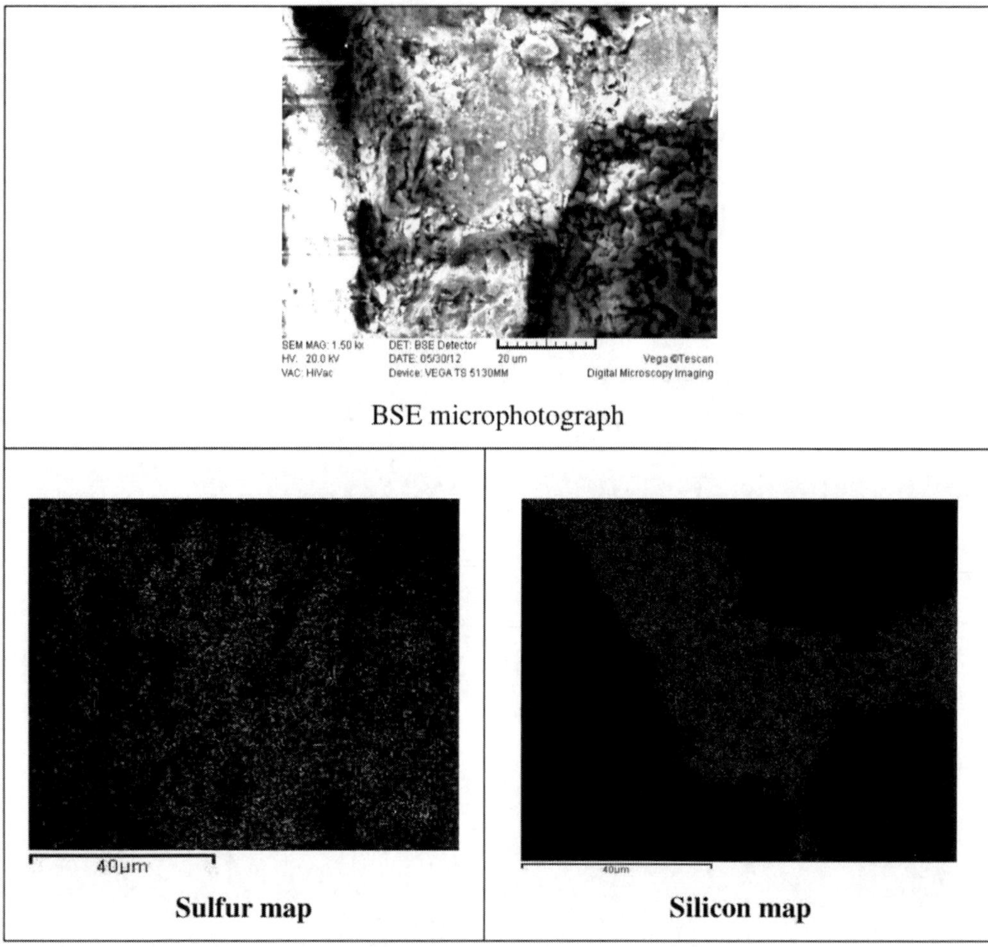

Figure 10. Maps of sulfur and silicon and corresponding BSE microphotograph of sulfur- polymer composite specimen after 21 days of treatment.

Energy dispersive X-ray analysis indicated the possibility of sulfur segregation in the sulfur- polymer composite specimens as a way of the material reacting to the used chemical agent. Although this analysis was done on a small area, it showed same tendency like the BSE analysis.

To determine the final distribution of the elements in the structure, one more possibility of SEM- elemental mapping was applied and a specimen was observed in the energy range of the selected element, sulfur and silicon. As the main ingredient of the aggregate, and thus present in the entire mass of the specimen, silicon was chosen as a referent element again.

The obtained maps for sulfur and silicon and the corresponding BSE microphotograph of sulfur- polymer composite specimen are presented in Figure 10.

Based on the presented maps, it can be concluded that qualitative differences in distribution of the chosen elements are present after 21 days of chemical agent influence on sulfur- polymer composite specimen. It is evident that after 21 days of chemical agent impact, initial sulfur segregation was happening. That indicated rearranging the ingredients within the entire volume of the specimen whereby concentration of sulfur occurred, which meant structure changes. This is a consequence of an external influence; in this case the effects of the chemical agent. The fact is that the rearrangement begins after a certain time of the agent action. The rearrangement of ingredients comes only when the interaction with the agent is achieved in the entire volume of the specimen. At that time, qualitative changes in the properties of the tested material are also occurring [33]. Monitoring the mechanical properties and image analysis showed that the complete interaction between the induced destruction agent and sulfur- polymer composite ingredients is achieved in the time interval from 14 to 21 days [3].

It should be mentioned that segregation in this case is only an indicator of sulfur rearrangement within a structure to increase its homogeneity.

3.2. SEM Analysis of Portland Cement Composite

SEM- pictures for Portland cement composite specimens are presented in Figure 11.

(a) (b)

Figure 11. SEM- pictures of Portland cement composite specimens: (a) untreated, (b) treated for 21 days.

The initial Portland cement composite structure is similar to the initial sulfur- polymer composite structure. Homogeneous grained structure with clearly distinguished components- aggregate and Portland cement binder can be noticed.

This material, after 21 days of treatment, exhibits a qualitative change to plate structure. That means that during the agent influence, agglomeration and production of plate formations on the sample surface are occurring. The plate structure indicates enhanced brittleness and generation of the layers that are not connected by strong bonds, as well as mechanical instability of the material. During the investigation, it was confirmed that the material was completely degraded. In terms of structural changes, the chemical agent by acting on components of Portland cement composite provoked creating instable plate structures, which led to complete degradation. This is in accordance with other results of the presented research.

Energy dispersive X-ray analysis of Portland cement composite specimens was not performed since this material is well-known, along with the fact that it was completely destructed after 21 days of acid treatment.

3.3. Overall Analysis- Comparison of the Results for Sulfur- Polymer Composite and Portland Cement Composite

Structural differences between sulfur- polymer composite and Portland cement composite specimens, both treated for 21 days are given in Figure 12.

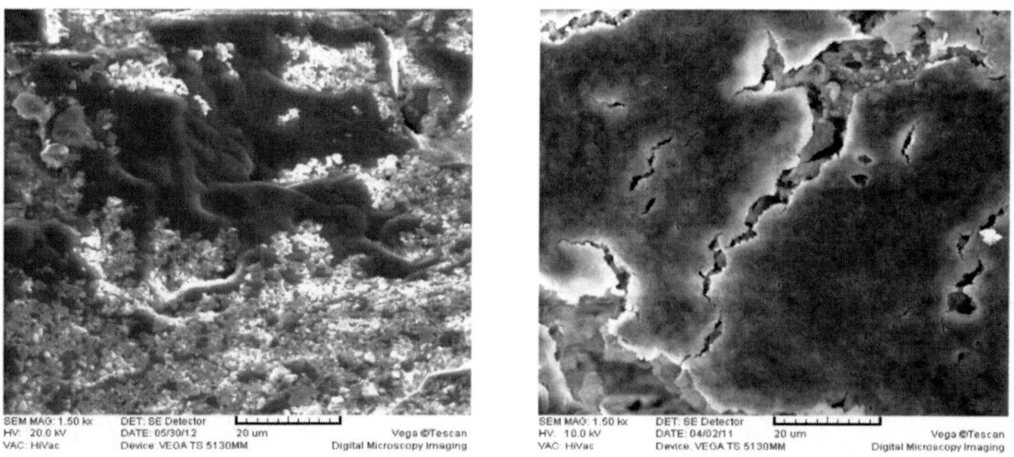

(a) (b)

Figure 12. SEM- pictures of specimens treated for 21 days: (a) sulfur- polymer composite, (b) Portland cement composite.

The analysis of SEM- pictures of sulfur- polymer composite and Portland cement composite specimens shows two different material changes provoked by the influence of the imposed chemical stress agent. In the case of sulfur- polymer composite, binder phase (sulfur) rearranged and conditionally homogenized the structure. As a result of the treatment, secondary bonding of the aggregate, which additionally homogenized the material was noticed. Therefore, correspondence between sulfur- polymer composite and prestressed concrete is logical, whereby introducing an external disturbance leads to a new quality of the material in terms of its exploitation. It can be concluded that chemical activation of sulfur- polymer composite, precisely sulfur as a binder, resulted in structural changes of the material. Unlike sulfur- polymer composite, the initial structure of Portland cement composite was completely degraded, whereby the binder phase was destroyed due to the chemical agent influence.

Conclusion

Sulfur- polymer composites have been developed with mechanical properties that can be absorbed by the market and low environmental impact. The viability of using sulfur obtained as a by-product of industrial processes has been demonstrated by the use of sulfur from a petroleum refinery. This material represents an interesting construction solution or alternative. Several applications have been developed in which sulfur- polymer composites can offer a high potential as construction materials (sulfur concrete, sulfur mortar and sulfur- modified asphalt), for solidification and encapsulation of different waste materials, for the production of eco-friendly polymer-sulfur composite coated controlled-release fertilizer and for corrosion protection of reinforcing steel and concrete.

Sulfur- polymer composite modification procedure is schematically presented in Figure 13.

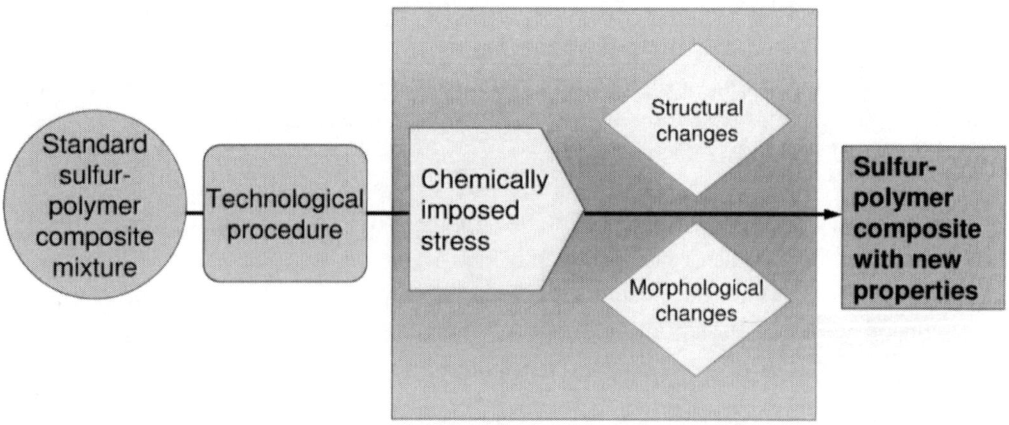

Figure 13. Sulfur- polymer composite modification procedure.

The focus of the presented research was directed to the analysis of changes in homogeneity during time.

By observing the structure of material specimens using increased resolution and by applying other available methods, it was possible to analyze volume changes in the material. Scanning electron microscopy as a high- resolution method was applied on unpolished and polished specimens. By analyzing the obtained microphotographs, changes as well as rearrangement of the structure during time were clearly identified. These changes were detected using secondary and back- scattered electrons as a source of visual information.

The obtained results indisputably proved that the investigated materials had responded differently to the impact of chemical agent. It is interesting that, based on the presented study, it can be concluded that the methodology of imposed chemical stress can be used in some cases to induce changes in the structure of materials and thus properties changes.

Noticeable changes in "resistance" to the impact of chemical agent during the exposure time are present on sulfur- polymer composite specimens. This can be explained by rearranging the structure and the increased influence of sulfur as a binder on the properties of sulfur- polymer composite after certain time. It is assumed that during total penetration of chemical agent, its interaction with sulfur throughout the volume is taking place. This interaction intensifies bonding properties of sulfur, which lead to structure compacting and the material becomes more resistant to destruction. Redistribution of sulfur in the specimen volume, which practically minimizes changes over time, also occurs.

These results indicated that the homogeneity of sulfur- polymer composite changed during the treatment time and after 21 days of the treatment homogeneity changed slower than before because of the slower structure rearrangement. It means that after 21 days of treatment, the effects of chemical agent action on the material were substantially weakened. A special priority of scanning electron microscope, energy dispersive X-ray elemental analysis (EDX) was applied. The observed changes in fractions of sulfur and silicon indicated the possibilities of sulfur segregation due to the effects of the chemical agent, which is in accordance with previously performed BSE analysis.

One of the methods, mapping, showed the concentration of sulfur in some locations. From a structural point of view, this means that the action of chemical agent leads to rapid rearrangement and that sulfur is the key ingredient that performs rearranging.

In the case of Portland cement composite, after 21 days of treatment, the existence of the starting, reference material is questionable. This means that the starting material does not exist any more in the original form, and that there is no solid material structure any more. For Portland cement composite, acting of the chemical agent is destructive and it completely destroys the basic structure of the material.

Changes in both systems were observed after 21 days of exposure due to the impact of the chemical agent. It can be assumed that after this period, total penetration of the chemical agent throughout the volume of the specimen is happening.

The presented SEM analyses using specific methods of investigating materials supported the conclusions obtained based on the previously applied test methods [3]. It is obvious that after 21 days of chemical agent acting, when the entire volume is under its influence, a concentration of sulfur was occurring. Sulfur was rearranged over specimen volume. It is clear that sulfur redistribution leads to improving its binding efficiency, which at the macro level presents increased homogeneity and improved mechanical properties.

The techniques (methods) used in this study adequately demonstrate the possibility of detecting chemical stress influence on the sulfur- polymer composite structures. The use of sulfur- polymer composite as a corrosion resistant construction material has been demonstrated. The material exhibited excellent mechanical properties compared with those of Portland cement composite and shown no signs of strength loss or degradation during the exposure to chemical stress. Industrial testing will be continued to establish durability and longevity of sulfur- polymer composite materials.

The presented analysis proves that for both materials changes in the structure on micro scale exist and they are result of interactions with the imposed chemical stress agent, which lead to different scenarios of sulfur- polymer composite and Portland cement composite life circles.

Further analysis should be focused on acting of another destruction system, which will selectively affect sulfur and confirm the proposed model.

ACKNOWLEDGMENTS

The described research was carried out as a part of the project funded by the Ministry of Science and Technological Development of Republic of Serbia under the Project Number 33007. The authors would like to extend their appreciation to the Ministry for this support.

REFERENCES

[1] Ye, E. Alternative Sulfur Management Solutions to Help Refiners Meet Clean Fuel and Environmental Challenges, The *Sulphur Institute's 10th Biennial, Sulfur Market Symposium*, Beijing, China, 2006.

[2] Spasic, A. M.; Mitrovic, M; Krstic, D. N. In *Finely Dispersed Particles: Micro-, Nano-, and Atto-Engineering*; Editors, Spasic, A. M; Hsu, J-P; CRC Press - Taylor & Francis Group: London, New York, 2006; pp 1-23.

[3] Vlahović, M. *Doctoral dissertation*. University of Belgrade, Faculty of Technology and Metallurgy, 2012.

[4] http://www.shell.com

[5] Gracia, V; Vàzquez, E; Cramona, S. Utilization of by-produced sulfur for the manufacture of unmodified sulfur concrete, *International RILEM Conference on the Use of Recycled Materials in Buildings and Structures*, Barcelona, Spain, 2004; 1054-1063.

[6] http://www.sulphurinstitute.org/pub/a0120f9d-b7f3-92f6-a60f-37552901cb41

[7] *Use of sulfur concrete in precast applications*, Document Prepared by Leading Carbon Ltd., 2011, http://v-c-s.org/sites/v-c-s.org/files/Shell%20Concrete%20Methodology_for%20public%20comment.pdf

[8] World Business Council for Sustainable Development report, *Cement Industry Energy and CO_2 Performance, Getting the Numbers Right*, 2009.

[9] Abdel-Mohsen, M. O.; El Gamal, M. *Environ. Geol.* 2012, 53, 159–175.

[10] Abdel-Mohsen, M. O.; El Gamal, M. *J. Hazard. Mater.* 2011, 192(2), 576-584.

[11] Lópeza, F. A.; Gázquez, M.; Alguacil, F. A.; Juan Bolívar, J. P; García-Díaz, I.; López-Coto, I. *J. Hazard. Mater.*. 2011, 192, 234-245.

[12] Kalb, P.D. ; Adams, J.W.; Milian, L.W. Sulfur Polymer Stabilization/Solidification (SPSS) *Treatment of Mixed-Waste Mercury Recovered from Environmental Restoration Activities at BNL*. 2001, Brookhaven Science Associates, Upton, Long Island, New York 11973, Formal Report.

[13] Ksiazek, M. *Composites Part B*. 2011, 42, 1084–1096.

[14] Diehl, L. Dicyclopentadiene-modified sulfur and its use as a binder, quoting sulfur concrete as an example, *New Uses for sulfur and Pyrites*. The Sulfur Institute, Madrid 1976; 202-214.

[15] Beaudoin, J. J.; Feldman, R. F. *Int. J. Cem. Compos. Lightweight Concrete*. 1984, 6(1), 13-18.

[16] Sullivan, T. A; McBee, W. C; Blue, D. D. In *New uses of sulfur: Advances in Chemistry Series* 140; Editor, West, J.R.; American Chemical Society, Washington, D.C., U. S., 1975; pp 55-74.

[17] Gregor, R; Hackl, A. In *New uses of Sulfur-II: Advances in Chemistry Series* 165; Editor, Bourne, D. J.; American Chemical Society, Washington, D.C., U. S., 1978; pp 54-78.

[18] *Drop forming method and sulphur polymer concrete process*, United States Patent 6228905. 2001.

[19] McBee, W. C.; Sullivan, T. A; Jong, B. W. Modified Sulfur Concrete Technology. *Proceedings SUFUR-81 International Conference on Sulfur*, Calgary 1984; 367-88.

[20] Blight, L; Currell, B.R.; Nash, B. J; Scott, R. A. M.; Stillo, C. In New uses of Sulfur-II: Advances in Chemistry Series 165; Editor, Bourne, D. J.; *American Chemical Society,* Washington, D.C., U. S., 1978; pp 13-30.

[21] Đorđević, N; Boljanac, T; Vidojković, V; Vlahović, M; Branković, A; Martinović, S. *Modification of secondary sulfur from oil refining process by dicyclopentadiene and application of obtained modified sulfur,* Patent P-2009/0436, 2010.

[22] Mohamed, A. M.O.; El Gamal, M. M. *Cem. Concr. Compos.* 2009, 31, 186-194.

[23] Mohamed, A. M. O.; El Gamal, M. M. *Sulfur concrete for the construction industry- A sustainable development approach.* J. Ross Publishing: U.S.; 2010.

[24] Vidojković, V; Đorđević, N; Branković, A; Martinović, S; Vlahović, M; Boljanac, T. *The processing procedure of obtaining sulfur concrete using elemental sulfur produced in the oil refining process and its application*, Patent-2009/0433, 2010.

[25] Methods of testing cement- Part 1: Determination of strength, *Serbian standard SRPS* EN 196-1:2008. 2008.

[26] Vlahović, M.; Jovanić, P.; Martinović, S.; Boljanac, T.; Volkov-Husović, T. *Composites Part B*, 2013, 44 (1), 458-466.

[27] Vlahović, M.; Savić, M.; Martinović, S.; Boljanac, T.; Volkov-Husović, T. *Mater. Des.* 2012, 34, 346-354.

[28] Vlahovic, M.; Martinovic, S.; Boljanac, T.; Volkov Husovic, T. Strength degradation modeling based on non-destructive testing, *43rd International October Conference on Mining and Metallurgy*, Kladovo, Serbia, 2011; 366-369.

[29] Vlahović, M.; Vidojković, V.; Martinović, S.; Boljanac, T.; Jančić-Heinemann, R.; Volkov-Husović, T. Corrosion resistance of sulfur concrete in acidic solution, *42nd International October Conference on Mining and Metallurgy*, Kladovo, Serbia, 2010; 102-105.

[30] Vlahović, M.; Boljanac, T.; Branković, A.; Vidojković, V.; Martinović, S.; Đorđević, N. *Hemijska industrija.* 2010, 64, 129-137.

[31] Vlahovic, M.; Martinovic, S.; Boljanac, T.; Jovanic, P.; Volkov-Husovic, T. Assesment of sulfur concrete corrosion resistance, *43rdInternational October Conference on Mining and Metallurgy*, Kladovo, Serbia, 2011; 370-373.

[32] Vlahovic, M.; Martinovic, S; Boljanac, T; Jovanic, P; Volkov Husovic, T. *Constr Build Mater.* 2011, 25, 3926–3934.

[33] Vlahović, M; Jovanić, P. In Microscopy Book Series- Number 5: Current microscopy contributions to advances in science and technology; Editor, Méndez-Vilas, A.; *Formatex Research Center, Spain, 2012; Vol.2,* pp 1171-1182.

In: New Developments in Polymer Composites Research
Editors: S. Laske and A. Witschnigg
ISBN: 978-1-62948-340-5

Chapter 11

EFFECT OF CONTACT TEMPERATURE ON THE TRIBOLOGICAL BEHAVIOR OF PTFE COMPOSITES

Marcello Conte**, *Bihotz Pinedo and Amaya Igartua
IK4-Tekniker – Tribology Unit, Spain

ABSTRACT

Polytetrafluoroethylene (PTFE) is one of the most widely used materials for tribological applications and its discovery was considered by STLE review to be one of the ten greatest events in Tribology history. In particular low friction and quasi-absence of sticking effect, high resistance to temperature and good compatibility make it of great interest.

Because of its chemical inertness, PTFE cannot be cross-linked like an elastomer, therefore it has no memory and its subjected to creep. Compounding fillers are used to control unwanted creep, as well as to improve wear, friction and other properties. It is important to keep in mind that PTFE fillers do not act like elastomer fillers, which become chemically bonded to the elastomer: the high shear modulus fillers are encapsulated and bound by the low shear modulus PTFE.

In this chapter an innovative way to study the PTFE composites tribological behaviour is proposed. Great attention is given to the material characteristics, such as its cristallinity and thermal properties and tribological tests are carried out. The use of Differential Scanning Calorimetry (DSC) and the frictional heating calculation tailored on the experimental tests allows a wider understanding of the tribolayer formation and re-generation and the different grades of wear.

Keywords: PTFE fillers, frictional heating, contact temperature

* E-mail address: marcello.conte@gmail.com (Corresponding author)

INTRODUCTION

Polymer composites are materials in which particles or fibers are dispersed in a polymer matrix i.e they are a combination of materials of different composition, where the individual constituents retain their separate identities. They consist of a matrix material, usually ductile and tough, and reinforcing materials, usually stiff and strong.The materials forming the composite act together to offer great advantages such as low weight, corrosion resistance, high fatigue strength, high creep resistance and faster assembly. Polymer composite is different depending on the type, shape, size and orientation of the reinforcing particles or fibers. Beside, all the properties depend on the degree of interfacial strength (adhesion) and the interparticle distance. A uniform distribution of fillers in the polymer matrix is desirable for avoiding cracks and brittle fractures [1].

Nowadays, reduction of lubricants is a goal of great importance in many tribological applications being their regeneration and disposal is governed by harsh acts which entail high added costs. In fact, they are estimated to be used 46% in the automotive and factory farm industry, 47% in the general production industry and 7% in the marine and aeronautical industry. Polytetrafluoroethylene (PTFE) is well known for its tribological characteristics, in particular its low friction, the quasi-absence of sticking effect, its resistance to high temperature and good compatibility with other materials. PTFE fibrils broke along their length and fragments run along the surface and/or are transferred to the counterface, orienting along the sliding direction. Such fragments fractionally cover the counterface filling in the roughness valleys to form a transfer film [2, 3].

Because of its chemical inertness, PTFE cannot be cross-linked like an elastomer, therefore it has no memory and is subject to creep: compounding fillers are used to control unwanted creep, as well as to improve wear, friction and other properties. In other words, PTFE fillers do not act like elastomer fillers, which become chemically bonded to the elastomer: the high shear modulus fillers are encapsulated and bound by the low shear modulus PTFE [4]. The most commonly used fillers for tribological applications are carbon, bronze, glass fibres, graphite and molybdenum disulfide in different percentages and sometimes combined between them for improving the wear resistance and the coefficient of friction. Of course other important properties are affected like abrasiveness (glass fibres and carbon), thermal conductivity (glass fibres), electrical conductivity (bronze), chemical inertness (bronze). Generally the fillers improve the wear resistance from 10 to 500 times and increase the thermal conductivity from 2 to 3 times [5, 6]. In a previous study [7] the authors have found the increment in thermal conductivity and diffusivity of PTFE composites improves their wear resistance to adhesion. A direct dependence of wear resistance from the thermal properties of PTFE composite was found by J. Khedkar et al. [8] and by the authors in [9]. Investigation on the effect of crystallinity on its fracture was presented by Brown et al. [10] finding that the formation of fibrils and associated increase in fracture toughness is restricted by the increase in crystallinity, that is, by diminishing of molecular mobility. In [11] Y. J. Shi et al. found that the friction coefficient of various PTFE composites are strongly dependent on filler crystal structure but the improvement of anti-wear properties is more dependent on filler shape. V. N. Aderikha et al. [12] found that friction of PTFE Carbon Black composites depends on changes in morphology and degree of crystallinity and on the ratio of adhesive activity of the filler surface and rigidity of the composite.

Y. Wang et al. [13], S. Bahadur [14] and E. Yang et al. [15] investigated the tribological properties of the transfer film of PTFE-based composites finding that its wear life is short because PTFE cannot form durable transfer film on the steel counterface. The different fillers morphologies were also studied.

Thus, a better understanding of the effects of fillers on PTFE tribological behavior is mandatory and moreover a general technique of characterization is needed. Relevant studies were carried out by K. Tanaka [16], A. Oshima et al. [17] and Q. J. Xue et al. [18]. Xiao-Xia Liu et al. [19] showed the effect of drawing process on PTFE's coefficient of friction. In the last years, important steps forward have been done also for PTFE nanocomposites [20-23].

Aim of this work is a contribution to the evaluation of the effects of fillers on the tribological behavior of PTFE by means of the energetic analysis of the tribotests. A comparative study between different kinds of filled PTFEs is proposed to have a map of PTFE materials to be used for tribological applications.

In particular, bronze, molybdenum disulfide (MoS_2), glass fibres and carbon fillers effects on tribological properties are compared and referred to pure PTFE. Wear specific energy formulation is used as key parameter for highlighting the ability of a material to resist wearing. Composites crystallinity is observed for a better understanding of the filler-matrix interactions.

Beside, it is also aim of this chapter to give a contribution to study the effect of contact temperature on the tribological behaviour of materials. When two materials in contact are sliding against each other a big part of the energy of friction is converted into heat resulting in local temperature rise called "flash temperature". This temperature rise affects the properties of the polymer and can even thermaly degrade it.

In this chapter, a frictional heating calculation method is proposed matching experimental tests configuration at laboratory scale and theory to overtake restrictive assumptions. In particular, temperature trend is calculated as function of measured friction in steady state regime. The method herein proposed is applied on virgin PTFE material.

METHODS

Mechanical and Thermal Characterization of PTFE Composites

In order to analyze the effect of different fillers on the tribological behaviour of PTFE it is necessary to perform a material characterization prior to the tribotests. For polymer characterization thermal and mechanical analysis techniques and microscopic analysis were used. Table 1 shows the PTFE polymers analized in this.

Glass fibers improve the compressive properties and creep resistance of the virgin PTFE and descreases the deformation of the polymer under load. Carbon fibers have similar effects to the glass fibers on PTFE. They add also the creep resistance , increases the hardness and decreases the deformation of the PTFE under load. Carbon filled composites show excellent wear properties and usually need a lower amount of carbon fillers than glass fibers to achieve the same effect. They are chemically inert so that it can be used in alkali and HF environments where glass fibers fail and have higher thermal conductivity than glass fibers compounds. However, the carbon tends to fail under high loads and high sliding velocities.

Bronze fillers aport high creep resistance and high thermal conductivity and can be easily machined. Adding bronze fillers can be useful for applications with high mechanical loads and high speed rubbing contacts. By contrast, they are not suitable for electrical applications and they tend to oxidise. Graphite apports good chemical resistance to corrosive environments and good initial wear and sliding characteristics both under dry and lubricated contact conditions. Graphite fillers are usually used in combination with other fillers. Adding molybdenum disulfide (MoS_2) fillers it is possible to increase the hardness and stiffness of the compound and to decrease the friction coefficient. MoS_2 is usually used in low percentages and in combination with other fillers.

Table 1. PTFE composites

Sample	
1	PTFE virgin (reference)
2	PTFE + 25 wt. % carbon
3	PTFE + 60 wt. % bronze
4	PTFE + 15 wt. % graphite
5	PTFE + 25 wt. % glass fibres
6	PTFE + 20 wt. % glass fibres + 5 wt. % graphite
7	PTFE + 15 wt.% glass fibres + 3 wt. % MoS_2

Hardness Shore D measurements were carried out at several locations for each sample and at the contact worn surface following the ISO R 868 Standard. The indentation load for each measurement was 50 N and hold time was 15 s. In Table 1 the comparative hardness value of the PTFE composites are reported. It is clear that addition of fillers to PTFE causes a significant increase of hardness (up to 15%). In particular the composite n. 3 (PTFE+60 wt. % Bronze) displays the maximum hardness, due to te presence of bronze. Scanning Electron Microscopy (SEM) was used for investigating the contact surface of each polymer composite (Figure 1). PTFE composites were chosen in order to evaluate the effect of their presence in weight and their different way to act. The samples were tested as received after cutting process. As can be seen in the images taken by Scanning Electron Microscopy (SEM) reported in Figure 1, surface structures are not homogeneous. In details, for PTFE composites, fillers cropping out the surface can be observed: carbon particle of different sizes can be seen in composite 2; bronze particles of composite 3 are mostly hidden by the bulk material; graphite flakes of composite 4 seem to be more merged with the bulk material; glass fibers for composites 5 to 7 are visible and oriented in different directions. Gaps between the glass fibers and the bulk material are visible.

Differential Scanning Calorimetry (DSC) was used to study the thermal transition of the polymers and to evaluate the crystallinity of the semi-crystalline materials. The dimension of the crystals depends on the mobility of the polymeric chains. This mobility is higher at higher temperatures, and is influenced by the content of fillers and their nature. The samples were warmed up and cooled down again from 30-350 °C with a cooling/heating rate of 10 °C/min in inert environment of N_2.

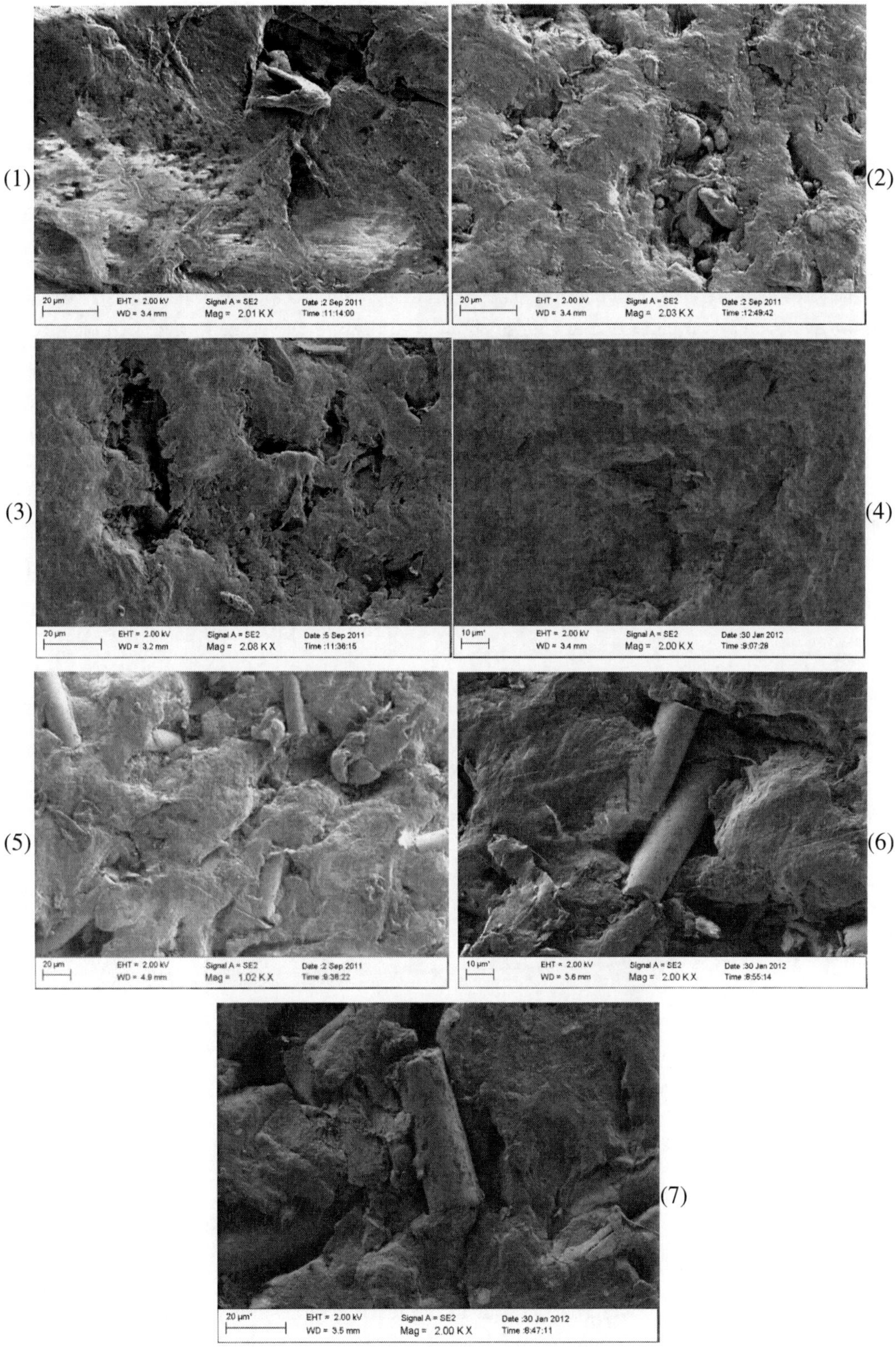

Figure 1. SEM of PTFE composites surfaces before tribological test.

The dimension of the crystals in fact depends on the mobility of the polymeric chains, easier at higher temperatures, and it is influenced by the content of fillers and their nature. DSC analysis results of the polymers are shown in Figure 2. All the polymers showed a melting point of about 330 °C.

The heat of fusion values of the samples, defined as the heat required to a unit mass of a solid at its melting point into a liquid without an increase in temperature, are shown in Table 1. Taking the virgin PTFE as reference, six commercial PTFE composites were tested in sliding contact against a quenched tempered CrVNiMo rod.

The PTFE samples are prepared by compression molding of powder mixture of PTFE and filler followed sinterization and cooling. The PTFE dispersion has a median particle size of 0.25 μm; the average carbon particle diameter is about 10-25 μm; the average bronze particle size is about 25 μm; the glass fibers have a diameter of about 5 μm and a length of about 90-120 μm; the graphite flakes are 25-50 μm size; the MoS_2 particles have 8-10 μm diameter. The density of the PTFE composites is in the range 2.15-2.20 g/cm^3, the density of composite 3 is about 3.8-4.0 g/cm^3. In Table 1, the seven samples are listed with the mass loss after the tribological test (Δm), the frictional energy E, the wear specific energy E_w, the Shore D hardness before and after the test, the heat of fusion ΔH_f and the coefficient of friction at regime μ.

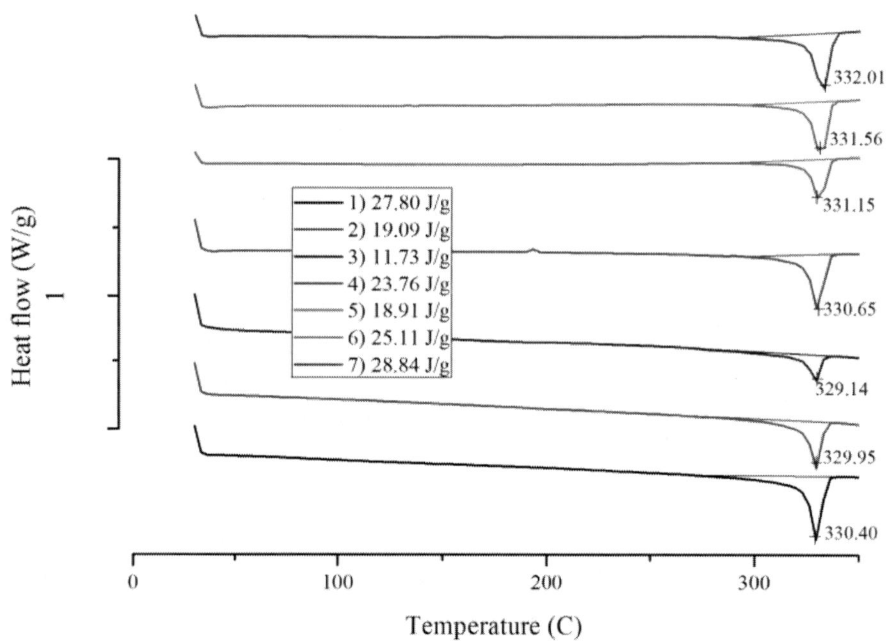

Figure 2. DSC analysis.

Tribological Characterization

Friction and wear tests were carried out in the test rig represented in Figure 3: a PTFE sample (1) is mounted on a frame (2) and loaded by an axially sliding rod of CrVNiMo (3) (15 mm diameter, 22 mm length, HRC > 60±1 and surface texture following standard D6425-02), supported and fixed by two holders (4) on which normal constant load is applied and transmitted to the axis of the rod. The holders and a couple of screws avoid rotation of the rod around its axis. Normal load was set on 50 N for the firsts 30 s and then at 100 N, 60 Hz frequency and 2 mm stroke for each sample. Coefficient of friction (CoF) was recorded at regime condition for 30 minutes; each test was repeated 2-3 times. Resulting friction coefficients are shown in Figure 4.

The CoF of the PTFE composites increases in the first 1/10 of the tests. Only the composites 1, 5 and 7 reach a stable value at regime while the others presented a constant increment (composites 5 and 6) or reached it in the last 1/3 of the test (composites 3 and 4). CoF of composite 2 maintain certain instability along the test.

The non-conservative friction force arises in response to the work needed to enable relative motion between two bodies, and the energy associated with this work is distributed differently. In the friction stack concept [24] a number of parallel layers represent potential locations in which frictional energy can be dissipated and several frictional energy dissipation zones are identified.

The energy that is transformed as a consequence of frictional contact can be dissipated (conversion to heat, vibration, material deformation, new surface creation) or stored into the tribosystem. Therefore, materials exhibiting the same friction coefficient can have different wear rates because the energy is partitioned differently between and within the materials.

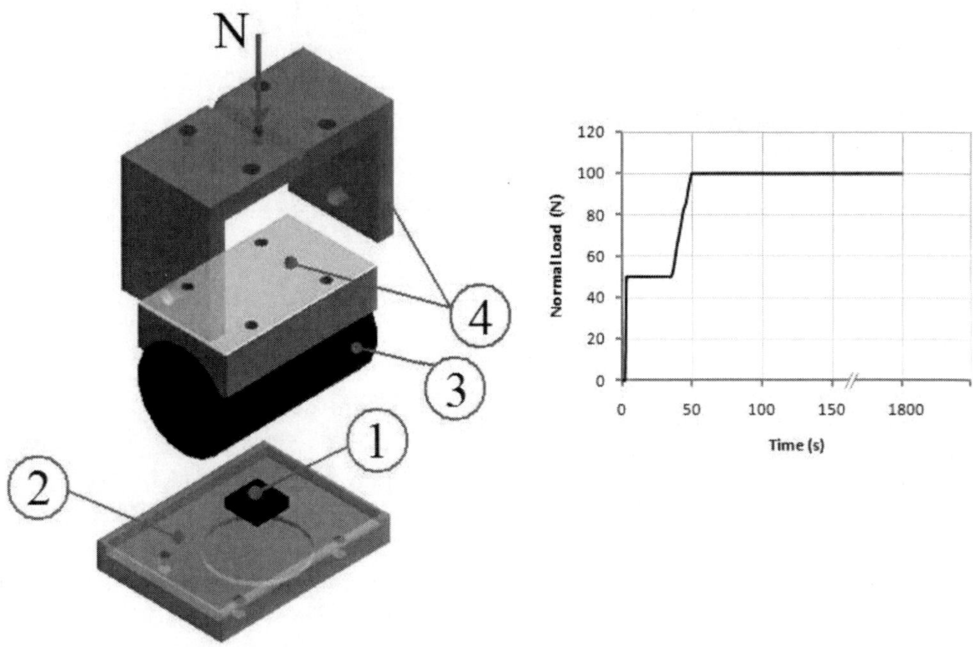

Figure 3. Tribological test configuration.

Both friction coefficient and wear rate are needed for a good understanding of the test. Specific wear energy (1) is a criterion, which takes both into account: it is the ratio of the friction work spent in the interface divided by the mass loss due to the wear.

$$E_W = \frac{E}{\Delta m} = \frac{v_m \cdot F_N \cdot \int_{t_i}^{t_f} \mu_{fr}(t) \cdot dt}{\Delta m} \qquad (1)$$

where v_m is the mean relative sliding velocity [m/s], F_N the normal load [N], μ_{fr} the coefficient of friction [-] and Δm the total mass loss [mg].

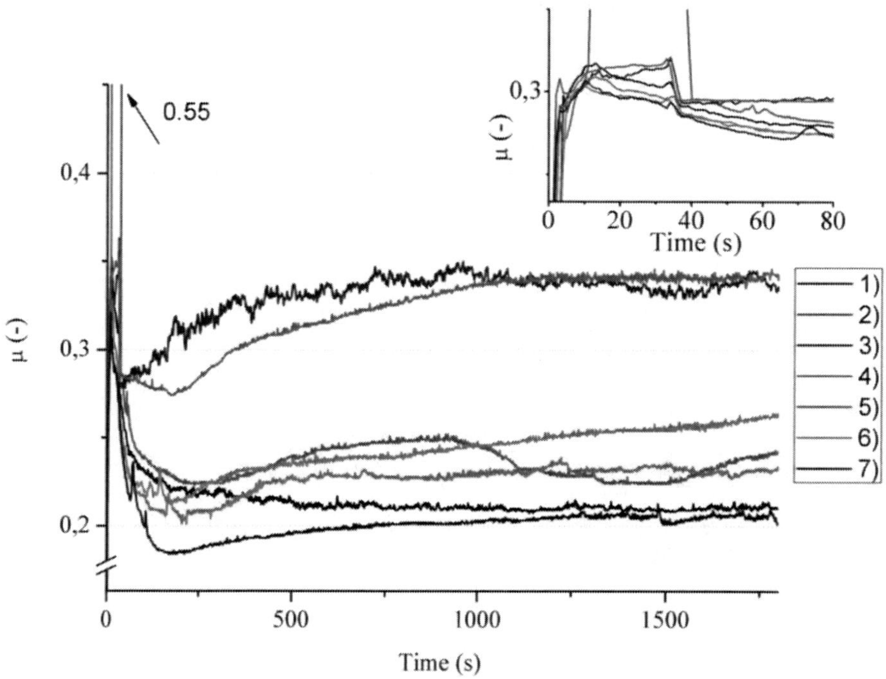

Figure 4. Frictional behavior of PTFE composites.

Unfortunately, also wear specific energy is not univocally determined, therefore, representation of frictional energy vs mass loss is herein considered a good way for comparing tribological characteristics also according to [25] and [26] considerations. Another way to illustrate results could be used. M. Jahangiri et al. [27] for example, defined the wear specific volume (wear volume/dissipated energy (mm^3/J)).

Both representations can be used as a substitute of the traditional wear rate because of the simultaneous involvement of several wear parameters and of its strong dependence on the wear mode.

All the filled PTFE composites are influenced by three groups of parameters: sliding contact mechanical and geometric conditions; bulk mechanical properties of the polymer (with dependence of temperature); role and properties of the transfer film.

At regime formation of a transfer film with high degree of lateral mobility of PTFE chains quickly induces molecular orientation in the sliding direction.

This phenomenon is also aided by PTFE smooth linear molecular profile and ability of the electron clouds of the large fluorine ions to screen bonding forces between neighboring molecules.

The measured dynamic friction coefficient is directly related to the molecular structure of the polymer. PTFE, which is composed of rigid rod-like molecules with a smooth molecular profile, shows a low coefficient of friction because of an easy slip between the aligned molecule chains at the interface. Basically the fillers are thought to provide a thin, coherent, securely attached transfer layer of the polymer species on the adjacent counterface. A stable transfer layer promotes low friction and low wear. During the sliding, adhesive wear is generated as consequence of PTFE molecular reorientation in the direction of sliding. Such reorientation occurs for the surface of PTFE in contact with the countersurface as well as for the transfer film that is deposited on the countersurface. As consequence, worn surface are more smoothed and packed than before the test (Figure 6). Tribological and hardness results as well as DSC results are reported in Table 2.

Figure 4 presents the trends of the friction coefficients along the distance. At regime, formation of a transfer films with high crystallinity on the counterparts causes the relatively steady friction coefficient in the sliding process even with the presence of fillers. As previously explained by the authors in [7] the transfer film and consequently the friction properties are greatly influenced by the prevailing interfacial temperature and also by the kinematics of the relative motion between the two bodies. It is well known how self lubricating nature of PTFE originates from the low activation energy for the motion of PTFE molecular bonds.

In the temperature range of 30-90°C, easily achieved in a frictional event [28], PTFE fails along weal intermolecular bonds due to the minor temperature rise which involves slippage of crystalline formations of the bond structure.

Table 2. PTFE composites

	Sample	Δm (mg)	E (MJ)	Ew (MJ/g)	Sh D Hardness before test	Sh D Hardness after test	Heat of fusion ΔH_f (J/g)	CoF μ (-)
1	PTFE Virgin (REF)	5.3	9.5	1.8	54	52	27.80	0.22
2	PTFE+25 wt. % Carbon	0.4	10.6	26.5	62	62	19.09	0.24
3	PTFE+60 wt. % Bronze	13.0	14.3	1.1	64	61	11.73	0.33
4	PTFE+15 wt. % Graphite	1.1	14.2	12.9	56	56	23.76	0.33
5	PTFE+25 wt. % Glass fibres	0.7	9.9	14.2	57	57	18.91	0.23
6	PTFE+20 wt. % Glass fibres+5 wt. % Graphite	0.6	10.7	17.8	57	57	25.11	0.24
7	PTFE+15 wt.% Glass fibres+3 wt. % MoS_2	0.4	8.8	22.1	60	60	28.84	0.19

In other words, temperature rise is quite relevant as consequence of low thermal conductivity and the energy dissipated when the carbon-to-fluorine bonds, which have high bond dissociation energy, are broken.

Two ranges of E_w were observed [19]: (i) $E_w < 5$ MJ/g: severe wear conditions; (ii) Ew> 5 MJ/g: high protective PTFE film formation on the counterface with constant rebuilding of the PTFE layer. In the first case, PTFE is transferred in the form of flakes of very small sizes during the initial period of friction. In the second case, the thickness of the transferred film is higher (augmented transfer film adhesion).

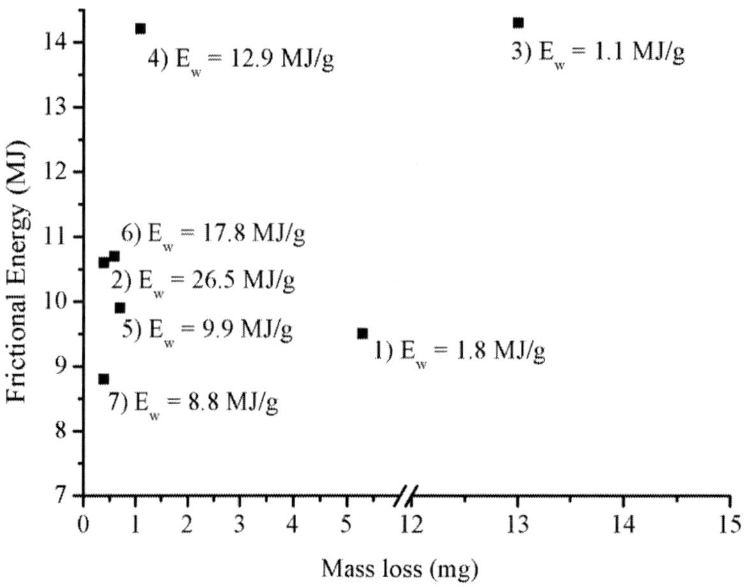

Figure 5. Wear specific energy.

In Figure 5 Frictional energy vs mass loss is shown in order to identify the best combination of low frictional energy, and mass loss. PTFE composites 5 and 7 correspond to these criteria. It is also shown how the addition of fillers improves the wear resistance of PTFE up to 15 times (PTFE+25% carbon). This behavior was attributed to the presence of hard particles of carbon in between the material matrix increasing the strength of the material. The increase of hardness value (see Table1) demonstrates that the carbon particles act as strengthening elements.

The use of a 15-25% by weight of a glass particles in a PTFE composite increases the wear resistance up to 7-12 times. On the other hand, the wear morphology of glass fiber-filled PTFE shows to exhibit crack nucleation at the polymer-fiber interface and crack propagation parallel to the surface and crack shearing to the surface resulting in flakes of polymer debris.

Comparing the composites 5, 6 and 7, 5% of Graphite is more efficient than 5% of glass fibers acting as a solid lubricant, and 3% of MoS2 is more efficient than 10% of glass fibers. In particular, the composite 7 reaches a good level of wear resistance due to the lamellar MoS_2 particles that prevent the contact between the glass fibers and the countersurface. This allows to have a more stable transfer film on the countersurface and to reach the lowest coefficient of friction.

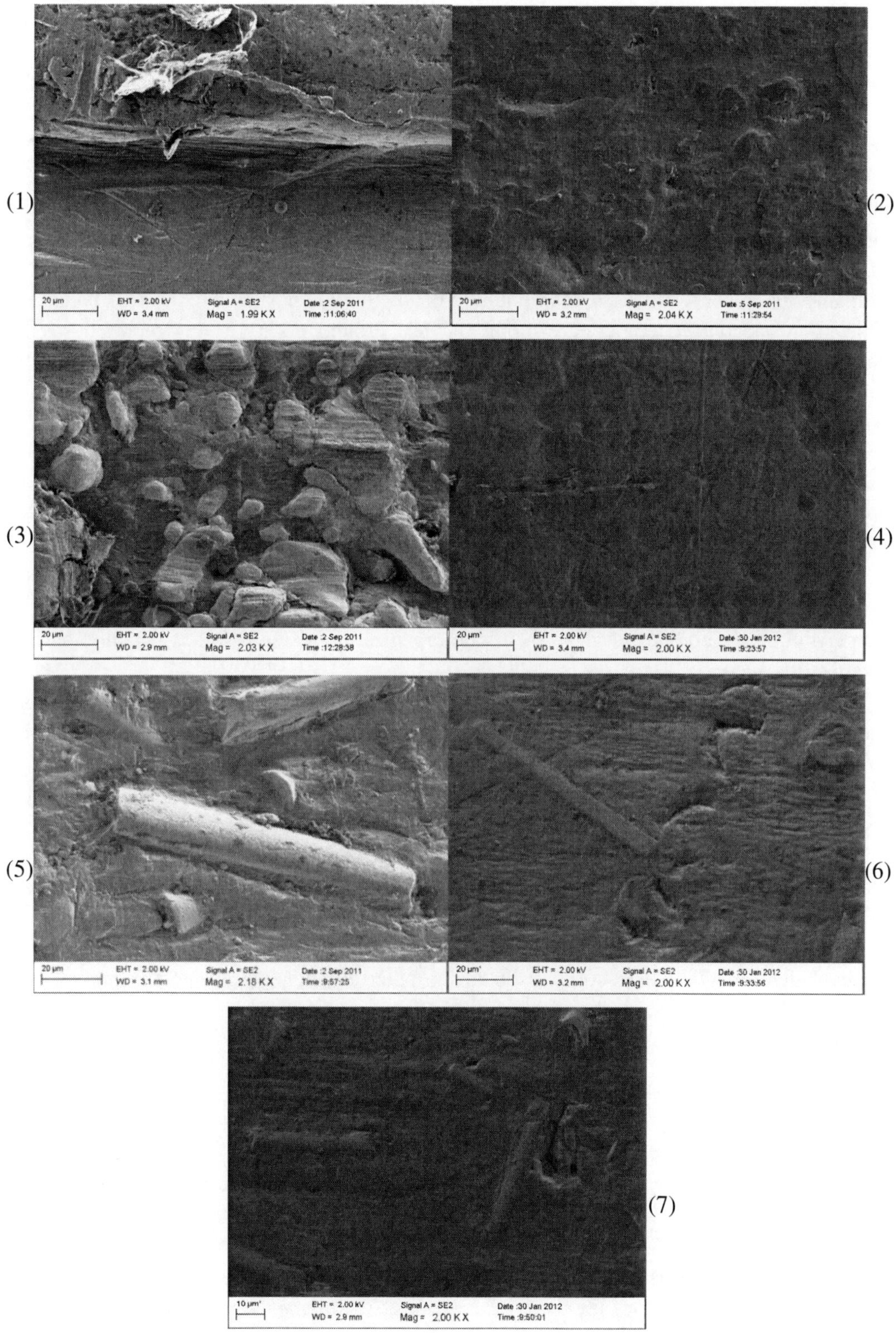

Figure 6. Variation of surface aspect after the test.

Measuring hardness at the worn surface, no relevant changes were observed in all the composites but the compound 3, the PTFE+60 wt.% Bronze, exhibits a decrement of hardness probably due to the reorganization of the composite structure. In Figure 6 it can be seen the worn surface of the PTFE composites. In particular, as expected, virgin PTFE (compound 1) shows plucked marks due to the transfer film formation and subsequent adhesion; even if the effect of black carbon was demonstrated in [13, 29] to be quite complex, it can be affirmed that the smallness of the black carbon dispersed particles (compound 2) avoid abrasion of the countersurface due to their self-lubricating property; the presence of MoS_2 on the glass fibers surface (compound 7) at the contact reduces the abrasiveness and their displacement seen when only glass fibers are used (compound 5). For the composites 1, 2 and 3, being the estimated mean contact temperature along the test around 60-70 °C (as reported by the authors in [7] and verified experimentally by Tzanakis in [30]), it is possible to say that the small crystals formed by nucleation at lower temperature are destroyed resulting in flakes separation. From the DSC analysis in fact it can be seen that crystallization of pure PTFE starts at low temperatures, therefore the crystals are less resistant (higher wear) than those grown up at higher temperatures. The presence of fillers improves the material resistance to adhesion wear mechanism changing the structure of the PTFE: the presence of fillers affects crystals nucleation depending of their nature and concentration.

For the composites 5, 6 and 7, the higher the presence of glass fiber, the higher the crystallization. When no more than 20 wt. % of macrofillers in the polymeric matrix is present, (like in the composites 1, 4, 6, 7), the organization of the polymer structure and its crystallization is easier.

DSC analysis allows the study of the friction process of energy balance, that is, to establish that friction conditions affect both total friction work and ratio between quantity of energy stored and the quantity of heat generated [7, 31].

One disadvantage of using hard phases in soft matrix is the possibility of transfer film removal and the aggressive abrasion of the hard counterface if the hard phases are released to the interface as a third body due to loosening of the bond between the filler and the polymer matrix: it was found that the particles used as fillers were rubbed off during the test highlighting the weakness between the transfer film and the countersurface.

Taking out the virgin PTFE, the viscoelastic deformation and its effect on friction force were considered negligible in comparison to the other components due to the presence of fillers (like glass fibers, graphite and bronze) able to improve the mechanical characteristics of PTFE. No appreciable plastic deformation was observed.

Contact Temperature Calculation Method

Frictional heating calculation in metal to metal contact has been deeply investigated and accordingly many reviews have shown it is still not well defined and the contact conditions vary in time and location [32, 33]. In [34] a comparison of some calculation model was presented showing high disagreement between them, due to two important concepts: the contact area to be considered and the variation of the materials properties with temperature. In polymer to metal contact, the first assumption does not affect calculations otherwise temperature reached at the asperities contact would overtake the melting point of the materials, according to some calculation methods but it was not found in any experimental test. On the other hand, the latter, like for example the coefficient of friction which enters calculation to the first power, is still of great importance. In metal to metal contact, flash temperature was estimated determining the phase transformations and microstructural changes in materials and relating them with the necessary temperature for these transformations [35, 36]. In polymer to metal contact, the authors [7] have used Differential Scanning Calorimetry (DSC) technique [41, 42, 43, 45] to evaluate the effects of fillers on the crystallinity ratio of some PTFE (Polytetrafluoroethilene) composites and observed relevant structural changes at the polymers contact surfaces. It was also observed in [44] how the crystallinity of transfer film was higher than the bulk material. According to Persson's investigation [38], due to the strong temperature dependence of viscoelastic properties of rubber-like material, local temperature calculation should be considered in order to explain the frictional behavior of the tribo-pair, so an estimation of flash temperature at the polymer to metal contact surface is herein proposed for a better understanding of the tribochemical and tribophysical phenomena occurring at the contact. Direct experimental measurement of contact temperature is not viable for different reasons [39, 40]. Taking into account the low diffusivity of this kind of materials, a high velocity thermo camera has been seen to be a good but expensive tool [30].

In this section, a frictional heating calculation method based on tailored experimental tests is presented. In particular, virgin PTFE was chosen as a case study of the method herein presented for three main reasons: 1) because of its high resistance to high temperature having a melting point of 330 °C and 2) because, the fluorine chains are not cross linked like in elastomers so the mechanical work done for overtaking the friction force can be considered quasi-entirely transformed in frictional heating [37]; 3) to have a correspondence with experimental tests found in literature [30].

The temperature at the interface between rubbing bodies is known as "surface conjunction temperature"; the calculation of this temperature is possible by applying the laws of energy conservation and heat transfer. In fact, most of the energy dissipated during the process of friction is converted into heat [46] resulting in local temperature rise called "flash temperature" that could be divided into transient and steady state flash temperature rise. As reported by Bowden and Tabor [47], in dry contacts the transient flash temperature may become larger than the steady state component. The flash temperature calculation method was formulated firstly by Block [48] and then improved by Jaeger [49] and Archard [36]. It provides a set of formulae for various velocity ranges and contact geometries and it is based on the following assumptions:

- A planar source of heat is considered;
- a steady state condition is attained;
- the thermal properties of the bodies do not change with temperature;
- the coefficient of friction is known and reaches a steady value.

A well defined friction dependent heat source at the contact between the two bodies in stationary condition is herein considered as suitable approximation of the actual situation. The component of energy lost in deformation of the bodies can be considered negligible due to the high mobility of the polymeric chains, and an appropriate test configuration (like high velocity reciprocating movement on a short stroke) could avoid the calculation of the Peclet number [46]. Furthermore, the experimental curve of the coefficient of friction can be used in the calculation, having as result a direct correlation with the work done by the friction force (Equation 2).

$$Q(t) = \mu(t) \cdot W \cdot v \tag{2}$$

where μ is the coefficient of friction, W the normal load and v the sliding velocity. The contact temperature can be then described also as function of the time by:

$$T_{fmax}(t) = \frac{2 \cdot Q(t)}{K \cdot A} \cdot \sqrt{\frac{2 \cdot \chi \cdot b}{\pi \cdot v}} \tag{3}$$

where K is the thermal conductivity, χ the thermal diffusivity, A the contact surface and $2b$ the contact width. Both K and χ are temperature dependent so, the maximum flash temperature is also function of temperature itself i.e., $T_{fmax} = T_{fmax}(t, T)$, and they should be evaluated experimentally. The contact area considered for calculations in equation (3) is the cylindrical sector of the contact zone at the end of the test, being not reliable to measure or to hypothesize its evolution over the time: it means that the calculated temperature was under-estimated. Equations (2) and (3), thus, describe the progressive evolution of the work done by the friction force and the temperature evolution at the contact. Maximum surface contact temperature, $T_{contact}$, is calculated considering equation (4):

$$T_{contact} = T_{bulk} + T_{fmax} \tag{4}$$

where T_{bulk} is the bulk temperature of the sample before entering the contact (25ºC).

The energy that is transformed as a consequence of frictional contact can be dissipated (conversion to heat, vibration, material deformation, new surface creation) or stored into the tribosystem. Specific wear energy is a criterion which takes coefficient of friction and wear rate into account: it is the ratio of the friction work spent in the interface along the test (between the initial time t_i and the time at the end of the test t_f) divided by the mass loss (Δm) due to the wear as expressed by equation (1). The amount of material removed from a surface is a function of the energy expended on it by the friction force.

Case Study: Virgin PTFE

As a case study of the contact temperature calculation method herein presented virgin PTFE has been chosen.

Polytetrafluoroethylene (PTFE) is a well known semicrystalline polymer having relevant tribological characteristics, in particular low friction and quasi-absence of sticking effect, high resistance to temperature and good compatibility. Compared to other polymers, PTFE has high density and high melting point (around 330°C). At atmospheric pressure it undergoes several phase changes from sub-ambient temperature to the melting point.

DSC analysis was carried out in order to characterize thermally the polymeric sample. For this purpose, the sample was heated up in the range of 30-350ºC with a ramp of 10ºC/min (Figure 7) using nitrogen as inert gas. Above 30°C the material converts into a pseudo hexagonal disordered phase from a partially ordered hexagonal phase so, thermal diffusivity temperature dependence above 25°C is small [43].

The thermal properties of virgin PTFE are almost independent of temperature between ambient temperature and the melting point (see Figure 8 and Figure 9, by kind permission of the authors of reference [43]). The values of thermal conductivity, K, and thermal diffusivity, χ, used for flash temperature and contact temperature calculations, according to the provider's communication, are shown in Table 2.

According to the provider's communications, the PTFE sample comes from a process of compression molding of the PTFE powder and subsequently sinterization and cooling. The PTFE dispersion has a median particle size of 0.25 μm.

For calculations the results of the tribological tests (test configuration shown in Figure 3) presented in the previous section were used.

Considering the thermal transition curve of the virgin PTFE shown in Figure 7, it is possible to see how the PTFE material can be affected by the heat flux generated at the contact.

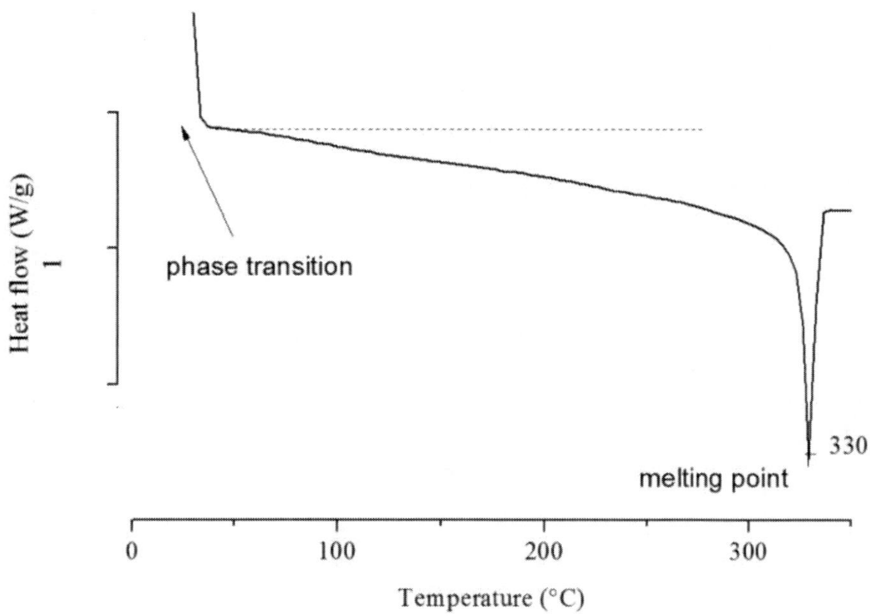

Figure 7. DSC analysis of virgin PTFE.

In particular, the small crystals formed in that range of temperature start wear processes. In fact, the small crystals formed at lower temperature are destroyed resulting in flakes separation being less stable.

Calculated contact temperature by means of equations (2) and (3) is around 60-70°C (Figure 10). These results match the experimental measurements carried out by Tzanakis [30]. PTFE fails along weak intermolecular bonds due to the minor temperature rise which involve slippage of crystalline formations of the bond structure. In other words, temperature rise is quite relevant as consequence of low thermal conductivity and the energy dissipated when the carbon-to-fluorine bonds, which have high bond dissociation energy, are broken.

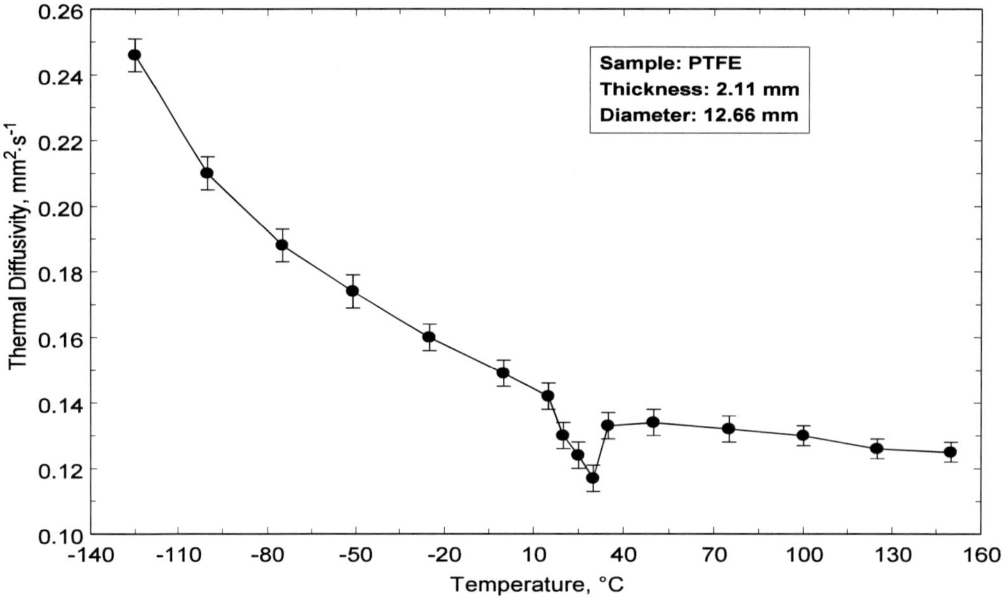

Figure 8. Virgin PTFE thermal diffusivity [43].

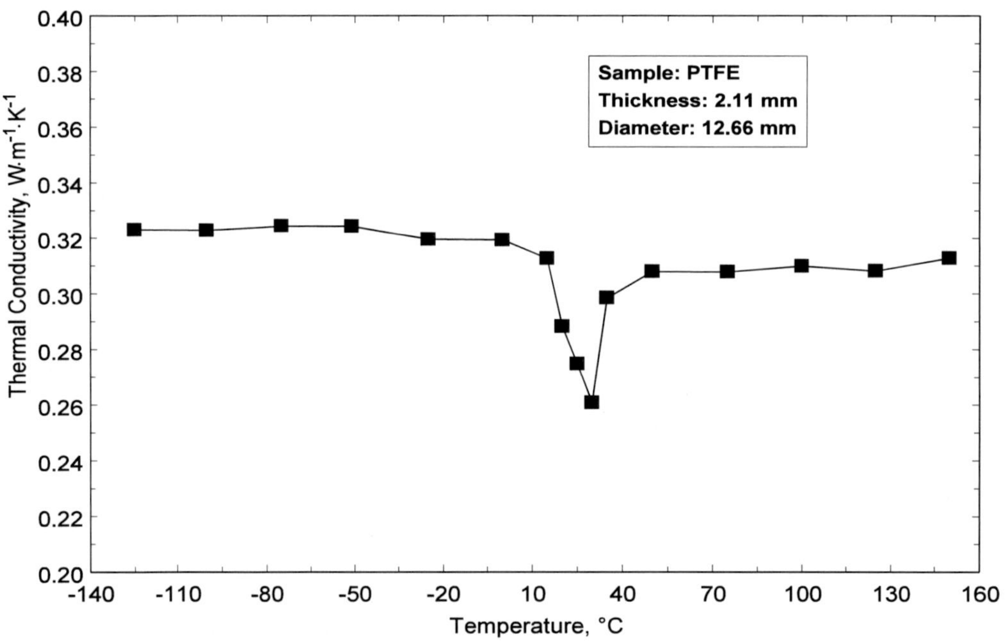

Figure 9. Virgin PTFE thermal conductivity [43].

Table 3. Thermal properties of virgin PTFE

Thermal conductivity K (W/mK)	0.24
Thermal diffusivity χ (m^2/s)	$9.24e^{-8}$
Melting temperature T_m (°C)	330
Heat of fusion H (J/g)	24.80

The work done by the friction force and the mass loss for the case of virgin PTFE are reported in Table 1. Wear specific energy was calculated according to equation (1).

In the initial stage of the test the high degree of lateral mobility of PTFE chains quickly induces molecular orientation in the sliding direction. This phenomenon is also aided by PTFE smooth linear molecular profile and ability of the electron clouds of the large fluorine ions to screen bonding forces between neighboring molecules. The measured dynamic friction coefficient is directly related to the molecular structure of the polymer. PTFE, which is composed of rigid rod-like molecules with a smooth molecular profile, shows a low coefficient of friction because of an easy slip between the aligned molecule chains at the interface.

A stable transfer layer promotes low friction and low wear. During the sliding, adhesive wear is generated as consequence of PTFE molecular reorientation in the direction of sliding. Such reorientation occurs for the surface of PTFE in contact with the countersurface as well as for the transfer film that is deposited on the countersurface.

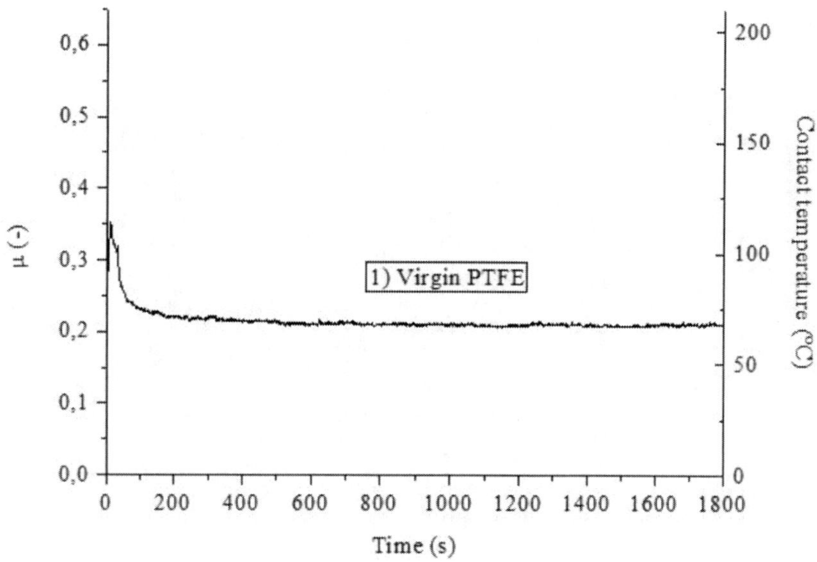

Figure 10. Coefficient of friction and calculated contact temperature evolution of the PTFE sample.

Figures 11 and 12 show the structure of the virgin PTFE sample before and after the tribological test, respectively. It can be observed that the sample shows after the test plucked marks due to the transfer film formation and subsequent adhesion.

CONCLUSION

External friction is a thermodynamically irreversible dissipative process composed of a synergism of strain, heating and mass transfer.

A comparative study based on the frictional energy analysis of PTFE composites was herein presented taking into account several aspects of the tribological behavior of the materials under analysis.

Figure 11. SEM caption of virgin PTFE surface before the tribological test.

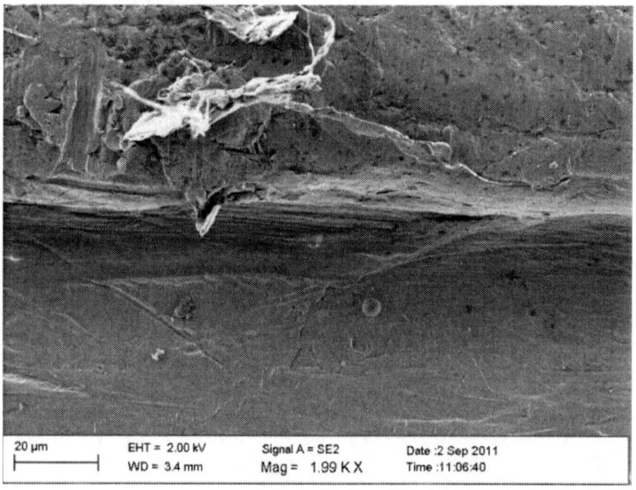

Figure 12. SEM caption of virgin PTFE contact surface after the tribological test.

When using hard-particle fillers to resist severe wear of PTFE, such fillers compromise PTFE's attractive low friction properties ($\mu<0.1$) but the wear of the composite in comparison with the virgin PTFE is further reduced because the load-carrying capacity of the composite is much higher than that of the matrix and thus any sub-surface fracture and yielding of the material are minimized due to the presence of the hard and strong second phases.

The use of both soft and hard phases in a polymer matrix enhances the self-lubricating and the load-carrying properties of the matrix improving the tribological properties of the PTFE.

A simplified frictional heating calculation method based on a tailored experimental test configuration was presented. This tailored experimental configuration allows considering:

- a planar stationary heat source at the contact;
- the evolution of the coefficient of friction along the time;
- the variation of the thermal properties of the materials along the experimental test.

The analytical calculation method was applied to a virgin PTFE polymer knowing all its thermal characteristics. A good matching with experimental measurements coming from literature was found.

The simplified calculation methods can be considered a useful tool to understand the real behavior of rubbing polymers in several applications like, for example, seals. When polymer composites are considered, thermal characteristics should be checked vs temperature and, furthermore, frictional heating calculations should be carried out when a hard phase present in the bulk material is rubbing against the countersurface.

ACKNOWLEDGMENTS

The authors would like to thank their colleagues Xana Fernandez, Miren Blanco and Gemma Mendoza for carrying out the tribological tests, the thermal analysis and SEM analysis respectively. Furthermore, specific acknowledgements are addressed to Dr. Juergen Blumm because his kind permission to reproduce Figure 8-9 allows a better understanding of the material behavior.

REFERENCES

[1] K. Friedrich, "Friction and wear of polymer composites"– *Composite materials Series*. Volume 1– Elsevier – ISBN: 0-444-42524-1.

[2] R. Flitney, "*Seals and sealing handbook*" – BH Pubblication – Elsevier – ISBN: 978-1-8561-7461-9.

[3] B. Fernandez-Diaz et al., "*Investigation of the friction and wear behaviour of HVOF and PVD coatings and PTFE tribopairs under lubricated reciprocating sliding*" – ECCM 2010 – Paris, France, May 16-21, 2010.

[4] S. K. Sinha, B. J. Briscoe, *Polymer Tribology*, ICP, ISBN-13 978-1-84816-202-0.

[5] D. M. Price et al., "Thermal conductivity of PTFE and PTFE composites" – *Proceedings of the 28th Conference of the North American Thermal Analysis Society*, Orlando, Florida, October 4-6, 2000.

[6] D. Gong et al., *Wear*, vol. 134, pp. 283–295, 1989.

[7] M. Conte et al., "Effect of surface temperature on tribological behavior of PTFE composites" - *10th International Conference on Surface Effects and Contact Mechanics*, 20-23 September 2012, Malta.

[8] J. Khedkar et al., *Wear*, vol. 252, pp. 361-369, 2002.

[9] M. Conte et al., "Frictional heating calculation based on tailored experimental tests" - *15th NordTrib conference*, 12-15 June 2012, Trondheim, Norway.

[10] E. N. Brown et al., *Materials science and engineering C*, vol. 26, pp. 1338-1343, 2006

[11] Y. J. Shi et al., *Tribology International*, vol. 40, pp. 1195-1203, 2007.

[12] V. N. Aderikha et al., *Wear*, vol. 268, pp. 1455-1464, 2010.

[13] Y. Wang et al., *Wear*, vol. 261, pp. 1359-1366, 2006.

[14] S. Bahadur, *Wear*, vol. 245, pp. 92-99, 2000.

[15] E. L. Yang et al. *Journal of Materials Research*, vol. 7, pp. 1-11, 1992.

[16] K. Tanaka, *Friction and wear of polymer composites*, Elsevier, vol. 1, pp. 137-174, 1986.
[17] A. Oshima et al., *Radiation Physics and Chemistry*, vol. 62, pp. 77-81, 2001.
[18] Q. J. Xue et al., *Journal of Applied Polymer Science*, vol. 69, pp. 1393-1402, 1998.
[19] Xiao-Xia Liu et al., *Wear*, vol. 262, pp. 1414-1418, 2007.
[20] J. R. Vail et al., *Wear*, vol. 267, pp. 619-624.
[21] S. E. McElwain et al., *Tribology Transactions*, vol. 51, pp. 247-253, 2008.
[22] Fei Li et al., *Wear*, vol 249, pp. 877-882, 2001.
[23] Shangguan Qian-quiana et al., *Wear*, vol. 260, pp. 1243-1247, 2006.
[24] P. J. Blau, "*Friction science and technology*" – New York: Marcel Dekker, 1986.
[25] P. J. Rae, D. M. Dattelbaum, *Polymer*, vol. 45, pp. 7615-7625, 2004.
[26] P. J. Rae, E. N. Brown, *Polymer*, vol. 46, pp. 8128-8140, 2005.
[27] M. Jahangiri et al., *Wear*, doi: 10.1016/j.wear.2011.08.023.
[28] B. N. J. Persson, *Journal of Physics: Condensed Matter*, vol. 18, pp. 7789-7823, 2006.
[29] Y. Takeichi et al., *Wear*, vol. 264, pp. 308-315, 2008.
[30] I. Tzanakis, M. Hadfield, "Evaluation of flash temperature of a composite elastomer tip seal with dry sliding conditions in contact with a high carbon steel plate" - *10th International Conference on Surface Effects and Contact Mechan*ics, 20-23 September 2012, Malta.
[31] B. I. Kostetsky, *Wear*, vol. 159, pp. 1-15, 1992.
[32] F. E. Kennedy, Thermal and thermomechanical effects in dry sliding, *Wear* 100 (1984) 453-476.
[33] F. Bucher, K. Knothe, A. Theiler, Normal and tangential contact problem of surface with measured roughness, *Wear* 253 (2002) 204-218.
[34] M. Kalin, Influence of flash temperatures on the tribological behaviour in low-speed sliding: a review, *Materials Science and Engineering: A* 374 (2004) 390-397.
[35] F. E. Kennedy, Single pass rub phenomena – Analysis and experiment, *Journal of Lubrication Technology* 104 (1982) 582-588.
[36] J. F. Archard, The temperature of rubbing surfaces, Wear 2 (1958-1959) 438-455.
[37] S. K. Sinha, B. J. Briscoe, *Polymer Tribology*, ICP, UK, 2009.
[38] B. N. J. Persson, Rubber friction: role of the flash temperature, *J. Phys.: Condens. Matter*, 18 (2006) 7789-7823.
[39] G. Sutter, N. Ranc, Flash temperature measurement during dry friction process at high sliding speed, *Wear* 268 (2010) 1237-1242.
[40] F. E. Kennedy, D. Frusescu, J. Li, Thin film thermocouple arrays for sliding surface temperature measurement, *Wear*, 207 (1997) 46-54.
[41] F. U. Tieyuan, Q. I. Yuchen, W. U. Suyun, C. Donglin, Equilibrium heat of fusión of PTFE, *Polymer Communications*, 4 (1985) 323-331.
[42] Y. Wang, F. Yan, Tribological properties of transfer films of PTFE-based composites, *Wear* 261 (2006) 1359-1366.
[43] J. Blumm, A. Lindemann, M. Meyer, C. Strasser, Characterization of PTFE using advanced thermal analysis techniques, *Int. Journal of Thermophysics*, 31(10) (2010) 1919-1927.
[44] E. N. Brown, P. J. Rae, E. B. Orler, G. T. Gray III, D. M. Dattelbaum, The effect of crystallinity on the fracture of PTFE, *Materials Science and engineering C* 26 (2006) 1338-1343.

[45] X. Q. Wang, D. R. Chen, J. C. Han, S. Y. Du, Crystallzation behavior of PTFE, *Journal of Applied polymer science* 83 (2002) 990-996.
[46] G. W. Stachoviak, A. W. Batchelor, Engineering Tribology, Elsevier, UK, 2005.
[47] F. P. Bowden, D. Tabor, *Friction and lubricating wear of solids* part 1, Clarendon press, UK, 1964.
[48] H. Block, Theoretical study of temperature rise at surfaces of actual contact under oiliness lubricating conditions, *General discussion on lubrication*, London 2 (1937) 222-235.
[49] J. C. Jaeger, Moving sources of heat and the temperature at sliding contact, *Proc. Royal Society*, N. S. W. 76 (1943) 203-224.

In: New Developments in Polymer Composites Research ISBN: 978-1-62948-340-5
Editors: S. Laske and A. Witschnigg

Chapter 12

INJECTION MOLDING COMPOUNDING OF POLYMER NANOCOMPOSITES

Markus G. Battisti [*] ***and Walter Friesenbichler***
Department of Polymer Engineering and Science – Chair of Injection Molding of Polymers, Leoben, Austria

ABSTRACT

In comparison to the conventional compounding process, where the compound has to be pelletized and fed into the injection molding machine for the second plasticizing process, the injection molding compounding combines these two processing steps which normally take place at the raw material manufacturer and the injection molder. The material compounding and the subsequent injection molding are done directly "by only one plasticizing process" by the use of a heated melt pipe and a melt accumulator. The molten material exits the extruder and is fed into a melt accumulator or into the injection molding machine, respectively. The accumulator serves as a buffer element between the continuously-operating extruder and the discontinuously-operating injection unit and so the final part is processed directly. In contrast to many other industrial installations this worldwide unique MUL Polymer NanoComposite Injection Molding Compounder (PNC-IMC) features some special devices, or to be more precise a special arrangement, for the production of polymer nanocomposites.

This chapter will give an overview of the work on the injection molding compounding technique at the Chair of Injection Molding of Polymers at the Montanuniversitaet Leoben. It will show the comparison of different processing and compounding techniques and their influence on the Young's modulus as well as the tensile strength. Both, conventional processing of polymer nanocomposites and processing in the worldwide unique Injection Molding Compounder (PNC-IMC) were evaluated. Additionally the effects of nanofillers on the thermal conductivity at various pressures and the warpage of glass-fiber reinforced polypropylene further reinforced with layered silicates were investigated.

These very first results give a good overview on both the possibilities and the limitations of the innovative concept of the PNC-IMC.

* E-mail address: markus.battisti@unileoben.ac.at; Address: Otto-Glöckel-Straße 2; 8700 Leoben (Corresponding author)

Keywords: Injection molding compounding, nanofiller, layered silicates, thermal conductivity, warpage

INTRODUCTION

The injection molding compounding (IMC) concept was firstly presented by Krauss Maffei at the K1998 for long glass-fiber reinforced thermoplastics. Its characteristic feature is a co-rotating twin-screw extruder, which is directly integrated into an injection molding system. The integration of a co-rotating twin-screw extruder in an injection molding machine opens a wide field of new applications for the direct processing of plastics. It gives the processor greater freedom during processing and greater flexibility when it comes to improving both the component structure and the quality of the molded part, as well as lower production costs. The combination of the injection molding with the continuous plastification offer further processing advantages – particularly where high throughput rates are important to manufacture products with high filler contents or to solve complex plastification and compounding problems. The continuous plastification process has a significant advantage over the traditional single-screw of the injection molding machine, which unfortunately reaches its limits in certain applications (Reitner, 2001, Sieverding et al., 2005).

Let's go back to the main principle of the injection molding compounding. The injection molding compounding process combines two processing steps, the material compounding which normally takes place at the raw material manufacturer and the injection molding process which usually is done at the injection molder. The final part is processed directly "by only one plasticizing process" and the material used has not to be granulated and plasticized again after the compounding step in the plasticizing unit of an injection molding machine. After compounding the molten material exits the extruder and is directly transferred into the injection unit by use of a heated melt pipe and a melt accumulator. After the specimen volume is reached, a conventional injection molding process starts. During the injection and the holding pressure phase of the injection molding process, the accumulator serves as a buffer element between the continuously-operating extruder and the discontinuously-operating injection molding machine (Reitner, 2001, Sieverding et al., 2005).

All presented systems till now have been designed for the processing of fibers (long glass-fibers, natural fibers). The IMC technology is excellently suited for the production of extensive thin-walled parts due to the low deformation speeds and smaller reduction of the fiber length. The greatest damage to the fibers occurs while the melt containing the fiber is filling the mold. The melting process greatly affects fiber length as well. There are major differences between injection molding and injection molding compounding. With the injection molding machine, the initial fiber length is restricted by the size of granulates. In contrast, the injection molding compounder melts the pure matrix material without fiber. The fibers are added to the melt later and are thus exposed to correspondingly less mechanical stress. Measurements revealed that the long glass-fibers amount added as long glass-fiber granulates in the final part is twice as high with the production of the IMC compared to a conventional processing in the injection molding machine and even three times as high by using continuous filaments (Bürkle et al., 2003). Another example is the dispersion of natural fibers which is much better and therefore increases the dynamic loading capacity of IMC molded parts. An additional advantage of natural fiber reinforced polymers in contrast to glass-fiber or carbon fiber reinforced polymers is, that the length of the fibers in the final part is not that important than for other fibers (Ruch et al., 2002).

Further benefits of processing in a "single-heat" are also lower working temperatures and the shorter residence time in the whole processing facilities. The material's properties are adjusted through the addition of additives, for instance, to improve bonding of the polar natural fibers to the nonpolar polyolefin, which is also necessary for layered silicates. Other common additives are UV stabilizers, impact modifiers and color masterbatches (Bürkle et al., 2009).

Another, especially nowadays very important factor is energy saving. The combination of compounding and molding in a single machine concept not only reduces the thermal load on the materials to be processed. It also raises the energy efficiency of the entire process (enthalpy change resulting from thermal and dissipation energy when considering the conversion of energy for material compounding). Compared to the conventional two-step process, the one-step process in the injection molding compounder reduces energy consumption by up to 40 %. Moreover, the more energy-efficient, continuously operating co-rotating twin-screw extruder replaces the energy-intensive step of plasticizing by means of a reciprocating screw (Bürkle et al., 2009).

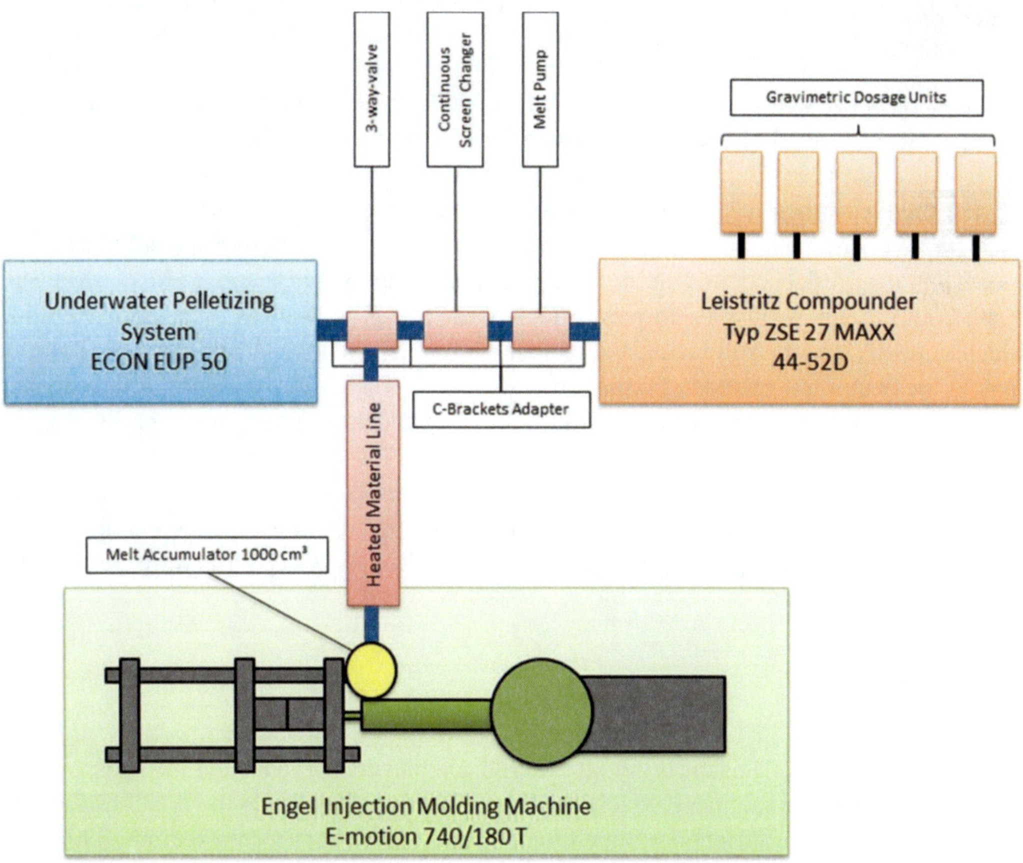

Figure 1. System schematic of the MUL Polymer NanoComposite Injection Molding Compounder (PNC-IMC).

Although there are already a lot of different applications implemented in the concept of the injection molding compounder there are many more regarding the advantages of injection molding in a "single-heat". One example is the easier production of articles from compounds that have a tendency to separation.

The MUL polymer nanocomposite injection molding compounder (PNC-IMC) described above(Figure 1) is the world's first special adapted IMC for manufacturing and injection molding of polymer nanocomposites. It consists of a ZSE 27 MAXX Leistritz-Compounder (Leistritz Extrusionstechnik GmbH, Nürnberg, Germany) and an 1800 kN fully-electric injection molding machine by Engel (Engel Austria GmbH, Schwertberg, Austria). A 3-way-valve and its special arrangement allows for using the IMC as "three-in-one" production line either for injection molding compounding, for injection molding or as a separately acting twin-screw compounder line with an underwater pelletizing system EUP 50, purchased by Econ GmbH (Weiszkirchen, Austria). Furthermore, this PNC-IMC can also be used for the homogenization of fillers in high temperature polymers (HTP) in a temperature range up to 400 °C.

Compounding by melt mixing is industrially the most attractive method for polymer nanocomposites preparation due to its technological simplicity. Due to the possibility to use common polymer processing machines it is much easier and even more pleasant to produce practical amounts of compounds. In comparison the laboratory-methods like "in situ" or "solution methods" are complicated and only capable for a limited amount of compound. Therefore these methods are not so interesting for many industry applications (Mignoni et al., 2011, Vaia et al., 1996).

For the preparation of polymer nanocomposites and the successful dispersion of silicate plates in a polymer melt both, high shear energy (usage of kneading blocks and high screw speed) and long residence time are necessary. Based on preliminary work at MUL (Kracalik et al., 2009) it is now possible to realize high shear energy as well as long residence time, which normally act oppositional in the extrusion process, by implementing a melt pump in front of the twin-screw extruder. Further possibilities to diversify the residence of the polymer melt in the compounder are the screw length of the extruder, and the processing in an IMC, which both is possible with the PNC-IMC described. The challenge is now, to find the optimal processing conditions or the process window respectively where the shear energy input and the residence time create the desired material properties, which are not always the same when we are thinking on high stiffness and/or high toughness and low warpage all at the same time.

EXPERIMENTAL

This paragraph explains the different material combinations, production methods, process parameters and test equipment which were used for research with the PNC-IMC. It is important to know that different research topics require different process settings, which are explained below.

The main polymer for the research with the PNC-IMC is on the one hand a virgin and on the other hand a glass-fiber reinforced polypropylene (PP). The PP homopolymer Bormed DM55pharm (MFR 2.8 g/10 min, 230 °C/2.16 kg) supplied by Borealis (Linz, Austria) was used for the investigations concerning Young's modulus, tensile strength and thermal conductivity. The semi-crystalline PP homopolymer RIALENE P100 SGF10plus ST (MFR 8.2 g/10 min, 230 °C/2.16 kg) supplied by VTS (Vienna, Austria) with 10 % special-glass-fiber was used mainly for the investigations concerning thermal conductivity and warpage. The used nanofiller for all tests was Nanofil 5 (montmorillonite intercalated with dimethyl distearyl ammonium chloride) supplied by Rockwood Clay Additives GmbH (Moosburg, Germany). To improve the bonding of the polar nanofiller to the nonpolar polyolefin the compatibilizer Scona TPPP 2112 GA (PP grafted with 1.2 wt% of maleic acid anhydride, PP-g-MA, MFR 6 g/10 min, 190 °C/2,16 kg) supplied by BYK Kometra (Schkopau, Germany) was used for both PP types.

First of all the various options for the production of the polypropylene nanocomposites compounds are explained in Table 1.In the “3-in-1 compounding process” all the components were fed simultaneously by separately controlled gravimetric dosage units. PP and compatibilizer were fed upstream through the main hopper, and the layered silicates are fed downstream at the 4th extruder barrel in distance of 16 D from the main hopper. After the compounding step the material can be processed with the one-step (highest energy saving potential) or the two-step-process.

Another possibility is production with the pre-stage masterbatch (MB) process. Here a MB of 50 wt.% organoclay and 50 wt.% compatibilizer is produced in the twin-screw extruder and then pelletized. The masterbatch pellets were fed into the twin-screw extruder again and then diluted with virgin PP to receive compounds with the desired wt.% of nanoclay and compatibilizer. It is important to know, that the “3-in-1 compounding process” with two-step processing has also potential of energy saving because there is no need to plasticize the material again after the additional MB process.

Table 1. Various options for the production of the polypropylene nanocomposites compounds

One-step process (pure IMC process)	• Production of test specimens or injection molded parts in a single-heat IMC-process (only one plasticizing and cooling process)
Two-step process	• Conventional compounding and pelletizing • Subsequent conventional injection molding of test specimens or injection molded parts (two plasticizing and cooling steps)
3-in-1 compounding process	• Direct compounding of the organoclay and the compatibilizer into the polymer; • Processing either with the one or the two-step process.
Masterbatch (MB) process	• Step 1: Production of the MB (50 wt% organoclay and (50 wt% compatibilizer) by conventional compounding • Step 2: Dilution of the MB in the compounder with PP to receive a compound with 5 or 10 wt% organoclay and compatibilizer • Step 3: Processing either with the one or the two-step process

Different process parameters like screw speed, counter-pressure of the melt pump and screw geometry can also be varied in a wide range. It is important to recognize, that the pressure gradient of the melt pump (Witte Pumps & Technology GmbH, Uetersen, Germany) is negative to increase the residence time and shear rate of the polymer compound leading to improved dispersion of the nanofiller (Witschnigg et al., 2009). The influence and the subsequent conclusions will be explained more precisely in the result paragraph for each experiment. As explained earlier, the residence time is the second very important factor for the production of polymer nanocomposites. The minimal residence time was measured as the time between color masterbatch granulate insertion into the hopper and color change of the outgoing molten string at the 3-way-valve and the injected part, respectively.

As test equipment for the mechanical properties a universal tensile testing machine Z010 (Zwick and Co. KG, Ulm, Germany) was used. This machine was used to carry out the tensile tests according to ISO 527-2 at standardized conditions (23 ± 2 °C/50 ± 5 % r.H.). The data were evaluated using the test Xpert II software. To guarantee the best and mainly the same processing conditions for all test specimens, test series were made before the test specimen's production. The optimized injection molding conditions for the production of the test specimens are presented in Table 2.

Table 2. Injection molding conditions for the production of the test specimens

Barrel temperature (°C) Z1-Z2-Z3-Z4-Z_N	**190-200-200-210-210**
Mold temperature (°C)	40
Holding pressure (bar)	800
Injection speed (cm^3/s)	75
Dosing speed (m/s)	0.075 or IMC

Thermal conductivity measurements in dependency of temperature and pressure, based on the "Line-Source-Method" according to ASTM D5930-97, were carried out in the temperature range from 80 to 240 °C using a high pressure capillary rheometer (Rheograph 2002, Göttfert, Germany) with a single channel system having a diameter of 15 mm. At the beginning of each measurement the high pressure capillary rheometer was heated up to the specified temperature for 60 minutes. A defined weight of specimen was filled in the cylinder and compressed tightly while filling. Afterwards the material was compressed with a piston for 10 minutes and then the measurement started at the defined pressure stage. At least 3 measurements for each temperature were made keeping12 to 13 minutes delay time between each measurement. While cooling down temperature to the next measurement level the delaytime between each temperature range was 20 to 30 minutes.

The degree of intercalation and exfoliation has been characterized by X-ray diffraction (XRD). The measurements were carried out using BrukerNanoSTAR (Bruker AXS, Karlsruhe, Germany) small angle X-ray scattering equipment. This system was equipped with a two-dimensional X-ray detector using a wavelength of 0.154 nm (CuKα). The scattering measurements were performed on plates with a thickness of 1 mm in transmission mode. To determine the gallery period a Gaussian peak shape was used and the particle shape was set to disk. There are two different methods in order to evaluate the degree of exfoliation. On the one hand the interlayer distance and on the other hand the area A under the silicate layer peak (see Figure 2). This area is proportional to the number of particles hit by the X-ray beam. The better the nanocomposite is exfoliated, the lower the probability of hitting particles and therefore the smaller the area is (Laske et al., 2012).

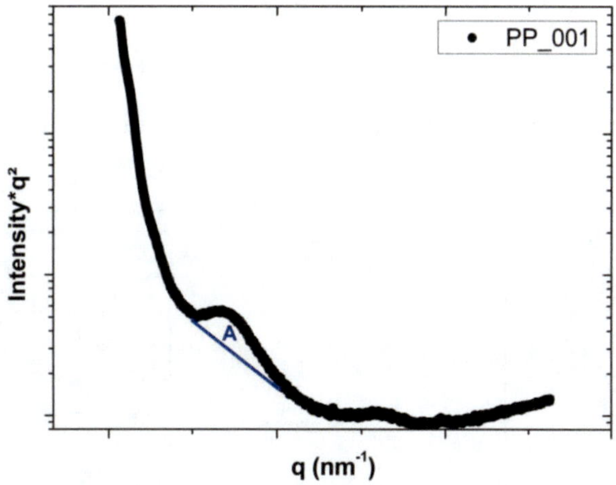

Figure 2. Calculation of degree of exfoliation using the area under the peak (A).

Results

First of all, results of the mechanical properties were presented. As explained earlier in this chapter the shear energy and the residence time are the two main influential variables. The possibility to vary the length of the extruder opened a large area for scientific development and research. For the first experiments 11 barrel or 13 barrel segments were used, where each barrel segment had a length of 4 D. All the components were fed by separately controlled gravimetric dosage units at an overall throughput rate of 6 kg/h. The formulation of the compounded polymer nanocomposites was constant at 90 wt% polypropylene, 5 wt% organoclay and 5 wt% compatibilizer produced with the "3-in-1compounding process". The used screw geometry can be seen in Figure 3.

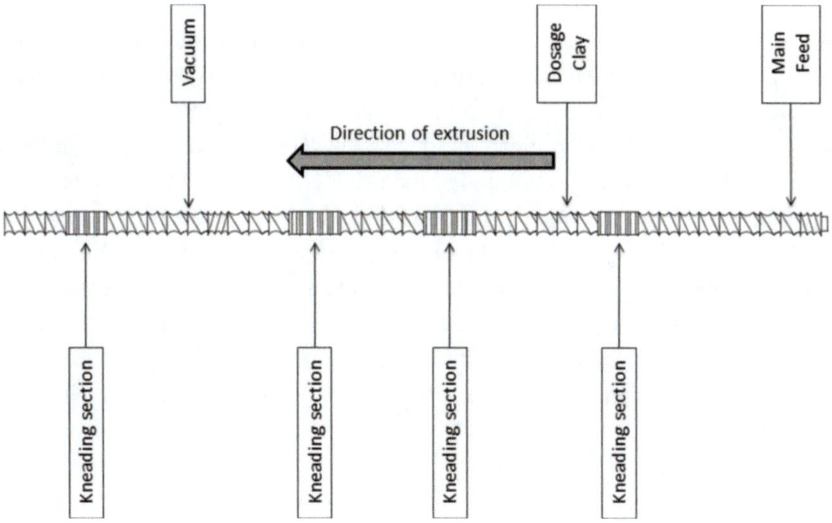

Figure 3. System illustration of the screw geometry with a length of 44 D.

The screw speed and the counter-pressure of the melt pump were varied according to Table 3 to observe the influence of different processing parameters.

Table 3. Design of Experiments for the prepared nanocomposites; extruder length as multiple of the screw diameter D.

Extruder Length (D)	Screw Speed (rpm)	Counter Pressure (bar)
44	50	0
		25
		50
		75
		100
	100	0
		25
		50
		75
		100
	150	0
		25
		50
		75
		100
52	50	50
		75
		100
	100	50
		75
		100
	150	50
		75
		100

The shear energy and residence time are the two main influential variables, which normally act oppositional to each other in the extrusion process. The effects of screw speed and counter-pressure on the Young's modulus for an extruder length of 44 D are plotted in Figure 4. Unfortunately there is no clear trend visible when comparing the counter pressure between 0 bar and 100 bar of the compounded material. Nevertheless it is noticeable that with only one exception at 0 bar the Young's modulus decreases with increasing screw speed. That means with higher screw speed and thereby higher shear force, the Young's modulus decreases. The values of the Young's modulus for the pure PP result from tensile tests of injection molded parts, which were produced under the same conditions like the compounds. It is striking to see, that the Young's modulus values decrease with higher counter pressure. A possible reason is the longer residence time in the compounder and the subsequent degradation of the material.

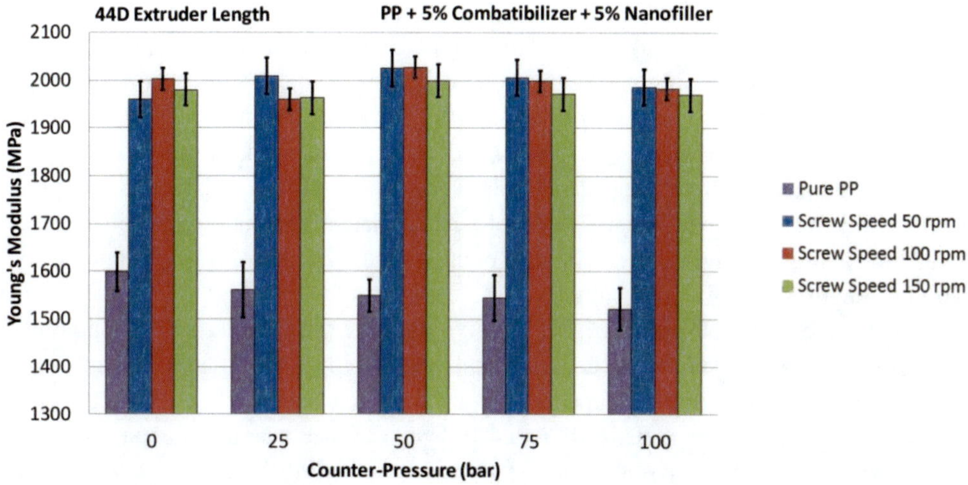

Figure 4. Effect of screw speed and counter-pressure on Young's modulus using an extruder length of 44 D.

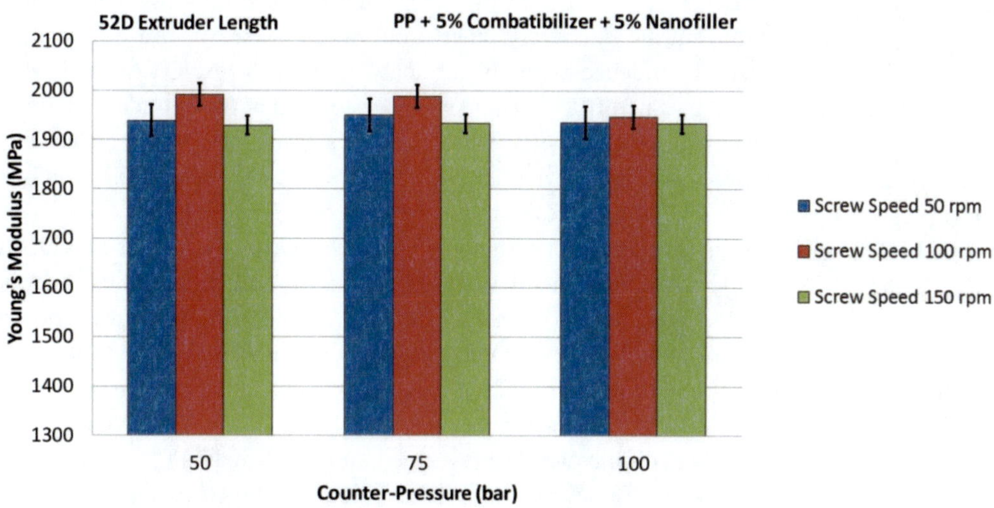

Figure 5. Effect of screw speed and counter-pressure on Young's modulus using an extruder length of 52 D.

Figure 5 shows the influence of screw speed and counter-pressure on the Young's modulus at an extruder length of 52 D. Due to the results of the 44 D extruder length Figure 5 only shows the Young's modulus values for 50 bar, 75 bar and 100 bar counter pressure. To investigate only the effects of longer residence time and not the shear energy input the additional 2 barrel segments with 8 D length were only used as lengthening of the metering zone with conveying elements (see screw geometry, Figure 2). In contrast to the results at the 44 D extruder length the processing conditions with a screw speed of 100 rpm results in the best Young's modulus values of the tested compound. Like in the experiments with the 44 D screw the 52 D screw shows no clear trend related to the counter-pressure. The residence time of the polymer melt varies between 260 s (100 bar), 245 s (75 bar) and 230 s (50 bar).

As one can see, the improvement in Young's modulus compared to virgin polypropylene was approximately 30% for the PNCs investigated, whereas the influence on the Young's modulus regarding processing conditions was really low. Same results occurred at the tensile strength values were the increase was about 10% but only minimal influenced through the different processing conditions. A consolidated view indicates that a longer residence time, caused by counter pressure (see also Table 5) and necessary for the diffusion process of intercalation and subsequent delamination of silicate platelets in the polymer matrix, seems to have only a minimal influence on the Young's modulus. The measurements with 52 D extruder length led to worse results than the experiments with 44 D. This means, that the additional use of the melt pump with a negative pressure gradient is not so important due to the additional long residence time caused by the relative small throughput and the long heated melt pipe of the chosen IMC concept.

To verify these results statistically an analysis of variance was made. Analysis of variance (ANOVA) is a collection of statistical models used to analyze the differences between group means and their associated procedures. To get not too deep into this topic only two important values which indicate the significance of a result will be shown. On the one hand the F-value, which examines the likelihood of a difference between two different observations, is calculated (threshold level > 1, the higher the better). On the other hand the post hoc Tukey range test, which compares every group mean with every other group mean to reveal, if a difference is significant, is analyzed. The confidence levels of all analyses with the ANOVA were set to 95%, which correspond to a α-value of 0.05. The ANOVA shows, that the extruder length (F-value = 50) is an important process parameter, which lead to significant different values also caused because of relative low standard deviations. The screw speeds (F-value = 2.74) and the counter pressures (F-value = 1.72) did not really influence the Young's modulus.

The next step was the investigation concerning the degree of exfoliation reached with the 44 D and the 52 D extruder. As mentioned earlier there are two different methods for the analysis of the XRD results. For these experiments the area A under the peak was chosen. Table 4 shows the different results for the 44 D and the 52 D extruder screw from 50 to 100 bar counter pressure and all tested screw speeds. It is striking to see, that there is no visible trend when comparing the different counter pressures. But if we take a closer look at the screw speed it is obvious, that with increased screw speed area A decreases for the 44 D and also the 52 D screw. Furthermore if one calculates the average of each screw speed for 44 D and 52 D screw length the values are more or less the same. To verify the results the ANOVA was made once more resulting in the screw speed as the only significant parameter for the degree of exfoliation in the compounds (F-value = 96). Counter pressure (F-value = 1.5) and extruder length (F-value = 0.07) do not have any effect.

Obviously the decrease in area A and as a consequence the degree of exfoliation only depends on the energy input of the rotating screw. It seems as if the degree of exfoliation is not affected by the residence time (counter pressure and extruder length). These results also explain that the decrease of the Young's modulus with increasing screw speed and increasing counter pressure (see Fig. 3 and Fig. 4) is only affected by the degradation of the material due to the higher residence time in the extruder.

Table 4. Area under the peak in dependency of counter pressure and screw speed at an extruder length of 44 D and 52 D

Extruder Length (D)	Area under peak (nm²)	Counter Pressure (bar)	Screw Speed (rpm)
44	562	50	50
	565	75	
	555	100	
	547	50	100
	526	75	
	538	100	
	453	50	150
	449	75	
	473	100	
52	559	50	50
	587	75	
	551	100	
	498	50	100
	582	75	
	455	100	
	511	50	150
	443	75	
	456	100	

Table 5 shows a residence time study for different pressure gradients and different throughput rates. The minimal residence time 1 was measured as the time between color masterbatch granulate insertion into the hopper and color change of the outgoing molten string at the 3-way-valve and the maximal residence time 2 was measured after no discoloration of the virgin material was noticeable. Some residence time values in Table 5 for the lowest throughput rate (4 kg/h) and/or the lowest screw speed (50 rpm) could not be measured due to exaggerated torque or insufficient counter pressure.

Table 6 shows selected results for the residence time depending on extruder length, counter pressure and throughput rate. The residence time increases by approximately 15 seconds when using the longer extruder screw but the difference is getting smaller when increasing the throughput rate.

The values in these two tables also partly explain the differences between the Young's modulus presented in Figure 4 and 5 in this paragraph and gives hints towards possible degradation of the material. Based on these results the construction of an electronically controlled 3-way-valve in order to realize high throughput rates up to 40 kg/h is planned. This valve should enable handling of high throughput rates and supplying material simultaneously for the injection molding compounder and the underwater pelletizing system. The pelletized compound further can be processed with a conventional processing machine and therefore a higher efficiency of the compounder is achievable.

Table 5. Residence time in dependency of the counter pressure, screw speed and the throughput rate at an extruder length of 44 D

	Counter pressure 0 bar					
screw speed (rpm)	50		100		150	
throughput rate (kg/h)	Time 1 (s)	Time 2 (s)	Time 1 (s)	Time 2 (s)	Time 1 (s)	Time 2 (s)
4	335	625	290	510	275	495
6	245	500	215	405	205	365
8	205	370	170	330	150	290
10	170	300	145	285	125	255

	Counter pressure 25 bar					
screw speed (rpm)	50		100		150	
throughput rate (kg/h)	Time 1 (s)	Time 2 (s)	Time 1 (s)	Time 2 (s)	Time 1 (s)	Time 2 (s)
4	355	625	300	510	275	490
6	250	485	220	410	205	365
8	205	370	170	330	160	290
10	175	305	145	285	130	255

	Counter pressure 50 bar					
screw speed (rpm)	50		100		150	
throughput rate (kg/h)	Time 1 (s)	Time 2 (s)	Time 1 (s)	Time 2 (s)	Time 1 (s)	Time 2 (s)
4	355	625	305	530	280	500
6	260	485	225	390	210	360
8	215	380	180	300	170	305
10	-	-	150	265	130	230

	Counter pressure 75 bar					
screw speed (rpm)	50		100		150	
throughput rate (kg/h)	Time 1 (s)	Time 2 (s)	Time 1 (s)	Time 2 (s)	Time 1 (s)	Time 2 (s)
4	360	625	315	465	-	-
6	265	495	225	330	210	355
8	225	380	180	300	170	280
10	-	-	150	165	135	250

	Counter pressure 100 bar					
screw speed (rpm)	50		100		150	
throughput rate (kg/h)	Time 1 (s)	Time 2 (s)	Time 1 (s)	Time 2 (s)	Time 1 (s)	Time 2 (s)
4	-	-	-	-	-	-
6	280	500	235	360	210	360
8	-	-	185	300	170	280
10	-	-	160	275	135	250

The second investigation concerning mechanical properties was carried out using the advantages of processing with the IMC. Figure 6 shows the difference of the Young´s modulus for variants of processing in a single-heat. It can be clearly seen, that the two-step process generates slightly higher Young's modulus and tensile strength values compared to processing in a single-heat. This can have various reasons. On the one hand the shear energy in the two-step process is higher than the dissipated energy in the one-step process which leads to a better exfoliation of the nanoclay in the polymer matrix (see decreasing area values in Table 4) based on the fact that the pelletized compound has to be molten again in the injection molding machine and so the injection screw dissipates shear energy again. On the other hand own measurements showed that the residence time of the single-heat process currently is 2.5 times higher than in the two-step process due to the PNC-IMC layout with its long melt conveying system. The residence time in the IMC depends on the cycle time of the injection molding machine and therefore it is variable to a certain extent. The ANOVA revealed that there are significant differences between the pure PP, the one-step process with 5 % nanofiller, the two-step process with 5 % nanofiller and the two-step process with 10 % nanofiller (F-value = 54) regarding the Young's modulus. The tensile strength is extremely significant (F-value = 1053) and only the one-step process with 5 % and the one-step process with 10 % nanofiller deliver non-significant values. In summery these experiments show that the two-step process generates clearly better mechanical properties of a compound than the one-step process.

Table 6. Residence time in dependency of the counter pressure, the throughput rate and the extruder length at a screw speed of 100 rpm

throughput rate (kg/h)	Counter pressure (bar)					
	50		75		100	
	Minimal Time (s)		Minimal Time (s)		Minimal Time (s)	
	44 D	52 D	44 D	52 D	44 D	52 D
4	305	320	315	330	-	-
6	225	230	225	240	235	250
8	180	190	180	195	185	200
10	150	160	150	165	160	170

Nevertheless the two interacting mechanisms of exfoliation and degradation are difficult to separate, thus, a lot of further work is necessary to decrease residence time on the one hand and to increase the degree of exfoliation without simultaneous material degradation on the other hand. Similar results were generated from other universities which also think that it may be advantageous to find a balance between mechanical stresses that necessitate a high level of viscosity and a diffusion process that requires rather low medium viscosity and enough time (Bousmina, 2006, Cho et al., 2011).

Table 7 displays another important result of these measurements in fact that the amount of compatibilizer and nanofiller has only little effect on the Young's modulus. The increase up to 10 wt.% improves the compounds stiffness less than 5 % which leads to the assumption, that there will be a plateau with increasing amount of nanofiller content (Laske et al., 2009).

The third experiment tries to visualize the influence of the pre-stage MB process. The effect of an additional masterbatch process for the preparation of polypropylene nanocomposites is plotted in Figure 7. Here "3-in-1 compounding process" is the abbreviation for mixing PP, nanofiller and compatibilizer in one compounding process. This compounding procedure is compared with the masterbatch pre-stage process, where a masterbatch of 50 wt.% organoclay and 50 wt.% compatibilizer was produced in the twin screw extruder and then pelletized. The masterbatch pellets were fed into the twin-screw extruder again and then diluted with PP to receive a compound of 5 wt.%nanofiller and 5 wt.% compatibilizer. This new compound was either transported directly into the injection molding machine (one-step process) or pelletized again and then injection-molded (two-step process).

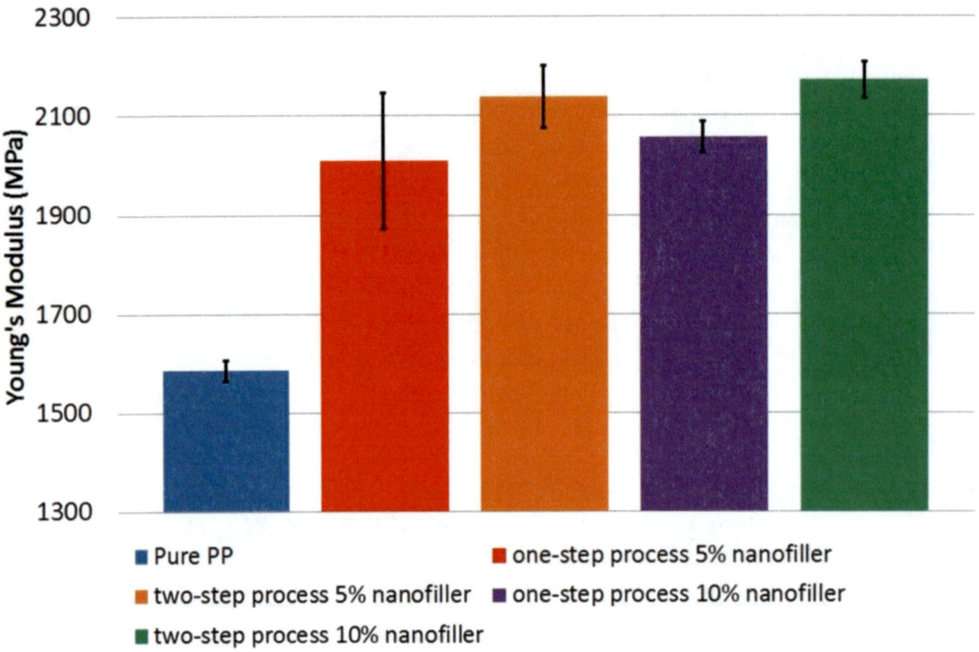

Figure 6. Effect of one-step and two-step process on Young's modulus using an extruder length of 52 D (zoom).

Table 7. Increase of Young's Modulus and Tensile Strength compared to the virgin unfilled PP DM55

Material	Young's Modulus (MPa)	Increase (%)	Tensile Strength (MPa)	Increase (%)
Pure PP	1590	-	33.6	-
one-step_5%	2010	26	36.8	9
two-step_5%	2140	35	37.8	13
one-step_10%	2060	30	36.7	9
two-step_10%	2170	36	38.3	14

On the one hand one can see again, that the two-step process leads to higher tensile strength and also generates higher Young's modulus values (Table 8) than processing in a single-heat and on the other hand the masterbatch process shows its great potential for the production of nanocomposites. The improvement in tensile strength compared to the virgin PP is not so impressive, but the increase in Young's modulus compared to virgin polypropylene is approximately 30 % for "3-in-1 compounding process", two-step process and one-step MB process whereas it is 40 % for the MB two-step process. That is not really a big difference but still noticeable. These results also were checked with ANOVA, where only the "3-in-1 compounding process" and the one-step MB process led to non-significant values concerning the Young's modulus. The other Young's modulus values showed as already assumed clear differences (F-value = 375). The tensile strength values are due to a really low standard deviation all significant (F-value = 322).

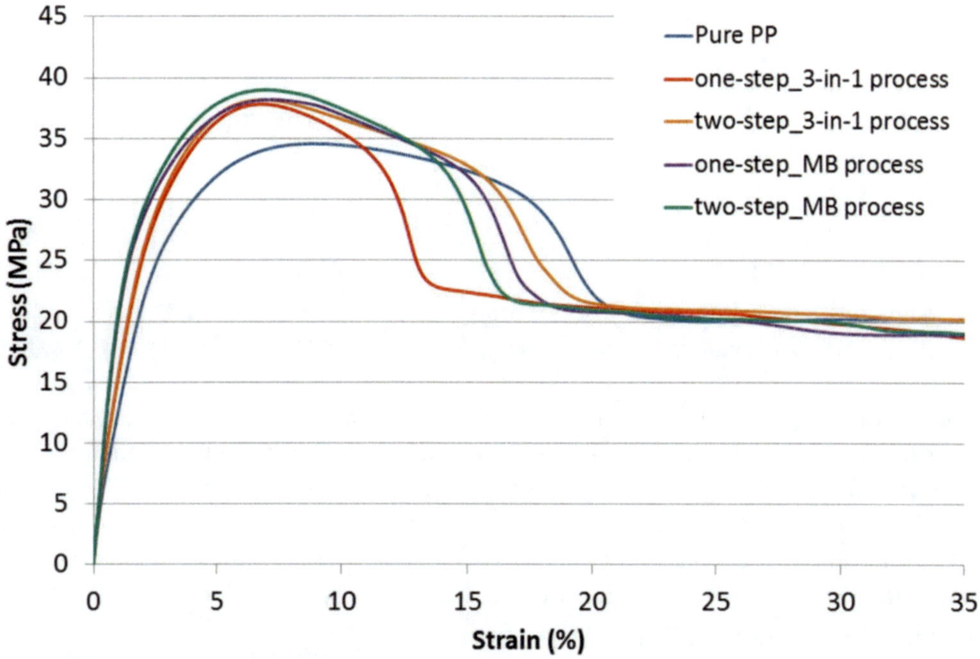

Figure 7. Effect of additional MB process in one-step and two-step processing on tensile strength using an extruder length of 44 D.

Table 8. Increase of Young's Modulus and Tensile Strength compared to the virgin unfilled PP DM55

Material	Young's Modulus (MPa)	Increase (%)	Tensile Strength (MPa)	Increase (%)
Pure PP	1590	-	33.6	-
one-step_3in1	2030	27	35.7	6
two-step_3in1	2040	28	36.3	8
one-step_MB	2080	31	36.9	10
two-step_MB	2190	38	37.7	12

In conclusion all investigations on Young's modulus and tensile strength have shown that the conventional compounding with the pre-stage MB process and the two-step process of the nanocomposite leads to the best values up to this day. Nevertheless, one should keep in mind that the "3-in-1" one-step process generates only 10 % lower Young's modulus values than the conventional process but with remarkable savings in energy (up to 40 %) and time.

These results have also an important aspect for the industry, because one of the biggest challenges still is the intercalation and subsequent delamination of silicate platelets in the polymer matrix. With the upstream masterbatch process stage it should be more successful to exploit the whole potential of nanofillers. Furthermore, a masterbatch is much easier to handle than a powder with a bulk weight of 210 g/l.

The second important research topic is the effect of nanofillers at various pressures and temperatures on the thermal conductivity of polymer nanocomposites. The thermal conductivity is the most commonly used property that helps to quantify the transport of heat through a material. In polymers the thermal conductivity is based on two different mechanisms. On the one hand heat is transferred through Van-der-Waals forces, and on the other hand, phonons are stimulated through covalent bondings. By definition energy is transported proportionally to the speed of sound and therefore thermal conductivity follows the relationship

$$\lambda = c_p \cdot \rho \cdot \upsilon \cdot l \qquad (1)$$

where the thermal conductivity λ is directly proportional to the specific heat capacity c_p, the density ρ, the speed of sound v, and the molecular separation l of the elastic waves (Oswald et al., 1996).

Due to the increase in density below the solidification of semi-crystalline thermoplastics, the thermal conductivity is higher in the solid state than in the melt. Furthermore, due to the free volume in the entangled polymer structure higher pressure increases the thermal conductivity too. Other factors such as temperature, morphology and molecular orientation are also affecting the thermal conductivity of polymers. Highly drawn semi-crystalline polymers like PP can have a much higher thermal conductivity in the direction of orientation of the polymer chains than perpendicular.

It is also important to know the thermal conductivity of filled compositions in order to model and analyze the heat transfer process during processing. Thermal conductivity can be influenced significantly by the addition of different fillers and additives since the thermal conductivity of most inorganic fillers is around 10 times higher than that of polymers. The increase of thermal conductivity depends on the type, on the amount as well as on the particle shape, the size and the distribution of the used fillers (Laske et al., 2012).

Table 9 displays the increase of thermal conductivity through the use of nanofillers for the one-step process and the two-step process as well as for the "3-in-1"and the MB formulations. An increase caused by 10 wt% nanofiller in the molten state of about 7.0 % compared to the molten virgin PP and about 10.2 % compared to the virgin solid PP can be seen. However, the process with the highest improve in thermal conductivity (one-step process, 10 % MB) leads to an increase in the molten state of about 10.8 % and in the solid state of 19.4 %. It is striking to see, that the pre-stage masterbatch process leads to significant higher thermal conductivity values compared to the "3-in-1" formulations. Furthermore, the longer residence time of the one-step process slightly improved the thermal conductivity of the compounds.

Table 9. Increase of thermal conductivity compared to the virgin unfilled PP DM55

Material	Melt state (%)	Solid state (%)
one-step_3in1_5 %	3.2	5.6
one-step_3in1_10 %	7.0	10.2
one-step_MB_5 %	7.6	12.7
one-step_MB_10 %	10.8	19.4
two-step_3in1_5 %	2.7	5.5
two-step_3in1_10 %	6.1	9.8
two-step_MB_5 %	7.4	13.4
two-step_MB_10 %	9.9	18.8

The increase in thermal conductivity and the influence of pressure on the thermal conductivity is shown in Figure 8. To keep the overview only 3 different materials produced with the "3-in-1 compounding process " with 0 wt%, 5 wt% and 10 wt% at 0 bar, 500 bar and 750 bar were plotted. The thermal conductivity strongly depends on the pressure applied. The increase in thermal conductivity at 0 bar from the melt state to the solid state was about 15 % and even 27.5 % at 750 bar measured for the virgin PP. The increase from 0 bar to 750 bar was about 10% in the melt state and about 25 % in the solid state for the three materials presented. It is very interesting to see, that the increase in thermal conductivity is significantly higher between the virgin PP and the compound with 5 % nanofiller than the increase with 5 % nanofiller up to 10 % nanofiller. So it can be assumed, that with further increase of the nanofiller content the influence will become less dominant.

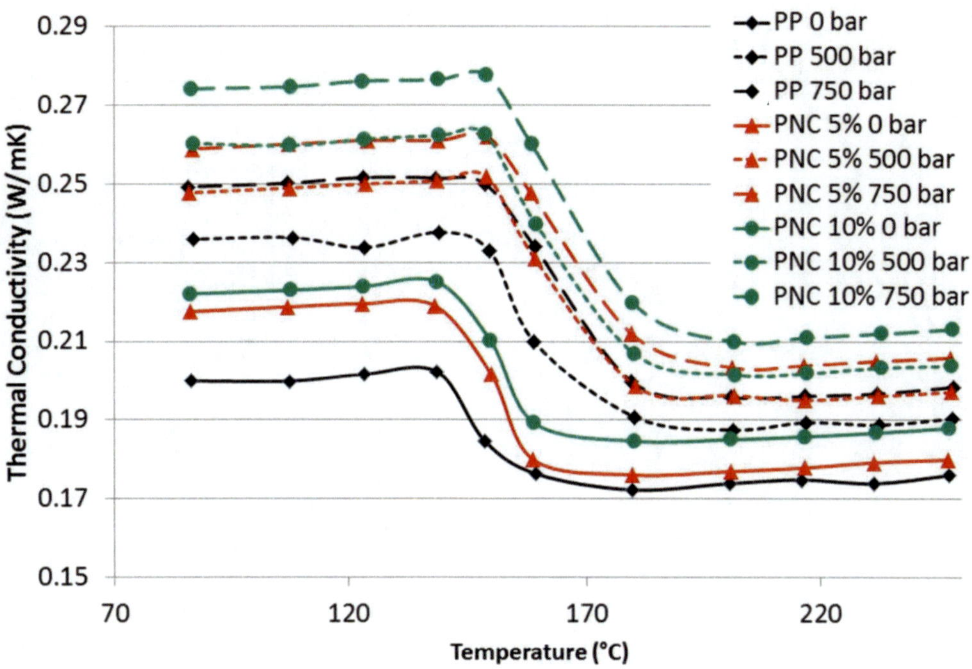

Figure 8. Effect of nanofillers and pressure on thermal conductivity of PP DM55.

These results are not really a new phenomenon, but should be considered in the production of injection molded parts regarding to cycle time reduction and cost savings. As it can be seen in the cooling time equation (2) the thermal conductivity is linked via the effective thermal diffusivity in inverse proportion to the cooling time so that an increase in thermal conductivity by 10 % leads to a reduction of the cooling period by approximately 10 %.

$$t_c = \frac{s^2}{a_{eff(T_w)} \cdot \pi^2} \cdot \ln\left[\left(\frac{4}{\pi}\right) \cdot \left(\frac{T_M - T_W}{T_E - T_W}\right)\right] \tag{2}$$

with t_c= cooling time, s = wall thickness, $a_{eff(Tw)}$ = effective thermal diffusivity, T_M = melt temperature, T_W = mean cavity wall temperature and T_E = demolding temperature.

Table 10 shows some calculated t_c values for plates with thicknesses of 2 and 4 mm, respectively. The thermal conductivity λ, specific heat capacity c_p and the density ρ were taken from the one-step "3-in-1 compounding process", at 750 bar and 80 °C. It can be clearly seen, that the saving potential is enormous, especially for thick-walled parts.

Table 10.Calculated cooling times and cooling time for 3 different materials at 750 bar and 80 °C and plates of 2 and 4 mm thickness

Material	Thickness (mm)	Cooling time (s)	Δt_c (%)
Virgin PP DM55	2	6.7	-
one-step_3in1_5 %		6.1	- 9.5
one-step_3in1_10 %		5.7	- 15.1
Virgin PP DM55	4	26.8	-
one-step_3in1_5 %		24.3	- 9.5
one-step_3in1_10 %		22.8	- 15.1

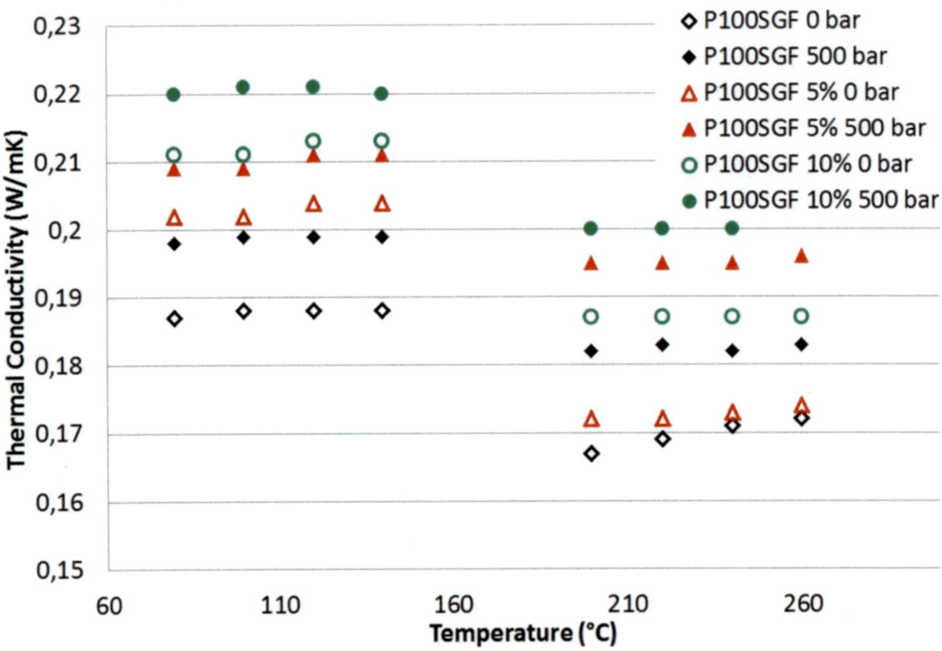

Figure 9. Effect of nanofillers and pressure on thermal conductivity of PP P100 SGF10plus ST.

A similar type of tests was made with the glass-fiber reinforced PP RIALENE P100 SGF10plus ST further reinforced with layered silicates. Additional to the theoretical calculated cooling time also an injection molding simulation of the cooling phase with Cadmould 3D-F (Simconkunststofftechnische Software GmbH, Würselen, Germany) was made. In contrast to the investigations with the PP DM55 for P100 SGF10plus, the pressure was only measured at 500 bar, due to the higher MFR and the consequently lower pressures in the injection molding machine. Especially notable is that the compounds with nanofillers increase clearer from the melt state to the solid state then the values of the virgin polymers at higher pressures (Figure 9). This phenomenon doesn't occur at the measured curves without glass-fibers. This difference can probably be explained with the different inner structure of the compounds caused by the glass-fibers. Another interesting aspect would be the thermal conductivity in the transition region which unfortunately is not really easy to measure (Kucher, 2009, Witschnigg, 2008).

The increase in thermal conductivity at 0 bar from the melt state to the solid state was about 10.6 % and 8.9 % at 500 bar measured for the virgin PP. The increase from 0 bar to 500 bar was about 7.5 % in the melt state and only 5.9 % in the solid state for the three materials presented. There is an increase in thermal conductivity caused by 10 wt.% nanofiller in molten state of about 10 % compared to the molten virgin PP P100 SGF and about 13 % compared to solid virgin PP P100 SGF (Table 11). Summarized it is obvious, that the increase in thermal conductivity by the use of layered silicates is more effective at virgin PP than at glass-fiber reinforced PP.

Table 11. Increase of thermal conductivity compared to the virgin glass-fiber reinforced PP P100 SGF10plus ST

Material	Thermal conductivity (W/mK)	
	Melt state	Solid state
P100SGF_0 bar	0.170	0.188
P100SGF_500 bar	0.183	0.199
P100SGF 5%_0 bar	0.173	0.203
P100SGF 5%_500 bar	0.195	0.210
P100SGF 10%_0 bar	0.187	0.212
P100SGF 10%_500 bar	0.201	0.221

Figure 10 shows the results concerning the reduction of the cycle time Δt_c theoretically calculated using equation (2) and the results of the simulation with Cadmould 3D-F. As can be seen, the theoretically calculated values are higher than the simulated results. Both show the same tendency although the gap between the results is getting larger with higher nanofiller amount. Till now it is not possible to decide beyond doubt, which values are closer to reality. Both calculation methods have the problem that they cannot take into account the pressure dependency. That is the reason why further investigations has to be done to study part cooling in injection molding with more complicated geometries taking into account also pressure dependency.

Another very important topic, especially for the application of parts in industry is warpage of glass-fiber reinforced polymers. This investigation shows some results concerning distortion tendencies in connection with the use of layered silicates in PP glass-fiber reinforced polymers. In polymers the warpage is based on two different mechanisms. Primarily it is attributed to residual thermal stresses associates with shrinkage and thermal contraction of the parts, but also flow-induced stresses generated during mold filling play a significant role (DePolo et al., 2006). To guarantee mainly the same conditions and make a valid comparison in our injection molding experiments the injection molding conditions were hold constant and so the flow-induced stresses generated during the mold filling should be similar.

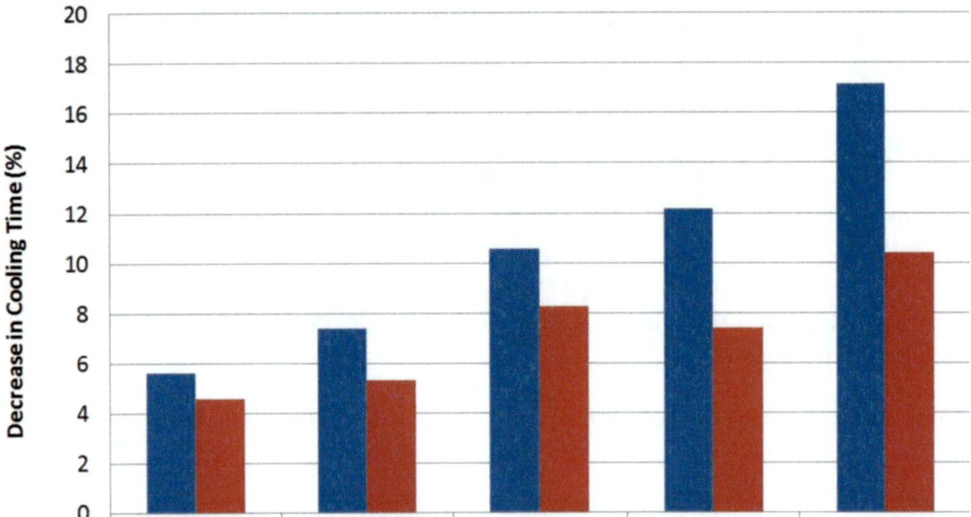

	P100SGF 500 bar	P100SGF 5% - 0 bar	P100SGF 5% - 500 bar	P100SGF 10% - 0 bar	P100SGF 10% - 500 bar
Theoretical	5.60	7.40	10.55	12.19	17.17
Simulation	4.55	5.30	8.25	7.37	10.37

Figure 10. Calculated cooling times decrease for 5 different materials for a plate with thickness 2 mm.

As can be seen in Figure 11, there is a big difference between the virgin glass-fiber reinforced PP (a) and the additionally layered silicate reinforced PP with a nanofiller content of 10 wt.% (b). The warpage is definitely lower for the compounded material, which can be explained by three possible reasons. On the one hand, possibly the anisotropic structure of the material with 10 wt.% glass-fibers could be more homogenized by the addition of layered silicates. On the other hand also the higher thermal conductivity and faster cooling of the molded parts could be responsible for the lower warpage of the injection molded plates. Third aspect is the cristallinity, which also has an influence on the warpage of the injection molded plates. There are a lot of different opinions concerning the effect of nanofillers acting as nucleation agent and most of the opinions are diverging (Tan et al., 2012, Ma et al., 2002, Svoboda et al., 2002). We also performed DSC measurements and found out, that cristallinity is more or less the same in all tested materials (Table 12). However, the crystallization kinetics should be changed towards formation of smaller sphärolithes due to faster crystallization. Which one of the three mechanisms is responsible for these results has to be verified in future and is therefore a new topic for further research. Of course there were made also experiments with a nanofiller content of 5 % which also improved the distortion tendencies of the injection molded parts, but not as impressive as by the compounds with 10 % nanofiller. The injection molded plate of the virgin PP P100 SGF10plus ST had a warpage of about 1.4 mm at the end of the plate. The addition of 5 % nanofiller decreased the warpage to 0.8 mm, which is an improvement of about 43 % and the addition of 10 % nanofiller decreased the warpage to 0.1 mm at the end of the plate which represents an improvement of impressive 93 %.

Figure 11. Injection molded plates of 2 mm thickness without (a) and with (b) 10 % nanofiller.

Table 12. Cristallinity of pure and reinforced PP P100 SGF10plus ST

Material	Cristallinity (%)
PP P100 SGF10plus	44,08
PP P100 SGF10plus + 5 % nanofiller	44,22
PP P100 SGF10plus + 10 % nanofiller	44,52

CONCLUSION

The preparation of PP polymer nanocomposites had been carried out with the intermeshing, co-rotating twin-screw extruder of the PNC-IMC in order to improve mechanical properties, thermal conductivity and warpage. On the one hand 11 or 13 barrel segments each with a length of 4 D were used for the compounding process and on the other hand the PNC-IMC was used for processing of injected parts in a single-heat, or to be precise a single plasticizing process, respectively.

As shown in this chapter, there are many variables influencing Young's modulus, tensile strength and thermal conductivity of polymer nanocomposites. Although it is possible to apply high shear forces and long residence time simultaneously the results show that a very long extruder (52D) and the additional use of a melt pump to generate counter pressure have a negative effect on the Young's modulus. Additionally higher screw speed is not absolutely essential for the increase of mechanical properties of polymer nanocomposites, but leads to better exfoliation of layered silicates in the polymer matrix.

More effective regarding to the improvement of Young's modulus is the use of the pre-stage masterbatch process, where a masterbatch of 50 wt.% nanofiller and 50 wt.% compatibilizer is produced and further diluted with PP.

The basic message of this chapter is that it is important to find a compromise between shear energy input and residence time. Shear energy is necessary for intercalation and subsequent delamination of silicate platelets in the polymer matrix, while residence time is the most important parameter for the diffusion process; however, too high residence time may be critical for material degradation. The results concerning the increase of thermal conductivity and the improved warpage are encouraging and motivate to carry out further intensive investigations based on industrial interest.

Currently the chosen setup and process parameter set of the PNC-IMC production line was able to reach almost the material property improvements of the two-step masterbatch process. Nevertheless, there are efforts to bring the one-step process beyond the two-step process. Therefore further research will be done on:

- modification of the screw geometry,
- perfect adaptation of the mixing ratio between nanofiller and compatibilizer,
- use of compatibilizers based on highly viscous PP-types,
- special treatment of the PNC melt between extruder and melt accumulator (flow induced improvement of intercalation and exfoliation),
- and electronic controlled 3-way-valve in order to realize high throughput rates up to 40 kg/h.

References

Bousmina, M., (2006). Study of intercalation and exfoliation processes in polymer nanocomposites, *Macromolecules*, vol. 39, no. (12): p. 4259-4263

Bürkle, E., Scheel, G., Darnedde, L., (2009). Energy-efficient processing of natural fiber-reinforced plastics. *Kunststoffe Plast Europe*, vol. 99, no (2): p. 25-29.

Bürkle, E., Sieverding, M., Mitzler, J., (2003). Injection Moulding of Long-glassfibre-reinforced PP. *Kunststoffe Plast Europe*, vol. 93, no (3): p. 17-19.

Cho, S., Hong, J.S., Lee, S.J., Ahn. K.H., Covas, J.A., Maia, J.M. (2011). Morphology and Rheology of Polypropylene/Polystyrene/Clay Nanocomposites in Batch and Continuous Melt Mixing Processes. *Macromolecular Materials and Engineering,* vol. 296, no. (3-4): p. 341-348.

DePolo, W.S., Baird, D.G., (2006). Flow-induced warpage of injection-molded TLCP fiber-reinforced polypropylene composites. *Polymer Composites*, vol. 27, no (3): p 239-248.

Kracalik, M., Laske, S., Gschweitl, M., Friesenbichler, W., Langecker, G.R., (2009). Advanced compounding: extrusion of polypropylene nanocompsites using the melt pump. *Journal of Applied Polymer Science*, vol. 113, no. (3): p 1422-1428.

Kucher, C., (2009). Untersuchungen zur Wärmeleitfähigkeit von Polypropylen mit mikroskaligen Füllstoffenbei hohen Drücken.*Master thesis at the institute of polymer processing at the Montanuniversitaet Leoben.*

Laske, S., Duretek, I., Witschnigg, A., Mattausch, H., Tscharnuter, D., Holzer, C., (2012). Influence of the degree of exfolation on the thermal conductivity of polypropylene nanocomposites. *Polymer Engineering & Science*, vol. 52, no. (8): p. 1749-1753.

Laske, S., Kracalik, M., Gschweitl, M., Feuchter, M., Maier, G., Pinter, G., Thomann, R., Friesenbichler, W., Langecker, G.R., (2009).Estimation of Reinforcement in Compatibilized Polypropylene Nanocomposites by Extensional Rheology, *Journal of Applied Polymer Science*, vol. 111, p. 2253-2259.

Ma, J., Zhanf, S., Qi, Z., Li, G., Hu, Y., (2002). Crystallization Behaviors of Polypropylene/Montmorillonite Nanocomposites, *Journal of Applied Polymer Science*, vol. 83, p. 1978-1985.

Mignoni, M.L., Silva, J.V.M., De Souza, M.O., Mauler, R.d.S., De Souza, R.F., Bernado-Gusmao, K., (2011). Polyehtylene-montmorillonite nanocomposites obtained by in situ polymerization of ethylene with nickel-diimine catalysts. *Journal of Applied Polymer Science*, vol. 122, no. (3): p. 2159-2165.

Oswald, T.A. et al. (1996). *Material Science of Polymers for Engineers.*Carl HanserVerlag, Munich/Vienna/New York.

Reitner, J., (2001). Making intelligent use of synergies.*Kunststoffe Plast Europe*, vol. 91, no. (9): p. 40-+.

Ruch, J., Fritz, H.G., Bürkle, E., Zimmet, R., (2002). Innovative Direktverarbeitung von Naturfasern. Flachsfaserverstärkung von Polypropylen. *Kunststoffe*, vol. 92, no. (2): p. 28-34.

Sieverding, M., Bürkle, E., Zimmet, R., (2005). IMC Technology opens up new fields of application. *Kunststoffe Plast Europe*, vol. 95, no. (8): p. 34-37.

Svoboda, P., Zeng, C., Wang, H., Lee, L.J., Tomasko, D.L., (2002). Morphology and Mechanical Properties of Polypropylene/Organoclay Nanocomposites. *Journal of Applied Polymer Science*, vol. 85, p. 1562-1570.

Tan, S., Tincer, T., (2012).A Study on Nanofiller Effects of Serpentine on Polypropylene Matrix.*Journal of Applied Polymer Science*, vol. 124, p. 1419-1425.

Vaia, Ra., Jandt, K.D., Kramer, E.J., Giannelis, E.P., (1996). Microstructural evolution of melt intercalated polymer-organically modified layered silicates nanocomposites. *Chemistry of Materials*, vol. 8, no. (11): p. 2628-2635.

Witschnigg, A., (2009). Messung der Wärmeleitfähigkeit in Abhängigkeit von Druck und Temperatur.*Student research project at the institute of polymer processing at the Montanuniversitaet Leoben.*

Witschnigg, A., Laske, S., Kracalik, M., Feuchter, M., Pinter, G., Maier, G., Märzinger, W., Hberkorn, M., Langecker, G.R., Holzer, C., (2009). In-Line Characterization of Polypropylene Nanocomposites Using FT-NIR. *Journal of Applied Polymer Science*, vol. 117, no. (5): p 3047-3053.

In: New Developments in Polymer Composites Research ISBN: 978-1-62948-340-5
Editors: S. Laske and A. Witschnigg

Chapter 13

Structural Organisation and Biomimesis of Nature's Polymer Composites

*Parvez Alam**
Centre for Functional Materials, Abo Akademi University, Turku, Finland

Abstract

In nature, there are many examples of complex hierarchical material designs where weak constituent polymers assemble into composites with exceptional mechanical properties. These structures are attracting considerable attention in the scientific and engineering communities as sources of inspiration for polymer composite design. Natural polymers or, biopolymers, may include proteins, polypeptides and polysaccharides. These biopolymers are the essential building blocks of hierarchical polymer composites in nature. This chapter elucidates a variety of different polymer composite structures from nature; how they are organised, and the specialised designs that provide for superior functionality. Recurring concepts in biological material structures will also be explained and hence, the chapter will draw upon the elements of structural biology in context to biomimetic and bioinspired composites science.

PACS: 82.35.Pq, 87.15.rp, 87.85.jf, 87.15.-v, 81.05.Qk

Keywords: Biological structures and biomimesis, biopolymers, biomimetics, mechanics, composites, biostructures

AMS Subject Classification: 74R10, 74C10, 74L15

1. Introduction

1.1. Bioinspiration

A holistic approach to determining the structure-property relationships of biological materials, most recently termed *biomateriomics* (Cranford and Buehler 2012), is a neces-

*E-mail address: parvez.alam@abo.fi (Corresponding Author). Parvez Alam, Adjunct Professor of Composite Materials and Biostructures, Centre for Functional Materials, Abo Akademi University, 20500, Turku, Finland.

sary precursor to biomimetic and bioinspired materials science. By analogy, a robot consisting of moving parts, a motor, a sensory system, a power supply and a processor functions only as a robot when all five of these criteria are present. There is no robotic function in any of the individual robotic parts, nor indeed will the robot function should any one of the parts be missing. Similarly, biological structures, or *biostructures*, when broken down into their constituent elements lack the superior properties they would otherwise exhibit when all the components function in unison as a single composite body. Notably, the majority of biostructures are polymer based composites and as such are dominated by secondary interactions. This being the case, it could be easy to assume that biostructures lack the qualities of high stiffness and strength. Quite to the contrary, many biostructures exhibit respectable mechanical properties, and when factored by density, establish their strength–to–weight superiority over many man–made materials (Chen et al. 2008). It is noted that the range of properties is high in natural materials, i.e. many natural polymer composites will display more modest properties of strength and stiffness; however, this is in itself expected since natural materials are suitably adapted to environmental and intrinsic factors and quite simply, not all these materials need to be mechanically stiff and strong given their application and utility in nature. Numerous natural polymer composites have an outstanding ability to absorb mechanical energy (termed toughness) and will fracture after experiencing considerable mechanical work. Figs 1 and 2 provide qualitative property maps for natural materials. Fig. 1 also compares the elastic modulus of natural materials to common engineering materials.

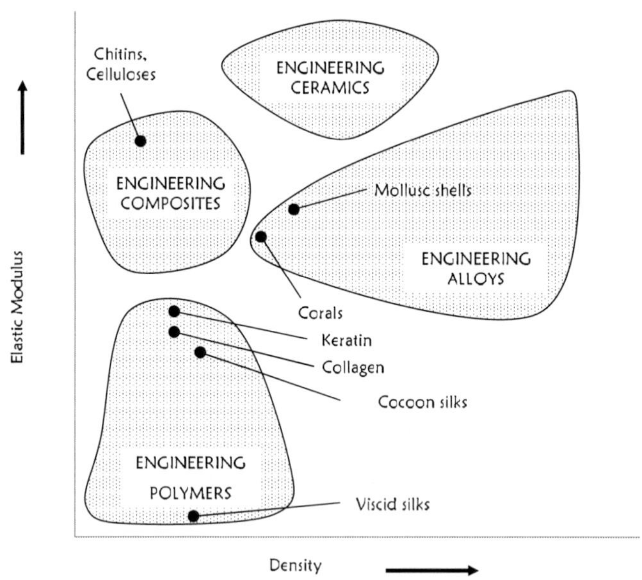

Figure 1. Elastic modulus/density relationships for engineered and natural materials (adapted from Ashby (1989) and Wegst and Ashby (1994)).

There are common mechanistic concepts by which certain biostructures are able to develop properties of high strength, toughness and fracture resistance. The main ones can be considered as:

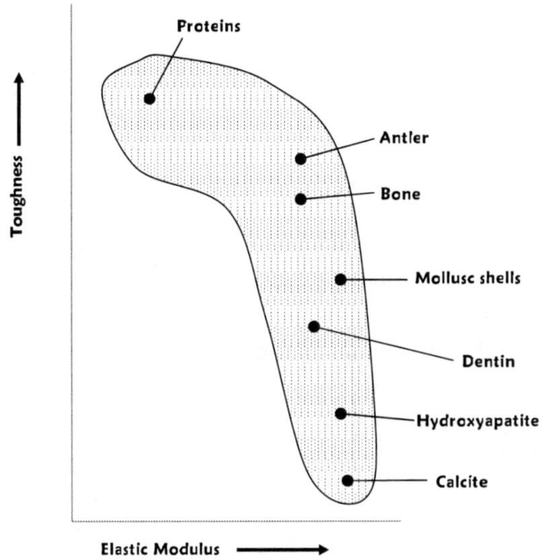

Figure 2. Toughness/elastic modulus relationships for a range of natural materials (adapted from Ashby (1989) and Wegst and Ashby (1994)).

- Structural hierarchy
- Crack retardation
- Orientation of composite components
- Hydrogen bonded networks
- Biological mineralisation
- Hard-soft segmentation.

Scientists in recent times acknowledge that these factors play a vital role in turning initially weak monomeric or polymeric building blocks into mechanically enhanced composite systems. As a consequence, a concerted worldwide effort is being undertaken towards mimicking, or taking benefit from such biostructures. This effort is collectively termed *biomimesis*, *biomimetics* or more generically, *bioinspiration*. Biomimetics can be subdivided into four interconncted stages.

1. Exploration and discovery
2. Understanding behind the science and function of a specific feature (or quality) in the natural organism under scrutiny
3. Mimicry of this feature, or the processes by which it develops, in man-made materials and structures
4. Application of the mimicked feature or process by identifying a niche area in science/technology where this feature could serve a distinct purpose.

Typical fields where biomimetics is currently applied include; robotics, artificial intelligence, aerodynamics, fluids and surfaces, smart systems and materials, biomedical applications (including molecular biomimetics) and in engineering composites.

1.2. The Building Blocks of Biostructures

Biopolymers and mineral compounds are the two main classes of material that typically conjoin to form composite biostructures. Both classes of material may exist as crystalline, semi crystalline or amorphous subunits that ultimately determine the material properties of the composite biostructure. Moreover, mineral compounds may form meta-stable polymorphs that can convert to a stable polymorph under certain conditions of pressure and temperature. The following sections describe these essential building blocks of composite biostructures.

1.2.1. Amino Acids

Amino acids are named as such because they contain a weak (carboxylic) acid group [-COOH] and an amine group [$-NH_2$]. They are absorbed by biomineralising organisms (e.g. corals), adsorbed to nucleating crystallites during the process of biomineralisation, involved in numerous metabolic processes and can conjoin to form long chain polymers (e.g. proteins). Each amino acid subtly differs with respect to their particular side chain. Resultantly, the molecular lengths, conformations and chemical behaviours of amino acids vary considerably. Figs. 3–6 show the chemical structures for twenty proteinogenic amino acids.

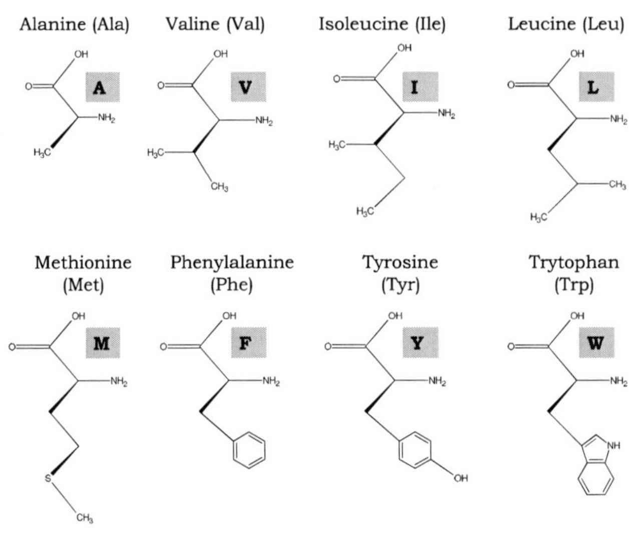

Figure 3. Proteinogenic amino acids with hydrophobic side chains.

Proteins are formed when amino acids form polypeptide chains. Amide bonds are formed when the carboxyl and amine groups from two amino acids react with one another in a condensation polymerisation reaction. A reaction as such, between two amino acid

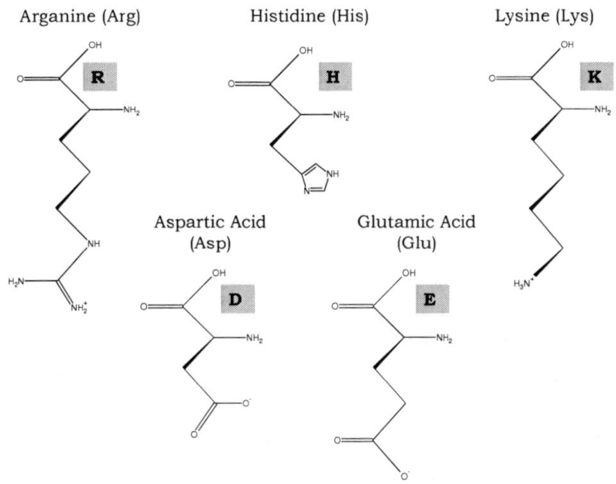

Figure 4. Proteinogenic amino acids with electrically charged side chains.

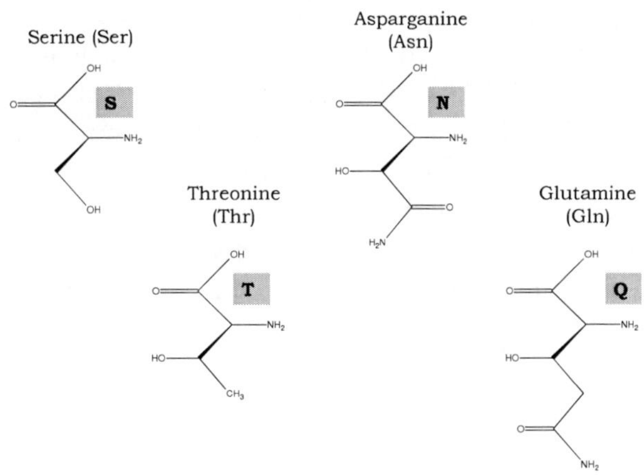

Figure 5. Proteinogenic amino acids with uncharged polar side chains.

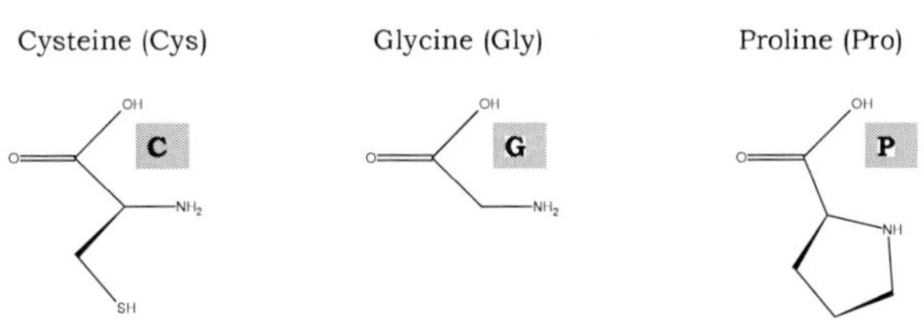

Figure 6. Proteinogenic amino acids not fitting into categories in Figs. 3-5.

molecules releases one molecule of water. Adjacent amino acids (or proteins) may interact through electrostatic and van der Waals forces, the strength of which depends on their molecular conformations, their proximal distances and the types of side chains available for interaction.

Structural biopolymers may be built up of polypeptide chains, polysaccharides or a combination of the two. Biopolymers such as collagen, silk, and keratin are built up of polypeptides, while celluloses and hemicelluloses are constructed from polysaccharides. Chitin is the anomaly in that it is formed of both polypeptide and polysaccharide chains. Celluloses and chitins are the two most abundant biopolymers on earth. While cellulose is most commonly found within the plant kingdom, chitin is a common structural component in the insect kingdom and in the materials of marine organisms. The following sections introduce these two ubiquitous biopolymers and a list of polypeptide based structural proteins.

1.2.2. Chitin

Chitin has the chemical formula $(C_8H_{13}O_5)_n$, and its structure is shown in Fig. 7. It is the second most common biopolymer in nature after cellulose. Chitin is a long chain derivative of glucose and has repeat units of N–acetylglucosamine. It is found in the exoskeletons of arthropods such as insects and crustaceans, and in the cell walls of fungi. Chitin is soluble in aqueous solutions of some acids and chitin chains tend to pack into nanofibrils with a diameter of *ca.* 2.5 nm. The nanofibrils connect into larger units known as microfibrils which have a diameter *ca.* 10 fold higher than single nanofibrils. Chitin forms strong but flexible crystalline structures since there are numerous hydrogen bonding units. The hydrogen bonds are not strong individually 5–20 kcalmol^{-1}, but when there are many of them they stabilise the network and have a higher possibility of reforming once broken. There are three forms of chitin in nature; α–chitin, which is the most abundant structure consists of both parallel and anti-parallel molecular strands packed tightly in an orthorhombic unit cell, β–chitin, which exists in parallel chains packed in a monoclinic unit cell, and γ–chitin, which is rarely found in nature and is regarded an intermediate crystalline form of α–chitin and β–chitin.

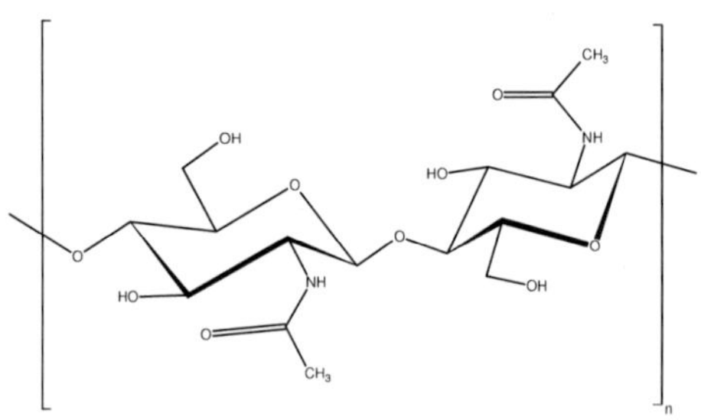

Figure 7. Molecular structure of chitin.

1.2.3. Cellulose

Cellulose is the most abundant biopolymer in nature. It has the chemical formula $(C_6H_{10}O_5)_n$ and its chemical structure is shown in Fig. 8. It is formed in the cell walls of green plants but is also produced by algae and certain bacteria. The most efficient cellulose producing bacterium is *Acetobacter xyllinum* and this bacterium is known to produce cellulose with up to 95 % crystallinity. Bacterial cellulose is secreted extracellularly as ribbon shaped fibrils that can reach 100 nm in length (3–8 nm thick). Each ribbon is further made up of microfibrils.

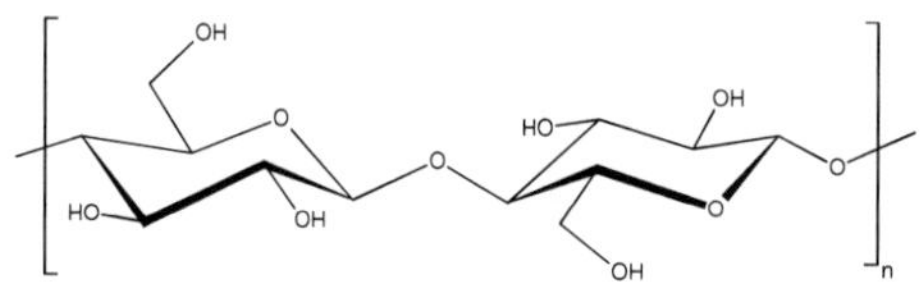

Figure 8. Cellulose repeat unit (cellubiose).

Table 1. Composition of some common structural proteins.

Protein	Composition
Keratins	Predominantly Arg, Cys, Gly with lower concentrations of His, Lys, Tyr, Trp and Phe (Block 1939).
Collagen	Predominantly Gly, Ala, Arg with lower concentrations of Leu, Val, Ser, Thr, Lys, Asp and Glu (Bowes et al. 1955)
Spongin	Collagen related varying in structure and amino acid composition according to Porifera, or marine sponge, species (Garrone 1985) and also as short chain collagens (Auoacheria et al. 2006)
MaSp1	poly-Ala blocks (Gly-Ala)$_n$ and Gly-Gly-X (Hu et al. 2006)
MaSp2	poly-Ala blocks (Gly-Ala)$_n$, Gly-Gly-X and Gly-Pro-Gly-X-Y (Hu et al. 2006) (X = Pro, Y = hydroxyl-Pro)
MiSp1	poly-Ala blocks, Gly-(Ala)$_n$ and Gly-Gly-X (Hu et al. 2006)
MiSp2	poly-Ala blocks, Gly-(Ala)$_n$ and Gly-Gly-X (Hu et al. 2006)

1.2.4. Structural Proteins

Structural proteins are named as such because they provide structural and scaffolding support. Table 1 describes some of the more common structural proteins.

1.2.5. Biominerals

Many natural materials incorporate hard biominerals within a polymeric matrix, which act essentially as reinforcing to the softer polymer domains. The most common mineral compounds include calcium carbonate, calcium phosphate, silica and hydroxyapatite. Each is briefly considered in Table 2.

The variety of forms in which biominerals exist creates considerable diversity in the properties of biostructures. Molecular interactions between polymeric phases of biostructures with biominerals will moreover vary with respect to the type of surface they encounter, which is itself a function of the crystal structure or mineral compound morphology. What makes biominerals different to pure minerals (listed in Table 2), is that they are in themselves, polymer composites. These structures consist of quantities of crystalline mineral compounds effectively glued together by organic polymer phases; such as proteins, polysaccharides and lipids (Wilt 2002).

1.3. Lessons from Nature

In the following sections, we will focus on particular features from polymer composite biostructures, from which inspiration can be derived to aid the design of high performance

Table 2. Common mineral compounds making up biominerals, their formulae, and example biostructures they are found in.

Biomineral	Generic chemical formula	Commonly existing as	Biostructure
Calcium carbonate	$CaCO_3$	*Aragonite* *Vaterite* *Calcite*	Corals Sponges Shells Crustaceans Molluscs
Calcium phosphate	$Ca_3(PO_4)_2$	*Monocalcium phosphate* $Ca(H_2PO_4)_2$ *Dicalcium phosphate* $CaHPO_4$ *Tricalcium phosphate* $(Ca_3(PO_4)_2$ *Hydroxyapatite* $Ca_5(PO_4)_3(OH)$	Bones Teeth Beaks Horns Antlers
Silica	SiO_2	*Amorphous (though local oxygen tetrahedral arrangement around the silicon atoms)* *Rhombohedral* *Hexagonal* *Orthorhombic* *Tetragonal* *Cubic* *Monoclinic*	Sponges Diatoms

man-made composites. We begin by looking at the characteristics determining the functionality of spider silk, which is an archetypal high mechanical energy absorbing material. This is followed by considering crack stopping and energy dissipation mechanisms of laminated materials such as are found in beetle forewings, wood, bone, mollusc shells, and armadillo and turtle carapaces. Finally various marine biostructures that mineralise intricate exoskeletons to resist flexural damage are detailed.

2. Spider Silk

Spiders are able to create different silks for different purpose from amongst seven different glands. Dragline silk is the primary structural silk used for web construction and its spidroins are secreted from the *ampullate (major)* gland. The *ampullate (minor)* gland produces silk that is used for temporary scaffolding during web construction. Silk produced from the *aciniform* gland is used for wrapping prey. The *aggregate* gland produces sticky silk glue, while the *piriform* gland produces silk that is used to adhere separate silk threads at attachment points. The *cylindriform* gland is responsible for egg sack silk and the *flagelliform* gland makes core fibres of sticky silk. Individual spiders do not have all seven glands but rather, males will usually have at least three of the glands and females have a minimum of four.

2.1. The Structure of Silk

Spidroins (Sp) of the major (Ma) and minor (Mi) ampullate glands are MaSp1 and MaSp2 (Major) and MiSp1 and MiSp2 (Minor). These spidroins have high molecular

Table 3. Amino acid composition in different types of silk (adapted from Lombardi and Kaplan (1990)).

Amino acid	***Bombyx Mori***	***Nephila clavipes* (dragline reeled)**	***Nephila clavipes* (glandular)**	***Argiope auranti* (dragline reeled)**	***Neoscona domhil-iorum* (dragline reeled)**
Gly	44.1	37.1	38.1	34.7	38.0
Ala	29.7	21.2	23.4	22.2	18.0
Ser	12.4	4.5	3.9	5.1	6.8
Tyr, Phe	7.5	10.2	4.3	3.8	3.7
Leu, Ile, Val, Asx, Glx	3.6	11.7	16.6	19.2	2.4
Thr	1.2	1.7	2.0	0.8	0.9
Arg	1.5	7.6	7.2	2.9	0.6
Trp	0.5				
Pro		4.5	3.9	6.4	11.2
His, Cys, Lys		1.0	1.9	0.8	0.9

weights (200–350 kDa or larger) and are special in that they are composed of predominantly alanine and glycine. The spidroins consist of long repeating amino acid chains (up to 500 amino acids in each repetition) with non-repeating conserved N and C terminal regions. Amino acid compositions (from Lombardi and Kaplan, 1990) are provided for a variety of silks in Table 3.

The MiSp1 and MiSp2 have a higher percentage of polyalanine units than the MaSp1 and MaSp2, which provide greater structural stability and extra support for temporary scaffolds. The polyalanine or polyglycine–alanine sequences form highly stable β–sheets as a consequence of protein folding and amino acid orientation during fibre extrusion. During spinning, liquid–crystal dope from the ampullate major gland is drawn through a tapered duct. The silk proteins resultantly elongate and align giving rise to hydrogen bonding that form readily due to the polypeptide structures. Though hydrogen bonds are weak (5–20 kcal/mol), when there are many forming an interconnected network, the structure becomes significantly more stable. In fact, β–sheet nanocrystals have the dual characteristics of stiffness and flexibility thanks to these H–bond arrays. The amorphous regions of dragline silk are rich in glycine. These exist as 3_1 helices with type II β–turns. By varying the spinning speed, the spider is thus able to control the final structural character of the silk according to requirement. Fibres spun at higher speeds tend to be highly crystalline as a result of greater alignment and thus the formation of higher levels of crystalline β–sheets. When a spider requires greater extensibility and energy dissipation, such as in a frame thread, it spins the fibres more slowly, which increases the ratio of amorphous polymer to crystalline polymer. Dragline silk itself at the macro level is reasoned to be a five–layer assembly (Sponner et al. 2007). From outside to inside these are the lipid layer, the glycol layer, the skin layer, the outer layer and the inner layer, Fig. 9. The main structural components of silk are found within the outer and inner (core) layers.

1. Lipid Layer: The lipid layer does not contain proteins. It may provide protection

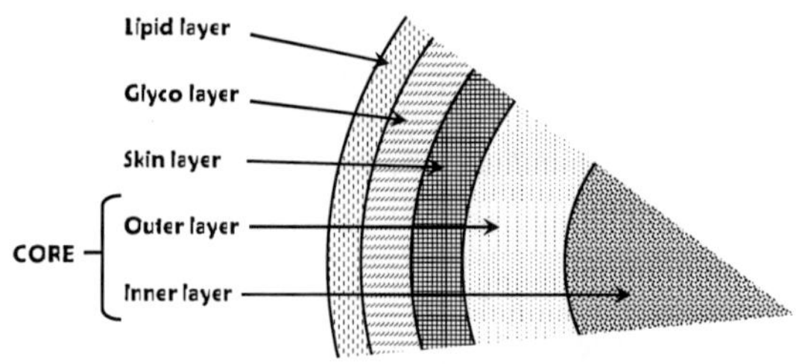

Figure 9. Layered structure of dragline silk (adapted from Sponner et al. 2007).

from the environment and micro–organisms. Its main purpose is to carry pheremones for sex and species recognition. Radial dimension, 10–20 nm.

2. Glyco Layer: Protection against microbial attack. Regulates water balance (and has therefore direct influence to the mechanical properties). May also act as a lubricant. Radial dimension, 40–100 nm.

3. Skin Layer: Contains silk proteins of equal or higher molecular weight than those the in spidroins and may provide some plasticity and support. High resistance to chemical agents suggests the layer protects the inner and outer layers from environmental impact. Radial dimension, 50–100 nm.

4. The core (inner and outer layers): The outer layer contains more load resisting β–sheet nanocrystals (see Fig. 10) than the inner layer of the core. The outer layer also has a higher degree of molecular orientation than the inner layer. The outer core provides rigidity and strength, while the inner core provides elasticity and strength. Outer layer radial dimension, 300–400 nm. Inner layer radial dimension, 1800–2300 nm.

We shall consider the core in greater detail since it is essentially the structural component of dragline silk. What is incredible about dragline silk is that its outstanding properties of toughness (> 150 MJm^{-3}) come from a material body made up solely from polymer (of amino acid building blocks). These amino acid units form into highly crystalline and semi–amorphous domains. The semi–amorphous domains are rich in glycine, and polyalanine segments within these domains are normally less ordered than in the polyalanine richer domains (Simmons et al. 1996). The ordered polyalanine rich regions align in an anti–parallel array and form β–sheet nanocrystals. Near–neighbour β–strands have a high affinity to pair up, which ultimately gives rise to β–sheet formation (Zhang et al. 2011). This *first come first pair* rule holds true only for β–strands in similar conformation and these strands tend not to pair up with coiled or α–helical polyalanine conformations. The β–sheet nanocrystals act essentially as the load bearing reinforcements in the soft semi–amorphous matrix and consist of poly(Gly–Ala) (Hakimi 2007, Du 2006) and poly–Ala repeat sequences (Simmons 1994). The β–sheets can be stacked (Gosline 1986) about the equatorial planes of the strand and are typically between 3 nm and 7 nm in length. The final β–sheet nanocrystals occupy between 15 % and 40 % of dry silk fibres (Gosline 1999, Simmons 1996, Gosline 1986). These fractions can be controlled by the spider and give rise to highly

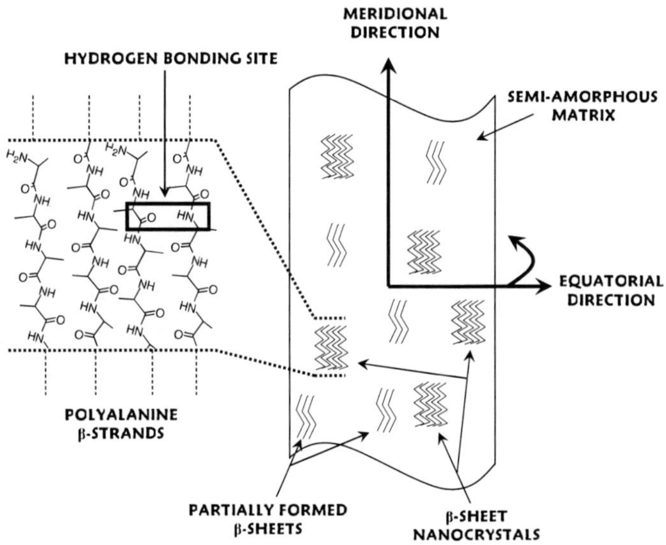

Figure 10. Antiparallel β–strands forming hydrogen bonds and β–sheets. The core materials consist of these sheets which provide the strength and a polyglycine rich semi-amorphous region.

variable mechanical properties (Gosline 1999, Denny 1976). One means by which the spider controls the β–sheet crystal fractions, sizes and orientations (with respect to the meridonial direction) is by altering the speed at which the silk thread is reeled (Vollrath 2001, Du 2006). Nanocrystals not in line with the meridional axis (Simmons 1996, Du 2006, Plaza 2012, Grubb 1997, Ulrich, 2008) do however orient to the meridional axis on deformation (Grubb 1999), which allows the silk further opportunity to resist load.

The crystalline polymeric β-sheet nanocrystals adhere primarily through hydrogen bonding (Hayashi 1999). Hydrogen bonds are in fact very weak bonds with strengths typically ranging from 5–20 kcalmol^{-1}. As mentioned earlier, when many hydrogen bonds work together however, they create a very stably bonded network that provides properties of both self healing and intra–crystalline flexibility. These two properties, give β–sheet nanocrystals the qualities of very high energy absorption and dissipation (Xu 2010). When a β–sheet nanocrystal starts absorbing large amounts of mechanical energy and comes close to a critical breaking stress, it can flex, redistribute the energy and absorb more energy. Moreover, the breaking of hydrogen bonds in a network consisting of multiple hydrogen bonds is not as detrimental to a crystal as is the breaking of a covalent bond since, depending on the rate of deformation, the hydrogen bond can in fact reform thus restoring the energy absorbing ability of the crystal. The size of a β–sheet nanocrystal may nevertheless improve or indeed lessen the mechanical properties of silk (Keten 2010, Cetinkaya 2011, Alam 2013) and these nanocrystals have orthotropic mechanical properties (Alam 2013, Cetinkaya 2011).

Another important aspect of spider silk is its ability to strain-harden. Fig. 11 shows typical stress–strain curves for both viscid and dragline silks. The microstructural factors giving rise to strain hardening may arise through unfolding mechanisms of β–sheet proteins yield-

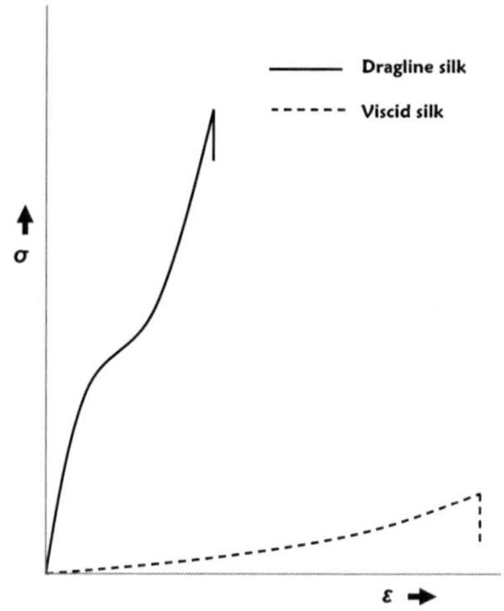

Figure 11. Example stress-strain profiles of dragline and viscid silks (adapted from De Tommasi et al. 2010).

ing greater resistance in the meridional axis. It is also suggested that strain hardening will occur through inter-β–sheet shear failure giving rise to smaller, higher toughness crystals (Keten 2010). This can be contested conceptually nevertheless, since β–sheet fragmentation has been shown to also contribute to strain–weakening (Du 2011).

2.2. Composites of Silk-Polymer

The comprehensive review of Hardy and Scheibel (2010) describes several processing methods for the manufacture of pure silk into films, foams, hydrogels, capsules, spheres and fibres. Silks moreover may alter the morphology of crystallising calcium carbonate (Cheng et al. 2008), since the amino acid based building blocks of silk position themselves between the crystallising segments, thus controlling the pattern of crystal growth and the final polymorph. In such cases, pure silks are foremost dissolved in concentrated solutions of inorganic or organic salts, in fluorinated solvents, ionic solvents or strong acids. Through dissolution, the hydrogen bonds are broken and the silk proteins are denatured, however, dissolution also allows the silk to be manufactured into a range of different structures. Manufacturing methods may include hand drawing, wet or dry spinning and electrospinning for the creation of fibres. Films may be formed through dip coating or spin coating, while hydrogels can be manufactured by exposure to an aqueous solution of polyethylene glycol (PEG) (M_w 8,000 to 10,000) solution (25-50 wt %), (Jin et al. 2007). Freeze drying such hydrogels gives rise to silk foams. Other methods for producing foams from silk hydrogels include gas foaming and salt leaching. Silk capsules are formed by adsorption of silk proteins to water in water–oil emulsions and silk spheres by electrospraying or precipitation reactions. Dissolution destroys the β–sheet proteins, however post–manufacturing treatment with alcohol or potassium phosphate encourages β–sheet proteins to reform (Hardy et al. 2008, Hardy and Scheibel 2009, Fu et al. 2009), thus recreating nanocrystalline reinforcing segments.

Biomimetic polymer composites using silk proteins as reinforcement are reported in combination with both biodegradable and slow–degradation polymer matrices. Slow–degrading matrices that have been used include nylon 66 (Liu et al 2004), polyacrylamide (Freddi et al. 1999), polyacrylonitril (Sun et al. 1997), polyethylene oxide (Jin et al. 2002) and polyvinyl alcohol (Sashina et al. 2007). Biodegradable polymer matrices reinforced with silk proteins include polyaspartic acid (Kim et al. 2008), poly–ϵ–caprolactone (Qiao et al. 2009) polycarbonate urethane (Dal Pra et al. 2003) and polylactic acid (Zhu et al. 2009).

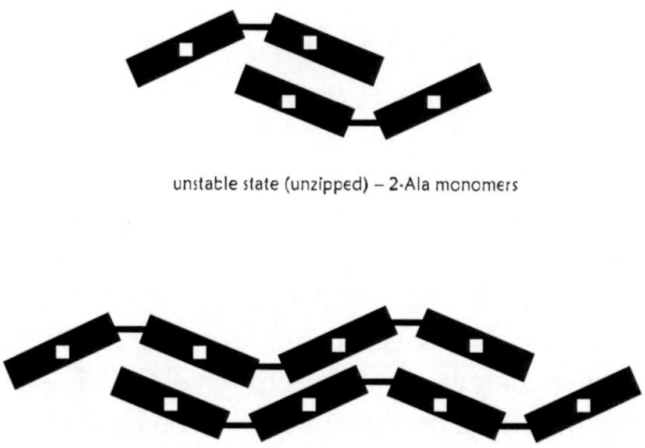

Figure 12. Poly–alanine segmentation in β–sheets have a minimal number of four monomer units for a stable and *zipped* conformation (adapted from Bratzel and Buehler 2012).

Despite the many attempts to mimic the high hydrogen bonded internal structures of silk through the incorporation of silk–like proteins, researchers are still far from imitating the high strength-high stretch characteristic of spider silk. Major guiding factors for the reinforcement of polymer matrices with silk proteins have thus far included biocompatibility, mechanical enhancement, price and good *specific* properties. In truth, a deeper understanding of the relationship between the molecular structure of silk and its macroscopic properties is necessary if closer mimicry is desired. As such, certain research papers have provided exceptional contributions that improve our current understanding of aspects of silk biomechanics and silk processing. Bratzel and Buehler (2012) for example used molecular dynamics simulations to identify critical lengths at which polyalanine chains will essentially become zipped in a β–sheet form. Polyalanine with less than four monomer repeat units are unstable, four monomer repeat units form metastable nanocrystals, and repeat units greater than four yield stable nanocrystals (assuming the polyalanine strands are aligned in parallel) Fig. 12.

The alignment of side chains along each polyalanine β–strand gives rise to extra molecular interlocking, which further enhances the flexural and torsional rigidity of the nanocrystal. The toughness of the nanocrystal itself is governed by a finely tuned ratio of intermolecular and intramolecular forces of attraction, which varies with respect to sheet stacking and β–strand length (Alam 2013). The formation of these nanocrystals is moreover likely to arise through the interlocking of polyalanine β–strands, rather than through folding mechanisms along the backbone since the latter requires higher extensional forces for β–strand alignment (Keten 2010). Smaller β–sheet nanocrystals have a greater reinforcing effect in the silk to an optimal size of *ca.* 3 nm (Keten 2010). This is because adjacent β–strands can only act cooperatively when they are in a stable and 'zipped' state (Bratzel and Buehler 2012). The mechanism of deformation for the smaller, stable nanocrystals is dominated by shear. As the nanocrystals increase in size, though they are also 'zipped', they have more torsional and flexional capabilities and under loading may be freed via high shear stresses that arise through flexure or torque.

3. Layered Biostructures

Layering is a key structural quality in a variety of biological organisms. In this section we will focus on understanding how biological organisms make use of laminates, and how material combinations affect the mechanics and failure mechanisms. Laminates are important where good transverse or off–axis properties are desirable alongside good axial properties. Laminates also have the advantage of distributing concentrated stresses along interfaces and can be critical for retarding fracture. Moreover they can reduce weight whilst increasing strength (e.g. insect forewings). We begin by describing the structural organisation of three very familiar materials, beetle forewings, wood and bone.

3.1. Beetle Forewings

The forewings of beetles have special, lightweight structures with good mechanical properties. The combination of light weight and strength is important since the beetles (a) fly and (b) require a protective outer cuticle from predatory attack. The forewings are made up of chitin fibres and protein, and are structured as laminated fibre reinforced composites, where these chitin fibres reinforce the softer protein matrix. This laminated sandwich structure has external and adjoined dual laminating cuticles about a near hollow lamina containing numerous trabecula (pillars). The inner lamina (closest to the body) are rich in chitin fibres, the outer laminae are rich in proteins. The trabeculae provide support to the external laminae whilst also minimising weight in the overall structure. Chitin fibre in the external lamina is circular in cross section and is sparsely distributed in the soft protein matrix. Chitin fibre in the inner lamina is square section and densely distributed in the matrix. The square section fibres give more support, Fig. 13, and fibre crosslinking provides additional lateral support.

3.2. Wood

Wood is a hierarchical material with both fibrous and laminated arrangements recurring at different length scales. The fundamental structural component of wood is crystalline cellulose. These units are united into elements known as microfibrils. The microfibrils make up the cell walls and are oriented in laminating layers within the cell wall structure. Hemicelluloses and lignins make up the rest of the cell wall. The primary wall, P, and middle lamella, M, are the non-structural regions of the cell wall. The middle lamella acts as an adhesive between other cell wall components. The secondary layers, S1, S2 and S3 are the structural components with the S2 layer most likely generating the greatest axial

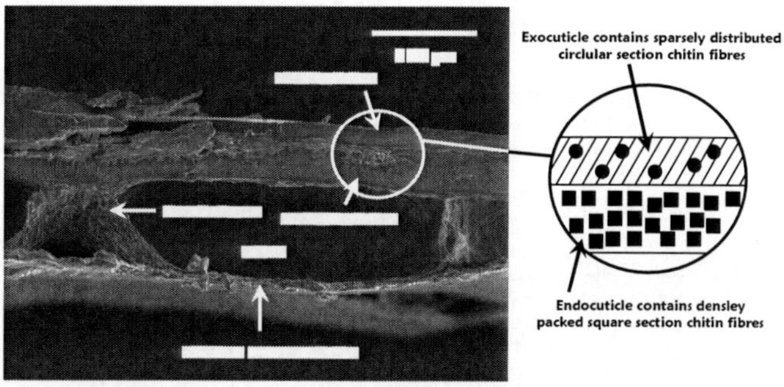

Figure 13. Beetle forewing: a fibre reinforced, laminated light–weight sandwich composite.

strength. These form cellular arrangements which subsequently become more or less dense through the seasons, and are termed latewood and earlywood respectively. These then form the clearly laminated structures we see in wood and sawn timbers. Fig. 14 illustrates the hierarchical structure of wood.

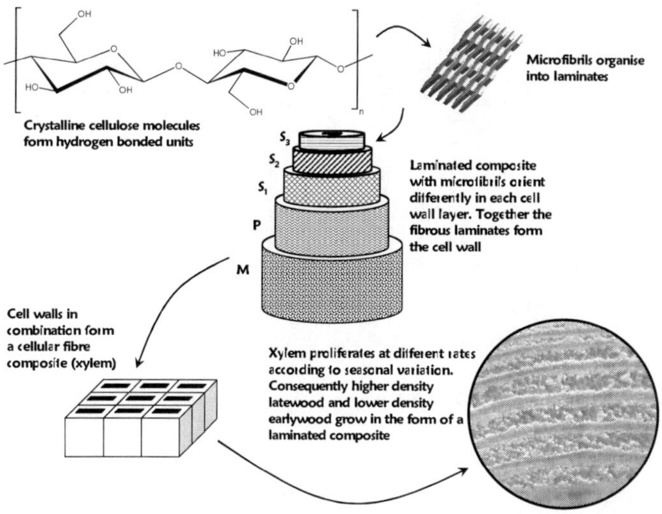

Figure 14. The hierarchical, fibrous and laminated structure of wood.

3.3. Bone

Bone is a material that naturally biomineralises. Similarly to wood, it is a hierarchically structured material with an immense capacity for bearing load. To understand the mechanical properties of bone, the component phases and the structural relations between each of them needs to be understood. These levels of structural organisation are (Rho et al. 1998):

1. Macrostructure: cancellous and cortical bone
2. Microstructure (10–500 μm): Haversian systems, osteons
3. Sub microstructure (1–10 μm): lamellae

4. Nanostructure (few hundred nm to 1 μm): fibrillar collagen and embedded mineral

5. Sub–nanostructure (<a few hundred nm): molecular structures of constituent elements such as mineral, collagen and non-collagenous organic proteins.

Fig. 15 illustrates how bone is structured hierarchically. Within the bony structure, there exist a number of substructures. These include:

- Cancellous bone: (also trabecular bone or spongy bone), which has a higher surface area than compact bone but is less dense, softer, weaker and less stiff

- Cortical bone: compact bone, the main function of which is to support the body. This bone is denser, stronger and harder than cancellous bone and has a Young's (tensile) modulus of 14–20 GPa (Reilly et al. 1974).

- Osteon–Haversian system: a fundamental unit of compact bone which is organised into cyclindrical sections. The Haversian canal contains the nerves and blood supply of the bone and is made up in a laminated structure of 3–8 laminae and consisting of mineralised collagen fibres. These lamella have a Young's (tensile) modulus of 26 GPa. The osteons themselves have a Young's (tensile) modulus of 22 GPa (Rho et al. 1999). The lamellae are *ca.* 3–7 μm thick. It is commonly believed that the collagen fibres in a lamella of an osteon are arranged in parallel within that lamella, with a change in the orientation angle from one lamella to the next, much like in plywood.

- Collagen molecule: is a polypeptide chain, a linear array of amino acids with repeat units of Gly–X–Y. Gly is glycine, X is predominately proline (Pro) and Y is usually hydroxyl–proline (Hyp). These molecules are coiled and form a triple helix, thus yielding the collagen molecule. Its molecular weight is 285 kDa.

- Collagen fibrils: in hydrated state have a Young's modulus of *ca.* 400 MPa (Sasaki and Odajima 1996).

- Collagen fibres: in hydrated state have a Young's modulus of 0.15–1 GPa (Bulter et al. 1986).

3.4. Fibre Reinforcement in Natural Biostructures

In each of the prior examples (beetle forewings, wood, bone), the structural organisation is essentially developed from a fibrous reinforced composite. The distribution of fibres within the matrix materials will depend on various factors. Fibre clustering is a reality and ideal fibre distribution seldom if never happens. Fibres with the same diameter, packed as closely as possible will tend towards a hexagonal close packing, Fig. 16(a). This tendency may predominate if the fibre fraction is sufficiently high, the fibres are unidirectional and the processing method is one that encourages close packing. The square shape of chitin fibres found in beetle forewings allows them to pack more tightly than circular cross section fibres. In man–made fibre composite processing; matrix material is injected into a close packed array of fibres. However, extremely high pressures would be required to retain

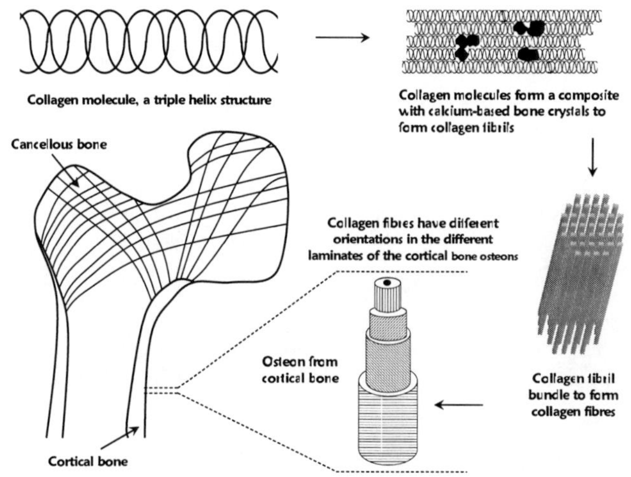

Figure 15. The hierarchical, fibrous and laminated structure of bone.

the arrangements found in natural materials. Such high pressures are not practicable and in nature, it is the combination of cellular activity with environmental pressure that drives their packing characteristics. Even with man-made uni–directional fibre packings, the fractions of fibre are not sufficiently high to facilitate precise hexagonal arrangements. Rather, fibres tend to cluster, become interlocked and, matrix rich regions with irregular fibre distributions manifest, Fig. 16(b). In fact, packing fractions for unidirectional (UD) fibre composites are seldom higher than 0.7.

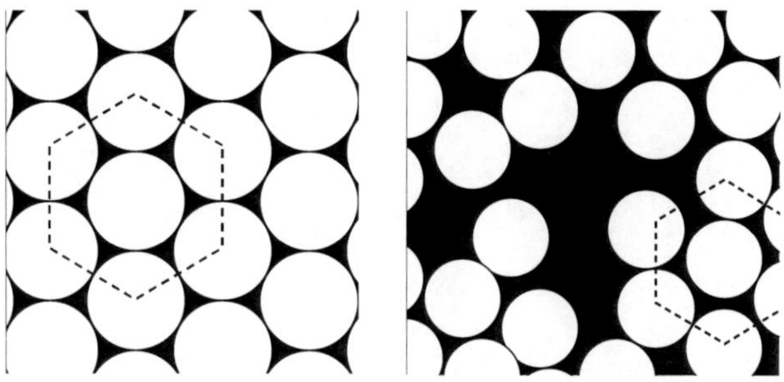

Figure 16. (a) Example of hexagonal close packed structure, (b) more realistic packing arrangement in UD fibre composites (occasional areas of near hexagonal close packing).

3.4.1. Fibre Models

When a material is loaded, it will experience a displacement of the bulk. Load and displacement are fundamental parameters in mechanics and are linearly related to stress and strain respectively. Load and displacement have essentially no generic meaning as they are related to a material with specific dimensions, which vary as a function of time. Stress, σ, and strain, ϵ, remove the constraint of *dimension* and are thus useful parameters for comparing materials of different dimension, Eqs. 1 and 2 respectively. In these equations,

F is the force, A refers to the area of cross section and L is the length of a structure. The Young's modulus, E, is taken to be the linearly proportional region of a stress–strain curve, Eq. 3.

$$\sigma = \frac{F}{A} \tag{1}$$

$$\varepsilon = \frac{\Delta L}{L} \tag{2}$$

$$E = \frac{\sigma}{\varepsilon} \tag{3}$$

When a material is loaded, it will experience a distortion perpendicular to the axis of loading. The ratio of the transverse, $_T$, and axial, $_A$, distortions (expressed as strain) gives rise to another material parameter called *Poisson's ratio*, Eq. 4. Normally in tension, materials will become thinner and in compression they will become thicker. Auxetic materials become thicker when stretched and thinner when compressed. Such materials also exist in nature, for example human skin, and are another important area of biomimetic investigation.

$$\nu = \frac{\varepsilon_T}{\varepsilon_A} \tag{4}$$

Stress transfer in natural composites is fundamentally the means by which the crack retarding potential of these materials can be understood. Moreover, it is the means by which we can understand the superior load bearing capabilities and toughness of many of these materials. When a reinforced system is loaded, the load is distributed and transferred along the interfaces of the materials from and to adjacent components.

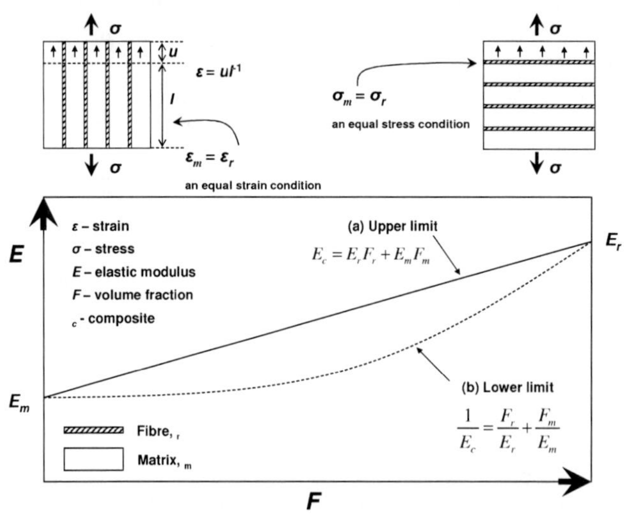

Figure 17. Stress transfer, unidirectional fibre composite extremes (Voigt and Reuss) and the fundamental relationship between composite elastic modulus and the fractions of component materials.

Understanding these upper and lower extreme limits, Fig. 17, allows us to understand that orientation of fibres (reinforcing components) away from the axis of loading will decrease the stiffness of the material reaching a minimum when the fibre is perpendicular to the axis of loading.

3.4.2. Fracture Mechanics and Crack Retardation

One of the key features of two (or more) phase hierarchical natural materials is their ability to retard crack growth. Fig. 18 shows a picture of the release of stored energy as a crack progressively grows. The critical strain energy release rate, G_c (kJm^{-2}) is a measure of how much energy can be retained within the body of material to the point of failure, Eq. 5.

$$G_c = \frac{\sigma^2 \pi a}{E} \tag{5}$$

At fast fracture (crack expanding at high speed leading to catastrophic failure) the following relationship can be derived, Eq. 6.

$$\sigma\sqrt{\pi a} = \sqrt{EG_c} \tag{6}$$

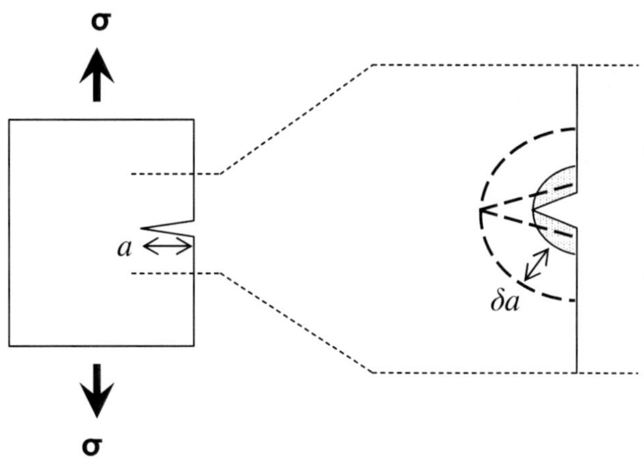

Figure 18. Crack tip and release of stored strain energy as crack progressively grows.

Here, a is the crack length, E the elastic modulus and σ the applied stress. The left hand side of Eq. 6 is related to variable parameters. It says that fast fracture occurs when a material at a particular applied stress also contains a crack that reaches a critical size a, or, that fast fracture occurs when a material with a crack of size a is subjected to a critical stress. The right hand side of the equation depends only on invariable materials properties (G_c being dependant on the material properties and hence invariable). Importantly therefore, it can be noted that the critical combination of σ and a at which point fast fracture initiates is a material constant. The stress intensity factor, K (usually MNm$^{-3/2}$), is the left hand side of this equation and the critical stress intensity factor, K_c (or fracture toughness, usually MNm$^{-3/2}$) is the right hand side of the equation, therefore:

$$K = \sigma\sqrt{\pi a} \tag{7}$$

$$K_c = \sqrt{EG_c} \tag{8}$$

$$K = K_c \tag{9}$$

Following crack initation, two modes of crack propagation may occur through the harder (crystalline, covalently bonded) and polymeric phases of biological structures. These include *ductile tearing* and *cleavage*. *Ductile tearing* is a situation where plastic flow occurs at the crack tip essentially blunting the tip and resulting in high energy consumption through plastic flow. The crack tip is not sharp due to the occurrence of ductile flow and voids that may arise in the plastic zone link up to further crack growth. *Cleavage* is a situation where atoms peel apart resulting in a crack tip that remains sharp (blunting does not occur). Materials that have cleavage dominated failure are usually brittle (like glasses and ceramics) and have very high yield strengths. Fig. 19 illustrates these modes of crack propagation.

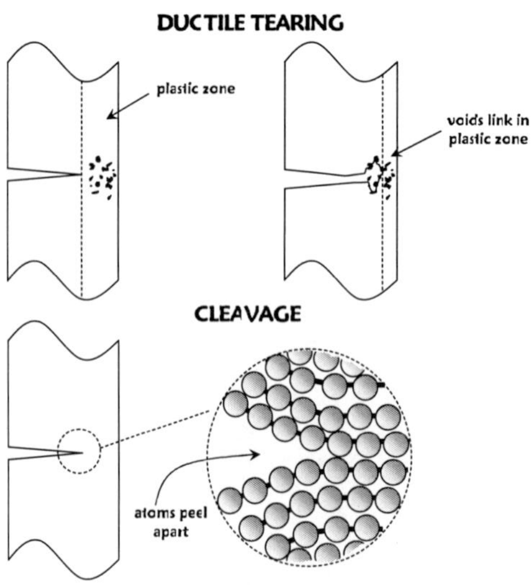

Figure 19. Modes of crack opening for ductile and brittle materials.

In materials comprised of distinctly different phases (harder phase and softer phase), the reinforcing phase will normally re–direct, stop or slow down crack propagation. This is one way by which fast fracture is delayed, and is also the means by which natural composites like spider silk and nacre realise high toughness. Crack propagation can give rise to fracture in the reinforcing material and/or debonding at the interfaces between the two phases, Fig. 20.

In the examples of wood and bone, the fibrous organisation differs between adjacent laminates. This results in a transformation of force vectors and increased global composite stability. Rather than that the xylem cell walls (in wood) and the osteons (of cortical

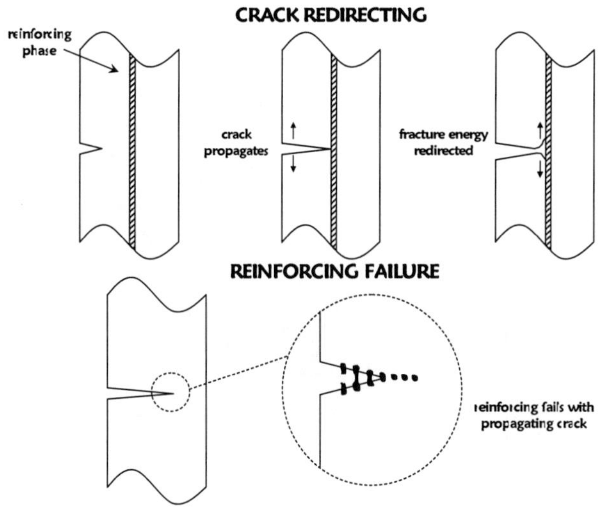

Figure 20. Crack blunting mechanisms in two phase materials.

bone) are stiff in only one direction, such stress transformations produce lowered unidirectional stiffness, but higher stiffness at a variety of angles. This in turn is beneficial when considering the plethora of loading conditions both wood and bone experience throughout their lifetimes. Calculations of these lamina stress and strain transformations are possible through an elastic transformation matrix (shown here in two dimensions), Eqs. 10 and 11. These transformations are necessary to consider for material design where the boundary axes will differ from the principal material coordinates, Fig. 21(a). If a single lamina with off axis UD fibres is loaded in the boundary coordinate system, lamina deformation is skewed, Fig. 21(b).

Typical curves showing the variation of material properties in the system (Fig. 21) as a function of the boundary to coordinate system angle, are shown in Fig. 21(c).

$$\left\{\begin{array}{c}\sigma_{xx}\\ \sigma_{y}\\ \sigma_{xy}\end{array}\right\}=\left[\begin{array}{ccc}m^2 & n^2 & -2mn\\ n^2 & m^2 & 2mn\\ mn & -mn & m^2-n^2\end{array}\right]\cdot\left\{\begin{array}{c}\sigma_{11}\\ \sigma_{22}\\ \sigma_{12}\end{array}\right\} \tag{10}$$

$$\left\{\begin{array}{c}\varepsilon_{xx}\\ \varepsilon_{y}\\ 2\varepsilon_{xy}\end{array}\right\}=\left[\begin{array}{ccc}m^2 & n^2 & -mn\\ n^2 & m^2 & mn\\ 2mn & -2mn & m^2-n^2\end{array}\right]\cdot\left\{\begin{array}{c}\varepsilon_{11}\\ \varepsilon_{22}\\ 2\varepsilon_{12}\end{array}\right\} \tag{11}$$

where $m = \cos\theta$ and $n = \sin\theta$, see Fig. 21.

In the following sections we consider the protective layered polymer-composite organisation of turtle and armadillo carapaces.

3.5. Turtle Shells

Turtles are one of the oldest extant vertebrates in the world (Krauss et al. 2009), the earliest carapace fossils of which are from the Jurassic period of the Mesozoic era. The turtle carapace is structurally optimised for both protection from predatory attack and weight

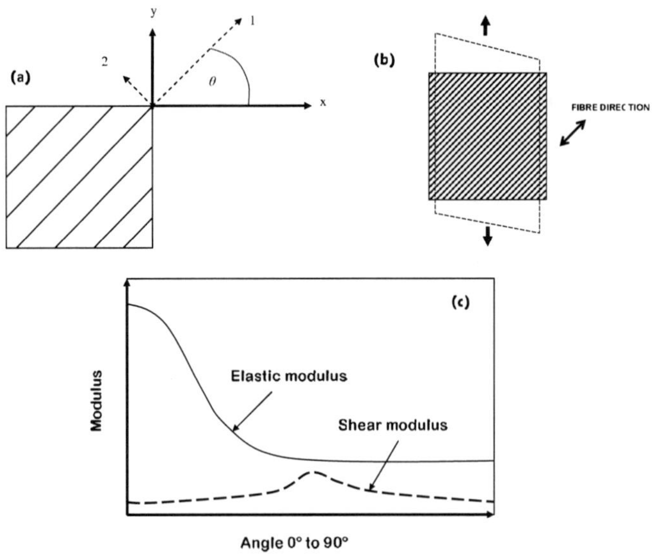

Figure 21. (a) orthotropic and boundary coordinate systems. Angle θ is integral in the transformation matrix (b) loading in boundary coordinate system of lamina where the orthotropic coordinate system is different (c) elastic and shear moduli trends as a function of increasing angle of orthotropic coordinate system relative to boundary coordinate system.

(Rhee et al 2009). The carapace is essentially the laminated calcified upper shell taking up *ca.* one third of the turtle body mass (Zhang et al. 2012), which aside from its protective value, may also regulate pH, reserve fats, wastes and water (Zhang et al. 2012, Balani et al. 2011). The carapace is segmented into eight neurals aligned along the centre of the shell, eight costals on each side of the neurals, twenty–two peripherals on the edges of the carapace (eleven on each side) and two suprapygals to the posterior of the turtle shell, Fig. 22. The marginal surfaces of shell segments are reinforced with bony extensions that serve to interlock adjacent segments creating a characteristic suture between shell segments. The shell bones are lined with scutes. These scutes are essentially plates made up of protective keratinous material (Scheyer 2007).

From a cross sectional perspective, the turtle carapace is a multiphase sandwich composite with a soft porous foam–like layer located between two dense bony layers, Fig. 23, (Rhee et al. 2009). The soft inner is fibrous and the pores in this layer are closed and distributed randomly. The soft layer has *ca.* 65 % pore space, whereas the external dense layers have *ca.* 7 % porosity (Xu and Zhang 1995). At the surface there are two waxy layers. The outermost waxy layer is *ca.* 10–15 μm thick and under this layer is a semi–waxy layer *ca.* 5–10 μm thick (Balani et al. 2011). These waxy layers provide minimal mechanical resistance but serve to (a) protect the outer casing of the shell from environmental and predatory damage and (b) reduce friction for easier slip from a predatory jaw. Since the carapace is essentially a composite of external dense bony layers sandwiching an internal porous bony network of fibrous foam, the sandwich as a whole is able to absorb considerable mechanical energy, which ensures the protection of the internal organs (Liang et al. 2012). In fact, on impact loading, very little force reaches the innermost (endocortical) dense bony layer since

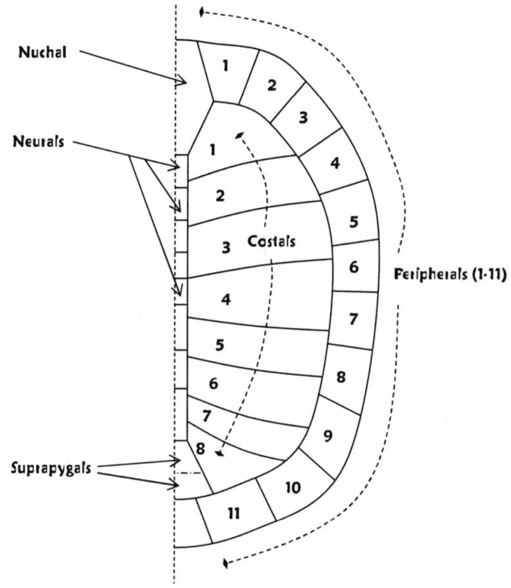

Figure 22. Fundamental structural organisation of a turtle carapace.

the porous layer absorbs most of it (Krauss et al. 2009). The mechanical purpose of the porous inner layer is to absorb and dissipate mechanical energy. This sandwich composite is able to withstand flexural loading since the stiffer, denser layers are farthermost from the axis of neutrality (which is in the lower stiffness porous layer). The relative densities of the endocortical (inner), middle and exocorticle (exterior) layers are 1520 kgm^{-3}, 1030 kgm^{-3} and 1470 kgm^{-3} respectively (Zhang et al. 2012). The 4th and 5th layers (see Fig. 23) are very soft and serve to further protect the internal organs as secondary shock absorbers for any mechanical energy that reaches the endocortical layer. More material in the 4th layer in particular, is able to return to its original state after deformation than in any other layer of the carapace (Balani et al. 2011).

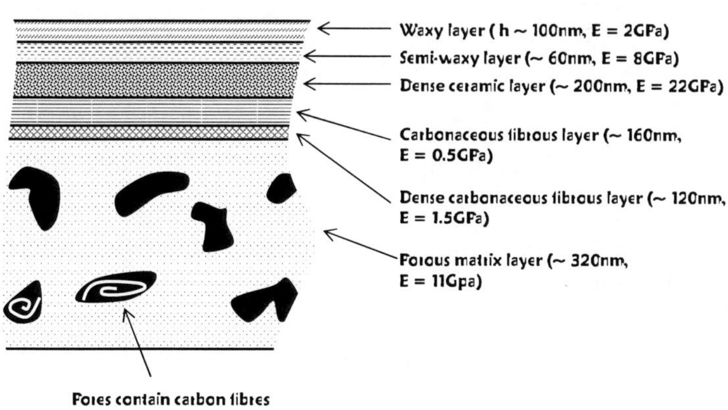

Figure 23. Schematic cross sectional layup of a turtle carapace (adapted from Balani et al. 2011).

Mechanical rigidity in a turtle shell is primarily a function of the properties of the bony plates and the mechanism that connects them (Rhee et al 2009). The shell, being essentially a bony casing of fused plates covered by a shield of keratinous scutes, is very stiff. Due to its high stiffness, mechanical loading that arises through respiration, locomotion, falls or predatory attack could give rise to microfractures. To avoid microfractures, the shell must be compliant such that it can retain rigidity and stiffness, whilst still allowing for a degree of flexibility. This flexibility becomes possible due to the presence of the inter–shell segment sutures. These softer sutures that exist between the bony shell–segments are uniquely shaped, 50–80 μm in width and 150–800mm in length (Krauss et al. 2009). The concentration of mineral decreases from the bony elements to these sutures and the sutures themselves are free of mineral compounds. As a result the sutures provide a source of flexibility between the bony segments. This structure helps to both absorb and dissipate mechanical energy across a larger area of the shell, thus decreasing the chances of brittle fast fracture within the bony segments. Due to the presence of sutures, turtle shells in flexure display non–linear characteristics of load by deformation, even though the material may still be within its limits of recoverably distortion.

3.6. Lateral Polymer Bridging of Armadillo Armour

In similitude to the turtle carapace, the top layer of the mid–section of an armadillo carapace is covered by a keratinous layer beneath which, are an organised array of triangular bony tiles connected by collagenous fibres. The collagen fibres are termed Sharpey's fibres and are the polymeric component of the composite responsible for binding the bony tiles. A structure–properties research conducted by Chen et al. (2010) reports that these unmineralised fibres diverge from the the centres of the tiles (Vickaryous and Hall 2006) and this arrangement is show in Fig. 24. In this figure, only the osteoderm is shown, which consists of primarily the Sharpey fibres and the bony plates. The polymeric Sharpey fibres, much like the sutures of the turtle carapace, allow for a degree of flexibility in an otherwise rigid carapace layer.

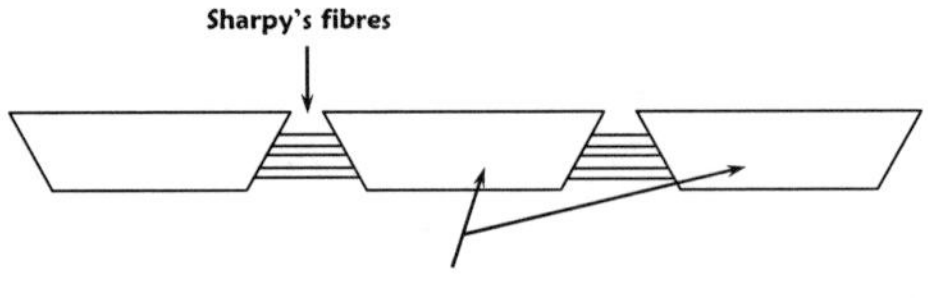

Figure 24. Laterally bound scales of armadillo armour.

Chen et al. (2010) report that armadillo osteoderms display two failure modes in tension, which are a function of their hydration levels. These failure modes are illustrated in Fig. 25. When the osteoderm is dehydrated, fracture propagates through the bony tiles whereas in hydrated form, the fracture propagates between the tiles and ruptures the Sharpey fibres. The stiffness and strength of dehydrated osteoderms ($E = 425$ MPa, $\sigma =$ 23 MPa) are higher than in the hydrated composite ($E = 150$ MPa, $\sigma = 13$ MPa), and this is minimally related to material embrittlement, or plasticisation, through a lack of, or sufficiency of, electrostatic forces of attraction. Importantly, the interplate shear strengths in the armadillo osteoderm clarifies that the polymer Sharpey fibres are primarily responsible for resisting load. In fact these fibres are able to resist almost as much force in shear as in tension. In an armadillo osteoderm composite therefore, the polymer input is critical to the overall mechanical performance of the osteoderm.

Figure 25. Fracture through armadillos tiles as a function of hydration content as described in Chen et al. (2011).

4. Marine Biostructures

4.1. Biomineralisation

A most important aspect of many marine biostructures is their ability to biomineralise. This is a process by which living organisms produce minerals. Organisms including; corals, sponges, diatoms, molluscs, shells and brittle stars, are often able to control the fractional content, shapes and/or the polymorphism of their biominerals (Falini et al. 1996, Gotliv et al. 2003, Aizenberg et al. 1995, Pokroy et al. 2006). An outcome of biomineralisation is the hardening up of soft living tissue. Biomineralisation can also serve to protect the soft tissue from possible environmental damage and predators. The most common minerals that are the product of biomineralisation are calcium carbonates, calcium phosphates and silicates. Fig. 26 shows a few examples of marine organisms that biomineralise.

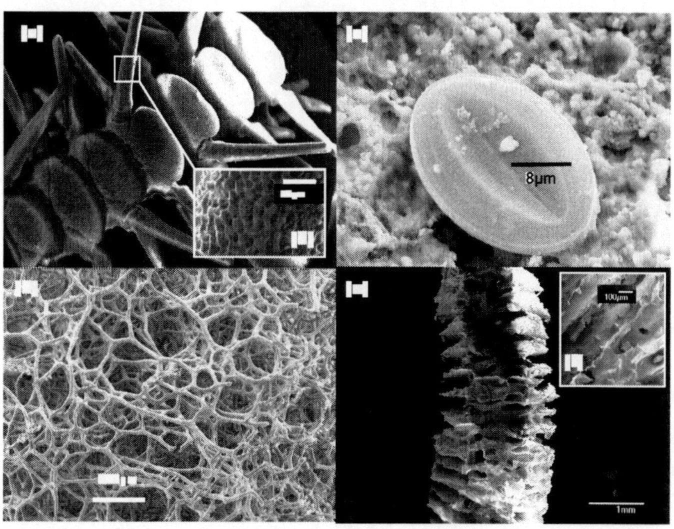

Figure 26. Examples of marine organisms that biomineralise (a) Echinoderm: *Ophiocoma* (b) a mono–raphid diatom (c) Porifera: *Sclerospongiae* and (d) a *Scleractinian* coral. Organisms collected at Pantai Drini, Java Trench, Indonesia during the months of January and October 2012.

The processes controlling biomineralisation are interesting to material scientists as mineralisation in itself has great potential in coating/composite technologies. There are two fundamental routes for biomineralisation. The first and more complex is *biologically induced* mineralisation where the process of mineralisation is metabolised by cells. In this form of mineralisation, the crystal morphology is highly intricate and often beautiful. The second route to biomineralisation is *biologically controlled* mineralisation where the process of mineralisation is environmentally controlled and dependent on the minerals available in the surroundings. These mineral compounds nucleate to cell walls and grow from the points of nucleation. The crystal morphologies are less controlled than in the biologically induced mineralisation process. Consequently the crystal morphologies are not as intricate as those produced through biologically induced biomineralisation, though they maintain good structural functionality as load bearing exoskeletons.

Biominerals are essentially composed of crystalline mineral componds bound together by biopolymers (Wilt 2002, Weiner and Dove 2003). The organic biopolymer phases in biominerals are most commonly proteins, polysaccharides and lipids. These phases predominantly bind at the edges and faces of mineral crystals. Other examples of mineral composites such as described here include bone, teeth and shells, which, in similitude to many corals are calcium phosphate or calcium carbonate crystals bound together by a nano–scale organic matrix (Lowenstam and Weiner 1989, Mann et al. 1989, Simkiss and Wilbur 1989, Weiner and Price 1986, Mann 1983). Typical elements found in biominerals therefore include hydrogen, carbon, oxygen, magnesium, silicon, phosphorous, calcium, manganese and iron. Approximately 50 % of crystals in nature contain calcium (Lownestam and Weiner 1989). The calcium cation holds importance in many organisms since it is used in a variety of metabolic processes (Lowenstam and Margulis 1980, Berridge et al. 1998). This mineral compound exists in three different polymorphs, calcite, aragonite and vaterite, the most stable of which is calcite while vaterite is a metastable form (Addidi et al 2003). About 60 % of biominerals consist of one phase containing water and/or hydroxyl groups (Weiner and Dove 2003). Silica containing minerals exist for example, only in their hydrated forms (Lowenstam and Weiner 1989). Calcium carbonate biominerals typically initiate in their hydrated form prior to transforming into a final polymorph (Beniash et al. 1997, Weiss et al. 2002, Addadi et al. 2003).

4.2. Nacre

Nacre is a material with incredibly high toughness that lines the inner shells of many molluscs. Table 4 shows the mechanical properties of nacreous material. The material of nacre is a result of a highly effective process of biomineralisation. Abalone shells are primary sources for nacre, which is essentially, a combination of soft organic materials and stiff minerals and is hierarchically structured, Fig. 27. At the macro level, abalone shells reveal two layers with different microstructures. At the millimetre length scale, two crystal forms can be distinguished in each layer, prismatic calcites (first layer) and aragonites (second layer). The first layer has the role of protection while the second layer dissipates mechanical energy and hence provides toughness (Sarikaya and Aksay 1995). The first layer is composed of predominantly calcite while the morphology of the calcium carbonate within the second layer is aragonite, which is less stable than calcite (Currey 1977). At the

Table 4. Mechanical properties of nacre. Sources: [a]Jackson et al. (1988), [b]Menig et al. (2000), [c]Wang et al. 2001.

Property	
Youngs Modulus (dry)	70 GPa[a]
Youngs Modulus (wet)	60 GPa[a]
Tensile Strength (dry)	170 MPa[a]
Tensile Strength (wet)	140 MPa[a]
Work of Fracture (varied ambient)	350–1240 Jm^{-2}[a]
Dynamic Compressive Strength (parallel to plates)	548 MPa[b]
Dynamic Compressive Strength (perpendicular to plates)	735 MPa[b]
Static Compressive Strength (parallel to plates)	235 MPa[b]
Static Compressive Strength (perpendicular to plates)	540 MPa[b]
Bending Strength (parallel to plates)	194 MPa[c]
Bending Strength (perpendicular to plates)	223 MPa[c]

scale of *ca.* 100 μm, growth bands can be seen in the structure which retard crack propagation. Between these growth bands at the scale of *ca.* 10 μm there are further, smaller bands, which when looked at more closely (at *ca.* 500 nm) can be seen to be formed of interpenetrated interlocking platy tiles, the surfaces of which furthermore have small < 10 nm scale nano-grains and between which there exists a protein-rich viscoplastic matrix (Barthelat 2007, Smith 1999). At every level of hierarchy, laminar structures serve to arrest crack propagation.

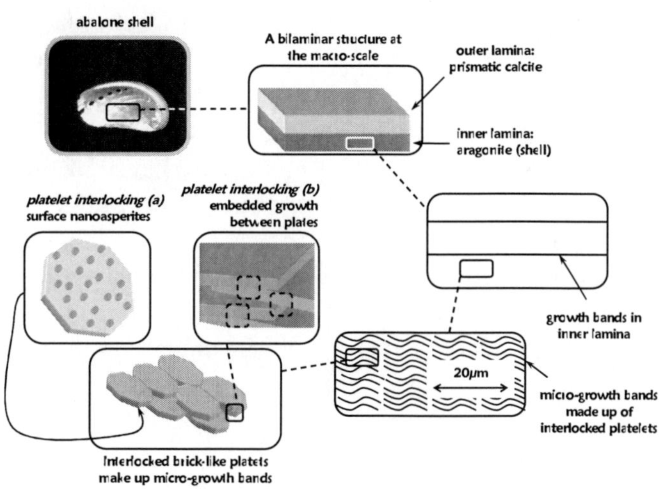

Figure 27. Hierarchical structural organisation of abalone.

Due to this complex hierarchical structure, and the fine blend of 95 % aragonite 5 % organics (proteins and polysaccharides), Sarakaya and Aksay (1995), this composite material is able to easily retard crack growth and as a consequence the toughness of this composite is very high (Curry and Taylor 1974). Organic scaffolds can be observed during the steady state growth of aragonitic tiles in nacre (Lin et al. 2008). Amino acids such as aspartic acid (Levi–Kalisman et al. 2001) containing carboxylic groups are primarily responsible for scaffolding by forming intimate contacts between the peptide chains and the inorganic compounds (Metzler et al. 2008), though within the inorganic phases there also exist intracrystalline glycoproteins (Levi–Kalisman et al. 2001). Aragonitic tiles interlocking during steady state growth and their interlocking mechanisms possibly arise through nucleation and growth around initial screw dislocations (Katti et al. 2005, Wise and Devillie 1971, Yao et al. 2006). Nacre has 106 screw dislocations per square centimetre (Katti et al. 2005). A screw dislocation can most easily be understood by imagining a cubical block which is cut in the centre but only part way through. One side of the block slips about the cut boundary and that represents the screw dislocation.

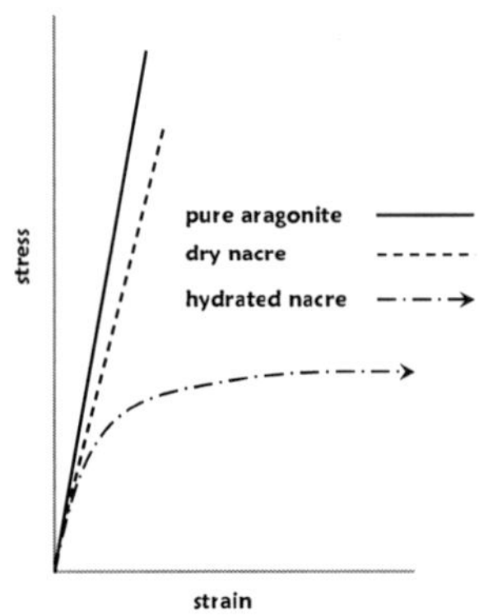

Figure 28. Stress-strain curves for nacre.

Nano–scale asperitic growth of mineral compounds on the surfaces of the aragonite plates further help to resist shear deformation of the plates and increase the mechanical energy that the abalone shells can bear. These asperites can in places, be direct mineral compound bridges that physically connect adjacent plates (Song et al. 2002a, Song et al. 2002b, Wang et al. 2001, Schaffer et al. 1997). The existence of these nano–asperitic growths and bridges means these structures are more like brick–bridge–mortar than simply brick and mortar structures (Song and Bai 2001, Song et al. 2002a, Song et al. 2003). These polymer–mineral composite shells therefore show minimally, six levels of hierarchy by which means they are able to raise the fracture toughness. There are three main mechanisms attributed to nacre platelet growth (Qiao et al. 2008), though the process may also have multiple stages (Mayer 2005):

1. By growth of single crystals
2. Through the coherent aggregation of nano-grains
3. By phase transformation from an amorphous carbonate to a metastable vateritic morphology

These platelets show characteristics of a brittle Hookean solid on deformation whether it be in a dry or wet state. It is the viscoplastic organic phase, that when hydrated, induces non–linear load–deformation characteristics in the nacreous material (Jackson et al. 1988, Smith et al. 1999). In a dehydrated state, the organic phases at the platelet interfaces

Table 5. Amino acid fractions in cDNA of red abalone.

Amino acid in cDNA	Fraction in cDNA (%)
Serine	16
Proline	14
Glycine	13
Cystine	9

lose many of the secondary bonding interactions, become brittle and the nacreous material behaves similarly to pure aragonite, Fig. 28.

4.2.1. Molecular Structures in Nacreous Material

The organic, polymeric matrix of nacre is made up of proteins, glycoproteins and chitin. More than 50 % of the organic phase proteins are made up of amino acids common to silk fibroins (glycine, alanine and serine), Weiner and Traub (1980). The combination of these silk–like proteins with chitin forms a skeletal basis for mineral deposition. Other proteins that are found in the polymeric organic phase of nacre are rich in aspartic acid or serine (Addadi and Weiner 1985, Weiner and Traub 1980). Both of these proteins guide the conformational structures of nacre and the aspartic acid rich proteins assume a β–sheet conformation (Addadi and Weiner 1985). Stereoselectivity occurs through the reaction of aspartic acid rich proteins with calcite, where carboxyl groups connect to the electrically charged (001) faces.

Morphological evolution from calcite to aragonite is to an extent, controlled by protein–crystal activity (Fritz et al. 1994, Belcher et al. 1996, Zaremba et al. 1996, Su et al. 2002). Proteins control the nucleation and orientation (Addadi et al. 1987) of calcite, which when oriented to e.g. the (104) face, will develop as nacreous aragonite. Specific amino acid motifs within proteins are critical for the successful mineralisation of nacre. Take the example of red abalone for which a biomineralising protein is cDNA (coding for Lustrin A). This protein has a highly modular configuration with high proportions of certain amino acids, Table 5. Within the protein there are ten cystine rich domains, which govern the mineralisation of nacre (Shen et al. 1997). The strength and hence rate of mineralisation in such chemical systems are a function of distance between the mineral ions and the protein functional groups (Ghosh et al. 2007).

Hydrated nacre, as we have discussed, is key to improved mechanical energy absorption and the encouragement of ductility within the nacreous polymer–mineral composite. Water in nacre at the molecular level, is present in three different forms. The first form is incorporated within the organic matrix and is partially bonded through electrostatic forces of attraction. In the second case, water is in contact with other water molecules and thus completely hydrogen bonded. The third case exists in pores within the aragonite bulk and at the interface between the organic polymer and aragonite. In the latter case, the water is chemisorbed to the surface of the aragonite (Verma et al. 2007). Water in contact with the mineral phases of nacre tends to remain stagnant even when deformation takes place. Contrarily, under conditions of deformation, water in the organic phases of nacre moves in conjunction with the movement of protein molecules (Ghosh et al. 2007).

In summary, the primary mechanisms by which energy is dissipated so effectively in nacre are:

- Crack deviation and branching leading to highly convoluted fracture paths
- Crack bridging by fibrous networks in the organic polymer matrix
- Failure process initiated by micro-cracking, rather than directly through fast fracture
- Frictional/shear resistance between plates through nano-asperity and mineral bridging
- Protein unfolding and breaking of cross-linked polymer in the organic phase
- Plasticisation of nacre in the presence of water
- Platelet interlocking
- Viscoelastic-plastic behaviour of the organic phase
- Resilience of the mineral platelets
- Rotational behaviour of the nano-grains during deformation.

4.2.2. Nacre Biomimesis

The successful fabrication and manufacture of nacre–mimetic materials remains a major challenge in nacre biomimesis. Though there are many engineering examples of platy mineral particle–polymer composites, none of these composites have the same hierarchical organisation or improved level of material toughness as seen in nacre relative to monolithic calcium carbonate. Synthetic nacre, as such should be formed as a combination of inorganic minerals and organic polymers (Oaki and Imai 2005, Oaki et al. 2006). A variety of approaches for manufacturing synthetic macre have been suggested and tested.

The bottom–up approach through covalent assembly seems to be one means to ensuring a well characterised hierarchical arrangement of biominerals (Verma et al. 2008). With this approach, inorganic crystals grow from an organic phase in a supersaturated solution. The properties of the inorganic phase accelerate or inhibit assemblage, alongside polymer molecular weights, concentration, and the density of functional groups along the backbone or side chains of the molecules (Tsortos and Nancollas 2002). Through such self assemblage methods, surfaces with high concentrations of functional groups can be developed, which in turn facilitate bioinspired material synthesis (Luz and Mano 2009). Polyacrylic acid (PAA) solvent has been used in a number of experiments as a means of manufacturing synthetic nacre (Oaki and Imai 2005, Oaki et al. 2006). The primary solutes being potassium sulphate (Oaki and Imai 2005) and calcium carbonate (Oaki et al. 2006). Crystallisation of potassium sulphate from a PAA solvent results in hierarchy, oriented assembly and nanostorage (the ability to host organic molecules), Oaki and Imai (2005). Using calcium carbonate actually promotes the generation of synthetic nacre through the physical bridging of nanocrystals in combination with organic polymer (Oaki et al. 2006). Similar bottom up mineral–polymer crystallisation methods using chitosan, polygalacturonic acid and hydroxyapatite evidence improved mechanical performance over individual component materials (Katti et al. 2008). Laminated microstructures can also be achieved using bottom–up methods through the dip–coating process using a solution of silica, surfactant and organic monomers. During evaporation silica–surfactant–monomer micelle–type structures assemble into liquid crystalline mesophases, while guiding the development of organic and inorganic precursors into the nano–scale laminated composite (Sellinger et al. 1998). Importantly, in this process, covalent bonds link together organic–inorganic phases at their interfaces. Clay particle with natural platy geometries are an obvious choice for inclusion into a nacre-mimetic composite (Bonderer et al. 2008). In this case, a bottom–up colloidal assemblage of clay platelets incorporated into a polymer matrix is a way of enhancing mechanical performance.

Electrophoretic deposition involves the application of an electric field to charged nanoparticles in suspension, which consequently compels nanoparticles to oppositely charged electrodes where they deposit (Luz and Mano 2009, Sarkar and Nicholson 1996, Boccaccini and Zhitomirsky 2002). This deposition method is a low cost yet efficient method by which nanoparticulate layers can be arranged as laminates. By applying this method to polyamic acid with clay nanoplatelets (Wang et al 2008), composite nacre mimetic films can be manufactured. Polyamic acid can moreover be synthesised from promellitic di–anhydride and 4,40–di–anminodiphenyl ether with different montmorillonite addition concentrations (Luz and Mano 2009). Montmorillonite is a monoclinic clay–type mineral from the smectite mineral group and is the most commonly used layered silicate in polymer nanocomposites because it has a high capacity for swelling (Brindley and Brown 1980). At the crystallographic level, montmorillonite is made up two tetrahedral layers of silicate sheet surround an octahedral gibbite sheet with shared oxygen atoms. Intercalation of the layered silicate is made possible through reaction with polyamic acid, which gives rise to an ordered and layered assemblage with significantly enhanced mechanical performance over the pure polymeric film (155 % higher stiffness, 40 % higher strength). Electrophoretic deposition has also successfully been used to combine montmorillonite with acrylamide (Long et al. 2007) and acrylic anodic electrophoretic resins (Lin et al. 2008).

A third route to the production of nacre–mimetic polymer composites is the layer–by–layer (LbL) method. This is essentially a time–consuming technique used predominantly for manufacturing thin films by deposition of oppositely charged materials from one layer to the next. This is done by submerging a charged substrate between two polyelectrolytes with opposite charges. This said, it is still considerably faster than dip–coating methods (Vertlib et al. 2008). The benefit of this method is that a large range of materials can be layered and the assemblage of the composite does not induce high mechanical stress, such that bio–macromolecular characteristics can be preserved (Crespilho et al. 2006). Though normally used to manufacture thin films, 5mm thick nacre-mimicking composites can be built up LbL through sequential immersion of a glassy substrate into polycationic and polyanionic solutions of montmorillonite clays (Luz and Mano 2009). The strength of such a composite is close to that of nacre while the Young's modulus is similar to lamellar bone. There is therefore, considerable mechanical potential for nacre–mimetic composites as bone replacement material. In view of establishing anti-bacterial properties using the LbL method, this can be achieved by positioning clay layers on starch stabilised silver nanoparticles (Podsialdo et al. 2005). Other ways of promoting bone growth can be through the application of coatings that encourage apatite fixation (Couto et al. 2009). The LbL method here united bioactive glass nanoparticles with chitosan resulting in a layered, nacreous coating. High strength and transparent thin nacreous films are formed using the LbL method by combining polyvinyl alcohol with sodium carbide–montmorillonite clays (Podsialdo et al. 2005, Podsialdo et al. 2008). The LbL method is sufficiently versatile to be used in combination with certain other techniques, such as with chemical bath deposition. In a study reported by Burghard et al. (2007), these methods were used conjointly to manufacture titanium oxide–organic polymer films with nacre–like microstructures. The temperature used in the chemical bath deposition method can be quite low (30°C) and is thus a similar temperature to that observed in natural organic–inorganic biomineralisation.

Template inhibition involves the deposition of minerals from solution to an ordered 2D structure (Li and Kaplan 2003). The starting point involves the deposition of soluble macromolecules onto an insoluble matrix. Using this method, thin films of calcium carbonate can be grown from polymeric solutions. In solutions of polyacrylic acid, the calcium carbonate grows as thin–film crystals. For layered nacre–like composite films, the same method and base materials (calcium carbonate/acid rich macromolecules) can be alternated with spin coating of polysaccharides like cellulose or chitin (Kato et al. 1998, Kato and Amamiya 1999, Hosoda and Kato 2001, Kato et al. 2000). Aragonite crystals in a chitosan matrix can be manufactured as thin-film nacre–like material by combining polyaspartic acid with magnesium ions (from magnesium carbide) and calcium carbonate in solution (Sugawara and Kato 2000). This solution is set down upon a chitosan matrix and layered structures are then possible by alternating the chitosan layer with the thin–films of aragonite–polymer crystals. The method of template inhibition can also be used to synthesise macroscopic and continuous films of calcium carbonate (Xu et al. 1998). The soluble inhibitor in this process is polyacrylic acid and mineral deposition is promoted through the use of amphiphilic porphyrin templates. The benefit of this particular nacre–like synthesis is that it can be used to manufacture thick samples (up to 0.6mm) of nano-laminated material at low temperatures ranging between 22°C and 48°C.

4.3. Sponges and Corals

Sponges (Porifera) and corals (Cnidaria) are organic living organisms (multi–cellular animals) which are able to biomineralise for increased structural rigidity. They tend to dwell at the bottom of the ocean and there are many different types of sponges and corals. Sponges and corals differ anatomically since that corals are a collection of genetically identical multicellular organisms (polyps). Physiologically, sponges acquire their energy through maximising the flow of sea water through their structure and filtering the food/nutrients. Corals however, gain considerable energy by catching photosynthetic unicellular algae (zooxanthellae) and hence, tend to be shallow water dwellers, usually not exceeding depths greater than 50–60 metres. Corals reproduce by and large sexually and 75 % of hermatypic corals (stony corals/reef building corals) are hermaphrodites, while 25 % of hermatypic corals have single sex colonies. Polyps in a coral head reproduce asexually. Most sponges are hermaphroditic but without gonads, reproducing via sperm and egg interactions. Fig. 29 shows the basic physiology of Porifera and Cnidaria.

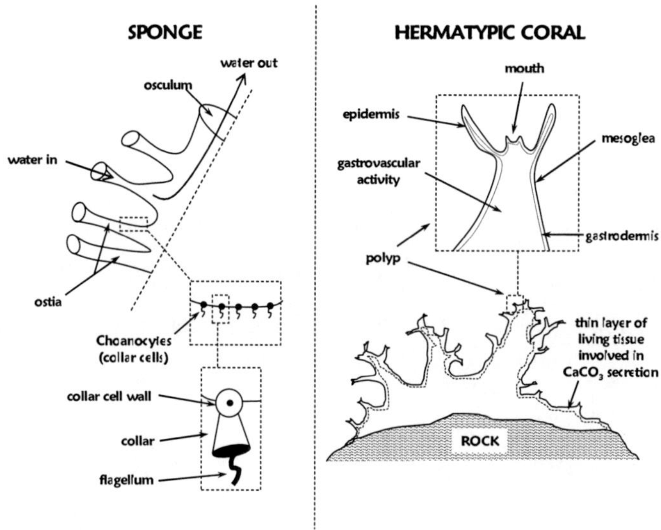

Figure 29. Basic structures of corals (Cnidaria) and sponges (Porifera).

4.3.1. Sponges

In sponges the process of mineralisation involves the creation of spicules, which are secreted from cells known as *sclerocytes*. These mineralised crystals reinforce, stiffen up and protect the *mesophyl*, which is essentially the endoskeleton of sponge tissue. The geometrical range for spicules is very large. Fig. 30 shows common spicule forms. Sponges can be classified through the following routes:

1. Skeletal characterisation including the chemical nature of spicules, the type and size of spicules, and the arrangement/pattern of the skeleton
2. External characteristics such as the shape of the sponge
3. The texture and feel of the sponge, whether fragile, brittle, tough or hard
4. Pigmentation
5. Sponge surface features within the membrane
6. Reproductive behaviour
7. Biochemical similarities
8. Histology
9. Ecology.

The material scientist will often prefer to classify according to the skeletal characterisation including the chemical nature of spicules, the type and size of spicules, and the arrangement/pattern of the skeleton. Therefore the following four classes of sponge are classified according to their skeletons/spicules:

- Class Calcarea: the skeleton consists of individual spicules of calcium carbonate
- Class Hexactinellida: (glassy sponges) members have spicules of silica (glass) fused in a continuous and often very beautiful latticework
- Class Demospongiae: the largest class, has unfused silica spicules, or a tough keratinous protein called spongin (rich in nitrogen and sulphur) or a combination of the two
- Class Sclerospongiae: the smallest class, their skeletons contain all three types of material i.e. silica, calcium carbonate and spongin.

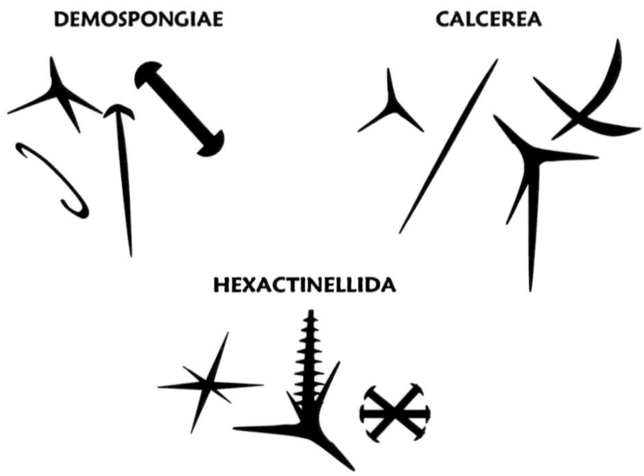

Figure 30. Examples of different sponge spicules exisiting in different classes.

The spicule sizes differ between Porifera species and as a function of environmental pressure (Garrone et al. 1981, Hartman 1981, Weaver et al. 2010). The two largest groups of Porifera, the demosponges and the hexactinellids, have spicule sizes that are to an extent, controlled by the method of spicule growth. The axial filaments in sponges provide the substrate structures onto which silica glass spicules biomineralise. In demosponges this filament, which is hexagonal in cross section (Simpson 1984, Simpson et al. 1985, Weaver 2003), is fully grown prior to spicule formation (Uriz et al. 2000). Contrarily, the square cross section axial filaments of hexactinellids (Reiswig 1971) grow continuously during spicule mineralisation (Leys 2003). Larger glassy spicules of hexactinellids can be laminated and this structure significantly improves their characteristics of strength and toughness over pure, synthetic glassy materials (Weaver et al. 2010).

4.3.2. Corals

Coral exoskeltons are often hierarchically structured composites consisting of biopolymers, a vast majority of which are amino–acid based (Mass et al. 2012), and mineral segments that rigidify, stabilise and protect the soft tissue of the underlying animal. Moreover, corals existing in colonies obtain extra mechanical stability from adjacent corals/polyps, which further decrease the chances of predatory attack, or damage through hydrostatic pressures and oceanic currents. Corals are by–and–large porous materials, much like sponges, and porosity is a defining factor for mechanical stability (Hamza et al. 2013), and is temperature sensitive in certain coral species (Caroselli et al. 2011). Whereas Porifera display a wide range of biomineral compounds, corals tend to form predominantly calcium carbonate based skeletons (Juraz–de la Rosa et al. 2012), the more common polymorph of which is aragonitic in corals (Dahan et al. 2003). The very large variety or corals (Carins 1999) and the characteristics of their microstructures lends them to a wide range of possible biomimetic applications. Some of the more common are for the purpose of mechanically compatible implants for bone or as chemically compatible media for oseoblastic proliferation. The porosity is particularly important in bone replacement engineering since colloidal blood must retain the ability to flow, and blood vessels should have space to grow and indeed reticulate. Many corals are further strengthened by the presence of nanograin biominerals, and hierarchical mineral nanobridging between biomineral segments further serves to reinforce the coral at the nanoscale (Przenioslo et al. 2008). Porosity nevertheless, seems to be the primary cause for decreased mechanical performance as is evidenced in Hamza et al. (2013). In their work, an increase in porosity of red coral *Corallium rubrum* from 50 % to 70 % reduces the Young's modulus from 3.44 GPa to 2.02 GPa, the compressive strength from 32.82 MPa to 24.34 MPa and the bending strength from 22.1 MPa to 15.3 MPa.

Allemand et al. (2004) describe four primary areas of influence in the process of biomineralisation in reef–building corals.

Morpho–functional anatomy:

The calicoblastic epithelium (calcination controlling) is the cause of the aragonite skeleton and is located under the polyps. Calicoblastic cells are 10–100 μm long and overlap one another. Skeletogenic cells are connected via desmosome junctions (localised cell structures that are specialised for cell to cell adhesion). The skeletogenic process occurs in a biologically controlled medium and is separated from the sea water by a few layers of cells 10–50 μm thick.

Biomineralisation – ions and organic molecules:

Biominerals are essentially composite materials of fractions – one is mineral the other is an organic matrix. Calcium and inorganic carbon have to be taken in by the coral from the ambient seawater to the sites of calcification. Protons are also eliminated in the mineralisation process. The fundamental formula is: $Ca^{2+} + HCO^{3-} \rightleftharpoons CaCO_3 + H^+$. Approximately 60-70 % of the carbon utilised in this process originates from respired CO_2. The enzyme carbonic anhydrase is also involved in calcification and essentially speeds up the following equilibrium: $CO_2 + H_2O \rightleftharpoons HCO^{3-} + H^+$. This enzyme is histochemically localised within the calicoblastic epithelium. The organic matrix has a firm role in biomineralisation. The organic matrix is responsible for nucleation, micro and macro regulation of crystal morphology and the mineral characteristics. The weight % of organic matrix in a dried coral skeleton is less than 0.01 % of the total but is crucial to the process of calcination. The organic matrix consists of amino acids, proteins, polysaccharides, glycosaminoglycanes, lipids and chitin (in some species).

Interactions between photosynthesis and calcification:

Mechanisms behind the way in which light exacerbates the rate of calcification are not well understood. Light is however known to enhance calcification considerably. The mechanisms for these interactions proposed so far include:

1. Modification of the dissolved inorganic carbon (DIC) equilibrium within the coral tissues caused by CO_2 uptake for photosynthesis – lowering of CO_2 partial pressures it the coral tissues thus favouring carbonate precipitation
2. OH^- production through photosynthesis – OH^- secretion through photosynthesis neutralising the H^+ protons and allowing further calcification
3. Production of oxygen through photosynthesis – high O_2 levels increase the respiratory rate and thus CO_2 production (again favouring carbonate precipitation)
4. Removal of inhibiting substances
5. Synthesis by symbionts of organic matrix molecules or precursors – amino acid contribution to DIC.

Environmental Control:

Changes in environmental parameters such as light, temperature, CO_2 partial pressure and dissolved nutrients directly affect the growth rates of corals and consequently, coral calcification.

4.3.3. Porosity

Porosity can be simply determined as $p = 1 - V_s$, where V_s is the fraction of solid and p is porosity. It is a physical feature that affects the mechanical properties and characteristics of materials such as sponges and corals. The main experimental methods for determining p include the direct methods such as weight assessment, intrusion methods such as mercury porosimetry, optical methods such as pictographical image analysis and gas absorption methods. In characterising pore space, parameters such as circularity, Feret's diameter and aspect ratio and tortuosity are of importance. Tortuosity, Eq. 12 (Epstein 1989) of the pore space reflects the solid state anfractuosity, which is inversely proportional to mechanical stiffness (Alam 2010a). In this equation, h is the total length through the measured section of material and h_e is the shortest length through the pore network through the same section. Tortuosity values of 1 signify that the pore network runs linearly through the material. If the value of tortuosity value rises from this value of unity, the running distance through the pore network is thus also increased and the pore network is non-linear. Circularity, Eq. 13, Feret's diameter and aspect ratio (all with respect to angle of loading) indicate the effective length of the pores, and the extent to which pores disprut the solid state continuum in the direction of loading. A final aspect that indicates the extent of heterogeneity in the solid state medium is the pore coordination number (Alam 2009).

$$\tau = \frac{h}{h_e} \tag{12}$$

$$C = 4\pi\left(A/P^2\right) \tag{13}$$

A is the area of a pore and P is the pore perimeter. Circularity is thus a measure of how circular a pore is and the value extents are 0 and 1. A value equal to 1 signifies a perfectly circular pore and as the value approaches 0, the pore aspect ratio increases. Corals may be considered in essence, porous particle–polymer composites. The particulate fractions being biominerals have some characteristics of ceramic materials. Alam (2010b) summarises predictive stiffness models that have been applied to multiphase porous particle–based composites like corals and sponges. The simplest porous models occupy Voigt bounds and the elastic modulus, E, of the material is considered inversely proportional to porosity, p, Eq. 14, where E_c is the composite elastic modulus. This model is far too simplistic though, and rarely does well to predict E.

$$E = E_c\,(1-p) \tag{14}$$

Non–linear models such as in Eq. 15 are more accurate and have been suggested by Phani et al. (1988), Maitra and Phani (1994) and Wagh et al. (1993). It is essentially an extended exponential of the linear form of Eq. 14 where a and b are fitting parameters.

$$E = E_c\,(1-ap)^b \tag{15}$$

A majority of fitting models for porous composites tend to use such fitting parameters, sometimes with a numerical exponent, Eq. 16, (Brown et al. 1964), and sometimes through the product of the composite elastic modulus and the natural exponential for when $p \leq 0.5$, Eq. 17, and for when $p \geq 0.5$, Eq. 18, (Rice 1993, Knudsen 1959).

$$E = E_c\left(1 - ap^{2/3}\right) \tag{16}$$

$$E = E_c \mathrm{e}^{-ap} \tag{17}$$

$$E = E_c \mathrm{e}^{-a(1-p)} \tag{18}$$

Janowski and Rossi (1967), Boccaccini et al. (1993) also suggest non–linear models, but using fitting parameters, a, that are a function of the pore aspect ratios, A_r, Eqs. 19 and 20 and $p \leq 0.5$ in each case.

$$E = E_c\,(1-ap) \tag{19}$$

$$E = E_c\left(1 - p^{2/3}\right)^{1.21a} \tag{20}$$

A model suggested by Hashin (1962)) incorporates a variable, k, which is related to the Poisson ratio of the composite, Eq. 21.

$$E = E_c\frac{(1-p)^n}{1+kp} \tag{21}$$

More advanced models are usually derivatives of the classical Voigt, Reuss and Halpin–Tsai models (Voigt 1889, Reuss 1929, Halpin-Tsai 1969) since these models do not predict well the elastic modulus for porous materials. The Voigt and Reuss models are provided in Fig. 17. The Halpin–Tsai model is shown in Eq. 22. In these models, the volume fractions, F, of matrix, $_m$, and reinforcement, $_r$, are coupled to the elastic moduli of the individual materials.

$$E = \frac{E_m\,(1+\xi\eta F_r)}{(1-\eta F)} \tag{22}$$

where

$$\eta = \frac{(E_r/E_m)-1}{(E_r/E_m)+\xi}$$

and

$$\xi = 1$$

in particulate reinforced materials like corals.

McAdam (1951) modified the Reuss model such that the final elastic modulus of the composite is linearly reduced as porosity increases, Eq. 23.

$$E = \left(\frac{F_r}{E_r} + \frac{(1-F_r)}{E_m}\right)^{-1} \cdot (1-p) \tag{23}$$

Bert (1985) modified the Halpin–Tsai model. In his model, Bert includes a composite term that considers the pore space and shape. This model, Eq. 24, assumes $K_0 = 2$ for spherical pores.

$$E = \frac{E_m\,(1+\xi\eta F_r)}{(1-\eta F_r)} \cdot \left(1 - \frac{p}{(1-F_r)}\right)^{K_0(1-F_r)} \tag{24}$$

Alam (2010a) suggests a series of modifications to the Voigt model in order to more comprehensively characterise the physical attributes of a porous two phase composite, Eq. 25. The modifications include an *effective* binder fraction, a stress transfer aspect ratio, s_t, an L_r/d_r, product for the reinforcing phase, an anfractuosity parameter, A, and an effective pore width ratio, $w_{p,\max} \cdot w^{-1}$. Here, L_r is the average length of the reinforcing phase in the loading direction and d_r is the diameter. $F_{c,eff}$ is an effective fraction of biopolymer connecting together the reinforcement phase. This model has so far provided the most accurate predictions when compared against other models for porous composites (Alam 2010a).

$$E = \left(E_r F_r \left(\frac{\bar{L}_r}{d_r} \cdot \frac{1}{s_t}\right) + E_m F_{c,eff}\right) \cdot \frac{1}{A} \cdot \left(1 - \frac{w_{p,\max}}{w}\right) \tag{25}$$

If there are no pores then $A = 1$ and $w_{p,\max} = 0$.

4.3.4. Biomimicry of Coral and Sponge Composite Structures

For the most part, biomimetics for the control of crystal morphology has involved observing the interactions between various types of polymer and mineralising via numerous crystallisation methods. The most typical methods for characterising the crystal polymorph that develops have been X-ray diffraction and Fourier Transform Infra-Red spectroscopy (Plavsic et al. 1999). Table 6 summarises some of the more notable research on calcium carbonate polymorphs that have been manufactured through the addition of polymers.

The examples in Table 6 describe biomimetic attempts to vary calcium carbonate polymorphs. The actual application of biomimetic crystals for engineering and technological benefit has however, not been studied in as much detail. Liu et al (2011), Kato (2000), Ajikumar et al. (2004), Stillfried et al. (2013) and Alam et al. (2013) all consider the growth of calcium carbonate polymorphs onto engineering fibre surfaces. Katos (2000) application of thin $CaCO_3$ crystal films to chitin surfaces was facilitated through dissolution of the calcium carbonate in polyacrylic acid (PAA). Kato reports that the OH and NH moieties found in chitin and chitosan molecules are able to interact with the carboxylate groups in PAA, leading to adsorption of the PAA macromolecules to chitin/chitosan surfaces. The PAA marcomolecules in turn have a high affinity to calcium ions, resulting in the nucleation of calcium carbonate to the surface of chitin. Using this method, thin crystallised laminated arrangements can be easily manufactured. Ajikumar and co–workers (2004) evidenced the importance of surface treating fibres for the growth of different crystal polymorph as fibre coating. Acid and alkali surface pre–treatments result in very different calcium carbonate morphologies. Moreover, using different organic acid polymer (poly–Asp, poly–Glu) as surface pre–treatments also controls the polymorph. Importantly, the combination of surface pre–treatment to the underlying fibre substrate is the final determinant of the calcium carbonate polymorph. Poly–Asp and poly–Glu applied to nylon–66 surfaces for example gives rise to a calcitic coating. Applying the same (poly–Asp and poly–Glu) polymers to demineralised eggshell membrane (an organic polymer), yields metastable vateritic and aragonitic rich coatings respectively. Liu and co–workers (2011) applied PAA–$CaCO_3$ coatings to electrospun cellulose acetate fibres. In such an instance, PAA colecules adsorb to to the cellulose acetate fibres through the interaction of OH moieties of the cellulose acetate molecules to the COOH groups in PAA. Redundant COOH groups in PAA then bind to the calcium ions, thus instigating the process of calcium carbonate nucleation and growth. In such an instance, nanoneedle–like calcitic aggregates form in a tight packing around the cellulose acetate fibres.

Though the research of Ajikumar et al. (2004), Liu et al. (2011) and Kato (2000) details how engineering fibres can be biomimetically mineralised using a variety of polymers in combination with calcium carbonate, none of these research efforts attempt to characterise the mechanical benefit of doing so. While Stillfried et al. (2013) researched mechanical improvement of mineralised flax fibres, Alam et al. (2013) considered the mechanical benefit of biomimetically mineralising flax fibres and applying them within a rubber matrix; as would be typical for engineering polymer composites. Alam and co–workers used glacial acetic acid to dissolve calcium carbonate prior to recrystallisation on a fibre surface. The work aimed to mimic blocky crystallite structures found on the surfaces of certain Sclerosponiae sponges (Porifera). Composites like these display numerous levels of structural

Table 6. Examples of altered calcium carbonate morphology through the addition and interaction of different polymers.

$CaCO_3$ morphology	Notes
Microscopic spherules	*Method:* Simple gas diffusion. *Additive:* Dimyristoylphosphatidylglycerol (DMPG) vesicles. Time and temperature have a pronounced effect on the properties, size and polymorph of the crystal (Liu et al. 2008).
Apple core type	*Method:* Addition reaction. *Additive:* Alanyl–alanine derived poly(isocyanide)s 1 and 2. The polymer additive is an efficient nucleation agent for crystal growth (Ren et al. 2011).
Rhombic spherical and spindly	*Method:* Titration reaction. *Additive:* Glycine. Proportion of vaterite increases as a function of additive concentration and time (Manoli et al. 2002).
Twinned peanut–like	*Method:* Simple solution synthesis. *Additive:* Double–hydrophilic block copolymers (PEG–b–PMAA) –DHBC. At higher concentrations of additive dumbbell-like protuberances occur and at higher concentrations still, the protuberances become quasi-spherical (Cölfen and Qi 2001).
Hollow spherical	*Method:* Mixed solution synthesis. *Additive:* DHBC with surfactant complex micelles. Unique vaterite discs and calcite pine cone shaped crystals (Qi et al. 2002).
Porous microspheres	*Method:* Simple precipitation reaction. *Additive:* Poly(styrene–alt–maleic acid) –PSMA. Formation of self organised hierarchically structured CaCO3 (Yu et al. 2004).
Microspheres	*Method:* Simple solution reaction. *Additive*: Doublyhydrophilic block copolypeptide poly (Ne–2(2–(2–methoxyethoxy)actetyl–L–lysine 100–b–poly(L–aspartate sodium salt). Fairly large mean diameter range of 30–50 μm (Yu et al. 2002) however provided these can be sparsely distributed about the surfaces of 100 μm organic fibres they may give rise betterment of mechanical properties within the framework of a composite.
Urchin–like	*Method:* Surfactant assisted. *Additive:* Sodium dodecylbenzenesulfonate (SDBS). Below temperatures of 80°C the structures are a mix of the three typical polymorphs of calcium carbonate (aragonite, vaterite and calcite) (Zhang et al. 2005).
Rods	*Method:* Microemulsion based synthesis. *Additive:* Cetyltrimethylammonium bromide (CTABγ 1–pentanol/cyclohexane). Morphological change from hexagonal vaterite crystals to prismatic rod shaped calcite on ripening (Liu and Yates 2006).
Spherules	*Method:* Mixed solution synthesis. *Additive:* Polypetide + Double–hydrophilic block copolymers (DHBCs. Microspheres average size 1–2 μm, which aggregate on ripening to 10–15 μm in diameter (Zhang et al. 2006).
Stacked rhombohedral and spherical	*Method:* Simple precipitation reaction. *Additive:* Nonionic pluronic amphiphilic triblock copolymer (F68). The concentration of additive distinctly alters the crystal morphology but not the crystalline form. Stacked rhombohedrals form at lower concentrations, turning into spherical precipitates at higher concentrations (Zhao et al. 2012).
Microspheres	*Method:* Simple precipitation reaction. *Additive:* Deoxyribonucleic acid (DNA). Creates unusual morphological profiles that are based on microspheres of rhombohedral and/or prismatic surface structures (Cheng et al. 2010).

hierarchy. The first is in the flax fibre itself, which is a porous structure. The next level of hierarchy results at the surface where biomimetic mineralisation takes place. The final level of hierarchy is at the macro–level, where the mineralised fibres are incorporated into a rubber matrix. Importantly, Alam and co–workers report that mere mineralisation is not enough to achieve mechanical benefit. In fact, thick crystal envelopes around engineering fibres may crack and consequently weaken the composite. Small mineral deposits however, work well to anchor the polymer matrix and increase the shear stresses required to debond the fibre from the matrix. As a consequence, small deposits of crystals on the fibre surface improve the strength and toughness properties of fibre reinforced composites and this therefore, is a fundamental consideration for the engineering design of biomimetically mineralised fibre/reinforced composites.

5. Conclusion

Natural biopolymer composites have advanced structural organisation from the molecular level to the macro level. Further, the methods employed by many biological organisms in the manufacture of such complex hierachical architectures are not trivial. Novel man-made polymer composites with excellent mechanical properties are being continually developed and improved by imitation of the sophisticated structural designs found in nature. Nevertheless, there is still considerable ground to cover. The sophisticated manufacturing methods and the highly controlled order of composite components in many biological materials remain challenging to imitate in biomimetics. This chapter has introduced a few organisms with biostructural organisation that may be inspiring to composites scientists, and has in each case, elucidated current research on the structure-properties relationships as well as on current biomimetics research.

References

[1] Addadi, L., Moradian, J., Shay, E., Maroudas, N. G., Weiner, S. (1987). A chemical model for the cooperation of sulphates and carboxylates in calcite crystal nucleation relevance to biomineralization. *Proceedings of the National Academy of Sciences of the United States of America*, 84, 2732-2736.

[2] Addadi, L., Raz, S., Weiner, S. (2003). Taking advantage of disorder: amorphous calcium carbonate and its roes in biomineralization. *Advanced Materials*, 15, 959-970.

[3] Aizenberg, J., Hanson, J., Ilan, M., Leiserowitz, L., Koetzle, T. F., Addadi, L., Weiner, S. (1995). Morphogenesis of calcite sponge spicules: a role for specialized proteins interacting with growing crystals. *The FASEB Journal*, 9, 262-268.

[4] Ajikumar, P. K., Lakshminarayanan, R., Valiyaveettil, S. (2004). Controlled deposition of thin films of calcium carbonate on natural and synthetic templates. *Crystal Growth and Design*, 4(2), 331-335.

[5] Alam, P. (2010a). A mixtures model for porous particle-polymer composites. *Mechanics Research Communications*, 37, 389-393 and (Erratum) 672.

[6] Alam, P. (2010b). Chapter 2. Porous particle-polymer composites. In: *Advances in Composite Materials Analysis of Natural and Man Made Materials*, Ed. Tesinova, P., Intech, 29-54.

[7] Alam, P. (2013). Biomimetic insights: structure-toughness relations in spider silk nanocrystals. *Advanced Materials Research*, 622-623, 1799-1802.

[8] Alam, P., Byholm, T., Kniivil, J., Sinervo, L., Toivakka, M. (2009). Calculating the permeability of model paper coating structures comprising incongruent particle shapes and sizes. *Microporous and Mesoporous Materials*, 117, 685-688

[9] Alam, P., Stillfried, D. G., Celli, J., Toivakka, M. (2013). Effects of fibre-surface morphology on the mechanical properties of Porifera-inspired rubber-matrix composites. *Applied Physics A: Materials Science and Processing*, 111(4), 1031-1036.

[10] Allemand, D., Ferrier-Pages, C., Furla, P., Houlbreque, F., Puverel, S., Reynaud, S. P., Tambutt, R., Tambutt, S., Zoccola, D. (1994). Biomineralisation in reef-building corals: from molecular mechanisms to environmental control. *General Palaeontology* (Palaeobiochemistry), 3, 453-467.

[11] Ashby, M. F. (1989). On engineering properties of materials, *Acta Metallurgica*, 37, 1273-1293.

[12] Auoacheria, A., Geourjon, C., Aghajari, N., Navratil, V., Deleage, G., Lethias, C., Exposito, J-Y. (2006). Insights into early extracellular matrix evolution: spongin short chain-collagen related proteins are homolgous to basement membrane type IV collagens and form a novel family widely distributed in invertebrates. *Molcualar Biology and Evolution*, 23(12), 2288-2302.

[13] Balani, K., Patel, R. R., Keshri, A. K., Lahiri, D., Agarwal, A. (2011). Multi-scale hierarchy of Chelydra serpentine: microstructure and mechanical properties of the turtle shell. *Journal of the Mechanical Behavior of Biomedical Materials*, 4(7), 1440-1451.

[14] Barthelat, F. (2007). Biomimetics for next generation materials. *Philosophical Transactions of the Royal Society A*, 365, 2907-2919.

[15] Belcher, A. M., Wu, X. H., Christensen, R. J., Hansma, P. K., Stucky, G. D., Morse, D. E. (1996). Control of crystal phase switching and orientation by soluble mollusc-shell proteins. *Nature*, 381, 56-58.

[16] Beniash, E., Aizenberg, J., Addadi, L., Weiner, S. (1997). Amorphous calcium carbonate transforms into calcite during sea-urchin larval spicule growth. *Proceedings of the Royal Society London B*, 264, 461-465.

[17] Berridge, M. J., Bootman, M. D., Lipp, P. (1998). Calcium A life and death signal. *Nature*, 395, 645-648.

[18] Bert, C (1985). Prediction of Elastic Moduli of Solids with Oriented Porosity. *Journal of Materials Science*, 20, 2220-2224.

[19] Block, R. J. (1939). The composition of keratins. *Journal of Biological Chemistry*, 181-196.

[20] Boccaccini, A., Ondracek, G., Mazilu, P., Windelberg, D. (1993). On the Effective Youngs Modulus of Elasticity for Porous Materials: Microstructure Modelling and Comparison between Calculated and Experimental Values. *Journal of the Mechanical Behavior of Materials*, 4, 119-128.

[21] Boccaccini, A. R., Zhitomirsky, I. (2002). Application of electrophoretic and electrolytic deposition techniques in ceramics processing. *Current Opinion in Solid State Materials Science*, 3, 251-260.

[22] Bonderer, L. J., Studart, A. R., Gauckler, L. J. (2008). Bioinspired design and assembly of platelet reinforced polymer films. *Science*, 319, 1069-1073.

[23] Bowes, J. H., Elliott, R. G., Moss, J. A. (1955). The composition of collagen and acid-soluble colagen of bovine skin. *Biochemistry*, 6r, 143-150.

[24] Bratzel, G., Buehler, M. J. (2012). Sequence-structure correlations in silk: Poly-Ala repeat of N. clavipes MaSp1 is naturally optimized at a critical length scale. *Journal of the Mechanical Behavior of Biomedical Materials*, 7, 30-40.

[25] Brindley, G. W., Brown, G. (1980). Crystal structures of clay and their X-ray identification. *Mineralogical Society*, London, UK.

[26] Brown, S.; Biddulph, R., Wilcox, P. (1964). A Strength-Porosity Relation Involving Different Pore Geometry and Orientation. *Journal of the American Ceramics Society*, 47, 320-322

[27] Burghard, Z., Tucic, A, Jeurgens, L. P. H., Hoffmann, R. C., Bill, J., Aldinger, F. (2007). Nanomechanical properties of bioinspired organic-inorganic composite films. *Advanced Materials*, 19, 970-974.

[28] Butler, D. L., Kay, M. D., Stouffer, D. C. (1986). Comparison of material properties in fasciclebone units from human patellar tendon and knee ligaments. *Journal of Biomechics*, 19, 425-432.

[29] Cairns, S. D. (1999). Species richness of recent Scleractinia. *Atoll Research Bulletin*, 459, 1-12.

[30] Caroselli, E., Prada, F., Pasquini, L., Marzano, F. N., Zaccanti, F., Falini, G., Levy, O., Dubinsky, Z., Goffredo, S. (2011). Environmental implications of skeletal microdensity and porosity variation in two scleractinian corals. *Zoology*, 255-264.

[31] Cetinkaya, M., Senbo, X., Grater, F. (2011). Effects of crystalline subunit on silk fiber mechanics. *Soft Matter*, 7, 8142-8148.

[32] Chen, I. H., Kiang, J. H., Correa, V., Lopez, M. I., Chen, P-Y., McKittrick, J., Meyers, M. A. (2010). Armadillo armour: mechanical testing and micro-structural evaluation. *Journal of the Mechanical Behavior of Biomedical Materials*, 4, 713-722.

[33] Chen, P-Y., Lin, A. Y. M., Lin, Y-S., Seki, Y., Stoked, A. G., Peyras, J., Olevskey, E. A., Meyers, M. A., Mcittrick, J. (2008). Structure and mechanical properties of selected biological materials. *Journal of the Mechanical Behavior of Biomedical Materials*, 1, 208-226.

[34] Cheng B., Cai W., Yu J. (2010); DNA-mediated morphosynthesis of calcium carbonate particles; *Journal of Colloid and Interface Science*, Vol. 352, pp. 4349.

[35] Cheng, C., Zhengzhong, S., Vollrath, F. (2008). Silk fibroin-regulated crystallization of calcium carbonate. *Advanced Functional Materials*, 18, 2172-2179.

[36] Couto, D. S., Alves, N. M., Mano, J. F. (2009). Multilayer coatings combining chitosan with bioactive glass nanoparticles. *Journal of Nanoscience and Nanotechnology*, 9, 1741-1748.

[37] Crandford, S. W., Buehler M. J. (2012). Biomateriomics. *Springer Series in Materials Science* 165, Springer, Dordrecht.

[38] Crespilho, F. N., Zucolotto, V., Olivera Jr, O. N., Nart, F. C. (2006). Electrochemistry of layer by layer films: a review. *International Journal of Electrochemical Science*, 1, 194-214.

[39] Currey, J. D. (1977). Mechanical properties of mother-of-pearl in tension. *Proceedings of the Royal Society of London: Series B-Biological Sciences*, 196(1125), 443-463

[40] Currey, J. D., Taylor, J. D. (1974). The mechanical behavior of some molluscan hard tissues. *Journal of Zoology*, 173, 395-406.

[41] Cölfen H. and Qi L. (2001); A systematic examination of the morphogenesis of calcium carbonate in the presence of a double-hydrophilic block copolymer; *Chemistry A European Journal*, Vol. 7, No 1, pp. 110-116.

[42] Dahan, D., Vago, R., Golan, Y. (2003). Skeletal architecture and microstructure of the calcifying coral Fungia cimplex. *Materials Science and Engineering C*, 23, 473-477.

[43] Dal Pra, I., Petrini, P., Charini, A., Bozzini, S., Fare, S. Armato, U. (2003). Silk fibroin-coated three-dimensional polyurethane scaffolds for tissue engineering : interactions with normal human fibroblasts. *Tissue Engineering*, 9, 1113-1121.

[44] Denny, M. (1976). The physical properties of spiders silk and their role in the deign of orb webs. *The Journal of Experimental Biology*, 65, 483-506.

[45] De Tommasi, D., Puglisi, G., Saccomandi, G. (2010). Damage, self healing and hysteresis in spider silk. *Biophysical Journal*, 98, 1941-1948.

[46] Du, N., Liu, X. Y., Narayanan, J., Li, L., Lim, M. L. M., Li, D. (2006). Design of superior spider silk: from nanostructure to mechanical properties. *Biophysical Journal*, 91, 4528-4535.

[47] Du, N., Yang, Z., Liu, X., Y., Li, Y., Xu, H. Y. (2011). Structural origin of the strain-hardening of spider silk. *Advanced Functional Materials*, 21, 772-778.

[48] Epstein, N. (1989). On Tortuosity and the Tortuosity Factor in Flow and Diffusion through Porous Media. *Chemical Engineering Science*, 44(3), 777-779.

[49] Falani, G., Albeck, S., Weiner, S., Addadi, L. (1996). Control of aragonite or calcite polymorphism by mollusk shell macromolecules. *Science*, 271, 67-69.

[50] Freddi, G., Tsukada, M., Beretta, S. (1999). Structure and physical properties of silk fibroin polyacrylamide blend films. *Journal of Applied Polymer Science*, 71, 1563-1571.

[51] Fritz, M., Belcher, A. M., Radmacher, M., Walters, D. A., Hansma, P. K., Stucky, G. D., Morse, D. E., Mann, S. (1994). Flat pearls from biofabrication of organized composites on inorganic substrates. *Nature*, 371, 49-51.

[52] Fu, C., Shao, Z., Vollrath, F. (2009). Animal silks: their structures, properties and artificial production. *Chemical Communications*, 6515-6529.

[53] Garrone, R., Simpson, T. L., Pottu, J. (1981). Ultrastructure and deposition of silica in sponges. In: *Silicon and siliceous structures in biological systems*. Eds. Simpson, T.L., Volcani, B.E. New York: Springer, 495525.

[54] Garrone, R. (1985) *NATO ASI Series A*. Ed. Bairati, A., Garrone, R., 93, 157-175, Plenum, NY.

[55] Ghosh, P., Katti, D. R., Katti, K. S. (2007). Mineral proximity influences mechanical response of proteins in biological mineral-protein hybrid systems. *Biomacromolecules*, 8, 851-856.

[56] Gosline, J. M., DeMont, M. E., Denny, M. W. (1986). The structure and properties of spider silk. *Endeavour*, 10, 37-43.

[57] Gosline, J. M., Guerette, P. A., Ortlepp, C. S., Savage, K. N. (1999). The mechanical design of spider silk: from fibroin sequence to mechanical function. *The Journal of Experimental Biology*, 202, 3295-3303.

[58] Grubb, D, T., Jelinski, L. W. (1997). Fiber morphology of spider silk: the effects of tensile deformation. *Macromolecules*, 30, 2860-2867.

[59] Grubb, D. T., Ji, G. (1999). Molecular chain orientation in supercontracted and re-extended spider silk. *International Journal of Biological Macromolecules*, 24, 203-210.

[60] Hakimi, O., Knight, D. P., Vollrath, F., Vadgama, P. (2007). Spider and mulberry silkworm silks as compatible biomaterials. *Composites Part B: Engineering*, 38, 324-337.

[61] Halpin, J., Tsai, S. (1969). Effects of Environmental Factors on Composite Materials. *AFRL-TR*, 67, 423.

[62] Hamza, S., Slimane, N., Azari. Z., Pluvinage, G. (2013). Structural and mechanical properties of the coral and nacre and the potentiality of their use as bone substitutes. *Applied Surface Science*, 264, 485-491.

[63] Hardy, J. G., Romer, L. M., Scheibel, T. R. (2008). Polymeric materials based on silk proteins. *Polymer*, 49, 4309-4327.

[64] Hardy, J. G., Scheibel, T. R. (2009). Production and processing of spider silk proteins. *Journal of Polymer Science Part A: Polymer Chemistry*, 47, 3957-3963.

[65] Hardy, J. G., Sheibel, T. R. (2010). Composite materials based on silk proteins. *Progress in Polymer Science*, 35, 1093-1115.

[66] Hartman, W.D. (1981). Form and distribution of silica in sponges. In: *Silicon and siliceous structures in biological systems.* Eds. Simpson, T. L., Volcani, B. E. New York: Springer, 453493.

[67] Hashin, Z. (1962). Elastic moduli of heterogeneous materials. *Journal of Applied Mechanics*, 29, 143-150.

[68] Hayashi, C. Y., Shipley, N. H., Lewis, R. V. (1999). Hypotheses that correlate the sequence, structure and mechanical properties of spider slk proteins. *International Journal of Macromolecules*, 24, 271-275.

[69] Hosoda, N., Kato, T. (2001). Thin-film formation of calcium carbonate crystals: effects of functional groups of matrix polymers. *Chemistry of Materials*, 13, 688-693.

[70] Hu, X., Vasanthavada, K., Kohler, K., McNary, S., Moore, A. M., Vierra, C. A. (2006). Molecular mechanisms of spider silk. *Cellular and Molecular Life Sciences*, 63, 1986-1999.

[71] Jackson, A. P., Vincent, J. F. V., Turner, R. M. (1988). The mechanical design of nacre. *Proceedings of the Royal Society of London Series B-Biological Sciences*, 234, 415-440.

[72] Janowski, K., Rossi, R. (1967). Elastic behaviour of MgO matrix composites. *Journal of the American Ceramics Society*, 50, 599-603.

[73] Jin, H-Y., Fridrikh, S. V., Rutledge, G. C., Kaplan, D. L. (2002). Electrospinning Bombyx mori silk with poly (ethylene oxide). *Biomacromolecules*, 3, 1233-1239.

[74] Jin, H-J., Kaplan, L. D., Kim, U-J., Park, J. (2007). *Concentrated aqueous silk fibroin solution and use thereof.* PATENT Publication Number EP1613796 A4.

[75] Jurez-de la Rosa, B. A., Muoz-Saldaa, J., Torres-Torres, D., Ardison, P. L., Alvarado-Gil, J. J. (2012). Nanoindentation characterization of the micro-lamellar arrangment of black coral skeleton. *Journal of Structural Biology*, 177, 349-357.

[76] Kamat, S., Su, X., Ballarini, R., Heuer, A. H. (2000). Structural basis for the frature toughness of the shell of the conch Strombus gigas. *Nature*, 405, 1036-1040.

[77] Kato, T. (2000). Polymer/calcium carbonate layered thin-film composites. *Advanced Materials*, 12(20), 1543-1546.

[78] Kato, T., Amamiya, T. (1999). A new approach to organic/inorganic composites. Thin film coating of $CaCO_3$ on a chitin fiber in the presence of acid-rich macromolecules. *Chemistry Letters*, 3, 199-200.

[79] Kato, T., Suzuki, T., Amamiya, T., Irie, T., Komiyama, M, Yui, H. (1998). Effects of macromolecules on the crystallization of $CaCO_3$ the formation of organic/inorganic composites. *Supramolecular Science*, 5, 411-415.

[80] Kato, T., Suzuki, T., Irie, T. (2000). Layered thin-film composite consisting of polymers and calcum carbonate: a novel organc/inorganic material with an organized structure. *Chemistry Letters.*, 2, 186-187.

[81] Katti, K. S., Katti, D. R., Mohanty, B. (2010). Biomimetic lessons learnt from nature. *Biomimetics Learning from Nature, Amitava Mukherjee* (Ed.), 193-216, InTech.

[82] Katti, K. S., Katti, D. R., Pradhan, S. M., Bhosle, A. (2005). Platelet interlocks are the key to toughness and strength in nacre. *Journal of Materials Research*, 20, 1097-1100.

[83] Katti, K. S., Verma, D., Katti, D. R., Mohanty, B. (2008). Mechanical response and multilevel structure of biomimetic hydroxyapatite/polygalacturonic/chitosan nanocomposites. *Materials Science and Engineering C*, 28, 399-405.

[84] Keten, S., Xu, Z., Ihle, B., Buehler, M. J. (2010). Nanoconfinement controls stiffness, strength and mechanical toughness of -sheet crystals in silk. *Nature Materials*, 9, 359-367.

[85] Kim, H. J., Kim, U. J., Kim, H. S., Li, C. M., Wada, M., Leisk, G. G., Kaplan, D. L. (2008). Bone tissue engineering with premineralized silk scaffolds. *Bone*, 42, 1226-1234.

[86] Knudsen, F. (1959). Dependence of Mechanical Strength of Brittle Polycrystalline Specimens on Porosity and Grain Size. *Journal of the American Ceramics Society*, 45, 452-453

[87] Krauss, S., Monsonegro-Ornan, E., Zelzer, E., Fratzl, P, Shahar, R. (2009). Mechanical function of a complex three-dimensional suture joining the body of bony elements in the shell of the red-eared slider turtle. *Advanced Materials*, 21(4), 407-412.

[88] Levi-Kalisman, Y., Falini, G., Addadi, L., Weiner, S. (2001). Structure of the nacreous organic matrix of a bivalvemollusc shell examined in the hydrated state using cryo-TEM. *Journal of Structural Biology*, 135, 8-17.

[89] Leys, S.P. (2003), Comparative study of spiculogenesis in demosponge and hexactinellid larvae. *Microscopy Research and Technique*, 62, 300311.

[90] Li, C., Kaplan, D. L. (2003). Biomimetic composites via molecular scale self-assembly and biomineralization. *Current Opinion in Solid State Materials Science*, 7, 265-271.

[91] Liang, Y., Ma, C, Wu, C. (2012). The structure of the turtle shell and the porous metal mimicking the structure of the turtle shell. *Advanced Materials Research*, 391-392, 395-399.

[92] Lin, A. Y., Chen, P-Y., Meyers, M. A. (2008). The growth of nacre in the abalone shell. *Acta Biomaterialia*, 4, 131-138.

[93] Lin, W., C. A. Wang, H. L. Le, Long, B., Huang, Y. (2008) Special assembly of laminated nanocomposite that mimics nacre. *Materials Science & Engineering C-Biomimetic and Supramolecular Systems*, 28, 1031-1037.

[94] Liu D., Yates M.Z (2006); Formation of rod-Shaped calcite crystals by microemulsion-based synthesis; *Langmuir*, Vol. 22, pp. 5566-5569.

[95] Liu, L., He, D., Wang, G-S., Yu, S-H. (2011). Bioinspired crystallization of $CaCO_3$ coatings on eletrospun cellulose acetate fiber scaffolds and correspondgin $CaCO_3$ microtube networks. *Langmuir*, 27, 7199-7206.

[96] Liu, X., Bai, H., Zhang, L. and Wang, E. (2008); Lipid-based strategies in inorganic nano-materials and biomineralisation study; *Advances in Planar Lipid Bilayers and Liposomes*, Vol. 7, pp. 203-220.

[97] Liu, Y., Shao, Z. Z., Zhou, P., Chen, X. (2004). Thermal and crystalline behaviour of silk fibroin/nylon 66 blend films. *Polymer*, 45, 7705-7710.

[98] Lombardi, S. J., Kaplan, D. L. (1990). The amino acid composition of major ampullate gland silk (dragline silk) of Nephila clavipes (Araneae, tetragnathedae). *Journal of Arachnology*, 18, 297-306.

[99] Long, B., Wang, C. A., Lin, W., Huang, Y., Sun, J. L. (2007). Polyacrylamide-clay nacre-like nanocomposites prepared by electrophoretic deposition. *Composites Science and Technology*, 67, 2770-2774.

[100] Lowenstam, H. A., Margulis, L. (1980). Calcium regulation and the appearance of calcareous skeletons in the fossil record. In: *The Mechanisms of Biomineralization in Animals and Plants*, Eds. Omori, M., Watabe, N.; Tokai University Press, Tokyo.

[101] Lowenstam, H. A., Weiner, S. (1989). *On biomineralization*. Oxford University Press, New York.

[102] Luz, G. M., Mano, J. F. (2009). Biomimetic design of materials and biomaterials inspired by the structure of nacre. *Philosophical Transactions of the Royal Society A*, 367, 1587-1605.

[103] Maitra, A., Phani, K. (1994). Ultrasonic Evaluation of Elastic Parameters of Sintered Powder Compacts. *Journal of Materials Science*, 29, 4415-4419

[104] Mann, S. (1983). Mineralization in biological systems. *Structure and Bonding*, 54, 125-174.

[105] Mann, S., Webb, J., Williams, R. J. P. (1989). *Biomineralisation: Chemical and biochemical perspectives*. Wiley-VCH Verlagsgesellschaft, Weinheim.

[106] Manoli F., Kanakisa J., Malkajb P., Dalasa E. (2002); The effect of aminoacids on the crystal growth of calcium carbonate; *Journal of Crystal Growth*, Vol. 236, pp. 363370.

[107] Mass, T., Drake, J. L., Haramaty, L., Rosenthal, Y., Schofield, O. M. E., Sherrell, R. M., Falkowski, P. G. (2012) Aragonite Precipitation by Proto-Polyps in Coral Cell Cultures. *PLoS ONE* 7(4): e35049.

[108] Mayer, G. (2005). Rigid biological systems as models for synthetic composites. *Science*, 310, 1144-1147.

[109] McAdam, G. (1951). Some relations of powder characteristics to the elastic modulus and shrinkage of sintered ferrous compacts. *Journal of the Iron Steel Institute*, 168, No.4, 346-358.

[110] Menig, R., Meyers, M. H., Meyers, M. A., Vecchio, K. S. (2000). Quasi-static and dynamic mechanical response of haliotis rufescens (abalone) shells. *Acta Materialia*, 48, 2383-2398.

[111] Metzler, R. A., Kim, W., Delak, K., Evans, J. S., Zhou, D., Beniash, E., Wilt, F., Abrecht, M., Chiou, J-W., Guo, J., Coppersmith, S. N., Gilbert, P. U. P. A. (2008). Probing the inorganic-mineral interface at the molecular level in model biominerals. *Langmuir*, 24, 26802687.

[112] Oaki, Y., Imai, H. (2005). The hierarchical architecture of nacre and its mimetic material. *Angewandte Chemie International Edition.*, 44, 6571-6575.

[113] Oaki, Y., Kotachi, A., Miura, T., Imai, H. (2006). Bridged nanocrystals in biominerals and their biomimetics: classical yet modern crystal growth on the nanoscale. *Advanced Functional Materials*, 16, 1633-1639.

[114] Phani, K., Niyogi, S., De, A. (1988). Porosity dependence of fracture mechanical properties of reaction sintered Si_3N_4. *Journal of Materials Science Letters*, 7, 1253-1256.

[115] Plavsic, B., Kobe, S., Orel, B. (1999). Identification of crystallization forms of CaCO3 with FTIR spectroscopy. Koveine, Zlitine, *Tehnologije*, 33(6), 517-521.

[116] Plaza, G. R., Prez-Rigueiro, J., Riekel, C., Perea, G. B., Agull-Rueda, F., Burghammer, M., Guinea, G. V., Elices, M. (2012). Relationship between microstructure and mechanical properties in spider silk fibers: identification of two regimes in the microstructural change. *Soft Matter*, 8, 6015-6026.

[117] Podsialdo, P., Kaushik, A. K., Shim, B. S., Agarwal, A., Tang, Z., Waas, A. M., Arruda, E. M., Kotov, N. A. (2008). Can natures designs be improved upon? High strength, transparent nacrelike nanocomposites with a double network of sacrificial crosslinks. *Journal of Physical Chemistry B*, 112, 14359-14363.

[118] Podsialdo, P., Paternel, S., Rouillard, J. M., Zhang, Z., Lee, J., Lee, J. W., Gulari, E., Kotov, N. A. (2005). Layer-by-layer assembly of nacre-like nanostructured composites with antimicrobial properties. *Langmuir*, 21, 11915-11921.

[119] Pokroy, B., Fitch, A. N., Marin, F., Kapon, M., Adir, N., Zolotoyabko, E. (2006). Anisotropic lattice distortions in biogenic calcite induced by intra-crstalline organic molecules. *Journal of Structural Biology*, 155, 96-103.

[120] Pokroy, B., Zolotoyabko, E. (2003). Microstructure of natural plywood-like ceramics: a study by high-resolution electron microscopy and energy-variable X-ray diffraction. *Journal of Materials Chemistry*, 13, 682-688.

[121] Przenioslo, R., Storlaski, J., Mazur, M., Brunelli, M. (2008). Hierarchically structured scleractinian coral biocrystals. *Journal of Structural Biology*, 161, 74-82.

[122] Qi L., Li J., Ma j. (2002); Biomimetic morphogenesis of calcium carbonate in mixed solutions of surfactants and double-hydrophilic block copolymers; *Advanced Materials*, Vol. 14, No. 4, pp. 300-303.

[123] Qiao, L., Feng, Q., Lu, S. (2008). In vitro growth of nacre-like tablet forming: from amorphous calcium carbonate nanostacks to hexagonal tablets. *Journal of Crystal Growth and Design*, 8, 1509-1514.

[124] Qiao, X. Y., Li, W., Sun, K., Xu, S., Chen, X. D. (2009). Nonisothermal crystallization behaviors of silk-fibroin-fiber-reinforced poly(epsilon-caprolactone) biocomposites. *Journal of Applied Polymer Science*, 111, 2908-2916.

[125] Reilly, D. T., Burstein, A. H. and Frankel, V. H. (1974). The elastic modulus of bone. *Journal of Biomechanics*, Vol. 7(3), pp. 271-275.

[126] Reiswig, H. M. (1971). The axial symmetry of sponge spicules and its phylogenetic significance. *Cahiers de Biologie Marine*, 12, 505514.

[127] Ren D., Feng Q., Bourrat X. (2011); Effects of additives and templates on calcium carbonate mineralization in vitro; *Micron*, Vol. 42, Issue 3, pp. 228245.

[128] Reuss A. (1929). Berechnung der fliebgrenze von mischkristallen auf grund der plastizittsbedingung fr einkristalle. *ZAMM*, 9 49-58.

[129] Rhee, H., Horstemeyer, M. F., Hwang, Y., Lim, H., El Kadiri, H., Trim, W. (2009). A study on the structure and mechanical behaviour of the Terrapene Carolina carapace: a pathway to design bio-inspired synthetic composites. *Materials Science and Engineering C*, 29(8), 2333-2339.

[130] Rho, J.Y., Roy, M. E., Tsui, T. Y., Pharr, G.M. (1999). Elastic properties of microstructural components of human bone tissue as measured by nanoindentation. *Journal of Biomedical Materials Research*, 45, 48-54.

[131] Rho, J. Y., Kuhn-Spearing, L., Zioupos, P. (1998). Mechanical properties and hierarchical structure of bone. *Medial Engineering Physics*, 20, 92-102.

[132] Rice, R. (1993). Evaluating porosity parameters for porosity-property relations. *Journal of the American Ceramics Society*, 76, 1801-1808.

[133] Sarkar, P., Nicholson, P. S. (1996). Electrophoretic deposition (EPD): mechanisms, kinetics, and application to ceramics. *Journal of the American Ceramics Society*, 79, 1987-2002.

[134] Sarikaya, M., Aksay, I. A. (1995). Biomimetics: design and processing of materials. Polymers and Complex Materials, *American Institute of Physics*, Woodbury, NY.

[135] Sasaki, N., Odajima, S. (1996). Elongation mechanism of collagen fibrils and force-strain relations of tendon at each level of structural hierarchy. *Journal of Biomechics*, 29, 1131-1136.

[136] Shahina, E. S., Novoselov, N. P., Vnuchkin, A. V., Golubikhin, A. Y. (2007). Preparation and properties of films of fibroin-polyvinyl alcohol blends from solutions in hexafluoroisopropanol. *Russian Journal of Applied Chemistry*, 80, 466-471.

[137] Sellinger, A., Weiss, P. M., Nguyen, A., Lu, Y., Assink, R. A., Gong, W., Brinker, C. J. (1998). Continuous self-assembly of organic-inorganic nanocomposite coatings that mimic nacre. *Nature*, 394, 256-260.

[138] Schaffer, T. E., Ionescu-Zanetti, C., Proksch, R., Fritz, M., Walters, D. A., Almqvist, N., Zaremba, C. M., Belcher, A. M., Smith, B. L., Stucky, G. D., Morse, D. E., Hansma, P. K. (1997). Does abalone nacre from by heteroepitxial nucleation or by growth through mineral bridges? *Chemistry of Materials*, 9, 1731-1740.

[139] Scheyer, T. M. (2007). Comparitive bone histology of the turtle shell (carapace and plastron): implications for turtle systematics, functional morphology, and turtle origins. *PhD Thesis*, Mathematisch-Naturwissenschaftliche Fakultt, University of Bonn, Germany.

[140] Shen, X. Y., Belcher, A. M., Hansma, P. K., Stucky, G. D., Morse, D. E. (1997). Molecular cloning and characterization of lustring A, a matrix protein from shell and pearl nacre of Haliotis rufescens. *Journal of Biological Chemistry*, 272, 32472-32481.

[141] Simkiss, K., Wilbur, K. (1989). *Biomineralization: Cell biology and mineral deposition*. Academic Press Inc., San Deigo.

[142] Simmons, A. H., Michal, C. A., Jelinski, L. W. (1996). Molecular orientation and two component nature of the crystalline fraction of spider dragline silk. *Science*, 271, 84-87.

[143] Simmons, A., Ray, E., Jelinski, L. (1994). Solid state 13C NMR of Nephila clavipes dragline silk establishes structure of crystalline regions. *Macromolecules*, 27, 5235-5237.

[144] Simpson, T. L. (1984). *The cell biology of sponges*. Springer, New York, 1660.

[145] Simpson, T. L., Langenbruch, P. F., Scalera-Liaci, L. (1985). Silica spicules and axial filaments of the marine sponge Stelletta grubii (Porifera, Demospongiae). *Zoomorphology*, 105, 375382.

[146] Smith, B. L., Schffer, T. E., Viani, M., Thompson, J. B., Frederick, N. A., Kindt, J., Belcher, A., Stucky, G. D., Morse, D. E., Hansma, P. K. (1999). Molecular mechanistic origin of the toughness of natural adhesives, fibres and composites. *Nature*, 399, 761-763.

[147] Song, F., Bai, L. (2001). Mineral bridges of nacre and its effects. *Acta Mechanica Sinica*, 17, 251-257.

[148] Song, F., Soh, A. K., Bai, Y. L. (2003). Structural and mechanical properties of the organic matrix layers of nacre. *Biomaterials*, 24, 3623-3631.

[149] Song, F., Zhang, X. H., Bai, Y. L. (2002a). Microstructure and characteristics in the organic matrix layers of nacre. *Journal of Materials Research*, 17, 1567-1570.

[150] Song, F., Zhang, X. H., Bai, Y. L. (2002b). Microstructure in a biointerface. *Journal of Materials Science Letters*, 21, 639-641.

[151] Sponner, A., Wolfram, V., Monajembashi, S., Unger, E., Grosse, F., Weisshart, K. (2007). Composition and hierarchical organisation of a spider silk. *PloS ONE*, 2(10), e998.

[152] Stillfried, D. G., Toivakka, M. and Alam, P. (2013). *Crispatotrochus*-mimicking coatings improve the flexural properties of organic fibres. *Journal of Materials Science*, 48(24), 8449-8453.

[153] Su, X., Belcher, A. M., Zaremba, C. M., Morse, D. E., Stucky, G. D., Heuer, A. H. (2002). Structural and microstructural characterization of the growth lines and prismatic microarchitecture in red abalone shell and the microstructures of abalone 'Flat pearls'. *Chemistry of Materials*, 14, 3106-3117.

[154] Sun, Y. Y., Shao, Z. Z., Ma, M. H., Hu, P., Liu, Y. S., Yu, T. Y. (1997). Acrylic polymer silk fibroin blend fibers. *Journal of Applied Polymer Science*, 65, 959-966.

[155] Sugawara, A., Kato, T. (2000). Aragonite $CaCO_3$ thin-film formation by cooperation of Mg_2C and organic polymer matrices. *Chemical Communications*, 5, 487-488.

[156] Tsortos, A., Nancollas G. (2002). The role of polycarboxylic acids in calcium carbonate mineralization. *Journal of Colloid and Interface Science*, 250, 159-167.

[157] Ulrich, S., Glisovic, A., Salditt, T., Zippelius, A. (2008). Diffraction from the β-sheet crystallites in spider silk. *The European Physical Journal E*, 27, 229-242.

[158] Uriz, M. J., Turon, X., Becerro, M. (2000). Silica deposition in Demosponges: spiculogenesis in Crambe crambe. *Cell Tissue Research*, 301, 299309.

[159] Verma, D., Katti, K. S., Katti, D. R. (2007). Nature of water in nacre: a 2D Fourier transform infrared spectroscopic study. *Spectrochimica Acta Part A-Molecular and Biomolecular Spectroscopy*, 67, 784-788.

[160] Verma, D., Katti, K. S., Katti, D. R. (2008). Effect of biopolymers on the structure of hydroxyapatite and interfacial interactions in biomimetically synthesized hydroxyapatite/biopolymer nanocomposites. *Annuls of Biomedical Engineering*, 36, 1024-1032.

[161] Vertlib, V., Bietiker, M., Plotze, M., Yezek, L., Spolenak, R., Puzrin, A. M. (2008). Fast assembly of bio-inspired nanocomposite films. *Journal of Materials Research*, 23, 1026-1035.

[162] Vickaryous, M. K., Hall, B. K. (2006). Osteoderm morphology and development in the nine-branded armadillo, Dasypus novemcinctus (Mammalian, Xenathra, Cingulata). *Journal of Morphology*, 267, 1273-1283.

[163] Voigt W. (1889). ber die beziehung zwischen den beiden elasticittsconstanten isotroper krper. *Annals of Physical Chemistry*, 38 573-587.

[164] Vollrath, F., Madsen, B., Shao, Z. (2001). The effect of spinning conditions on the mechanics of spiders dragline silk. *Proceeedings of the Royal Society B*, 268, 2339-2346.

[165] Wagh, A.; Singh, J., Poeppel, R. (1993). Dependence of ceramic fracture properties on porosity. *Journal of Materials Science*, 28, 3589-3593.

[166] Wang, C. A., Long, B., Lin, W., Huang, Y, Sun, J. (2008). Poly(amic acid)-clay nacre-like composites prepared by electrophoretic deposition. *Journal of Materials Research*, 23, 1706-1712.

[167] Wang, R. Z., Suo, Z., Evans, A. G., Yao, N., Aksay, I. A. (2001). Deformation mechanisms in nacre. *Journal of Materials Research*, 16, 2485-2493.

[168] Weaver, J. C., Milliron, G. W., Allen, P., Miserez, A., Rawal, A., Garay, J., Thurner, P. J., Seto, J., Mayzel, B., Friesen, L. J., Chmelka, B. F., Fratzl, P., Aizenberg, J., Dauphin, Y., Kisailus, D., Morse, D. E. (2010). Unifying design strategies in demosponge and hexactinellid skeletal systems. *The Journal of Adhesion*, 86, 72-95.

[169] Weaver, J. C., Morse, D. E. (2003), Molecular biology of demosponge axial filaments and their roles in biosilicification. *Microscopy Research and Technique*, 62, 356367.

[170] Wegst, U. G. K. and Ashby, M. F. (1994). The mechanical efficiency of natural materials. *Philosophical Magazine*, 21, 2167-2181.

[171] Weiner, S., Dove, P. M. (2003). Chapter 1: An overview of biomineralization processes and the problem of the vital effect. In: *Biomineralization* Eds. Dove, P. M., De Yoreo, J. J., Weiner, S.; Reviews in Mineralogy and Geochemistry, Volume 54.,

[172] Weiner, S., Price, P. (1986). Disaggregation of bone into crystals. *Calcified Tissue International*, 39(6), 365-375.

[173] Weiner, S., Traub, W. (1980). X-ray diffraction study of the insoluble organic matrix of mollusc shells. *FEBS Letters*, 111, 311-316.

[174] Weiss, I. M., Tuross, N., Addadi, L., Weiner, S. (2002). Mollusk larval shell formation: amorphous calcium carbonate is a precursor for aragonite. *The Journal of Experimental Zoology*, 293, 478-491.

[175] Wilt, F. H. (2002). Biomineralization of the spicules of sea urchin embryos. *Zoological Science*, 19(3), 253-261.

[176] Wise, S. W., Devillie, J. (1971). Scanning electron microscopy of molluscan shell ultrastructures screw dislocations in pelecypod nacre. *Transactions of the American Microscopical Society*, 90, 376-380.

[177] Xu, G. Yao, N., Aksay, I. Groves, J. (1998). Biomimetic synthesis of macroscopic-scale calcium carbonate thin films. Evidence for a multistep assembly process. *Journal of the American Chemical Society*, 120, 11977-11985.

[178] Xu, Y., Zhang, L. (1995). Mechanical properties and microstructure of tortoise shell. *Fuhe Cailiao Xuebao/Acta Materiae Compositae Sinica*, 12(3), 53-57.

[179] Xu, Z., Buehler, M. J. (2010). Mechanical energy transfer and dissipation in fibrous -sheet rich proteins. *Physical Review E*, 81, 061910(6).

[180] Yao, N., Epstein, A., Akey, A. (2006). Crystal growth via spiral motion in abalone shell nacre. *Journal of Materials Research*, 21, 1939-1946.

[181] Yu J., Yu J.C., Zhang L., Wanga X., Wua L. (2004); Facile fabrication and characterization of hierarchically porous calcium carbonate microspheres; *Chemical Communications*, pp. 2414-2415.

[182] Yu S., Clfen H., Hartmann J., Antonietti M. (2002); Biomimetic crystallization of calcium carbonate spherules with controlled surface structures and sizes by double-hydrophilic block copolymers; *Advanced Materials*, Vol. 12, No 8, pp. 541-545.

[183] Zaremba, C. M., Belcher, A. M., Fritz, M., Li, Y. L., Mann, S., Hansma, P. K., Morse, D. E., Speck, J. S., Stucky, G. D. (1996). Critical transitions in the biofabrication of abaone shells and flat pearls. *Chemistry of Materials*, 8, 679-690.

[184] Zhang, L., Zhang, N., Ruan, J. S., Zhang, T. (2011). Studies on the rules of -strand alignment in a protein β-sheet structure. *Journal of Theoretical Biology*, 285(1), 69-76.

[185] Zhang, W., Wu, C., Zhang, C., Chen, Z. (2012). Numerical study of the mechanical response of the turtle shell. *Journal of Bionic Engineering*, 9(3), 330-335.

[186] Zhang X., Zhang Z., Yan Y. (2005); A facile surfactant-assisted approach to the synthesis of urchin-shaped aragonite micropatterns; *Journal of Crystal Growth*, Vol. 274, pp. 550554.

[187] Zhang Z., Gao D., Zhao H., Xie C., Guan G., Wang D., Yu S. (2006); Biomimetic assembly of polypeptide-stabilized CaCO3 nanoparticles; *Journal of Physical Chemistry B*; Vol. 110, pp. 8613-8618.

[188] Zhao, Y., Wang, X., Jiao, J. and Yu, L. (2012); The preparation of calcium carbonate crystals in pluronic F68 solution; *Journal of Molecular Liquids*, Vol. 169, pp. 144-151.

[189] Zhu, H. L., Feng, X. X., Zhang, H. P., Guo, Y. H., Zhang, J. Z., Chen, J. Y. (2009). Structural characteristics and properties of silk fibroin/poly(lactic acid) bend films. *Journal of Biomaterials Science Polymer Edition*, 20, 1259-1274.

EDITORS' CONTACT INFORMATION

Dr. Stephan Laske
Asst. Prof. Dipl.-Ing. Dr. Mont.
Montanuniversitaet Leoben
Department Polymer Engineering and Science
Polymer Processing
Otto Gloeckel-Strasse 2
8700 Leoben, Austria
Tel: +43/3842/402-3506
Fax: +43/3842/402-3502
Stephan.Laske@unileoben.ac.at

Andreas Witschnigg
Dipl.-Ing.
Montanuniversitaet Leoben
Department Polymer Engineering and Science
Polymer Processing
Otto Gloeckel-Strasse 2
8700 Leoben, Austria
Tel: +43/3842/402-3528
Fax: +43/3842/402-3502
Andreas.Witschnigg@unileoben.ac.at

INDEX

B

C

D

E

F

G

H

I

J

K

L

M

N

O

P

Q

R

S

T

U

V

W

X

Y

Z